EVOLUTIONARY BIOLOGY OF THE ANURANS

EVOLUTIONARY BIOLOGY OF THE ANURANS

Contemporary Research on Major Problems

Edited by James L. Vial

University of Missouri Press
Columbia, 1973

ISBN 0-8262-0134-2

University of Missouri Press, Columbia, Missouri 65201

Library of Congress Catalog Number 72-87836

Printed and bound in the United States of America

PREFACE

Because amphibians exhibit certain general characteristics, particularly a biphasic life cycle, external fertilization, and a unique occupancy of the transitional aquatic-terrestrial environment, they have proven to be excellent organisms for the study of many basic biological phenomena. During the forty-one years since the publication of G. K. Noble's classic reference, *The Biology of the Amphibia* (1931), there has been an almost exponential increase of information about these vertebrates. Of amphibians, the frogs and toads are an incomparably successful and dynamic group. The nearly 2,000 known species display a remarkable range of adaptations that has enabled them to occupy an array of diverse habitats in most areas of the earth. Studies of anurans in both the field and the laboratory have been a source of much of our current knowledge of embryology, physiology, genetics, ecology, population dynamics, systematics, evolution, and biogeography.

This volume is the progeny of a symposium presented during the 13th annual meetings of the Society for the Study of Amphibians and Reptiles, held on the Kansas City campus of the University of Missouri from August 27 to August 30, 1970. During the concluding sessions of the 1969 SSAR meetings, the idea for a symposium on anuran biology arose from my discussions with Dr. Kraig Adler (Cornell University) and Dr. William E. Duellman (University of Kansas). We met several times during the following weeks to develop a program with a unifying theme that, we had concluded, could be best embodied in an evolutionary approach. After composing lists of topics and of the outstanding authorities in each subject, we tried to balance the selection of participants between those who have been long established and those who have achieved more recent recognition. The purpose of mixing "old guards" and "young turks" was to increase the opportunity for an incisive, critical evaluation of the evidence.

The response from the invited specialists greatly reinforced our original optimism about the amount of interest the program would receive. They all shared our opinion that such a synthesis was overdue, and each labored for nearly a year, exchanging and revising early drafts in preparation for the meetings. However, because the time limitations imposed by the formal sessions restricted the length of the material, the authors were allowed further opportunity to revise and elaborate upon their manuscripts until press time in 1972. Thus, the papers as published here have a recency and a greater body of evidence than the original presentations. Except for a minor rearrangement in sequence, the organization of this book follows that of the symposium.

The salient problems in anuran evolution and those areas of investigation likely to produce significant resolutions are identified in Blair's introductory paper.

In Part I, Estes and Reig interpret early evolutionary lineages; from their meticulous re-examination of Mesozoic proanuran and anuran fossils, they derive support for the lissamphibian hypothesis of common origin for frogs, salamanders, and caecilians. Variation and trends in the evolution of skeletal morphology are assessed by Trueb. Her evidence suggests a greater adaptive lability in osteological features than has been generally recognized. Lynch reviews the substance of previous anuran classifications and, after presenting extensive and detailed character analyses, proposes an alternative phylogeny to delineate the archaic, transitional, and advanced frog families. The application of modern biochemical techniques to determine genetic relationships is described and evaluated by Guttman.

In Part II, Rabb compares variation and progressive complexities in components of reproductive behavior among anuran taxa, emphasizing the roles of hormonal and neurological mechanisms. Salthe and Duellman find simultaneous relationships in body size, clutch size, ovum size, and mode of reproduction, all of which impose adaptive limitations on the evolution of reproductive strategies. Starrett focuses upon functional anatomy of the branchial chambers, chondrocranium, and musculature to diagnose the significant evolutionary features of larval morphology. Four distinctive developmental patterns provide a framework for her definitions of the anuran suborders. Diverse modes of tadpole social behavior are recognized by Wassersug as responses to environmental variables, which he interprets in terms of adaptive and evolutionary trends.

In Part III, Schiøtz examines the ecological role of mating calls in the context of social interactions and sound environments. He concludes that voices are important determinants of evolutionary processes but of little use in phylogenetics. Straughan's complementary study deals with mating calls as functional units of communications systems that exhibit modular characteristics subject to selective pressures. An operational definition and application of the techniques of ecological genetics is provided by Merrell, and chromosomal mutations are used by Bogart to postulate derivations from ancestral karyotypes. Savage reconstructs the historical origins of modern frog distribution from his analyses of concurrent events in continental drift, paleoecological opportunities, and macroevolution.

Each of the above parts, which represent the program's sessions, is followed by an open discussion in which the contributors and moderator are identified by surname only. Other speakers are identified by their full names and professional affiliations at the point of their initial statements and subsequently by surnames only. The editing of these tape-recorded sessions has been necessarily severe. I have, however, tried to leave intact the informal atmosphere of the discussions, as well as to retain the innuendoes and syntax of the speakers.

During the preparation and presentation of the symposium, several students in the Biology Department at the University of Missouri–Kansas City, particularly Thomas J. Berger, James R. Stewart, Donald D. Smith, Cathy Ewart, Cathy Hunt, and Roger Jensen, offered their generous assistance. I also benefited by the experience of Dr. William W. Milstead, who organized and edited a noteworthy symposium on lizard ecology in 1965. Dr. Harold C. Burdick, Acting Chairman of the Biology Department, provided funds for staff assistance and communication expenses. Paul Anderson and Linda Folkman of the Division for Continuing Education managed most of the procedural details during the meetings.

Dean Edwin J. Westermann and Associate Dean Henry A. Mitchell of the UMKC College of Arts and Sciences supported the project by subsidizing the travel expenses for the foreign participants, arranging for the availability of facilities, and making possible my reduced teaching commitment during some of the months required for editing. Provost Wesley J. Dale also arranged some financial support for travel of foreign participants.

I have been aided greatly in the editorial work by two persons who deserve special recognition: Kay Earnshaw, who did the difficult transcription of discussion sessions and typed the many manuscript drafts that were necessary, and Thomas J. Berger, who has been of continued assistance in overseeing many editorial details.

To all of these, and others not individually mentioned, I wish to express my sincere gratitude. At a time when scientific specialists must search through an extensive body of literature in order to maintain professional competence and biologists with more general interests encounter equal difficulty in finding other than vintage references, I hope this volume will accomplish the intended objectives of providing a contemporary review, re-examination, and synthesis of those fields that significantly relate to anuran evolution.

J.L.V.
University of Missouri–Kansas City
August, 1972

Participants in the Symposium on Evolutionary Biology of the Anurans. *Seated, left to right:* Arne Schiøtz (Danmarks Akvarium, Charlottenlund, Denmark), Sheldon I. Guttman (Miami University, Ohio), Priscilla H. Starrett (University of Southern California), W. Frank Blair (University of Texas–Austin), Linda Trueb (University of Kansas), Stanley N. Salthe (Brooklyn College of the City University of New York), James P. Bogart (Louisiana Tech University). *Standing, left to right:* Jay M. Savage (University of Southern California), John D. Lynch (University of Nebraska), George B. Rabb (Chicago Zoological Society), Ian R. Straughan (James Cook University of North Queensland, Australia), Dean E. Metter (University of Missouri–Columbia), Richard Estes (Boston University), Robert F. Inger (Field Museum of Natural History), Richard J. Wassersug (University of California–Berkeley), David J. Merrell (University of Minnesota), William E. Duellman (University of Kansas), Osvaldo A. Reig (Universidad Austral de Chile, Valdivia, Chile). *Not shown:* Joseph A. Tihen (University of Notre Dame).

CONTENTS

EVOLUTIONARY BIOLOGY OF THE ANURANS

INTRODUCTION

MAJOR PROBLEMS IN ANURAN EVOLUTION

W. Frank Blair

INTRODUCTION

I am taking two approaches in this introductory statement. One is addressed to the accuracy of our present knowledge of the evolutionary history and degree of diversification (speciation) of the anurans. This approach, of course, raises the question of the accuracy of present schemes that represent presumed phyletic relationships of the anurans at all levels up to the highest taxa. The second approach is addressed to the use of anurans as materials by which we can elucidate evolutionary mechanisms and principles. The two approaches are not mutually exclusive and can be discussed as parallel, rather than isolated, entities.

With respect to evolutionary history, we are concerned with such questions as the following: What has been the area of origin of major evolutionary breakthroughs leading to the establishment of new phyletic lines—to new higher taxa in our schemes of classification? What has been the history of expansion of these lines into other areas? What has been the time scale within which these events have occurred? There are obviously related questions. What has been the ancestral group? How much has our understanding of phylogenetic relations been clouded by convergent evolution? How much by differential rates of evolution within various subgroups? How much by differential rates of change in different biological systems?

With respect to evolutionary mechanisms, we can also ask of our anuran material a number of significant questions: What is the real meaning of ecological diversity? Why are some genera monotypic and seemingly have been monotypic or nearly so throughout their histories, while others split into many species? Why is there greater diversity in the tropics than in extratropical regions? Why does diversity of taxa decrease and numbers of individuals increase as one ascends a tropical mountain range? How are major specializations initiated and perfected? What are the mechanisms of speciation? Why do evolutionary strategies differ so much between taxa, such as *Rana* and *Bufo*? Are these differences due to different ecological roles of the two genera, or do they relate back to basic differences in their genetic constitutions? Some of these questions have moderately satisfactory answers; some have no answers at all.

In raising these questions and in suggesting possible approaches that might provide answers, I am going to stress the multidisciplinary approach, particularly for the first set of questions. In so doing, I am going to call on my own experience with the genus *Bufo*. For that genus, the multidisciplinary, team approach has provided answers that are so superior to those provided by more limited methodologies that I would question purported answers based on data from a single biological system, such as osteology, myology, or blood proteins. Differential rates of evolution of different systems within the same species and convergent evolution within the same systems of different species can make the data from a single kind of analysis quite misleading. For example, the heavy deposit of dermal bone in the skull of *B. punctatus* might lead to the conclusion that this species is only distantly related to such thin-skulled toads as those of the *boreas* group. However, the data from hybridization, chromosomes, skin venoms, and vocal mechanisms all show close affinity of *B. punctatus* with these thin-skulled toads. The heavy skull is a specialization probably related to the burrowing habits of these xeric-adapted toads.

The anurans are especially rich in biological systems the comparative study of which can provide answers to many of the questions we have posed. The most useful of these areas of study are the following:

1. *External Morphology*. This is prone to reflect adaptive pressures and hence most apt to show similarities that result from convergence rather than phylogenetic affinity.

2. *Osteology*. Cranial osteology is useful, but intraspecific and intrapopulational variation in the cranium and in other skeletal elements is considerably greater than in mammals.

3. *Myology*.

4. *Karyology*. The number of chromosomes represents one significant attribute. Position of the centromere and the location of heterochromatic areas—secondary constrictions—are also important.

5. *Genetic Compatibility*. Artificial hybridization gives highly significant results with respect to phylogenetic relationships. These relationships range through a spectrum from complete incompatibility to fertility of F_1 hybrids in both reciprocals of interspecific crosses. The significance of these results varies from taxon to taxon. Some species of *Bufo* have retained the ability to produce viable hybrids through millions of years of separation (Blair, 1970), but in the genus *Rana* there appears to be rapid evolution of genetic incompatibility (Moore, 1949; Porter, 1969). In the latter genus, hybridization serves only to delimit species groups, while in the toads, and in at least Hylidae and Pelobatidae, much more distant relationships can be identified.

6. *Vocal Apparatus and Vocalizations*. The combined approach of W. F. Martin (1972) of studying functional morphology as well as characteristics of the vocalizations has advantages over the analysis of the vocalizations alone, which was the method of Bogert (1960), Blair (1958), and many others. Discrimination experiments, using gravid females and tape-recorded calls, provide information about species distinctness (Blair and Littlejohn, 1960; Gerhardt, 1969; and others).

7. *Venoms*. Skin venoms (Cei and Erspamer, 1966; Low, 1972; and others) have provided valuable information about phyletic lines. Two-dimensional paper chromatography has proved to be successful for separation and identification of these compounds.

8. *Blood*. Electrophoresis has been a successful technique for separation of

blood proteins in anurans (Guttman, 1972; and others). Serological tests have also been used (Cei, 1965).

PHYLOGENETIC QUESTIONS

What are the relationships of the major evolutionary lines of anurans?

The classification of the major taxa of anurans still in general use today is largely one that resulted from the morphological studies of such persons as Parker (1934) and from the somewhat more broadly based conclusions of Noble (1931). Recent investigations of other biological systems—for example, the studies of larvae by Orton (1957), and karyology by Morescalchi (1967) and Bogart (1972*b*)—have cast doubt on the classifications that have been based heavily, if not solely, on morphology.

Vertebrate systematists have traditionally placed great weight on osteological characters in assessing relationships. However, such emphasis now seems unwarranted in the anurans. Parker's unlikely classification of the Microhylidae has been shown by Nelson (1966) to stem heavily from his failure to recognize parallel, but independent, reduction in osteological structures. The great variance in skull type in the *B. spinulosus* group of South American toads (R. F. Martin, 1972) demonstrates the need for caution in the use of osteological characters alone.

Illustrative of the probable flaws in the present system of classification are the findings of Morescalchi (1967) and Bogart (1972*b*) that ceratophrynids (South America) and pelobatids (North America and Eurasia) are karyotypically, as well as ecologically, very similar.

Results of my own hybridization studies (Blair, 1972) strongly imply that *Melanophryniscus* is not only a bufonid but a *Bufo,* and W. F. Martin's (1972) work with vocal structures supports the idea that it is a primitive member of this genus. Barrio (1970) and Barrio and Rubel (1970), in studies of karyotypes and vocalizations of the genera *Pseudis* and *Lysapsus,* show the close relationship of these frogs and make me wonder about the suitability of recognizing both genera.

My conclusion is that a major overhaul of the classification of the higher categories of the anurans is overdue. Any new classification should be based on evidence from all biological systems amenable to analysis.

What are the main sources of error in making phylogenetic conclusions?

It seems hardly necessary to emphasize the dangers of error inherent in conclusions drawn from comparisons of a single biological system. For example, if one looked solely at chromosomes as studied by Bogart (1968) he might separate off many of the African *Bufo* from all other members of the genus because of the difference in chromosome number (2n equal of 20 versus 2n equal of 22 in all other species studied). However, hybridization data, as well as osteological characters, point to an affinity between these African toads and the broad-skulled group of toads that inhabits the neotropics.

Different rates of evolutionary change, or we might say different evolutionary strategies, in different taxa can also lead to error unless the existence of these differences is recognized. For example, the evolution of genetic incompatibility is very different in *Rana* and *Bufo* (Blair, 1970). In the latter, where genetic compatibility has been retained by species that have been isolated from one another for millions of years, hybridization experiments provide important information about the degree of affinity of even distantly related species. However, one cannot extrapolate to *Rana,* where incompatibility may evolve in a few thousand years (Porter, 1969) and where it is impossible to produce hybrids between species beyond the limits of the species group. The differences between these two genera are not ones of different degrees of affinity between species groups but of different evolutionary strategies.

Another source of error resides in the rather obvious fact that different biological systems within a species or higher taxon can be expected to evolve at different rates. Systems under the greatest selection pressures should show the greatest divergence; consequently, there should be a wide spectrum of response in the degree and rate of divergence in the various biological systems. The *B. marmoreus* group of toads and related species offer a case in point. The two members of this group, *B. marmoreus* and *B. perplexus,* are very similar in morphology, in their skin secretions, and in mating calls (Porter and Porter, 1967). However, *B. marmoreus,* presumably under the pressures of stream breeding, has departed from the normal *Bufo* pattern of egg laying and lays single, adhesive eggs, while *B. perplexus* lays eggs in jelly strands, as do most species of *Bufo*. The two species produced fertile F_1 hybrids. A related species, *B. punctatus,* differs considerably from these two in appearance. It lays eggs like those of *B. marmoreus* but produced only sterile hybrids or none in crosses with the two members of the *marmoreus* group.

Another potential source of error is our inability to distinguish between convergent characters and those attributable to common ancestry. There are remarkably impressive examples of presumed convergence in the anurans of different continents, for instance, rhacophorids and hylids. However, any grand generalizations about convergent evolution in anurans should be deferred until the degree of affinity of presumably convergent taxa has been established to the fullest extent of our presently existing methods for measuring and comparing biological systems.

I have the feeling that much of the past thinking about the discreteness of anuran—and other animal—taxa on different continents has been influenced by a widely held misconception about the inability of these animals to cross from continent to continent. The work with *Bufo* that my associates and I have done suggests that the toads of South America, North America, and Eurasia evolved virtually as if these land masses were all one. The evidence also points to the ability of the toads to move from continent to continent with great ease through Tertiary and Quaternary times. If toads could do this, many other kinds of animals should have been able to do so. Furthermore, new evidence about continental drift now makes it seem likely that the more primitive of anuran lines moved apart with their respective land masses or at most crossed water barriers that were much less extensive than presently existing ones.

In sum, geographical isolation should have slight weight in assessment of phylogenetic relations within a long existing group such as the anurans.

BIOGEOGRAPHICAL QUESTIONS

What has been the area of origin?

The fossil record for anurans is so incomplete that it has not been possible to establish with certainty the area of origin of the major taxa. The best that we can do is to marshal all possible lines of evidence that can be derived from the biogeography, ecological adaptations, and character states of living forms to augment the fragmentary fossil record in making the best estimate about the area of origin.

The *Bufo* work (Blair, 1972; Bogart, 1972*b*; Cei, Erspamer, and Roseghini, 1972; Guttman, 1972; Low, 1972; R. F. Martin, 1972; W. F. Martin, 1972; Reig, 1972) illustrates the kinds of information that can be utilized in order to increase the probability of a correct conclusion. In this instance the evidence points strongly to the New World—South America or southern North America—as the area of origin.

The fossil record, such as it is, is consistent with this conclusion. There are several major points:

(1) The oldest known fossil anurans are from the Jurassic of South America and thus establish the existence of frogs on the present South American continent long before the origin of *Bufo* (Reig, 1961).

(2) The leptodactylid frogs, from which the bufonids evolved, were present in South America in the early Tertiary, and unequivocal fossils of these frogs have been found only in the New World and Australia.

(3) The oldest known fossil *Bufo* are from the New World.

Keeping in mind the inadequacies of the anuran fossil record, the existing information is most logically interpreted as supporting a Gondwana center of anuran evolution as hypothesized by Reig (1960) and Casamiquela (1961).

Turning from fossils to the evidence that can be marshaled from the examination of many biological systems in living material, we find the Gondwana hypothesis and the idea of a New World origin for *Bufo* not only supported but greatly strengthened. The major evidence follows:

(1) In South America, there are living intermediate types, including *Macrogenioglottus* and ceratophrynids, that provide a transition from leptodactylid to bufonid morphology and karyotype (Reig, 1972; Bogart, 1972*a*).

(2) In South America and southwestern North America, there are living *Bufo* that are intermediate in morphology and crossability between the two major lines into which *Bufo* separated no later than in Miocene time and that now are represented in Eurasia and Africa, where no such intermediate forms have been found.

(3) In South America, there is such a species as *Bufo crucifer*, which has five of the six secondary chromosomal constrictions that are found in both evolutionary lines of *Bufo* and which has the only transferrin that is found in both lines.

What has been the pattern of spread?

The multidisciplinary approach to studying anurans can be used to establish the existence of relationships between taxa that occur today in quite distant parts of the world and hence can provide evidence about the history of the spread of the major group to its present distribution.

The *Bufo* studies (Blair, 1972) illustrate the effectiveness of this approach. The question is, Are the narrow-skulled toads of South America, North America, and Eurasia narrow-skulled because of affinity or because of convergence? The evidence from multiple sources points clearly to affinity. Main points of evidence are the following:

(1) All are primarily "cold-adapted," living at middle-to-high latitudes or at relatively high altitudes.

(2) There is a trend toward loss of the mating call, especially in New World members (*spinulosus* group, *bocourti* and *boreas* group).

(3) The relict *B. canorus* of California has retained a mating call very similar to that of the Eurasian *B. viridis*.

(4) There is a strong tendency toward sexual dimorphism.

(5) The testes are of distinctive shape (oval) and tend to have black melanin covers.

(6) Distinctive skin secretions—"Indole-20" (Low, 1972) and bufoviridin

(Cei, Erspamer, and Roseghini, 1972)—have been found only in these toads and in branches from this line.

(7) Hybridization data add strong support.

When all this evidence is collated, a dispersal of narrow-skulled toads from South America, through North America to Asia by way of a Bering Land Bridge, ultimately to Europe and North Africa is strongly indicated.

SPECIES DIVERSITY

The existence of latitudinal gradients in species diversity is one of the challenging phenomena that, to my mind, still defies adequate explanation either in ecological or in evolutionary terms. As reviewed by Pianka (1966), there are at least six distinct hypotheses, which might be referred to as (1) the time theory, (2) the theory of spatial heterogeneity, (3) the competition hypothesis, (4) the predation hypothesis, (5) the theory of climatic stability, and (6) the productivity hypothesis.

Some of these theories could bear much more rigorous testing than they have yet received, and anurans provide a virtually unexploited source of materials for such testing. For example, the theory of spatial heterogeneity rests on the idea that the greater species diversity of animals in tropical environments relates to the greater structural complexity of the environment (especially of the vegetation in tropical forests), but it does not explain the high species diversity of the plants. Comparisons of the niche diversity of evolutionary lines of anurans that have remained terrestrial in the tropical forest with those that have evolved the arboreal habit—and hence have the opportunity to differentiate with respect to the vertical, as well as horizontal, heterogeneity of the vegetation—could test this theory.

That there may be merit in such an approach is evident from just the crude data summarized in Table 1. Here, I have used Cochran and Goin's (1970) listing of Colombian frogs to compare primarily terrestrial groups with arboreal. Much greater species diversity is evident for the latter group. This is, of course, a very crude comparison that cannot partition out the various bases for the diversity, such as altitudinal variations, geographic isolation of montane populations, and the like. Comparisons should be made for the anuran components of single ecosystems and in areas of limited size.

Table 1. Diversity in Primarily Terrestrial and Primarily Arboreal Lineages of Tropical Frogs

Taxa	Family	Genera	Species
Bufonoid (Terrestrial)[a]	5	10	48
Hylidae (Arboreal)	1	12	81
Leptodactyloid (Terrestrial)[b]	1	9	19
Leptodactyloid (Arboreal)[c]	2	3	39
Total Terrestrial	6	19	67
Total Arboreal	3	15	120

[a]Includes Dendrobatidae, Pseudidae, Bufonidae, Atelopodidae, Microhylidae.
[b]All leptodactyloid genera, except *Eleutherodactylus*, *Centrolene*, and *Centrolenella*.
[c]Includes *Eleutherodactylus*, *Centrolene*, *Centrolenella*.
Source: Cochran and Goin (1970).

MECHANISMS OF SPECIATION

Most of our information about the process of speciation depends on work that has been done in the middle latitudes, where sweeping and repetitive climatic change seems to have been a major factor in promoting geographical speciation (see Blair, 1965). Some recent authors—for example, Haffer (1969) and Mayr (1969)—argue that this same mechanism accounts for the species diversity in the tropical forest, postulating major fragmentation of the rainforest into residual pockets that could fit the midlatitude model for geographic speciation.

It may be true that the midlatitude model can account for all speciation in the tropics, but I would be much happier if we had intensive studies of speciating populations in the tropical forest comparable to those available for midlatitude animals. Anurans provide excellent material for such studies.

The key question is, of course, whether there are features of the complex tropical forest and of the distribution of segments of a population that would permit evolutionarily significant limitation of gene exchange between subpopulations without requiring the massive climatic shifts proposed by Mayr, Haffer, and others.

We know so little about ecological life histories and demographic features of tropical anurans that a great extension of such studies will be necessary before we can provide answers to these questions.

REFERENCES

Barrio, A. 1970. Caracteres del canto nupcial de los pseudidos (Amphibia, Anura). Physis 29:511-515.

Barrio, A., and Delia Pistol de Rubel. 1970. Caracteristicas del cariotipo de los pseudidos (Amphibia, Anura). Physis 29:505-510.

Blair, W. F. 1958. Mating call in the speciation of anuran amphibians. Am. Naturalist 92:27-51.

________. 1965. Amphibian speciation, p. 543-556. *In* H. E. Wright and D. G. Frey, [ed.], The Quaternary of the United States. Princeton Univ. Press, Princeton.

________. 1970. Genetically fixed characters and evolutionary divergence. Am. Zool. 10(1):41-46.

________. 1972. Evidence from hybridization, p. 196-232. *In* W. F. Blair, [ed.], Evolution in the genus *Bufo*. Univ. Texas Press, Austin.

Blair, W. F., and M. J. Littlejohn. 1960. Stage of speciation of two allopatric populations of the chorus frogs *(Pseudacris)*. Evolution 14:82-87.

Bogart, J. P. 1968. Chromosome number difference in the amphibian genus *Bufo:* The *Bufo regularis* species group. Evolution 22(1):42-45.

________. 1972*a*. Afinidades entre los generos de anuros de las familias Pelobatidae y Ceratophrynidae. Act. del II Jornadas Argentinas de Zool. In press.

________. 1972*b*. Karyotypes, p. 171-195. *In* W. F. Blair, [ed.], Evolution in the genus *Bufo*. Univ. Texas Press, Austin.

Bogert, C. M. 1960. The influence of sound on the behavior of amphibians and reptiles. *In* W. E. Lanyon and W. N. Tavolga, [ed.], Animal sounds and communication. Am. Inst. Biol. Sci. Publ. 7:137-320.

Casamiquela, R. M. 1961. Un pipoideo fósil de Patagonia. Rev. Mus. La Plata, Sec. Paleontol. (Nueva Ser.) 4:71-123.

Cei, J. M. 1965. The relationships of some ceratophrynid and leptodactylid genera as indicated by precipitin tests. Herpetologica 20:217-224.

Cei, J. M., and V. Erspamer. 1966. Biochemical taxonomy of South American amphibians by means of skin amines and polypeptides. Copeia 1:74-78.

Cei, J. M., V. Erspamer, and M. Roseghini. 1972. Biogenic amines, p. 233-243. *In* W. F. Blair, [ed.], Evolution in the genus *Bufo*. Univ. Texas Press, Austin.

Cochran, D. M., and C. J. Goin. 1970. Frogs of Colombia. U.S. Natl. Mus. Bull. 288. Smithsonian Inst. Press, Washington, D.C. 665 p.

Gerhardt, H. C. 1969. Selected aspects of the reproductive biology of some southeastern United States hylid frogs. Ph.D. Dissertation, Univ. Texas, Austin.

Guttman, S. I. 1972. Blood proteins, p. 265-278. *In* W. F. Blair, [ed.], Evolution in the genus *Bufo*. Univ. Texas Press, Austin.

Haffer, Jurgen. 1969. Speciation in Amazonian forest birds. Science 165:131-137.

Low, B. S. 1972. Evidence from parotoid gland secretions, p. 224-264. *In* W. F. Blair, [ed.], Evolution in the genus *Bufo*. Univ. Texas Press, Austin.

Martin, R. F. 1972. Evidence from osteology, p. 37-70. *In* W. F. Blair, [ed.], Evolution in the genus *Bufo*. Univ. Texas Press, Austin.

Martin, W. F. 1972. Evolution of vocalization in the genus *Bufo*, p. 279-309. *In* W. F. Blair, [ed.], Evolution in the genus *Bufo*. Univ. Texas Press, Austin.

Mayr, Ernst. 1969. Bird speciation in the tropics, p. 1-17. *In* R. H. Lowe-McConnell, [ed.], Speciation in tropical environments. Academic Press, New York.

Moore, J. A. 1949. Patterns of evolution in the genus *Rana*, p. 315-338. *In* G. L. Jepsen, G. G. Simpson, and E. Mayr, [ed.], Genetics, paleontology and evolution. Princeton Univ. Press, Princeton.

Morescalchi, A. 1967. The close karyological affinities between *Ceratophrys* and *Pelobates* (Amphibia, Salientia). Experientia 23(1071):1-4.

Nelson, C. E. 1966. The evolution of frogs of the family Microhylidae in North America. Ph.D. Dissertation, Univ. Texas, Austin.

Noble, G. K. 1931. The biology of the amphibia. McGraw-Hill Book Co., New York. 577 p.

Orton, G. L. 1957. The bearing of larval evolution on some problems in frog classification. Systematic Zool. 6(2):79-86.

Parker, H. W. 1934. A monograph of the frogs of the family Microhylidae. British Museum (Nat. Hist.), London. 208 p.

Pianka, E. 1966. Latitudinal gradients in species diversity: a review of concepts. Am. Naturalist 100:33-46.

Porter, K. R. 1969. Evolutionary status of the Rocky Mountain population of wood frogs. Evolution 23:163-170.

Porter, K. R., and W. F. Porter. 1967. Venom comparisons and relationships of twenty species of New World toads (genus *Bufo*). Copeia 2:298-307.

Reig, O. A. 1960. Lineamentos generales de la historia biogeographica de los anuros. Act. y Trab. I^ro Congr. Sudam. Zool. 1:271-278.

_______. 1961. Noticia sobre un nuevo anuro fósil de Jurásico de Santa Cruz (Patagonia). Ameghiniana 2:73-78.

_______. 1972. *Macrogenioglottus* and the South American bufonid toads, p. 14-36. *In* W. F. Blair, [ed.], Evolution in the genus *Bufo*. Univ. Texas Press, Austin.

PART I

OSTEOLOGICAL AND BIOCHEMICAL CONSIDERATIONS

1

THE EARLY FOSSIL RECORD OF FROGS A REVIEW OF THE EVIDENCE

Richard Estes and Osvaldo A. Reig

An examination of the present state of knowledge about the evolutionary biology of the Anura is appropriately begun with a critical survey of whatever information the fossil record may offer about the origin and early evolution of the group. The problem of the origin of frogs as a major taxon is a matter of obvious theoretical interest for the modern theory of evolution since the latter does not yet include a clarification of the mechanisms involved in the emergence of a major group (cf. Reig, 1970).

While the unusual adaptations, behavior, and wide distribution of frogs has made them especially interesting to the systematist or evolutionist, similarity of body form, lack of meristic characters, and the presence of parallel trends has caused problems in the interpretation of frog classification and phylogeny. Current work on frog phylogeny has made little use of the existing fossil record, principally because character states used both historically (zonosternal construction, attachment of intervertebral bodies, musculature) and recently (karyotype, larval type, serum proteins) are difficult or impossible to determine in fossil material. Moreover, there has been considerable difference of opinion as to the significance and phylogenetic weight, if any, that should be accorded these features. Lacking knowledge of these and other aspects of soft anatomy, it is not surprising that few of the known fossil frogs have been used in, or have contributed much to frog classification. It is clear that much work is to be done before frogs may be used to clarify mechanisms involved in the origin of higher categories.

Although it is frequently expected that paleontological evidence may afford sound conclusions for various unsettled questions about the evolutionary biology of a given taxon, such an expectation will not necessarily be satisfied in the present paper. Only a few Mesozoic fossil frogs are known, unfortunately, and even these few appear closely related to modern groups. After the exciting discovery of a "proanuran" *(Triadobatrachus)* in the Triassic of Madagascar during the 1930's, and of true anurans in the Jurassic of Patagonia (Argentina) in the 1950's, only moderate progress has been made during the last few years. It is thus apparent that the early evolution of frogs is not yet represented in the fossil record and that only Triassic or late Paleozoic fossils are likely to clarify this early evolution.

Nevertheless, it is obviously important to have as clear and as detailed knowledge of the fossil record as possible. Therefore, such progress as has been made is worthy of review, and more important, the significant Mesozoic frogs from Argentina have been the topic of some discussion, such that critical restudy of the actual specimens was necessary. Estes was able to undertake this restudy, and by using techniques not available at the time of the original description of the Patagonian fossil frogs, was able to obtain new evidence that made it possible to settle most of the controversial points about the significance of *Notobatrachus* and *Vieraella,* the two most important known fossil genera connected with the early history of the Anura. Since Reig was responsible for the first descriptions and interpretations of these genera, it appeared fruitful to join efforts in a revisionary work based on the new evidence. These efforts have permitted full agreement, and we are pleased to offer some mutual accord in an area in which scholars have seemed eager to look for disagreement.

During the preparation of this paper, we have been concerned about the fact that paleontological controversies often arise because some requirements of the modern scientific method have been overlooked. Thus, prior commitment with particular positions, with special schools of thought, or with allegedly well founded theories, has compelled interpretation of facts beyond what is authorized by these facts. As with the Greek philosophers of the time of Plato, ideas may seem to be preferred over evidence, fostering an attitude to "save the phenomena," which has been considered a characteristic tenet of classical, in contrast to modern, science (Wightman, 1950). In this view, the enquiry begins with a generalization, and if the particulars are in agreement with the generalization, those particulars are considered to be "saved." From this position, an effort to compel the phenomena to fit the previous views involves only a single step—a step that is too often made. The convenient phrase in modern science is, rather, to "look at singulars in search of evidence for universals" (Bunge, 1959:74). Thus, evidence is to be used to build new interpretations only when it is necessary, or to modify or confirm previous interpretations when partial or full agreement actually exists.

We feel, therefore, that our task is primarily to increase the accuracy of known data through careful reanalysis, trying to avoid any commitment with previous positions. This should not be interpreted as a tribute to sheer empiricism, however, but as a recognition that at the present state of affairs, it is healthy and necessary to postpone extensive interpretation until there is previous analysis of what the facts authorize us to infer. Extensive phylogenetic or biogeographic inference based on the fossil record of frogs, therefore, belongs in the realm of "soft" conclusions, as compared with such conclusions drawn from the record of many other vertebrate groups. This paper is thus in great part an analysis of known facts; interpretation of these facts will remain as far as possible from unfounded or forced speculation. We have oriented this discussion towards two main goals: (1) a review of Mesozoic (especially Jurassic) anurans, with an attempt to resolve some of the controversy that has arisen about them, and (2) an attempt to see if any early trends in frog evolution can be observed. Although little new evidence is available, the origin of frogs, the significance of the early Triassic amphibian *Triadobatrachus,* and the source of the Lissamphibia are briefly discussed as a part of the second goal.

JURASSIC FROGS

From the early Triassic (Scythian) to the early Jurassic (Liassic) there is a gap of about 40 million years without any indication in the fossil record of the history of the Salientia (Proanura and Anura). This is unfortunate, as it is most probable that in this span of time (between *Triadobatrachus* and *Vieraella*), significant steps in the history of frogs took place.

Few Jurassic frogs are known, and few of these are well preserved. Only three described taxa are represented by well-preserved, complete specimens. Two of these are from Patagonia (Argentina): *Vieraella herbstii,* the earliest known frog (early Jurassic), and *Notobatrachus degiustoi* (middle Jurassic). Since a major part of this paper deals with a reinterpretation of these two forms, they will be discussed first. The third form represented by well-preserved complete material is *Eodiscoglossus santonjae,* from the late Jurassic or early Cretaceous of Spain (Hecht, 1970).

Vieraella herbstii

The specimen on which this species is based was discovered by Dr. R. Herbst and J. C. Viera in the plant-bearing Roca Blanca Formation, early Jurassic of Santa Cruz Province, Argentina, and was briefly described and discussed by Reig (1961) on the basis of the single specimen, the impression of the dorsal surface of a small frog. The age of the deposit was established by Herbst (1961) as Liassic. Since the discovery of *Vieraella,* stratigraphic work on the Roca Blanca and related formations has continued, and Dr. S. Archangelsky informs us that the early Jurassic date is firm (personal communication 1970). Moreover, the Roca Blanca Formation is overlain by the Chon Aike Formation, for which a Potassium-Argon date (Cazenueve, 1965) confirms a middle Jurassic age (160.7×10^6 years), reinforcing the estimate of the underlying Roca Blanca Formation as early Jurassic. The possibility of a middle Jurassic age mentioned by Casamiquela (1965*b*) can now be dismissed.

In the summer of 1965, R. Casamiquela, S. Archangelsky, and R. Herbst made the remarkable discovery of the counterpart of the type specimen at the original locality. This was described in detail, and the original specimen was reevaluated by Casamiquela (1965*b*). The animal is small, and its remains are badly crushed, rendering analysis difficult and several points of anatomical interpretation controversial. The present study was based on the original specimens and on high-fidelity latex molds made from them. Since the material is preserved only as an imprint, the molds produce a positive image that is more easily interpreted. The figures were prepared from the molds, so that statements here of "right" or "left" refer to morphological right or left rather than referring to the fossil imprint (cf. Casamiquela, 1965*b,* Figures 1, 2).

Only a brief description of the specimen is given here; more detailed discussion appears in Appendix 1. Sections of this appendix referring to individual points are indicated by numbers in parentheses given below in the description.

Description

A small frog with snout-vent length about 28 mm; nine presacral vertebrae (I,1a), of which the third and fourth (and probably also the second) bore free ribs (I,4); mode of centrum development probably perichordal, condition of intervertebral body unknown (I,1b); atlas wider than centra of other vertebrae, no atlantal transverse processes; neural arches not imbricate anteriorly; transverse processes of vertebrae 5–9 relatively long, not diminishing in size posteriorly, some with expansions perhaps indicating the presence of fused ribs; sacral diapophyses broken but probably simple.

Clavicles apparently curved (I,2c); scapula relatively long, probably as long or longer than the clavicles (I,2a); cleithrum not forked; coracoid with slightly dilated medial end (I,2b); humerus with epicondyles little distinguished from the ball, the latter unossified (I,5); radius and ulna fused, a longitudinal groove present as in most anurans, and with olecranon process well developed (I,3); at least seven carpals present, centrale 2–3 apparently fused, little or no torsion of

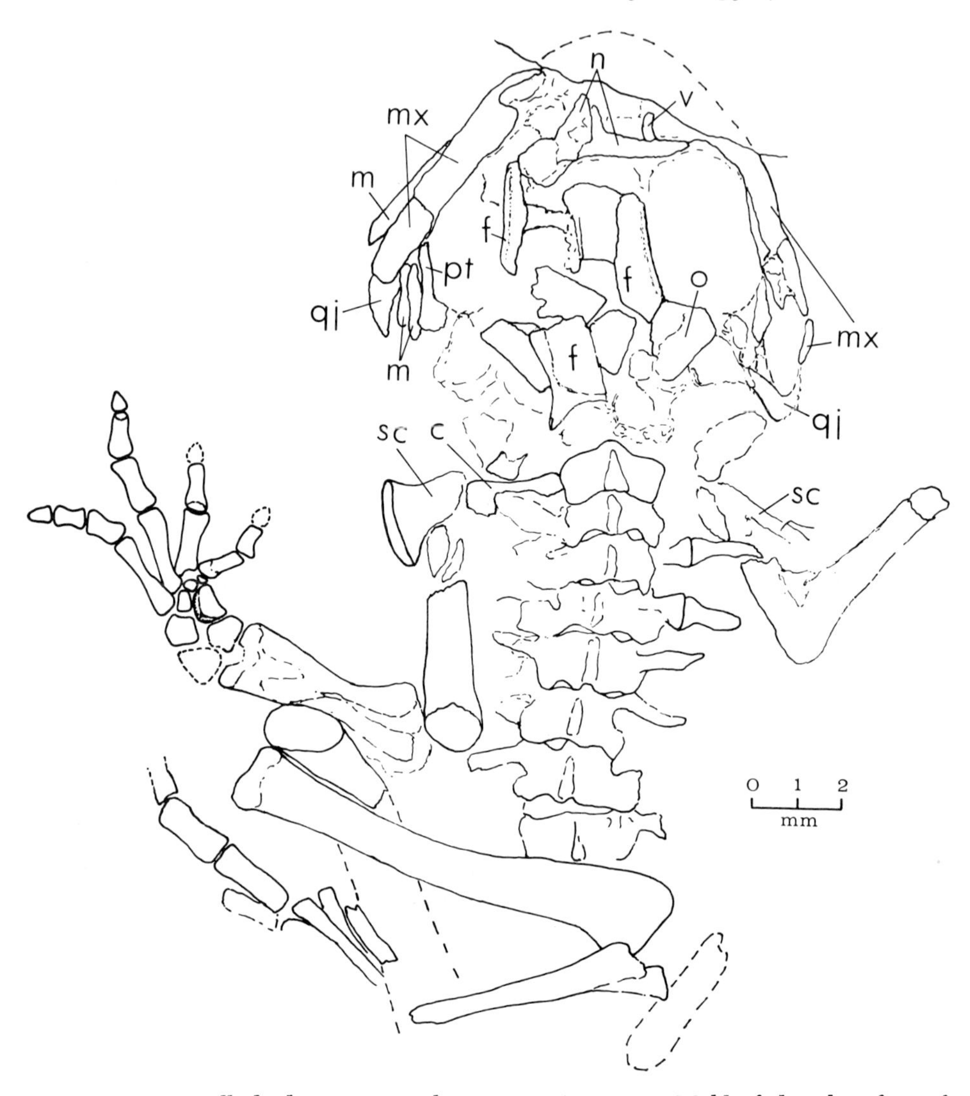

Figure 1–1. *Vieraella herbstii* Reig, early Jurassic, Argentina. Mold of dorsal surface of type specimen, M.L.P. 64-VII-15-1. *Dotted line*, areas restored from counterpart, except for distal phalanges of digits 1 and 2, which are not preserved. Thickness of line is directly proportional to clarity of the structure on the specimen. Abbreviations used for this figure, and those following: c, cleithrum; co, coracoid; cl, clavicle; f, frontoparietal; m, mandible; mx, maxilla; n, nasal; o, otic capsule; pa, parasphenoid; pt, pterygoid; qj, quadratojugal; sc, scapula; t, thyroid process of hyolaryngeal apparatus; v, vomer.

the brachiocarpal joint (I,6); prepollex absent or not preserved; manus tetradactyl, phalangeal formula 2?–2?–3–3.

Pelvis unknown, only a few fragments preserved; femur probably shorter than (or no more than equal to) length of tibiofibula; tibia and fibula fused; tibiale and fibulare apparently separate, pes poorly preserved.

Skull wider than long; a quadratojugal apparently present (II,1c); frontoparietals paired, with straight lateral borders and with only an anterior fontanelle, a posterolateral process of frontoparietal present on otic capsule (II,1e); auditory

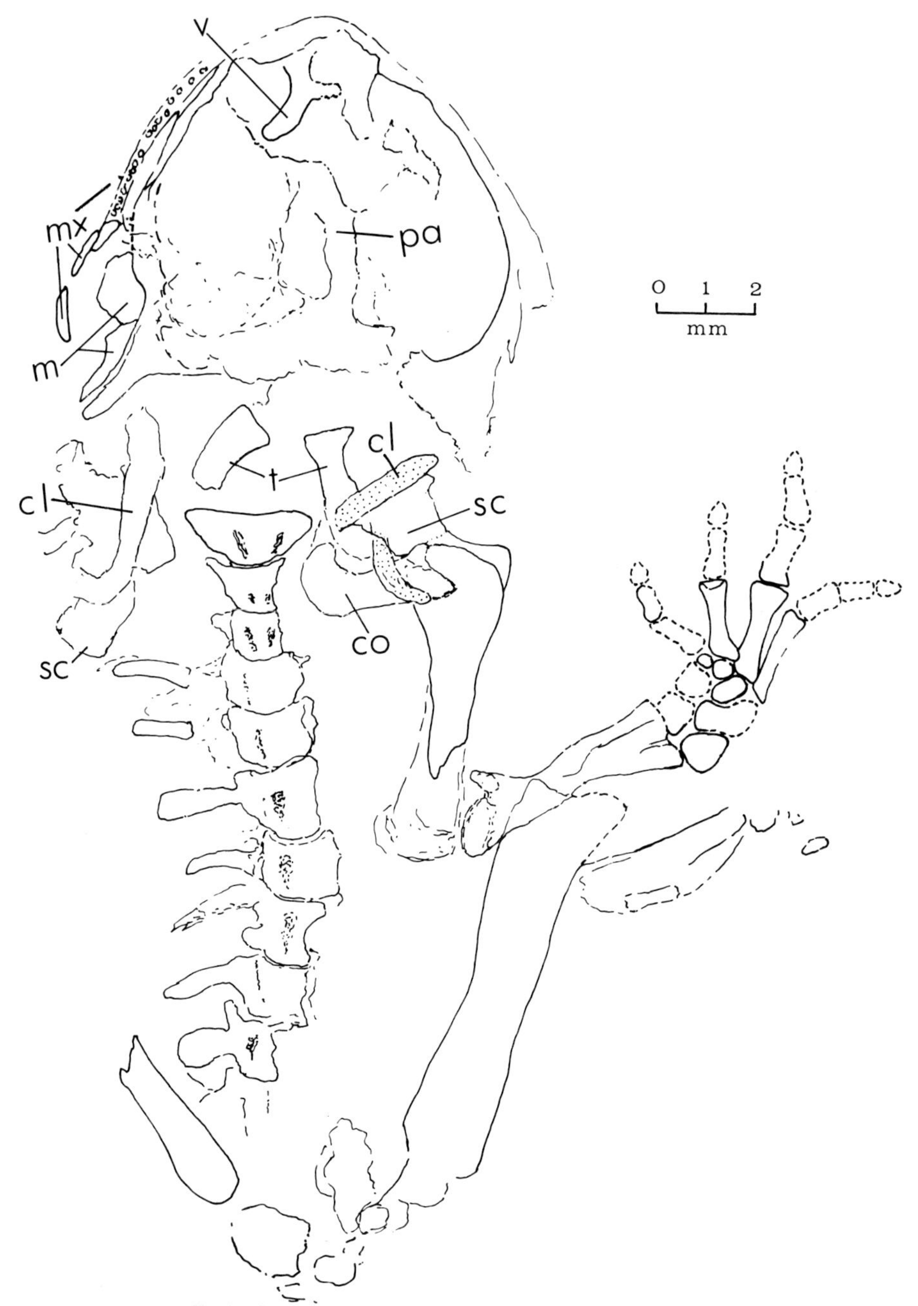

Figure 1–2. *Vieraella herbstii* Reig, early Jurassic, Argentina. Mold of ventral surface of counterpart of type specimen, P.V.L. 2488. For abbreviations, see Figure 1–1. The relative position of the two stippled areas of clavicle is indicated by the transverse line on the restoration on the right of Figure 1–3.

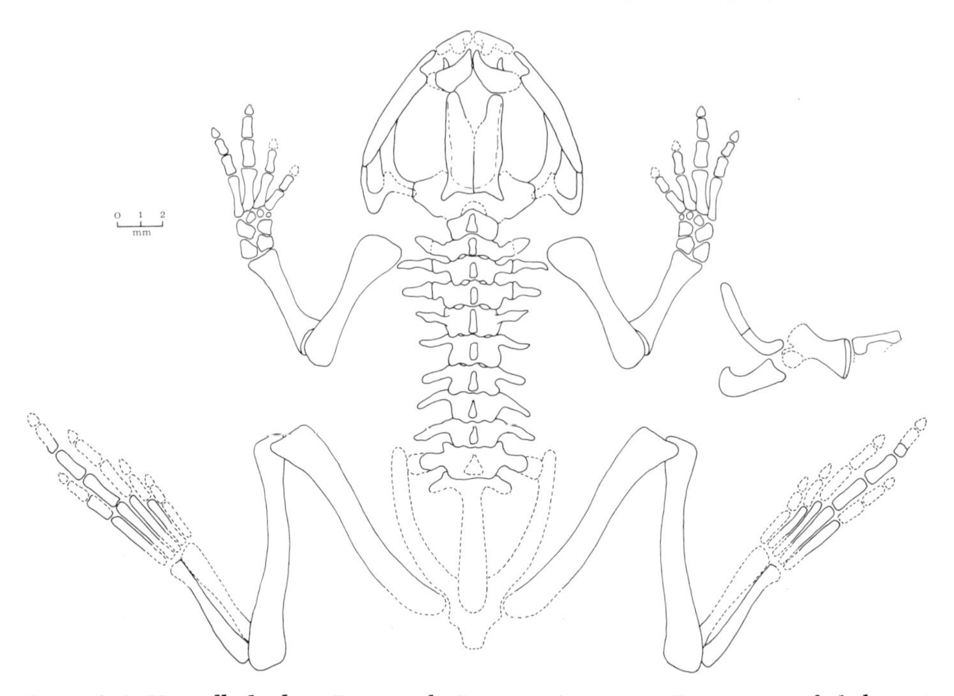

Figure 1–3. *Vieraella herbstii* Reig, early Jurassic, Argentina. Restoration of skeleton in dorsal view; shoulder girdle in ventral view, flattened so that all bones are in the same plane. *Dotted lines,* restoration; *transverse line* on drawing of clavicle, position of two pieces stippled on Figure 1–2.

capsules somewhat rounded and of limited lateral extent (II,1d); squamosal present but poorly preserved (II,1g); nasals poorly preserved but probably paired; a septomaxilla perhaps present (II,1a); vomer paired, sickle shaped, with a medial (perhaps tooth-bearing) extension (II,1a,2a); maxilla toothed; mandible with wide, posterior expansion for muscle attachment; mandible poorly preserved but probably formed only of dentary, prearticular, and mentomeckelian (II,2b); pterygoids of normal anuran type; parasphenoid apparently normal but poorly preserved; an ossified sphenethmoid probably present but represented only by a few poorly preserved fragments; thyroid processes of hyolaryngeal apparatus ossified and well developed.

Reig (1961) suggested the possibility that *Vieraella* could be considered as the only member of a new family Vieraellidae on the basis of an alleged association of character states that was not found in ascaphids, discoglossids, and *Notobatrachus,* the taxa that he considered as most closely related to the early Jurassic form. He insisted, however, that such a decision was only tentative, in view of the imperfect nature of the specimen. The referred association of character states was: (1) small scapula, (2) lack of prepollex, (3) similar number of phalanges in digits 3–4, (4) lack of lateral exposure of sphenethmoid, (5) presence of quadratojugal, and (8) absence of a frontal process of the maxilla. Most of these character states are here reinterpreted on the basis of new evidence (see appendixes 1 and 2; the exception is the presence of the quadratojugal).

Hecht (1963) stated that *Vieraella* was "an essentially modern frog which bears only one primitive character, the presence of ribs." By modern, we presume that he meant only that it closely resembled living families rather than being an ancestral form. Casamiquela (1965*b*), based on the counterpart and restudy of the original specimen, interpreted *Vieraella* as a member of the family Notobatrachidae, but our study does not support such a specific relationship. The following are the similarities of *Vieraella* to *Notobatrachus* according to Casamiquela, followed in each case by our comments:

1. GENERAL CONTOUR OF CRANIUM. This is equally similar to that of discoglossids, ascaphids, and many other frogs.

2. MORPHOLOGY OF THE "PREVOMER." The vomers of the two animals as we interpret them are different elements from those referred by Casamiquela. While they are also similar to each other in our interpretation, this is a similarity shared with ascaphids and a number of other frogs. Further, we cannot agree that the vomer is single (Casamiquela, 1965*b*); it is poorly preserved medially but appears paired as far as can be seen.

3. CONFORMATION OF THE MANDIBLE. We do not accept Casamiquela's interpretation of the presence of "splenial and angular" separated by a transverse suture, either in *Notobatrachus* or in *Vieraella*. The supposed suture is a break in the bone (Appendix 1, II,6).

4. CONSTRUCTION OF THE PECTORAL GIRDLE. Our interpretation of this structure (Appendix 1, I,2) is quite different from that seen in *Notobatrachus*. Even if we are in error about the curvature of the clavicles, the lack of coracoid expansion and, perhaps, the length of scapula still indicate differences between the two forms.

5. NUMBER AND CHARACTER OF THE PRESACRAL VERTEBRAE. The presence of nine presacrals is also found in living ascaphids, is not an uncommon variant in discoglossids (Kluge and Farris, 1969), and also occurs in pelobatids and some other groups (Kluge, 1966). Moreover, the relative size, appearance, and direction of the posterior transverse processes is different in the two Jurassic forms.

6. MORPHOLOGY OF THE SACRAL DIAPOPHYSES. The length of the single preserved diapophysis indicates that it is broken but was probably simple. A simple sacral diapophysis is probably primitive because it is present in all Jurassic forms in which this element is known (cf. Tihen, 1965; Trueb, 1973, this volume), and in any case occurs in a variety of unrelated frogs.

7. PRESENCE OF LATERAL DIAPOPHYSES ON THE FIRST POSTSACRAL VERTEBRA. Again, this is a feature common to primitive frogs.

8. PRESENCE OF FREE RIBS. This, too, is a feature common to primitive frogs.

9. MORPHOLOGY OF THE ISCHIA. The two deep holes in the sediment interpreted by Casamiquela as paired ischia (1965*b*, Figure 2) probably are not accurately identified because the faint and irregular impressions in this area give no real hint of the actual structure of any pelvic element.

10. MESIAL CONTACT OF ILIA WITH THE SACRAL DIAPOPHYSES. As described in items 6 and 9.

What, then, can be said about the relationships of *Vieraella*? Our view is that this early Jurassic frog is most similar to the Ascaphidae and that it shows some similarity to Discoglossidae, as well as possessing characters that are shared with some more specialized frogs. The combination of nine probably perichordal centra, probable *Leiopelma*-like scapula, *Leiopelma*-like vomer, unforked cleithrum, with the primitive character states of free ribs, carpus lacking torsion and with at least seven bones present, and simple sacral diapophyses suggest a primitive frog that could be referred to the Ascaphidae. The rather *Bombina*-like skull roof and ear capsules, in combination with the wide atlas (Ritland, 1955*a*), indicate similarity to the discoglossids. The presumed perichordal centra are shared with ascaphids, *Discoglossus* (in modified form), and a number of other frogs, but the primitive carpus is known only in ascaphids and discoglossids. A quadratojugal occurs in families as divergent as discoglossids and ranids, and its presence is a primitive character state.

The number of primitive character states that *Vieraella* possesses, in combination with some specific resemblances noted above, makes it possible to compare it only with the Ascaphidae and Discoglossidae. The presence of nine presumably perichordal presacral vertebrae, if weighted, would favor reference to the Ascaphidae more than such widely distributed and generalized character states as curved clavicles and presence of a quadratojugal would deny it. The wide atlas may be the only strictly discoglossid feature, but it is a weak resemblance more than counterbalanced by the elongate scapula, which is entirely unlike the very short structure uniformly present in discoglossids.

At present we do not believe that a separate family Vieraellidae is justified; the divergence from other ascaphids that occurs in *Vieraella* is too minor. It is possible to expand the definition of the Ascaphidae to include a form with the primitive character states of complete maxillary arch and curved clavicles. The difference between the arcuate clavicles of most frogs and the relatively straight ones of *Leiopelma* and *Ascaphus* is only a matter of degree that is, in any case, partially bridged by *Notobatrachus* (Figure 1–7). Although relationship suggested on the common presence of primitive character states is suspect, we do not see an alternative since sufficient morphological indication of family divergence is lacking.

In summary, *Vieraella* makes a good structural ancestor for both discoglossids and ascaphids. Whether its nine presacrals are a variant in the only known specimen or whether they are the normal condition for the genus is immaterial in this regard. The rather *Bombina*-like aspect of the skull is perhaps significant for this suggested intermediate position of *Vieraella*, since *Bombina* has been considered the most primitive, ascaphid-like of the discoglossids (Slabbert and Maree, 1945; Ritland, 1955*b*).

While we have not been able to confirm the close relationship of *Vieraella* to *Notobatrachus* that was suggested by Casamiquela (1965*b*), we have placed the two genera in the same family so that in effect the difference in interpretation is not so great as it might initially seem.

Notobatrachus degiustoi

This species was briefly described and discussed in a preliminary manner by Reig (Stipanicic and Reig, 1955) in a short paper signaling the discovery of frogs in Jurassic sediments in Patagonia. Later, Reig (1957) gave a detailed description and evaluation of the relationships of this frog, which represented then the earliest, most completely preserved, and most numerous fossil anuran. Additional material was described later by Casamiquela (1961*a*), supplemented by extensive disquisitions on its significance.

Notobatrachus is known from several complete and very well preserved imprints (i.e., molds from which all bone has disappeared), as well as many additional incomplete remains, all coming from three different localities of a single geological unit, the La Matilde Formation, of the province of Santa Cruz, Argentina. The deposits are rich in fossil plants, and Stipanicic (1957), on the basis of the paleobotanical evidence and data from regional geology, gave good reasons for placing the frog and plant-bearing sediments in the upper part of the middle Jurassic or the lowest part of the upper Jurassic. The recent Potassium-Argon date of the underlying Chon Aike Formation as middle Jurassic (Cazenueve, 1965) suggests that the second alternative is more likely. Moreover, Dr. S. Archangelsky, the paleobotanist of the Museo de La Plata, has recently informed us (*in litt.* 1970) that the fossil flora of the La Matilde Formation may be synchronous with Jurassic floras of Antarctica and southern Chile that are better placed in the late Jurassic (Tithonian). Therefore, all current evidence seems to indicate that we must now consider *Notobatrachus* as of late Jurassic age.

Hecht and Ruibal (1958) reviewed Stipanicic and Reig (1955), and Eaton (1959), Hecht (1963), Griffiths (1963) and Casamiquela (1965*b*) have all discussed various features of the species. Reig (1957, 1958) had placed *Notobatrachus* in a new family Notobatrachidae, a taxon formally accepted by Casamiquela (1961*a*, 1965*b*) and Kuhn (1961, 1962), and not directly contradicted by any authors other than Griffiths (1963), who referred it to the Ascaphidae.

We have reinterpreted a number of character states present in *Notobatrachus*. Our reasons for making these changes are given in Appendix 2; the numbers in parentheses in the following brief description of *Notobatrachus* indicate the part of the appendix in which the character is discussed.

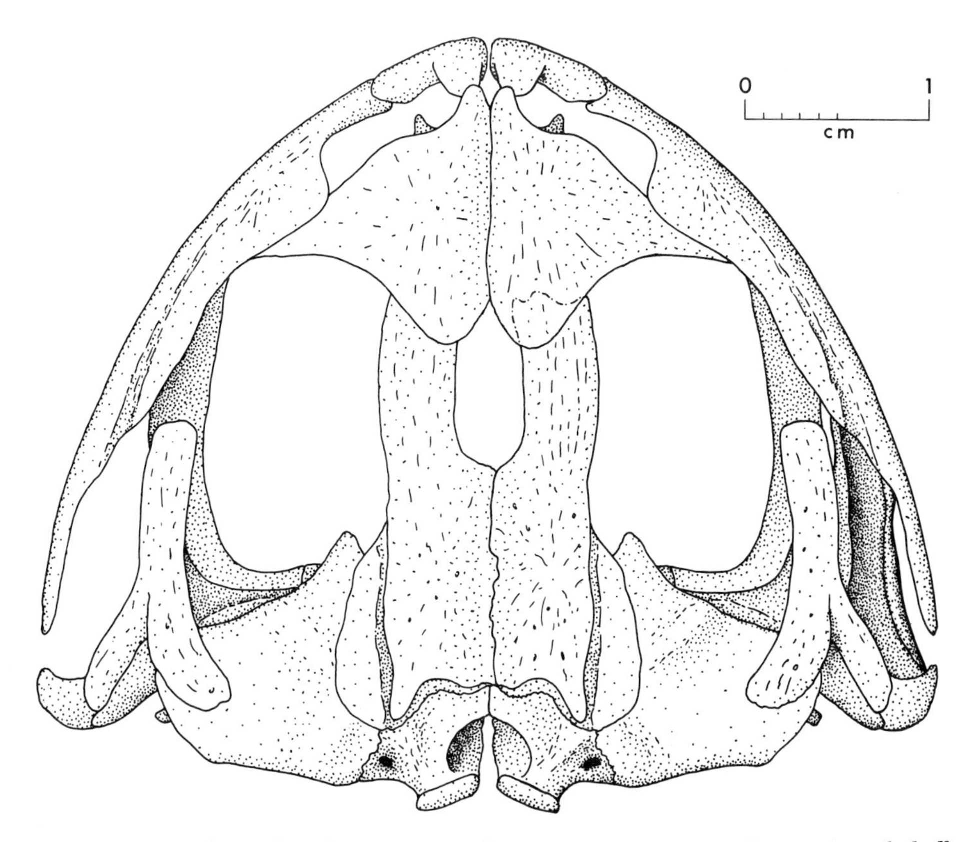

Figure 1–4. *Notobatrachus degiustoi* Reig, late Jurassic, Argentina. Restoration of skull in dorsal view; mandible removed on left side.

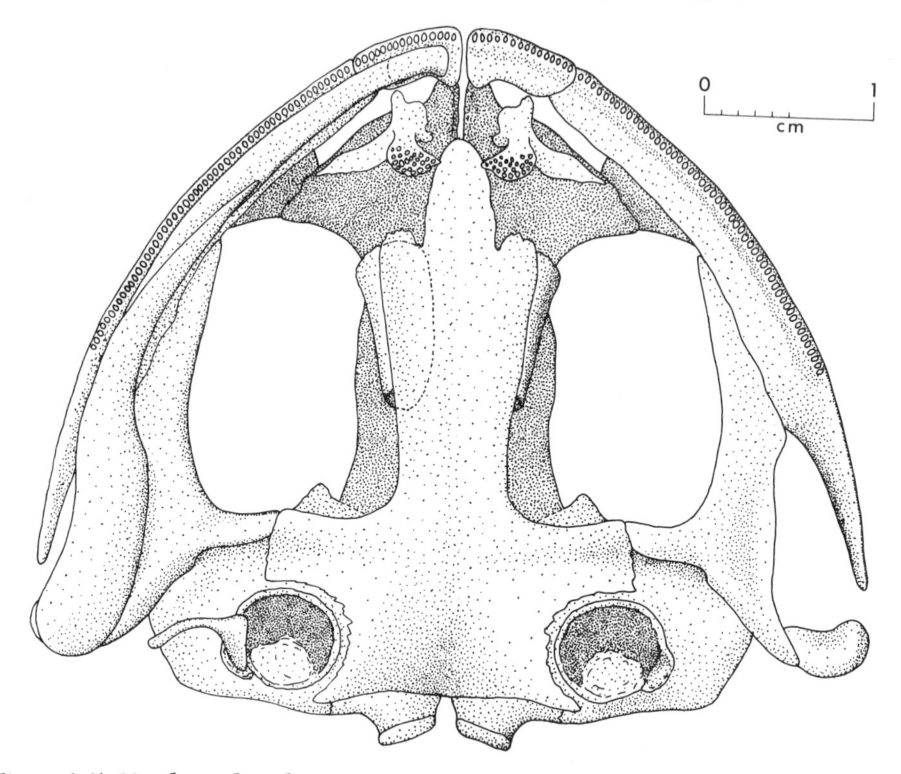

Figure 1–5. *Notobatrachus degiustoi* Reig, late Jurassic, Argentina. Restoration of skull in ventral view; mandible and columella removed on left side.

Description.

A large ascaphid frog with adult snout-vent lengths ranging from 120 to 150 mm. Skull wider than long; premaxillae with nasal processes meeting on midline; about 15 premaxillary teeth (III,6); maxilla long, reaching nearly to quadrate, about 50 maxillary teeth; teeth pedicellate (III,5); nasals broad, reaching maxillae and frontoparietals; anteromedially, narrow thickened premaxillary processes of nasals overlie (and perhaps meet) nasal processes of premaxillae (III,6); frontoparietals separated by fontanelle anteriorly and by suture posteriorly (II,2); groove on frontoparietal for occipital artery usually open dorsally but may be partially or wholly covered with a ridgelike encrustation of dermal bone; flanges of frontoparietal overlapping prootics laterally and exoccipitals posteriorly, forming a flat, sharp-edged dermal roof; squamosal T-shaped, with prominent ventral process to quadrate, flattened prootic process lying on prootic (no depression or articulation surface is visible), and anterior maxillary process that ends freely, failing to contact maxilla (II,5); quadratojugal absent (III,11); sphenethmoid paired, but the two halves may fuse anteriorly in large specimens (II,1); no dermal sphenethmoid (internasal) in fontanelle region (II,2); palatine absent (III,8); vomer toothed, with anterolateral processes surrounding choana (II,4); parasphenoid long anteriorly, reaching well between vomers, cultriform process acuminate medially and with blunt, more posterior lateral processes underlying sphenethmoids; parasphenoid corpus with anteriorly directed anterior borders ending in blunt squared-off lateral processes that reach pterygoid, and with posterior processes bordering fenestra ovalis medially (II,3; III,7); pterygoids meet-

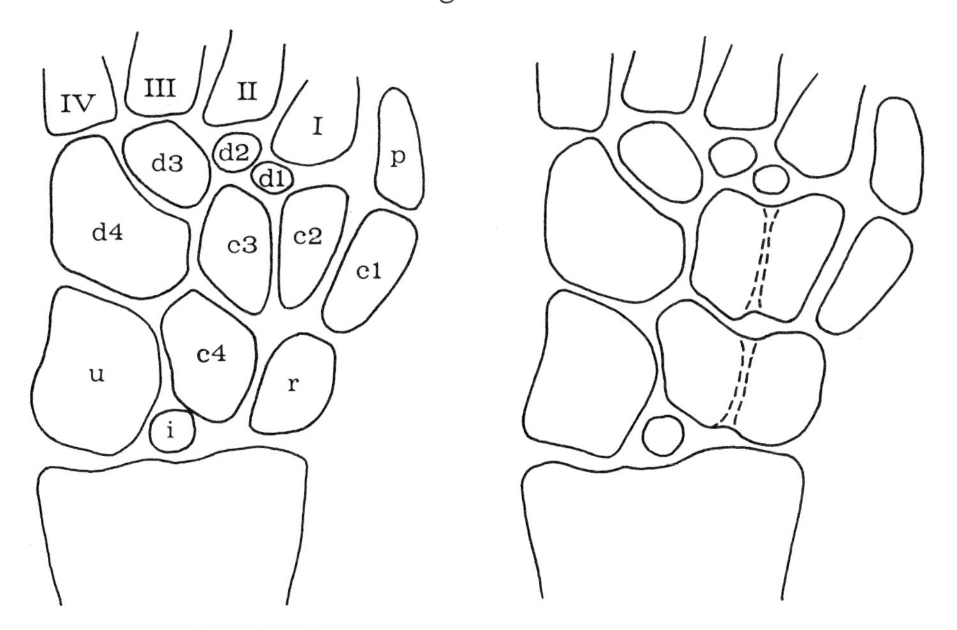

Figure 1–6. *Notobatrachus degiustoi* Reig, late Jurassic, Argentina. Ventral view of carpus; not to scale. *Left:* M.A.C.N. 17721b, modified from 17720, showing unfused radiale and centralia 2–4. *Right:* Composite of M.A.C.N. 17720 and P.V.L. 2194 using base of M.A.C.N. 17721b, showing fusions of radiale-centrale 4, and centrale 2–centrale 3. cl–4, centralia 1–4; dl–4, distal carpals 1–4; i, intermedium; p, prepollex; u, ulnare; I–IV, digits 1–4; r, radiale.

ing parasphenoids, prootics, and quadrates (III,9); probably a well-developed cartilaginous basal articulation on the braincase (III,9); prootics expanded, with large, ventrally facing fenestra ovalis; plectrum present and probably also operculum (I,3); quadrates well forward from plane of condyles (III,10); exoccipitals separated on midline ventrally but apparently joined dorsally, a large "vagus" foramen present posterolaterally; thyroid processes of hyolaryngeal apparatus ossified and well developed.

Mandible composed only of mentomeckelians, thin flat toothless dentaries (I,4); and long curved prearticulars (II,6); prearticular with deep dorsal channel for soft structures and unossified articular; prearticular with rugose posteromedial shelf for adductor muscle attachment, and strong, steep-sided posterolateral crest (III,11).

Nine perichordal presacral vertebrae (I,1); neural arches prominent, imbricate, with lateral crests for muscle attachment and prominent neural spines, the latter occasionally with flattened caps of dermal bone, especially on the first three or four vertebrae; atlas lacking ribs or transverse processes; ribs always present on vertebrae 2–5 and occasionally on 6 (III,1); urostyle with one separate postsacral vertebra in about four out of five specimens; urostyle relatively short and broad (0.46–0.51 of the remaining length of the vertebral column), dorsal surface of urostyle on same plane as neural arch of first postsacral vertebra (III,2).

Shoulder girdle probably arciferal (I,2); clavicle nearly perpendicular to midline, proximally expanded (I,2c); coracoid with extremely broad distal end such that the usual anuran anterior border is restricted to an acute proximal notch (I,2a); scapula with partes acromialis and glenoidales separated by a well-defined cleft (I,2b; III,12); scapula relatively short; scapula-clavicle ratio (based

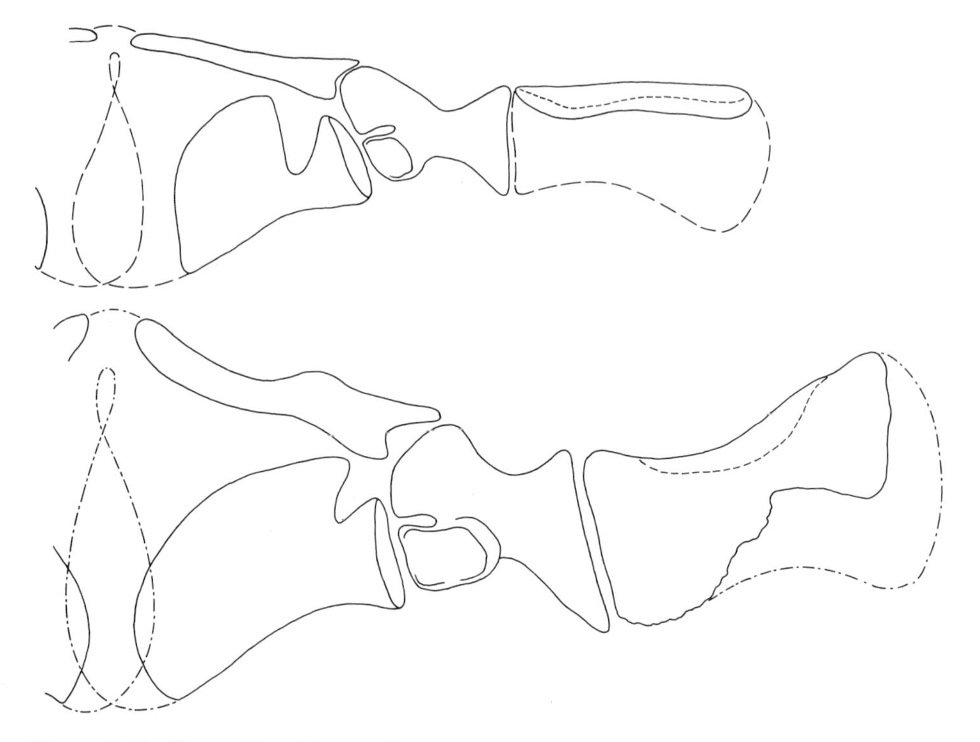

Figure 1–7. *Above:* Shoulder girdle of Recent *Leiopelma hamiltoni,* M.C.Z. 8521, in ventral view, flattened so that all bones are in the same plane; not to scale. *Long dashes,* epicoracoid and suprascapular cartilages; *short dashes,* dorsal margin of cleithrum. *Below:* Restoration of shoulder girdle of *Notobatrachus degiustoi* in ventral view, flattened so that all bones are in the same plane; not to scale. *Dotted and dashed line,* restoration of epicoracoid and suprascapular cartilages (after *Leiopelma*); *short dashes,* dorsal margin of cleithrum.

on maximum length of clavicle and anterior length of scapula) 1.5–1.7 (III,12); suprascapular cartilage not calcified; cleithrum large, clasping suprascapular cartilage anteriorly and articulating proximally with dorsal border of scapula (I,2d).

Humerus robust, epicondyles large, equally developed, fossa cubitalis ventralis absent, humeral ball unossified but apparently relatively small (III,3); humeral crests well developed, especially the distal medial and lateral crests on the dorsal side (probable sexual dimorphism in these crests exists, those of the male being better developed).

Radioulna shorter than humerus; olecranon process present but weak; shaft bilobed as in most frogs, indicating approximate boundaries of radial and ulnar portions; in most specimens a small notch separates radial and ulnar portions distally.

Carpus with nine to eleven bones present (III,4); intermedium separate; radiale and centrale 4 fused or not, centralia 1 and 2 fused or not; no carpal torsion; manus normal, phalangeal formula 2–2–3–3; prepollex present but lacking phalanges.

Ilium relatively short, shaft thick; supra-acetabular and subacetabular expansions approximately equal in size; no obvious crests or tubercles on ilium for

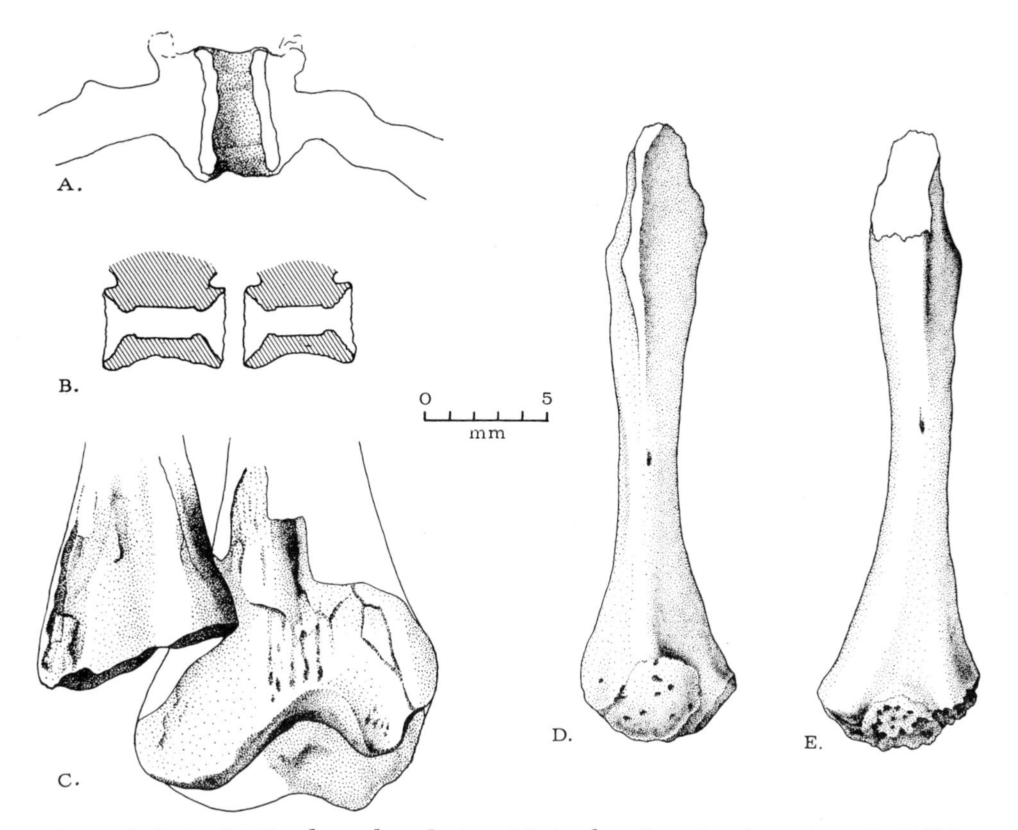

Figure 1–8. A.–C., *Notobatrachus degiustoi* Reig, late Jurassic, Argentina: A., P.V.L. 251, ventral view of sixth vertebra in section, showing imprint of intervertebral cartilage and notochord; B., the same, sagittal section of fifth and sixth vertebrae in natural position; C., distal left humerus, unnumbered M.L.P. specimen in ventral view, with proximal radioulna. Note unossified ball, prominent epicondyles, and lack of fossa cubitalis ventralis. D.–E., *Palaeobatrachus* sp., P.U. 17656, late Oligocene–early Miocene, France: D., left humerus in ventral view; E., the same, dorsal view. Note relatively small ball and olecranon scar, and lack of fossa cubitalis ventralis.

muscle attachment; ischia paired and only loosely connected on their flattened medial surfaces; pubes not visible, probably cartilaginous.

Femur and tibiofibula large, robust; tibiofibula slightly shorter than femur; tibiale and fibulare unfused; no ossified tarsus; foot normal, phalangeal formula 2–2–3–4–3.

Reig (1957) discussed the resemblance of *Notobatrachus* to other groups of frogs. On the basis of (1) nine perichordal presacral vertebrae, (2) free ribs, (3) small uncleft scapula, (4) clavicles perpendicular to midline, (5) expansion of medial border of coracoid, (6) lack of fusion of ischia and ilia, (7) phalangeal formula, and (8) carpal number, he recognized a close resemblance among *Notobatrachus, Leiopelma,* and *Ascaphus.* Of the above character states, we now know that the uncleft scapula was a misinterpretation, and that the scapula is also cleft in *Leiopelma* (III, 2). Reig (1957) concluded, however, that *Notobatrachus* could not be included in the Ascaphidae (Leiopelmatidae of his paper) because of the following character states: (1) probable firmisternal shoulder girdle, (2) coracoid with convex anterior border, and (3) toothed lower jaw. In another

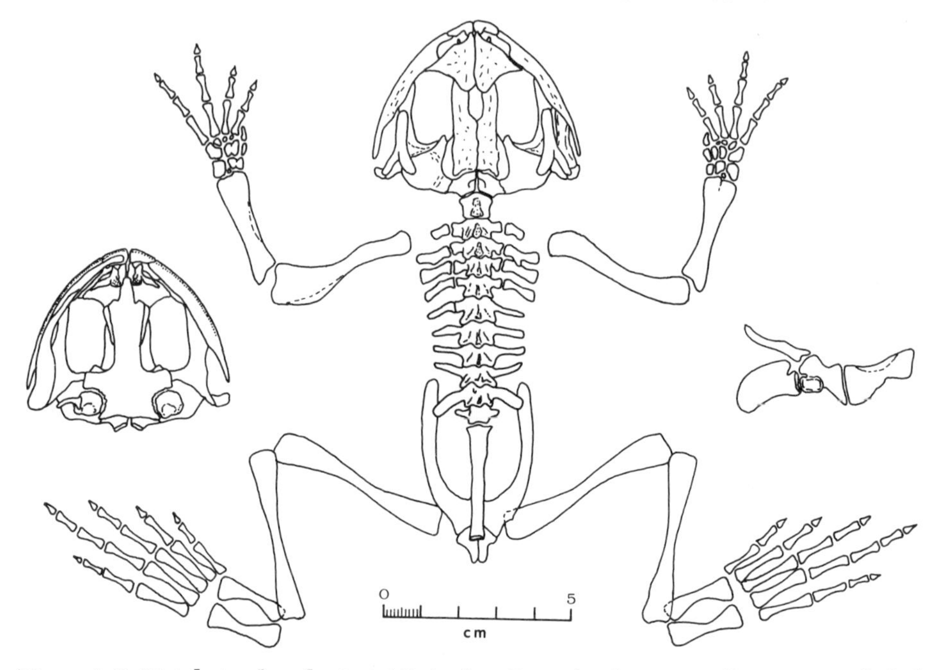

Figure 1–9. *Notobatrachus degiustoi* Reig, late Jurassic, Argentina. Restoration of skeleton in dorsal view, skull and shoulder girdle in ventral view; mandible and columella removed on left. Presumed male humeri and radioulna shown on left; *broken line,* muscle crests. Presumed female humerus and radioulna on right. Both stages shown in carpal fusion (see Figure 1–6).

paper (Reig, 1958) he added (4) greater number and better development of free ribs. From those four character states alleged by Reig to be distinctive, we now accept that (1) and probably (3) were misinterpretations (II,2,4), and that (2) and (4) are not sufficient to justify family separation. The value of the Notobatrachidae is therefore questionable, and we must explore whether it is not wiser, as already supported by Griffiths (1963), to place *Notobatrachus* in the Ascaphidae.

Most of the character states mentioned by Reig as indicative of a close resemblance of *Notobatrachus* to *Leiopelma* and *Ascaphus* are primitive rather than indicative of specific similarity to ascaphids. Regardless of this, we believe that some of these character states may be used, in combination with others summarized by Ritland (1955*a*, 1955*b*) and the shoulder-girdle similarities to *Leiopelma,* to allow allocation of *Notobatrachus* to the family Ascaphidae as the best hypothesis. The importance of these character states for family placement is enhanced by their association—they form a total morphological picture that precludes relationship to any other known family and obviates the separation of *Notobatrachus* in a family of its own. As mentioned above, in referring *Notobatrachus* to the Ascaphidae we agree with Griffiths, although we believe that he overemphasized the phylogenetic weight of the presence of nine presacral vertebrae in considering anurans diphyletic (Griffiths, 1963; cf. Kluge and Farris, 1969).

Notobatrachus possesses the following character states in common with ascaphids that in combination confirm reference to that family:

1. Nine presacral vertebrae,
2. Vertebrae perichordal, with notochord almost certainly continuous throughout column and into urostyle,

3. Sacral diapophysis simple,
4. Squamosal with long, squared-off anterior process that ends freely behind orbit,
5. Quadratojugal absent,
6. Short thick ilia and urostyle, with dorsoposterior urostyle surface in same plane as neural arch of urostylar vertebra,
7. Clavicles essentially straight, nearly perpendicular to midline and with proximal bifurcation,
8. Similarity of shoulder-girdle elements to those of ascaphids, implying an arciferal condition,
9. Coracoid expanded medially, restricting size of fenestra between coracoid and clavicle, and
10. Humeral ball relatively small, humeral epicondyles equally developed.

Of these character states, 4 and 9 are closer resemblances to *Leiopelma,* while 6 is more similar to *Ascaphus.* The following character states distinguish *Notobatrachus* from other ascaphids:

1. Well-developed middle ear, with plectrum and probably also operculum present,
2. Nasal processes of premaxillae meeting on midline,
3. Ventral position of fenestra ovalis, the latter quite large,
4. Presence of separate intermedium and all four centralia in the carpus,
5. Expansion of coracoid medially much greater than in *Leiopelma,* and
6. Occasional presence of ribs on sixth vertebra.

Of these character states, 1, 4, and 6 can be interpreted as indicating that *Notobatrachus* shows greater primitiveness than either of the two living genera of the Ascaphidae, which is to be expected in such an early member of the family. Casamiquela, however, considered the position of the suspensorium (in front of the plane of the condyles) to be a specialized feature of *Notobatrachus,* in contrast to the presumed reverse situation in *Leiopelma,* yet his argument in support of this conclusion is not convincing (1961*a*). Inference of primitiveness at the intrageneric level from the position of the suspensorium is negated by the fact that within a single living genus of anurans *(Lepidobatrachus),* this character varies greatly for each of the included species (Reig and Cei, 1963); our material shows that this is the case in *Leiopelma* as well.

The assessment of greater primitiveness of *Notobatrachus* within the ascaphids does not allow the conclusion that it is a potential ancestral frog, as previously sustained by Reig (1957). On the contrary, in spite of the great age of *Notobatrachus* we find little or no indication in its structure of closer relationship to any other family of frogs. For instance, the presence of a middle ear in *Notobatrachus* diminishes the separation of Ascaphidae from other frog families in general, but of course is not a resemblance to any one family specifically. Moreover, it is clear that the presence of nine presacral vertebrae of ascaphids has been fixed for a long period of time. Actually the origin and first diversification of frogs must have occurred much earlier than either *Notobatrachus* or the even older *Vieraella,* probably during (or before) the Triassic, a time that has not yet yielded any fossil anuran.

It is possible that the shape of the posterior corpus of the parasphenoid (with its four major processes partially surrounding the fenestrae ovales), the far anterior extent of the parasphenoid, and the rather expanded inner ear region of the prootic may indicate affinity of Pipidae with Ascaphidae through *Notobatrachus* or a related form. Some Cretaceous pipids from Israel (Nevo, 1968, Figure 10, especially 10b; see also this paper, Figure 1–11) show a parasphenoid shape that is reminiscent of that seen in *Notobatrachus,* although in loss of the quadratojugal the latter is already specialized over the earliest pipids. It is also possible that this rather labyrinthodont-like form of the parasphenoid is primitive for frogs, and that the expansion of the ear capsule in *Notobatrachus* parallels that of pipids. A further possibility is that the ventral position of the fenestra ovalis is a specialization and that its encroachment on the parasphenoid has produced the resultant shape. It is difficult at this point to choose one alternative over another.

Eodiscoglossus santonjae

This late Jurassic or early Cretaceous discoglossid from Spain, the earliest record of this family, was first figured by Piveteau (1955), and named *Eodiscoglossus santonjae* by Villalta (in Melendez, 1957). Reig (1957) briefly discussed this animal, while Hecht (1963) noted some of its distinctive features and has now given a full description (Hecht, 1970). The specimen comes from the famous lithographic limestones at Montsech, in the province of Lérida, Spain, an area noted for its preservation of complete invertebrates, fishes, and reptiles. As noted by Seiffert (1972), the locality is on the Jurassic-Cretaceous boundary, rather than late Jurassic as previously believed (Hecht, 1970).

Description.

A small discoglossid frog with a snout-vent length of approximately 28 mm. Premaxillae and maxillae apparently edentulous; frontoparietal with apparently

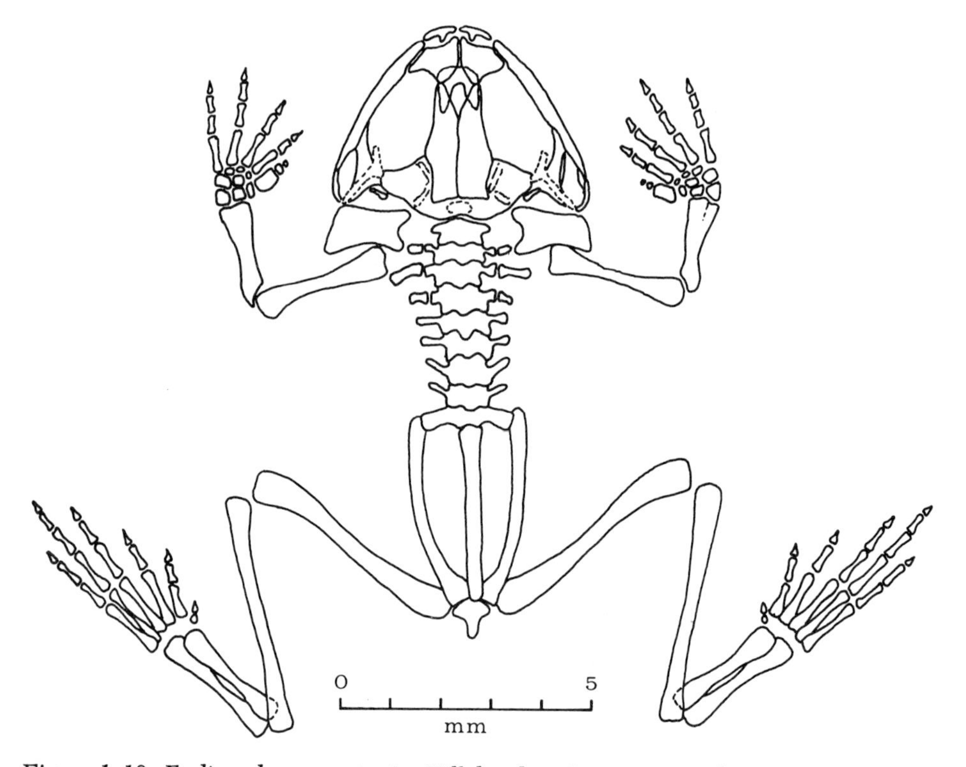

Figure 1–10. *Eodiscoglossus santonjae* Villalta, late Jurassic or early Cretaceous, Spain. Restoration of skeleton in dorsal view. Palate and ventral shoulder girdle unknown. Except for sphenethmoid and otic capsules, outlines of skull bones are partially restored. Mandible removed on left side.

a median fontanelle at least; otic capsules well developed, bluntly rounded; plectrum present; squamosals small, apparently T-shaped.

Eight presacral vertebrae, their method of articulation not preserved; free ribs on vertebrae 2–4; neural arches imbricate; sacral diapophyses simple, unexpanded; urostyle apparently lacking transverse processes anteriorly.

Cleithrum large, forked dorsally; scapula poorly preserved but evidently short; coracoid simple; clavicles probably arcuate; humerus and radioulna normal; at least four proximal and medial carpals present, distal series not apparent; carpal torsion probably not present; a nuptial pad present on pollex, and second and third digits; phalangeal formula 2–2–3–3.

Ilium with dorsal blade; iliac symphysis (synchondrosis) perhaps present; femur, tibiofibula, and unfused tibiale-fibulare normal, distal tarsals unossified; prehallux formed of two bones; phalangeal formula 2–2–3–4–3.

Hecht (1970) states that "major characters" appear in the skull and vertebral column that "demonstrate its affinity with living genera" of discoglossids. Those he cites, such as imbricate neural arches, presence of ribs on vertebrae 2–4, presence of plectrum, and simple sacral diapophyses, are all primitive or variable characters without special resemblance to Discoglossidae.

Lack of carpal torsion indicates a primitive condition also, one limited among living frogs to Ascaphidae and Discoglossidae. The presence of eight presacrals suggests the latter family as well as the remaining frog families. As Hecht (1970) states, the presence of nuptial pads on the first three digits of the hand is like that of living *Discoglossus* and *Bombina*, but nuptial pads on these three fingers are also present on some bufonids and ranids. The iliac synchondrosis (if present) occurs in the discoglossid *Barbourula*, but also appears in palaeobatrachids and pipids.

The cluster of primitive characters possessed by *Eodiscoglossus*, including the lack of carpal torsion and the presence of eight presacral vertebrae, indicates that among living families, *Eodiscoglossus* may be placed in the Discoglossidae without expanding the family definition and even without knowledge of the presence of opisthocoelous vertebrae or the shape of the scapula. The subjectively discoglossid-like appearance of the skull, especially the short sphenethmoid and blunt otic capsules, also tends to support this view. The family identification is probably correct and no alternative one can be suggested, but unquestionable reference would require additional data.

Other Jurassic frogs

The remaining Jurassic frog fossils are much less informative and much less secure in their identification.

Montsechobatrachus gaudryi

The imprint of this frog, also from the late Jurassic or early Cretaceous of Montsech, Spain, is poorly preserved, and as Hecht (1963) states, all that can be said about the specimen is that there is "little reason for a generic or specific designation for this poorly preserved cast."

The presence of palaeobatrachids in the late Jurassic or early Cretaceous of Spain noted below in the section on pipoids (Seiffert, 1972) indicates the possibility that the original identification of this animal as *Palaeobatrachus gaudryi* by Vidal (1902) could be correct, at least to family. This possibility will probably never be confirmed, however, if the specimen of *Montsechobatrachus* is as indeterminate as Hecht indicates. Only careful study by modern techniques (such as the black-light illumination used by Hecht to study *Eodiscoglossus*) will perhaps resolve this problem.

Eobatrachus agilis

Eobatrachus agilis is represented by a single humerus from the late Jurassic Morrison Formation of Wyoming. A few other skeletal elements present and originally referred to this species cannot be placed there without considerable doubt, for *Comobatrachus aenigmatis* (see below) also occurs at the same locality. These frog remains are the only ones known from the North American Jurassic. They were originally described by Marsh (1887), revised by Moodie (1912, 1914), and discussed by Gregory (*in litt.,* quoted by Reig, 1957). The most recent revision is by Hecht and Estes (1960), who recognized the presence of two frogs in the deposit. The latter study indicated that among living frog families, the type humerus of *Eobatrachus* resembled that of pipids most closely, especially in the shape of the fossa cubitalis ventralis; yet the ball is relatively larger than it is in that group. Reig (1957), on the basis of an atlas and a urostyle illustrated and not discussed by Moodie (1914), suggested that *Eobatrachus* might be placed in the Discoglossidae. These latter specimens have not been found either by Gregory (quoted by Reig, 1957) or by Hecht and Estes (1960), and as redefined by the latter, *Eobatrachus agilis* includes with certainty only the type specimen. No family allocation was given to *Eobatrachus* by Hecht and Estes, only an identification as "?Aglossa." Hecht (1963) stated that "if *Eobatrachus* must be assigned to any family, it is most likely closer to the Pipidae," and later in the same paper, he indicated that *Eobatrachus* was "clearly aglossan in affinity." Nevo (1968:294) referred to *Eobatrachus* as "Pipidae?" It is thus clear that *Eobatrachus* has been brought closer to the Pipidae (in reference, if not in morphology) than was the original intent of Hecht and Estes (1960).

Casamiquela (1965*b*), in discussing the early record of pipids and referring to *Eobatrachus,* noted that it was dangerous to refer an isolated humerus to the Aglossa, because of the well-known similarity of individual long bones throughout anurans. In addition, he cited as even more questionable the presumed use by Hecht (1963) of this information for phylogenetic and zoogeographic conclusions. Hecht's conclusions regarding *Eobatrachus* (1963) were (1) that two divergent types of frogs had already evolved in North America by Jurassic time (the other type being *Comobatrachus*), and (2) that the "clearly aglossan" affinities of *Eobatrachus* did not support the belief by Casamiquela (1961*b*) that fossil pipid frogs in South America indicated a center of evolution of frogs on southern continents. Hecht's first contention is clearly correct; the second, because of the tentative reference of *Eobatrachus* to the Aglossa, is less defendable. Clearly the pipids have been in South America at least since the later Cretaceous, but the fragmentary record and presence of early Cretaceous pipids in Israel makes it easy to agree with Nevo (1968) that "the crucial discoveries which will elucidate centers of origin of pipids and other early frog lineages are yet to made . . . ."

Obviously, then, too much has been made of the very fragmentary fossil known as *Eobatrachus*. The probable presence of palaeobatrachids in the Cretaceous of North America noted further in the text and the known presence of palaeobatrachids in the Jurassic of Europe suggests that the pipoid resemblance of *Eobatrachus* could be explained by relationship to Palaeobatrachidae rather than to Pipidae. Known humeri of palaeobatrachids are distinctive, however, and unlike that of *Eobatrachus* (Grazzini and Młynarski, 1969) in absence of any vestige of a fossa cubitalis ventralis. In summary, while *Eobatrachus* may be a pipoid, it cannot at present be referred either to Pipidae or Palaeobatrachidae.

Comobatrachus aenigmatis

This name is based on a humerus from the late Jurassic Morrison Formation of Wyoming, described by Hecht and Estes (1960) and based on a specimen originally referred to *Eobatrachus agilis*. Hecht and Estes note that while the relationships of this humerus seem to be with the more advanced frogs, it is not possible even to suggest a family to which it might be referred. The poorly developed epicondyles and the large, well-developed ball suggest an advanced family. All that can be added now to consideration of this form is that *Vieraella* also appears (at least so far as its poorly preserved humerus allows comparison) to lack perceptible development of epicondyles and seems to have a large (although unossified) ball. Until more material is available, this comparison has no special value, except to raise the possibility that *Comobatrachus* is not necessarily an advanced frog.

Stremmeia scabra

Two fused metapodials from the Jurassic of Africa were called reptilian by Stremme (1916) but were referred to the anurans by Nopcsa (1930), who gave them the name *Stremmeia*. We follow the interpretation of Hecht (1963) that the original description of these remains as reptilian was correct.

Were ranids present in the Jurassic?

Seiffert (1969*a*) has described a tiny element from the late Jurassic of France as an omosternum of a firmisternal frog. On the diagram in his Figure 2, he places this find in, or on the line leading to, the Ranidae.

It is, of course, questionable to refer a single fossil element of a frog skeleton to a living family unless that element is diagnostic. Seiffert has attempted to demonstrate that the omosternum is a diagnostic element and that his identification of the specimen is correct. The specimen is 1.2 mm long and 0.8 mm wide. We have not seen the specimen, but the figure appears to be quite detailed. Comparison of a large series of skeletons of firmisternal and arciferal frogs indicates that this element could as easily be the sternal style of an arciferal form, or the medial end of a coracoid of several different families, as an omosternum. Most ranid omosterna are forked posteriorly with two clavicular articulation surfaces separated by a distinct notch, unlike the Jurassic specimen, in which these surfaces are nearly or actually confluent; however, in some Recent ranids the notch is small and the similarity is quite close. Seiffert states (1969*a*) that the omosternum is "found in a rod-like development with bony peri- and enchondral middle and basal pieces only in the firmisternal family Ranidae." A cleared and stained specimen of the primitive arciferal pelobatid *Leptobrachium hasselti* (M.C.Z. 73935) has a prominent bony omosternum of similar sort except that the posterior end is not as enlarged as in ranids. Lynch (1971) cites the presence of omosterna in many arciferal Leptodactylinae. This fact alone indicates that the omosternum must be studied with greater care before it can be used in a diagnostic manner. At the moment, we believe that if the Jurassic element is anuran at all—and this is not in our view confirmed—Seiffert's identification and the results derived from it are an overinterpretation. For such an identification to be made on material of Jurassic age, there should be evidence derived from well-preserved, numerous, diagnostic remains from other parts of the skeleton. For this reason, we believe that until such diagnostic and complete material is available from the same or similar Jurassic localities, this record should be set aside in discussions of frog phylogeny.

A late Mesozoic palaeobatrachid

Seiffert (1972) has described a record of the pipoid family Palaeobatrachidae from the lithographic limestones at Montsech, northern Spain. The locality is on the Jurassic-Cretaceous boundary and extends the stratigraphic range of the group significantly, as discussed further in this paper.

THE FOSSIL RECORD OF PIPOID FROGS

The earliest well-known pipoids are referable to the family Pipidae and were described by Nevo (1968) from the early Cretaceous of Israel (Figure 1-11). Earlier identifications of these animals suggested that they were referable to other families, or were only in part pipoid (Hecht, 1963; Griffiths, 1963; Nevo, 1956). Nevo's paper (1968) shows definitively that all are referable to the Pipidae, and he interprets these remains as forming two genera and three species: *Thoraciliacus rostriceps, Cordicephalus gracilis,* and *C. longicostatus.* The two genera are different in many significant statistical details, as well as in the more obvious presence of the rostrum and extremely long ilia in *Thoraciliacus. Cordicephalus* is more delicately constructed than *Thoraciliacus,* and the two species of the former differ in minor skull details, size, and a number of statistically significant measurements.

Nevo (1968) suggested that *Cordicephalus* might be related to *Xenopus,* and principally on the basis of the presumed functional synsacrum and lack of fusion of presacrals, he compared *Thoraciliacus* to the African *Eoxenopoides* (Cretaceous or Tertiary) and the South American (Argentine) *Saltenia* (late Cretaceous) and *Shelania* (Paleocene or Eocene). Lack of fusion of presacrals is probably a common primitive character for these taxa. The extreme elongation of the ilia in *Thoraciliacus* is clear, but this feature is much less developed in the African and South American forms. The articulation of sacrum midway along the length of the iliac shaft (medioiliac condition), rather than on the tips of the ilia (acroiliac condition), is also present in *Xenopus,* thus vitiating the supposed relationship with *Eoxenopoides, Saltenia,* and *Shelania.* As Whiting (1961) has shown, there is considerable variation in this area as a result of the sliding union between sacrum and ilium.

Reig (1959) described *Saltenia ibanezi* as an early Cretaceous pipoid from Argentina. Many remains of this frog occur in the "Areniscas Inferiores" (=Pirgua Formation), a member of the "Salta System," which has been usually considered part of the lower Cretaceous by most Argentine geologists. Recent advances in regional geology and new fossil discoveries in the area suggest that this date has to be reconsidered, as advanced by Nevo (1968, *in litt.* from Casamiquela). *Saltenia* is now being studied by Miss Ana María Báez of the Universidad de Buenos Aires, and she has kindly reviewed for us (*in litt.*, 1970) the present status of age determinations for the Pirgua Formation. Leanza (1969) suggested that a lower Cretaceous age is to be maintained. Bonaparte and Bossi (1967), however, have found titanosaurid sauropod remains in two localities of this formation; these suggest a late Senonian or possibly early Maastrichtian age. There is also good indication, however, of an earlier age within the Senonian. Geographic correlations with the more clearly dated Puca Group of Bolivia show that the Yacoraite Formation, which overlies the Pirgua Formation in Salta, is correlated with the Bolivian El Molina Formation, dated as Campanian-Maastrichtian on the basic of charophytes, fossil fishes, and ammonites (Branisa and Hoffstetter, 1966). The Pirgua Formation is thus more probably correlated with the earlier Senonian (Santonian-Coniacian), a conclusion also reached by Moreno (1969).

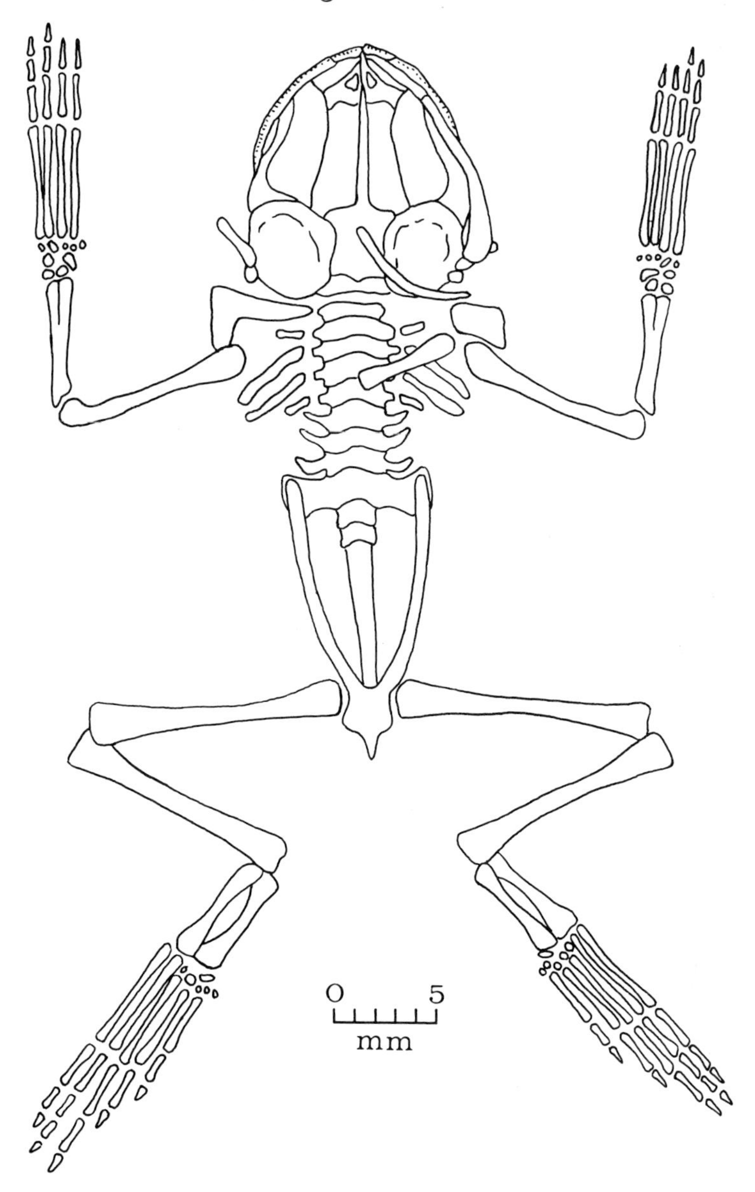

Figure 1–11. *Cordicephalus gracilis* Nevo, early Cretaceous, Israel. Restoration of skeleton in ventral view; right mandible removed, right ventral shoulder girdle removed to show ribs, vertebrae, and cleithrum. After Nevo (1968), with minor changes from the type specimen, H.U.J.Z. F–165.

Saltenia may better be considered a late Cretaceous form as a result of these data.

Reig (1959) described *Saltenia* as a probably procoelous aglossan with eight presacrals and three pairs of free ribs. Hundreds of specimens of this form are now known, and are being studied currently by Miss Báez, as noted above. We note here only that it appears to be a characteristic pipid with, so far as can now be seen, opisthocoelous vertebrae (of which the first two may be fused), an azygous frontoparietal, presence of both columella and operculum, large otic capsules, fused sacrum and urostyle, fused tibiale-fibulare, and three pairs of ribs (Estes has seen specimens in which these ribs appeared to be both fused and unfused). Casamiquela (1961*b*) and Kuhn (1962) both named a new family Eoxenopoididae for the African Cretaceous (or early Tertiary) *Eoxenopoides. Shelania,* an early Tertiary pipid described in the same paper of Casamiquela (1961*b*; see also 1960, 1965*a*) was also placed in this family. Both of these authors were apparently unaware that this family had already been named by Parodi Bustos et al. (1960; cf. Casamiquela, 1965*a*). Parodi Bustos et al. (1960) described as *Eoxenopoides? saltensis* the remains of the same taxon described in the previous year by Reig (1959) as *Saltenia ibanezi.* In later papers, Parodi Bustos and Kraglievich (1960) and Parodi Bustos (1962) insisted that this form should be placed in the genus *Eoxenopoides.* This generic problem will presumably be resolved in the current study by Miss Báez. A study of the skeletal differences of the various South American pipoid genera allegedly synonymous with *Pipa,* and further study of the Tertiary *Shelania,* will be necessary before definite conclusions may be reached.

Paleocene pipids from Brazil

In order to make this discussion as complete as possible, the presence of pipids in the Paleocene of São José de Itaboraí, Brazil, must be mentioned. These specimens, part of a diverse herpetafauna now being studied by the senior author, consist of uncrushed, nearly complete skulls and disarticulated postcranial elements. The material has yet to be studied in detail, but one astonishing fact is clear: the difference between these fossils and comparable elements of *Xenopus* is trivial. At the moment, a generic distinction could rest only on differences of time and geography from the Recent African genus—grounds that we do not consider valid, lacking morphological evidence.

The Palaeobatrachidae—Northern Hemisphere pipoids?

The Palaeobatrachidae is the only known extinct family of frogs. It is represented by numerous specimens from the Eocene through Pliocene of middle and western Europe (Špinar, 1967*a*; Grazzini and Młynarski, 1969), a Jurassic specimen from Spain has recently been discovered (Seiffert, 1972), and a possible Cretaceous record from North America is mentioned below. Following the original suggestions of Wolterstorff and Cope, Reig (1958) placed this group near the pipids, a view also held by Kuhn (1962) and Casamiquela (1961*b*). This relationship is discussed by Špinar (1966, 1967*b*, 1970), who has clarified the composition and relationships of the Palaeobatrachidae. Hecht (1963) and others have noted that several unrelated groups were included in this taxon by earlier workers and that the material is poorly preserved and difficult to interpret because of compression. Špinar's studies have shown that the family is valid and distinctive, and the other groups previously included in this group are shown to belong to pelobatids and perhaps ranids.

In addition to numerous adults, *Palaeobatrachus* is also represented by larvae at various stages of development. Both adults and larvae are notable in the preservation of numerous indications of soft anatomy, including nerves, blood vessels, glands, melanophores, and various cartilaginous structures. The sample is large enough, including hundreds of specimens, so that well-preserved examples of all character states are available. Following Špinar (1966 and *in litt.* 1967–1970), we believe that the palaeobatrachids differ from the pipids primarily in the possession of a synsacrum, procoelous vertebrae, and mentomeckelian bones. Figure 1–12 is modified from Špinar (1970, Figure 3) and is a restoration of

Palaeobatrachus grandipes, the most common form. A monograph of palaeobatrachids has been completed by Špinar (1972). The following are the most significant descriptive features of *Palaeobatrachus.*

Description.

Pipoid frogs about 50–70 mm in snout-vent length. Skull wider than long; otic capsules subcircular, expanded; both plectrum and operculum present; sphenethmoid single, large; parasphenoid lanceolate, corpus little distinct from cultriform process; small, toothed vomers present; palatines absent; pterygoid covering eustachian tube ventroposteriorly; os parahyoideum V-shaped, with short posterior projection; frontoparietal usually azygous; squamosal T-shaped at its upper end, with well-developed ventral process extending posteriorly from quadrate region to plectrum; quadratojugal vestigial or absent; maxilla elongate, toothed along the anterior half, and with strong anterior process overlapping premaxilla; premaxilla small, toothed, with prominent nasal process; teeth prominent, elongate, nonpedicellate; nasal semicircular, with faint dermal bone sculpture; septomaxilla present; mandible formed of prearticular, edentulous dentary, and well-developed mentomeckelian bones.

Vertebrae procoelous, epichordal; first two vertebrae fused, vertebrae 3–6 free, vertebrae 7–9 (or 8–9) fused into a synsacrum, the tenth vertebra seldom discrete; sacral diapophyses moderately expanded; urostyle ontogenetically composed of 4–6 vertebrae, usually one and sometimes two of these may be visible in the adult; urostyle with double condyle articulating with synsacrum; neural arches of vertebrae imbricate; free ribs in larvae on vertebrae 2–6, these more or less fused in adult but often still visible.

Pectoral girdle probably arciferal; cleithrum large, forked; scapula short, uncleft; clavicles arcuate, overlapping scapulae anteriorly; clavicle-scapula ratio 2.3–2.8; coracoid with prominent lateral expansion to meet scapula and clavicle (processus rostriformis of Špinar); humerus lacking fossa cubitalis ventralis, ball relatively small, olecranon scar reduced, epicondyles moderately developed; radioulna fused, olecranon process present; carpus usually formed of eight bones but in some specimens intermedium present in addition; metacarpals elongated; phalangeal formula 2–2–3–3.

Ilia, ischia, and pubes all paired and ossified; iliac symphysis (synchondrosis) present; major part of acetabulum on ilium; femur long, tibiofibula shorter than femur; tibiale and fibulare separate; distal tarsal of prehallux present but other distal tarsals absent; metatarsals quite elongate; phalangeal formula 2–2–3–4–4; prehallux with metatarsal and two phalanges.

Tadpole with labial cartilages and labial tentacles (the paired spiracles cited as present by Casamiquela, 1961*b*, cannot be determined on the specimens *fide* Špinar *in litt.* 1970).

The Palaeobatrachidae is undoubtedly a distinctive and valid family of frogs. In having procoelous vertebrae, well-developed mentomeckelian bones, and a synsacrum, the group is clearly distinguished from the Pipidae. Yet the lanceolate shape of the parasphenoid, azygous frontoparietal, forward extension of plectrum, squamosal with dorsal and ventral processes enclosing plectrum, enlarged otic capsules, ossified pubis, elongated metacarpals, and nonpedicellate teeth all indicate that the relationships of the palaeobatrachids are ultimately with the pipids and that the two families are reasonably to be joined in the suborder Pipoidea. Špinar (1967*a*, 1967*b*, 1972) recognizes the link with the pipids but places the palaeobatrachids in their own suborder nearer to the pelobatids, principally on the basis of the procoelous vertebrae. We agree with him that the

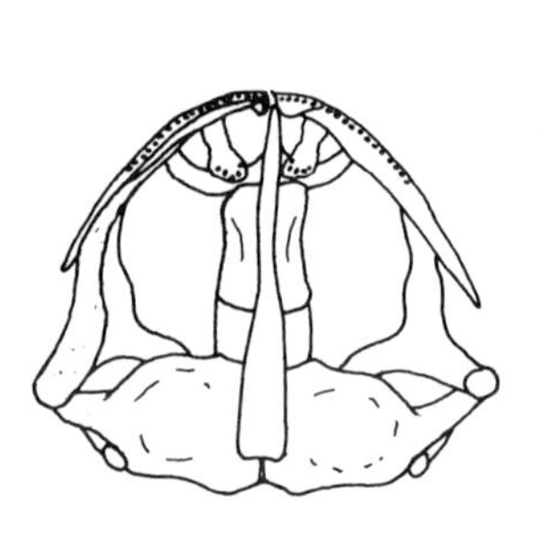

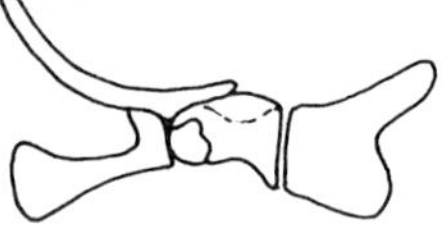

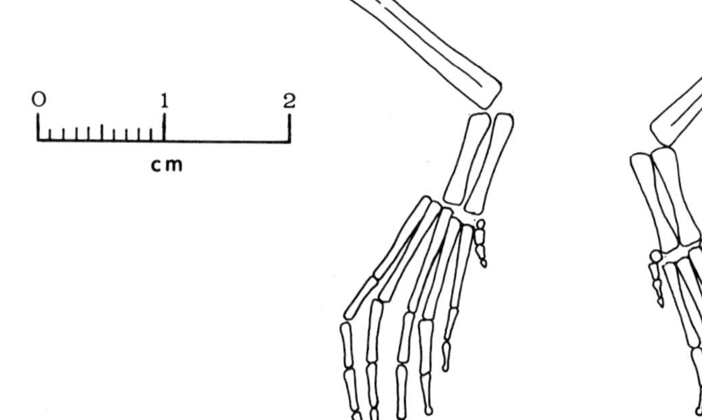

Figure 1–12. *Palaeobatrachus grandipes* (Giebel), Oligocene, Czechoslovakia. Restoration of skeleton of young metamorphosed animal in dorsal view. Skull and shoulder girdle in ventral view, left mandible removed. After Špinar (1967, 1970). Note synsacrum; *shading,* intervertebral spaces.

palaeobatrachids are a *morphological* link, at least, between the pipoids and the more advanced frogs; yet the total adaptation of palaeobatrachids is so pipid-like that we feel their systematic position is nearest to the latter family. Recent work on frog vertebrae (e.g., Kluge and Farris, 1969) has shown that the difference between opisthocoely and procoely is dependent only upon the end of the definitive centrum to which the intervertebral body fuses and that epichordal procoely (as in palaeobatrachids) and perichordal procoely (as in pelobatids) are not necessarily homologous. Further, we have found minute mentomeckelian ossifications in a few specimens of *Xenopus* in spite of their tonguelessness (thus, tonguelessness and lack of mentomeckelians are derived conditions in pipids, and palaeobatrachoids while perhaps pipoid were not aglossan). The synsacrum of palaeobatrachids is a distinctive familial specialization but cannot in our view justify subordinal status. Our difference of opinion from that of Špinar, however, is not a fundamental one; the phylogenetic diagram given in Špinar (1967*b*) clusters pipids, rhinophrynids, and palaeobatrachids relatively close together. We would simply include them in the same superfamily, implying a closer basal phylogenetic connection than does Špinar.

A specimen of a palaeobatrachid has recently been found in the late Jurassic or early Cretaceous of Spain and named *Neusibatrachus wilferti* (Seiffert, 1972). The specimen is about 21 mm snout-vent length, is essentially complete and well preserved, and differs from the Tertiary palaeobatrachids in (1) presence of a single sacral vertebra rather than a synsacrum, (2) presence of a quadratojugal, (3) free ribs present on at least some vertebrae, and (4) absence of a processus rostriformis on the coracoid. Many other smaller differences occur, but the animal is an undoubted palaeobatrachid conforming well to the description of the family given above, although more primitive in the above features. Seiffert believes *Neusibatrachus* to be ancestral to the Ranidae, as well as to the later palaeobatrachids. His description compares the animal only with these two families, and the list of presumed similarities between *Neusibatrachus* and Ranidae (Seiffert, 1972) indicates that his actual comparisons were limited. He cites the high number of of maxillary teeth (over 40) and toothed vomer as ranid features of *Neusibatrachus*, yet similar maxillary tooth counts are found in, for example, discoglossids, pelobatids, and leptodactylids; toothed vomers may occur in *Leiopelma* and pelobatids, as well as in *Palaeobatrachus* itself. Other supposed common features of ranids and *Neusibatrachus* are also misinterpreted and include a long maxilla (actually about the same length as in *Palaeobatrachus*, leptodactylids, and some other frogs), long processes of pterygoid (not significantly longer than those of *Palaeobatrachus* and many other frogs), absence of synsacrum (a condition unique in Tertiary palaeobatrachids), and absence of diapophyses on the urostyle (absent also on the majority of fossil and Recent frogs). More significant, however, is the stated presence of a diplasiocoelous vertebral column. The sacral vertebra is "excavated posteriorly for the . . . urostyle; moreover it possesses two cranial articulation surfaces for . . . the eighth vertebra as well." As the other vertebrae are procoelous the eighth is presumed to be amphicoelous. It is difficult to interpret this situation, for Seiffert does not picture the area in adequate detail. Even if diplasiocoely were confirmed, little else would support a ranid relationship. In our view, *Neusibatrachus* is a definitive palaeobatrachid with, as described, no clear indication of relationship to other families.

There is some evidence that palaeobatrachids were present in North America. The late Cretaceous specimen from Wyoming that was referred by Estes (1964, Figures 28 and 29) to *cf. Barbourula*, a living genus of discoglossids from the Philippines, was an unexpected find both temporally and geographically. An additional specimen is now known from the late Cretaceous of Montana (Estes et al., 1969). Pipids also have the unusual iliac synchondrosis present in these specimens, and Hecht and Hoffstetter (1962) also found it in presumed discoglossid remains from the Oligocene of Belgium. Hecht (1970) has now indicated its possible presence in the Jurassic *Eodiscoglossus.* More recent study of the Belgian Oligocene specimens reveals that they are actually referable to the Palaeobatrachidae rather than the Discoglossidae (Hecht, personal communication 1970). The detailed morphology of the Cretaceous specimens from Montana and Wyoming is closer to the Belgian specimens (Estes et al., 1969) than to the discoglossid *Barbourula;* in fact, the North American and Belgian specimens are probably congeneric. Further, the questionable discoglossid humerus from the late Paleocene Cernay fauna of France (Estes, Hecht, and Hoffstetter, 1967, Figure 6A-D) is also a palaeobatrachid probably belonging to the same genus. All these specimens, originally confused with discoglossids until the palaeobatrachids became better known (Grazzini and Młynarski, 1969; personal observations of Špinar's material by Hecht and by Estes), now seem to demonstrate that palaeobatrachids were present in the Cretaceous of North America, in addition to their late Mesozoic and Tertiary record in Europe. Thus the palaeobatrachid line is an ancient one.

BRIEF REVIEW OF THE FOSSIL RECORD OF FROGS

Ascaphidae. *Vieraella* and *Notobatrachus* have been referred above to this family. The questionable ascaphids cited by Estes (1964, Figure 27) are more likely to be from a small individual of the pelobatid present in the Lance Formation (*?Eopelobates,* see Estes, 1970*a*). The family has no fossil record, then, other than that from the early and late Jurassic of Patagonia.

Discoglossidae. In addition to *Eodiscoglossus* from the late Jurassic or early Cretaceous of Spain, Estes (1969) and Estes, Berberian, and Meszoely (1969) have noted the presence of presumed discoglossids from the late Cretaceous of North America. Lacking more complete material, the family reference is of course open to question, yet the best known form *(Scotiophryne)* shows some resemblance to known Tertiary discoglossids from Europe (Estes, 1969; see also Friant, 1960, and Estes, 1970*a*). As noted earlier, *cf. Barbourula* from the late Cretaceous Lance Formation of Wyoming is with great probability referable to the Palaeobatrachidae.

Pipidae. The earliest known pipids are described by Nevo (1968) from the early Cretaceous of Israel. By the late Cretaceous, they are known from both South America and Africa (Haughton, 1931; Reig, 1959). There are no other known Mesozoic occurrences, and Cenozoic records are limited to South America (Casamiquela, 1961*b*).

Palaeobatrachidae. The primary records of this extinct family are from the Eocene through Pliocene of Europe (Špinar, 1967*a*; Grazzini and Młynarski, 1969). A late Jurassic or early Cretaceous record has been described recently (Seiffert, 1972).

Rhinophrynidae. Unpublished Paleocene material of the Eocene *Eorhinophrynus* (Hecht, 1959) from Wyoming is being studied by the senior author. Holman (1963) cites the living genus *Rhinophrynus* from the Oligocene of Saskatchewan.

Pelobatidae. Estes (1970*a*) reviewed early pelobatids and revised the primitive genus *Eopelobates.* The latter resembles the living megophryines most closely but approaches pelobatines in some features. The material tentatively referred to as "near Hylidae?" by Estes (1964, Figure 31a-b), from the Cretaceous of Wyoming, is probably referable to *Eopelobates.* The family is thus probably known from the Cretaceous and without question from the Eocene.

Pelodytidae. This family is unknown before the Miocene (Taylor, 1941) of North America; additional specimens of the fossil genus are now being studied by Dr. Ted Cavender.

Leptodactylidae. Unstudied forms close to living genera are now being described by the senior author. These specimens extend the South American record of this family back to the Paleocene of Brazil and indicate that at least one living genus may be represented, supplementing the Oligocene records of *Eupsophus* (Schaeffer, 1949) and *Calyptocephalella* (Hecht, 1963) from Argentina. The leptodactylid status of the early Cretaceous fossils from Texas (Zangerl and Denison, 1950) is still in doubt (Estes, 1970*b*), and the late Cretaceous specimen from Wyoming cited by Estes (1964, Figure 32) as "family *incertae sedis,* near Leptodactylidae?" has now been referred to the discoglossid *Scotiophryne* (Estes, 1969). Lynch (1971) discusses other records of this family.

Bufonidae. Members of living species groups of *Bufo* are now being studied from the Paleocene of Brazil. The presence of the Old World *calamita* group of *Bufo* suggested as being present in the Oligocene of Argentina by Tihen (1962) has been confirmed by the presence of new material that is being studied by the senior author. Outside of South America, valid bufonid fossils are not known prior to Miocene or perhaps the Oligocene (Hecht, 1963).

Ceratophrynidae. Reig (1972) notes that this family goes back with certainty only to the Pliocene of Argentina; the Miocene *Wawelia* (also from Argentina) is referable to this family (Lynch, 1971).

Hylidae. Holman (1968) has recently described *Hyla* from the Oligocene of Saskatchewan, and Paleocene hylids are now being studied from Brazil.

Ranidae. No certain fossils of this group antedate the Oligocene (see above; and Estes, 1970*b*).

Microhylidae. The earliest substantiated record of this group is in the Miocene of Florida (see Holman, 1967).

No other frog families are represented as fossils.

TRIADOBATRACHUS AND FROG ORIGINS

Protobatrachus massinoti Piveteau (1937), from the early Triassic of Madagascar, was interpreted as an early ancestor of the frogs. Piveteau referred the unique specimen to an adult individual. Kuhn (1962) showed that the name *Protobatrachus* was preoccupied and replaced it with *Triadobatrachus.* The proper family name of the group is still Protobatrachidae Piveteau, however, contrary to Kuhn (1962) and Romer (1966), because the International Rules of Zoological Nomenclature prohibit changing the name of a family-group taxon in the event that the type genus is synonymized (Stoll et al., 1964, Art. 40). Triadobatrachidae Kuhn must thus be rejected.

The ancestral significance of *Triadobatrachus* was extended by Watson (1941) back to the Pennsylvanian *Miobatrachus* (now a synonym of *Amphibamus*), a dissorophid labyrinthodont. This interpretation was accepted by Reig (1957) and defended by Casamiquela (1961*a*, 1965*b*). Griffiths (1956) suggested that the specimen might better be interpreted as a tadpole in the later stages of metamorphosis rather than as an adult, without disagreeing with Piveteau as to its significance as an ancestral form to the Anura. Griffiths later (1963) continued to support the ancestral position, further discussed the foundation of his larval hypothesis, and also pointed out that the specimen might also be interpreted as an adult if it was considered aquatic. Hecht (1960, 1962) cast doubt on the annectent nature of *Triadobatrachus,* and disagreed with several points of Piveteau's interpretation, some of which we comment on below; areas not discussed reflect our accord with Hecht's reinterpretation. The senior author was able to study an excellent latex cast of the specimen in Paris, as a result of the kindness of Dr. D. Heyler.

Piveteau characterized *Triadobatrachus* as like frogs in three fundamental character states: (1) large frontoparietal, (2) anteriorly-elongate ilium, and (3) elongation of the proximal tarsals. Of the other frog-like character states that he cited, we consider only the following significant, either because of reinterpretation by Hecht or ourselves, or because they are difficult or impossible to determine in the specimen. These other character states are (4) sphenethmoid ossified only anteriorly, (5) large, frog-like parasphenoid, (6) shortened presacral column, and (7) short tail. Some of these features and others not included in the list deserve reconsideration.

Lower jaw. Hecht (1962) noted the presence of a lower jaw in *Triadobatrachus,* and Griffiths (1963) described its structure briefly. The presumed jaw is indicated by a curved imprint on each side, extending posteriorly from the quadrate to the fourth vertebra (Piveteau, 1937, Plate 1, Figure 1a). This thin, curved structure may be a lower jaw reflected back against the thorax, yet if this delicate imprint is restored to a closed position it reaches only to the middle of the sphenethmoid—rather short for a mandible. If it is a reflected jaw, then its visible portion (well shown on the cast) is the dorsal edge of the mandible, apparently indicating absence of mandibular teeth (Hecht did not believe that the presence or absence of teeth could be demonstrated). Griffiths (1963) believed the jaw to be composed of two elements on each side. If so, they could be homologized with the dentary and prearticular of frogs; if this and the toothless condition could be confirmed, they would be significant frog-like features of *Triadobatrachus.*

Frontoparietal. Hecht (1962) notes that the frontoparietal may be paired, rather than single as Piveteau believed, and that the position of the frontoparietal wings posterior to the otic capsules was unusual for frogs. This could be the result of the wings being dermal structures added on dorsal to the temporal musculature as in some Recent frogs. We do not believe this to be a frog-like feature of *Triadobatrachus;* however, the single unit of the dorsal skull table, whether azygous or paired, is a significant character state resembling frogs and fundamentally different from any labyrinthodont.

Nasal and prefrontal. We agree with Hecht that Piveteau's identification of nasals is open to question. On the cast, this area does not appear paired as Piveteau restored it, but rather as a raised azygous region with a rough surface; perhaps it represents sphenethmoid. The areas indicated by Piveteau as prefrontals could possibly be the posterior extensions of paired nasals as in many frogs, or perhaps could also represent frontal processes of maxillae.

Parasphenoid. While the parasphenoid is like that of branchiosaurs as Hecht (1962) has noted, there is also similarity to this element in *Notobatrachus,* at least so far as the broad, rectangular corpus of the bone is concerned. In combination with the other features it is perhaps significantly frog-like.

Caudal vertebrae. Piveteau indicated the presence of three caudal vertebrae, while Hecht saw six. If the sacrum is taken as the vertebra with the largest sacral diapophyses, there are six clear vertebral bodies behind the sacrum, each smaller than the one previous to it.

Ilium and sacral diapophyses. Hecht (1962) states that the anterior extension of the ilium is "a major adaptation for jumping" in frogs, yet it is equally important in symmetrical propulsive movements in swimming. The relatively longest ilia in frogs occur in the aquatic pipids. As noted by Griffiths (1963), the ilium is very frog-like in *Triadobatrachus* both in size and in having an elongated anterior ilial shaft with a large dorsal prominence (see Estes and Tihen, 1964). We will return to this below.

Tarsal elements. Hecht (1962) raised the possibility that Piveteau's identification of the proximal set of tarsal bones as sesamoids was in error. Hecht further suggested that these small bones might be actually proximal tarsals and that either of two possibilities might then be true: (1) that our presently accepted homologies of these bones as tibiale and fibulare are incorrect; or (2) that *Triadobatrachus* is convergent on modern frogs in tarsal elongation. We have what seems to us an even simpler explanation: If a late Jurassic form like *Notobatrachus* has a complete and primitive carpus, why may not primitive tarsal elements be present in *Triadobatrachus?* If the elements in question are identified as an intermedium and centrale 4, then their position more or less between the lateral tarsals is natural. Folding of the limb (especially if it is a young animal) and postmortem compression could easily alter their position slightly; only a small rotation of the bones as preserved would be necessary to bring the tibiale and fibulare into normal position with them.

One final point about the tarsus must be made. The tibiale and fibulare are often elongated in primitive amphibians, more so than the other tarsals. In fact, the much-heralded "frog-like" elongation of these bones in *Triadobatrachus* is little more than that already present in a number of Paleozoic groups, for example, trematopsid dissorophoids. We can thus consider the elongation of tibiale and fibulare in Paleozoic amphibians as a possible preadaptation for the frog condition; by reducing intermedium and centrale 4 and further enlarging tibiale and fibulare relative to the distal tarsals, the prefrog condition could easily be reached. *Triadobatrachus* is in this sense not incompatible with such a prefrog stage.

We have two other suggestions as to interpretation of *Triadobatrachus* that were not noted by Hecht or Piveteau. First, it is possible that the element identified by Piveteau as opisthotic could actually be an operculum of lissamphibian type. We do not here commit ourselves to the homology of these two bones as suggested by Eaton (1959), but we simply point out that the fenestra ovalis region of *Triadobatrachus* is very like that of frogs—more so when the cast is examined than from the figures of Piveteau.

The ribs of *Triadobatrachus* were interpreted by Piveteau as single-headed structures found on all vertebrae except the atlas and caudals. From the cast, the attachment of these ribs appears deep and somewhat bilobed as, for instance, in hynobiid salamanders. Confirmation of this view seems to come from what appears to us to be an atlantal rib, not interpreted as such by Piveteau but clear on the cast and the figure (Piveteau, 1937, Plate 1, Figure 1a; cf. this paper, Figure 1–13). There may be here a condition in which the most anterior ribs are strongly bicapitate while those farther back on the body tend toward approximation of the two rib-heads; such a condition is found in many salamanders. In view of the embryological differences in rib formation between living salamanders and frogs (Eaton, 1959), we cannot suggest homology between the condition in *Triadobatrachus* and that of either living group.

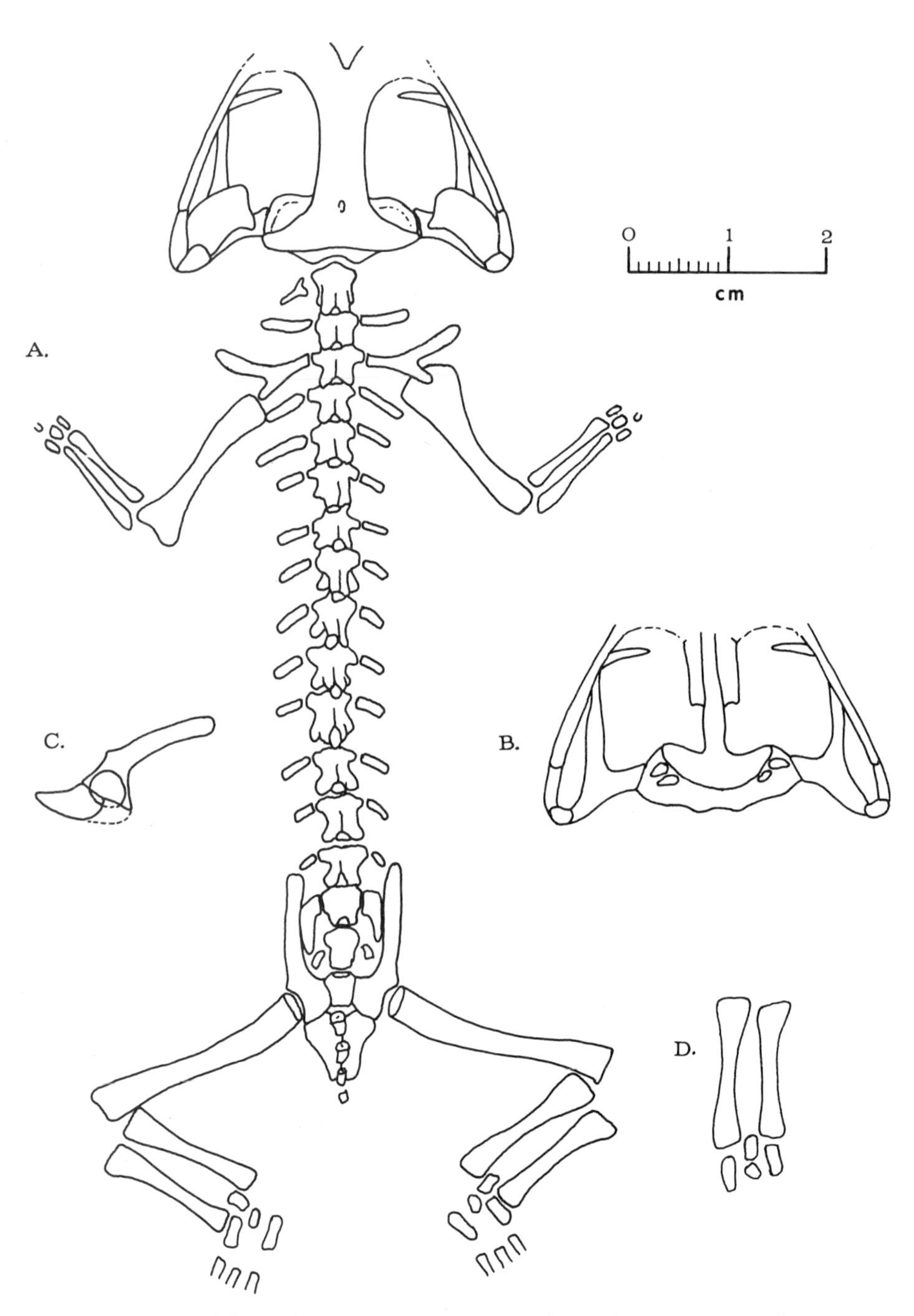

Figure 1–13. *Triadobatrachus massinoti* (Piveteau), early Triassic, Madagascar. A., skeleton in dorsal view. Left humerus seen directly from dorsal surface, right humerus seen obliquely more or less as in life. Right ischium placed in natural position, otherwise unrestored. Manus, pes, and anterior end of skull unknown; shoulder girdle omitted because of difficulty of interpretation. B., skull in ventral view. C., pelvis restored in right lateral view. D., tibia, fibula, and suggested restoration of proximal tarsals. Based on photographs in Piveteau (1937) and notes made from cast; all to same scale.

We do not accept, therefore, the statement of Hecht (1963) that the prefrog state in "the evolution of the frogs is again unknown." Although *Triadobatrachus* as a genus may not be on the direct line to anuran ancestry, it has too many similarities to frogs to be dismissed as convergent. In this view we concur with Griffiths (1963). If the vertebral column of *Triadobatrachus* were shortened from an already reduced number of fourteen presacrals to nine or ten, the total morphological picture would be undeniably frog-like. As stressed by Piveteau (1937) and Reig (1957), one may conclude from *Triadobatrachus* that the transition from a generalized Paleozoic amphibian to a true anuran involved evolution of skull morphology at a more rapid rate than evolution of the postcranial skeleton. We are far, however, from understanding the evolutionary forces involved in the modification of either area. A decision as to whether *Triadobatrachus* represented an early stage in the main line of salientian evolution (see the same conclusion in Casamiquela, 1965*b*) or whether as yet unknown proanurans were more significant, must await knowledge of related forms.

One final, more speculative matter remains to be discussed relative to *Triadobatrachus*. Considerable discussion of the possible origin of the jumping mechanism of frogs has occurred recently and has been ably reviewed by Gans and Parsons (1966). Three possible habitats in which this adaptation might have taken place have been suggested: aquatic, terrestrial, and riparian. The existing fossil record does not permit a definitive choice, but if *Triadobatrachus* is, as we believe, related to true frogs, then a review of evidence relating to the habits of this animal may be important to consider. The interpretation of Griffiths (1956, 1963) that the specimen of *Triadobatrachus* represents a metamorphosing tadpole is an interesting one. The presumed lack of a functional sacrum, the presence of vertebral segments in the tail, the separate forearm and shank bones, the absence of upper jaw structures, the possible presence of a *Stirnorgan* (Griffiths, 1963), and the separation of the lower jaw bones were character states suggested by Griffiths in confirmation of his view. We are not convinced that discussing *Triadobatrachus* in terms of the tadpole-metamorphosis transition seen in true frogs is entirely justified, yet the specimen of *Triadobatrachus* could have been the young stage of a form in which the sacral diapophysis would have fused to the sacrum in definitive adult condition. The sacral ribs of *Triadobatrachus* are actually rather well developed structures, larger than as restored by Piveteau (1937, cf. text Figure 2b, Plate II, Figure 1; and this paper, Figure 1–13). Moreover, the attachment of the sacral ribs to the sacrum is extensive and intimate even if unfused. If this animal was young, and if fusion took place in the adult, the sacral region would be a firm, well-developed structure with an elongated surface articulating with the ilium. The detailed frog-like appearance of the ilium as well as its anterior extension implies that either jumping or propulsive swimming movements were necessary to the animal; such an anterior extension and detailed morphology of the ilium are unique in frogs and *Triadobatrachus*.

In the late stages of development of recent aquatic anuran tadpoles, the sacral diapophyses are unossified and the sacral centrum little different from those in front of it. In these larvae the limbs, musculature, and vertebral column are used for symmetrical swimming movements before the tail is entirely resorbed, whatever their function may be in the adult. In *Triadobatrachus*, the length of the vertebral column suggests that it is more probable that symmetrical swimming movements were practiced than that this structure functioned in jumping. In this we agree with Hecht (1962), although if the sacral diapophyses were fused or more firmly connected to the sacrum in the adult, it is possible that this same propulsive movement could have been used in a riparian situation as a *primitive* sort of "jump" to help the animal regain the water more easily if a predator were present. At that stage of evolution any such thrust would confer an advantage to the animal, however imperfect it might be compared with the jump of a true anuran. The very frog-like skull of *Triadobatrachus* suggests that considerable brain evolution beyond the labyrinthodont condition might have occurred and that the neuromuscular and sensory coordination necessary for symmetrical limb movement could have been present.

Triadobatrachus, then, appears to us to be an aquatic animal, probably a young stage, that was able to use symmetrical propulsive swimming movements. If it is as closely related to the ancestry of true anurans as we believe, then the simplest hypothesis is to presume an aquatic origin and habitat for proanurans and perhaps for true anurans as well. The cumulative effects of directional selection for short body form in such an aquatic stock could have led to the development of a jumping mechanism in riparian forms that gave rise to true anurans. It is also possible that the beginning radiation of true anurans involved an early dichotomy in which an ascaphoid stock took advantage of the riparian habitat while a pipoid stock emphasized and refined the aquatic adaptation. Although we tend to emphasize the importance of swimming in early frogs, both larval and adult, our views are not incompatible in any serious way with those of Gans and Parsons (1966). These authors have persuasively argued that the riparian habitat could have been the one in which jumping per se evolved; we consider this view plausible and simply consider it to have been a phenomenon of the early riparian radiation of true anurans whose ancestors were aquatic.

Casamiquela (1965*b*), in a paper not seen by Gans and Parsons (1966), argued heatedly that Griffiths' (1963) belief in an aquatic ancestry for frogs was, among other things, "en flagrante contradicción con algunas premisas fundamentales de la biología." We cannot agree with this extreme view. Although to us the Permian trace fossils presumed by Griffiths (1963) to represent the handprints of a swimming proanuran are overinterpreted and not demonstrably of salientian origin, Griffiths' work hardly fulfills the accusation noted above. If one opposes Griffiths in this particular view, it is perhaps more a matter of taste than of biological principles.

The sketch restoration of *Triadobatrachus* given by Romer (1966, Figure 146) is inaccurate with respect to proportions; the skull is given natural size while the body length is one-half natural size; individual bone outlines are generalized and in some cases inaccurate. The restoration given here (Figure 1–13) has been made from photographs in Piveteau (1937) and notes based on a latex cast of the specimen.

THE ANCESTRY OF THE SALIENTIA

The problem of the origin of the Salientia (Proanura and Anura) is an integral part of the continuing discussion on the polyphyletic or monophyletic origin of the class Amphibia, a discussion that has been recently reviewed by Schmalhausen (1958, 1964), Szarski (1962), Reig (1964), and Thomson (1964). After demonstration by Williams (1959) that the Gadovian theory of vertebral arcualia was not applicable to tetrapods and the increasing knowledge about the morphological and biological features of the sarcopterygian-amphibian transition (Thomson, 1969; Thomson and Vaughn, 1968), little place remains, in our view, for acceptance of the hypothesis that separate labyrinthodont-salientian and lepospondyl-urodelan lines can be traced back to different groups of sarcopterygians (cf. Jarvik, 1962). It is now generally agreed that the best-founded hypothesis of amphibian ancestry is that all known fossil and living amphibians had common rhipidistian ancestors; as Carroll (1967) has noted, the ancestry might have been within a single family or superfamily and hence monophyletic at that level, rather than being derived from different superfamilies as Jarvik has suggested.

This hypothesis does not afford an answer, however, to the question of placement of salientian origins within the manifold radiation of the Paleozoic amphibians. In this connection, the Watsonian view of a labyrinthodont ancestry for the Anura (Watson, 1941) deserves careful consideration, in spite of reinterpretation of *Amphibamus* (Gregory, 1950; see, however, Griffiths, 1963). Gregory synonymized *Miobatrachus,* a form directly implicated by Watson in frog ancestry, with *Amphibamus* and included the latter in the rhachitomous family Dissorophidae, a view recently confirmed by Carroll (1964) and De Mar (1968). The dissorophids stand, therefore, as a group to be considered in the search for salientian ancestry. The same conclusion was reached by Estes (1965) from the point of view of the lissamphibian hypothesis.

The lissamphibian hypothesis rejects the view that the anuran line on the one hand and a urodelan-apodan line on the other hand are derived from two separate Paleozoic amphibian subclasses. This view has been carefully analyzed by Parsons and Williams (1963), who have revived the Haeckelian concept of the Lissamphibia and extended it, in the sense of Gadow, to include the orders Anura, Urodela, and Apoda, which are presumed to have had a common amphibian ancestor. Eaton (1959) had previously supported a close relationship between anurans and urodelans, but he regarded the apodans as a separate group. Reig (1964) was also inclined to exclude the apodans from close relationship with frogs and salamanders, and he pointed out that although the evidence seemed to indicate common origin of anurans and urodelans, these relationships ought not yet to be taken as conclusive.

Recent biochemical work on this subject has been interesting but contradictory. In a significant study, Salthe and Kaplan (1966) concluded that the molecular structure of the LDH enzyme in 45 species of frogs and salamanders supported a common ancestry for these two groups. Unfortunately, apodans were not included in their study, and it will be of considerable interest to determine the LDH enzyme structure in this group. On the basis of cytophotometric estimates of the DNA value in 33 species of amphibians, Beçak et al. (1970) found that apodans and urodelans share a high DNA value (about 400 per cent that of mammals), while in diploid anurans this value ranges from 43 per cent to 147 per cent. They concluded that the two former groups might have been derived from a common ancestral stock, while frogs might have originated from a labyrinthodont stock more closely related to the ancestry of reptiles. Similar work by Goin and Goin (1968), however, indicates that only salamanders had such high DNA levels; their data show apodans and frogs with levels about the same as mammals, while salamanders and sarcopterygians are on the order of 500 per cent that of mammals. Aside from the contradictory data from the two laboratories, the representation of apodans in all of these studies is too small to formulate useful results at this time. Such biochemical work may ultimately be useful in answering questions about relationships of modern amphibians, but greater precision and wider sampling is needed (see, for example, Szarski, 1970).

Karyological data for vertebrates has been reviewed recently by Morescalchi (1970), who has shown that the most primitive frogs and salamanders, and the few apodans studied to date, all have similar and presumably primitive karyotypes (high chromosome number, abundant acrocentric and microchromosomes), as opposed to conditions in specialized amphibians (lower numbers and presence of metacentric elements). Morescalchi thus suggested (1970) that the lissamphibian hypothesis appeared to be supported by karyological data. As for the biochemical studies noted above, however, sufficient data on apodans are still lacking. Interesting and suggestive as the karyological data are, wider sampling is necessary before hypotheses on the direction of chromosome evolution in amphibians may be used to support the concept of the Lissamphibia.

The fossil record of modern amphibians does not shed light on the question of the reality of the Lissamphibia. The early Triassic salientian *Triadobatrachus* is, in our view, already well advanced in the direction of anurans, yet neither this proanuran nor the earliest true anurans show any particular clues to relationship with any other modern group of amphibians. Vaughn (1965) has described *Lasalia cutlerensis,* based on frog-like vertebrae from the lower Permian of Utah. Although the elongated transverse processes of these vertebrae do resemble those of frogs, the holochordal centrum is unlike that of most frogs. Vaughn has suggested that holochordy might be the primitive state if *Lasalia* were related to frogs; the presence of perichordal centra in *Triadobatrachus* and *Notobatrachus,* as well as the analysis of Recent forms by Kluge and Farris (1969), indicates that perichordy is primitive. At present, we see no reason to ally *Lasalia* with frogs. The apodans have no described fossil record, and well-defined salamanders appear in the late Jurassic (Hecht and Estes, 1960; Seiffert, 1969*b*). Vaughn (1963) described a urodele-like vertebra from the lower Permian of New Mexico; this was named *Vaughniella urodeloides* by Kuhn (1964). Vaughn is correct in stating that the vertebra has a salamander-like haemal arch, but the specimen is poorly preserved and interpretation difficult. It is a caudal vertebra, and the morphology of presacral vertebrae should be known before salamander relationships are suggested. Thus, in all cases the scanty or controversial nature of the available Mesozoic or late Permian material tells much less about possible interrelationships of salamanders, anurans, and apodans than does study of their living relatives.

Needless to say, the value of the lissamphibian hypothesis would be immeasurably strengthened if a "protolissamphibian" could be identified in the fossil record. After their careful morphological analysis, Parsons and Williams (1963) were unable to arrive at conclusive identification of such a common ancestor within the known groups of Paleozoic tetrapods. The later analysis by Estes (1965) led to no better result beyond demonstrating that "the closest *resemblances* between modern and Paleozoic amphibians are with the temnospondyls (specifically the Dissorophidae) rather than with the lepospondyls." Estes (1965) pointed out, however, that "it is certainly true that no individual temnospondyls show tendencies toward skull bone reduction or the development of pedicellate teeth, both of which should be demonstrated in any immediately ancestral form or primitive lissamphibian." More recently, the origin of anurans from rhachitomous temnospondyls has been supported again by Shishkin (1970) on the basis of the structure and development of the anuran head. Shishkin denies relationship of urodeles and apodans to frogs, however, and believes the former two groups to be lepospondyl derivatives. He also feels that there has been a recent tendency in the literature to move from noting possible evidence for a close relationship of modern amphibians to an acceptance of the lissamphibian hypothesis, without much evidence in addition to that given by Parsons and Williams (1963), and he cautions against the use of the "hypothetical protolissamphibian" of Parsons and Williams in attempts to clarify early relationships of modern amphibians.

Nevertheless, an animal has recently been described that fulfills in more ways than any other Paleozoic amphibian the requirements for a "protolissamphibian." This animal, *Doleserpeton annectens* Bolt (1969), from the early Permian of Oklahoma, is related to the labyrinthodont family Dissorophidae but was placed in a monotypic family. *Doleserpeton* was not discussed by Shishkin (1970); presumably the latter's paper was in press by the time he could have received it. In view of the Permian age of *Doleserpeton,* it is interesting to note that Salthe and Kaplan (1966) concluded on the basis of their data on protein evolution, that the Permian is the period in which frogs and salamanders would have been expected to diverge. Parsons and Williams (1963) listed a series of character states that might be expected to occur in a common ancestor of the Lissamphibia and that could be recognized in a fossil; Bolt (1969), correctly, added bicuspid teeth, giving a total of twenty (see also Lehman, 1968, for discussion of this character). *Doleserpeton* has fifteen of these twenty character states, a number greater than any other Paleozoic group but not substantially so. Parsons and Williams (1963) also listed features that linked the three modern orders most intimately. Of these, unfortunately only two were osteological and thus possible to include in a list obtainable from fossils; these were (1) the presence of pedicellate teeth and (2) the presence of an operculum-plectrum complex of the middle ear. Again, we may add to this list the bicuspid tooth crown (Bolt, 1969).

The particularly significant feature of *Doleserpeton,* aside from its minute size (skull 12–20 mm long), is the possession of pedicellate bicuspid teeth that are formed of a distinct crown and pedicel. It thus has two of the three important character states noted above. The other, the operculum-plectrum complex, cannot be determined in *Doleserpeton* and has not been demonstrated in any other described Paleozoic amphibian. Estes (1965) mentioned that this complex seemed to be present in a dissorophid in process of description by Eaton, at least if a strongly reduced opisthotic forming a flat, oval piece of bone posterior to the stapes is actually to be interpreted as an operculum in this specimen. The interpretation is, however, uncertain, and the actual evidence about it not yet published. Detailed studies of dissorophid skull anatomy by Carroll (1964) and De Mar (1968) do not afford evidence of an operculum in other representatives of the family.

Aside from the pedicellate teeth, *Doleserpeton* is no closer to a "protolissamphibian" than are other dissorophoids. Although it has lost the ectopterygoid, it differs strikingly from modern amphibians in lack of reduction of the bones of the posterior skull region, a lack shared with other dissorophoids. Also, while no contrary data are known, the value of pedicellate teeth as an index of phylogenetic relationship cannot be decided unequivocally because analogous structures occur in some teleosts (Parsons and Williams, 1963). However, the teleost teeth lack a pedicel and are loosely attached to the jaw itself, unlike the distinctive pedicelly of lissamphibians. Parsons and Williams (1962) had considered the alternative hypothesis that this peculiar structure, shared by the vast majority of living amphibians and in detailed construction unknown elsewhere (except now in *Doleserpeton*), might be the result of parallelism. The review of the histological literature made by these authors shows that we are only at the beginning of a sound knowledge of the dental microstructure of the amphibians. Therefore, without disregarding the importance of pedicellate teeth for the lissamphibian hypothesis, it is clear that additional work is necessary before pedicellate teeth are considered as a necessary protolissamphibian character state, or (perhaps more important) that the presence of such teeth in a labyrinthodont is unequivocal evidence of ancestry for any of the modern orders. Besides this, it is also clear that *Doleserpeton* has no features that specifically suggest relationship to frogs. If it is at all related to any lissamphibian group, it is closest to the salamanders, which is not surprising in view of the more primitive structure of the latter group.

Estes (1965) concluded that dissorophids were the most plausible lissamphibian relatives although "they certainly cannot be cited as the group of origin at this precarious state of knowledge." Cox (1967), on the other hand, suggested that the gymnarthrid microsaurs were plausible ancestors, stating that "it seems very likely that the modern amphibians are descended from [them], or from some similar, closely related group of microsaurs." Commenting on this conclusion, Estes (1967) noted that the similarities given between the two groups by Cox were "faint, sometimes negative, and in some cases spurious resemblances." These statements make clear the necessary distinction between phenetic similarity and phylogenetic relationship, and while there is little to suggest that microsaurs or any other lepospondyl group were ancestral to Lissamphibia, dissorophoids may be supported with greater probability as progenitors. However, the known dissorophoids suggest a cautious view of their potentiality to be direct ancestors. Bolt (1969), after placing *Doleserpeton* in a family of its own, allied it with dissorophids and trematopsids in the superfamily Dissorophoidea. Vaughn (1965) has described additional evidence for relationship of these two families, supporting the previous suggestion of Romer (1947). De Mar (1968) found that dissorophids and trematopsids demonstrate an increased terrestrial adaptation during their Permian radiation. Even the Pennsylvanian dissorophids (principally *Amphibamus*) seem not to have been primarily aquatic animals (Carroll, 1964), and *Doleserpeton* "was possibly adapted more completely to terrestrial existence than any other known labyrinthodont" (Bolt, 1969:891). *Micropholis,* a small early Triassic labyrinthodont from the *Procolophon* zone of South Africa, was questionably referred to the Dissorophidae by Romer (1966), but Carroll (personal communication 1970) believes that it is without doubt dissorophid. According to Carroll, this animal, which has vestigial armor and enormously enlarged interpterygoid vacuities, represents either the continuation of aquatic Pennsylvanian dissorophids or a reinvasion of the aquatic environment by late terrestrial forms. Cutaneous respiration is found in all modern amphibians and probably occurred in their ancestors as well, since Thomson (1969) has suggested its presence in the sarcopterygians ancestral to the class Amphibia as a whole. In this regard, the increasing terrestriality of dissorophoid evolution, except for *Micropholis,* appears to have departed from the main line of lissamphibian ancestry, for the progressive development of armor in dissorophids seems incompatible with this method of gas exchange. Cox (1967) has attempted to show that cutaneous respiration arose as a terrestrial adaptation within Amphibia. While the lissamphibian features of *Doleserpeton* might be used to support this view, the work of Thomson (1969) and the fact that the greatest percentage of cutaneous respiration occurs in living aquatic amphibians (Foxon, 1964) seem to support origin from primarily aquatic forms. If dissorophids were actually involved in lissamphibian ancestry, then the latter group must have either retained a larval cutaneous respiration from terrestrial forms by paedomorphosis—a phenomenon still operating in salamander evolution (Wake, 1966)—or derived it from aquatic dissorophoids. If the latter, a choice between primitive Pennsylvanian forms or their later derivatives might depend on whether *Micropholis* was primarily or secondarily aquatic.

In summary, no more precise answer to the question of salientian relationships can be given at present than that they were probably derived from temnospondyl labyrinthodonts, that the group of origin may have been related to primitive dissorophoids, and that they probably form a natural group with the salamanders and, less clearly, with the apodans. New evidence must precede further speculation.

SUMMARY AND CONCLUSIONS

The Mesozoic fossil record of frogs shows that the Ascaphidae and Discoglossidae are ancient lineages. The earliest known frog, *Vieraella,* from the early Jurassic of Argentina, appears to be a member of the Ascaphidae in spite of its great age. Although it shows somewhat more similarity to the discoglossids than do the Recent ascaphids, the character states that indicate this similarity are neither numerous nor definitive, including only the wide atlas and rather *Bombina*-like skull roof. To us, these features simply support and somewhat strengthen the close relationship between these two families that has already been demonstrated (see, for example, Ritland, 1955*a*, 1955*b*). Although *Vieraella* probably possesses more primitive character states than does any other known frog, and thus makes a plausible structural ancestor for at least the Discoglossidae and perhaps also the Pelobatidae, it cannot be made a central figure in frog evolution without more complete knowledge of other early Jurassic frogs. At the least, however, *Vieraella* extends the ascaphoid line back to the early Jurassic. The late Jurassic *Notobatrachus*, clearly an ascaphid, shows that early members of the Ascaphidae had a well-developed middle ear, in contrast to the living genera *Ascaphus* and *Leiopelma*. *Notobatrachus* is further unusual and more primitive, apparently, than *Vieraella* in having a primitive carpus with free intermedium and (at least in some growth stages) unfused radiale and centrale 4, as well as centralia 2 and 3. In shoulder girdle construction and shape of squamosal, *Notobatrachus* is specialized toward *Leiopelma* and indicates an ancient split between the two living ascaphids.

A body of rapidly accumulating data now suggests that the hypothesis of continental drift is a viable one (Darlington, 1965; McElhinny and Luck, 1970; Elliot et al., 1970; Dickinson, 1971; Kitching et al., 1971; Keast, 1971; Colbert, 1971) and that major northern ("Laurasian") and southern ("Gondwanan") predrift continental nuclei can be identified (Hurley and Rand, 1969). Although the hypothesis of continental drift is now accepted by a significant number of geologists and biologists, many questions remain to be answered. Nevertheless, the brief discussion of distribution below (and the more extensive ones by Lynch, 1971, and Savage, this volume) shows that the known fossil frog data can be explained in terms of the drift hypothesis although it does not directly support or confirm it.

The presence of two fossil genera of ascaphids in the Mesozoic of South America makes it possible to explain the presence of *Ascaphus* on plausible zoogeographic grounds. Although interchange between North America and South America has always been relatively limited, the fact that it has occurred demonstrates that *Ascaphus* or its ancestors could have come from the south, perhaps in the late Mesozoic or early Cenozoic when other groups dispersed (see, for example, Estes, 1970*b* for some lower vertebrate data). Although some workers (Savage, this volume) have placed *Ascaphus* and *Leiopelma* in different families, there is in fact little evidence that such a separation is justified or is in fact useful. *Notobatrachus* displays some evidence of intermediacy between the two living genera and as Ritland (1955*a*, 1955*b*) has shown, the modern forms are similar in many features that suggest relationship, in spite of their wide geographic separation. Whether any of these similarities are derived, rather than primitive, must await further study. The northern occurrence of *Ascaphus* does not seriously weaken the fossil evidence that ascaphids as a group originated in and radiated from the southern hemisphere.

In contrast, the Discoglossidae is essentially a northern group, in both present and fossil distribution. The recent description of *Eodiscoglossus*, from the late Jurassic or early Cretaceous of Spain (Hecht, 1970), and the probable presence of the group in the Cretaceous and early Cenozoic of North America (Estes, Berberian, and Meszoely, 1969), indicates an antiquity for discoglossids almost equal to that of ascaphids. The fossil record of the two families thus demonstrates a long geologic history for what Lynch (this volume) has termed the superfamily Ascaphoidea, with the Ascaphidae essentially a southern hemisphere (Gondwanan) group, the Discoglossidae having a complementary Holarctic distribution.

The only other group of frogs with an almost equally ancient history is the superfamily Pipoidea (see Lynch, this volume). Well-defined members of the family Pipidae appear in the early Cretaceous of Israel (Nevo, 1968), and in the late Cretaceous of South America and Africa (Haughton, 1931; Reig, 1959). The Israeli pipids may be considered a Gondwanan occurrence (although a peripheral one) on geological grounds, and certainly no other unquestionable pipids have been found on the northern continents. On the other hand, the Palaeobatrachidae, as known, is entirely a northern family. The late Jurassic or early Cretaceous palaeobatrachid mentioned above and (if properly referred) the late Cretaceous palaeobatrachid from Wyoming indicate that palaeobatrachids may be considered an ancient Holarctic family of pipoids. This provides, for the Pipoidea, a parallel case to the Ascaphoidea described above: an essentially Gondwanan family Pipidae known from the early Cretaceous and probably of at least Jurassic antiquity, and a complementary Holarctic group Palaeobatrachidae of comparable age.

The earlier suggestion that *Notobatrachus* shows some similarity to the Pipidae, and the fact that the fossil record shows that the most ancient frog fossils are referable only to Ascaphoidea and Pipoidea, tends to support the suggestion of Morescalchi (1968) that these groups may be related because of karyological similarities lacking in higher frogs. These paleontological and karyological data (although slim) are also consistent with the computer-generated dendrogram of frog relationships given by Kluge and Farris (1969).

The described fossil record of other families is at present limited to the Cenozoic, with a few occurrences in the late Cretaceous. When now-undescribed material from the Cretaceous of Brazil has been studied, it is probable that the record of the Leptodactylidae or a closely related group will be extended at least back to the middle Cretaceous. Fossils probably referable to the Pelobatidae occur in the late Cretaceous (Estes, 1970*a*) and future finds of this group earlier in the Cretaceous may be expected. It is thus possible to predict that the northern hemisphere pelobatids and their probable Gondwanan relatives, the leptodactylids, will eventually also form complementary groups having an antiquity similar or nearly equal to that of the Ascaphoidea and Pipoidea. But the history of the remaining groups of frogs is not at present predictable from the fossil record.

Darlington (1957) has suggested that the center of origin of frogs was in the area now referred to as the Old World Tropics. This view has been supported by Reig (1957, 1960, 1961, 1968) and Casamiquela (1961*a*, 1965*b*) to the extent that a southern origin is concerned, while Hecht (1963) and Nevo (1968) have indicated that the record is too incomplete to determine the area of origin. Although we believe that the Madagascan *Triadobatrachus* belongs to a group that was ancestral to frogs, its relationship to the latter is too distant to permit use of its southern hemisphere Triassic occurrence as zoogeographic evidence. The presently known Pennsylvanian and Permian distribution of the labyrinthodont family Dissorophidae is Holarctic (*Micropholis* reached southern Africa in the lower Triassic); if the Lissamphibia is a valid group descended from them, then salamanders at least may be of northern origin, as their fossil and Recent distribution also indicates. However appealing it might be to suggest that the frogs are a complementary Gondwanan group, the phylogenetic connections of salamanders and frogs are still undocumented by fossils, and the record indicates only that some primitive frogs are northern while other families, equally primitive, are southern. Clearly it is premature to predict whether frogs are ultimately of Gondwanan or Holarctic origin.

The analysis of salientian origins has been hindered by lack of data. The only fossil indicating a frog-like form that is substantially more primitive than anurans is *Triadobatrachus,* whose interpretation as adult or larva, ancestor or convergent form, has not been entirely resolved. The frog-like appearance of the skull, both in its possession of a frontoparietal and in its general aspect, combines with the frog-like pelvis and reduction of presacral vertebrae to indicate that the skeleton of this animal resembles that of frogs very closely. Our interpretation is that *Triadobatrachus* is a representative of an ancestral proanuran stock. It may represent a young stage but is probably not far from adult. We believe it to have been aquatic and capable of symmetrical propulsive swimming movements. Unfortunately, there is no special resemblance of *Triadobatrachus* to any Paleozoic amphibian group.

The hypothesis of common origin of frogs, salamanders, and apodans is supported by recent work. The subclass Lissamphibia should be maintained for the present, but it will be necessary to obtain more data to confirm the relationship; data is especially needed from the apodans. Nevertheless, the lissamphibian hypothesis has assembled more detailed, relevant, positive evidence than any other suggestion has been able to contradict.

The presence of bicuspid pedicellate teeth of lissamphibian type in the recently described Permian dissorophoid *Doleserpeton* strengthens recent suggestions that these labyrinthodonts form a group from which the modern amphibians may be derived. We have presumed that cutaneous respiration is an aquatic adaptation that extends back through the Amphibia—at least through the young stages—to originate within Sarcopterygii. The terrestrial habits of almost all dissorophoids thus pose a problem unless lissamphibians were of paedomorphic origin (*sensu* Wake, 1966). Such a view could be supported, since modern amphibians (particularly salamanders) are marked by paedomorphic trends. The recent suggestion that *Micropholis* is a dissorophid and that it was aquatic makes it also possible that, if descended from dissorophoids, Lissamphibia need not have been originally paedomorphic. The families Doleserpetontidae and Dissorophidae have a greater number of resemblances to Lissamphibia than do any other groups of Paleozoic amphibians, yet neither group shows the reduction of dermal skull elements that would be expected in an ancestor. The differences between dissorophoids and lissamphibians are still profound, and while a dissorophoid origin for the latter is perhaps the most fertile hypothesis at present, we are still a long way from detailed knowledge of lissamphibian, much less anuran, ancestry.

Acknowledgements

It is a special pleasure to thank our scientific colleagues in Argentina, whose cooperation in obtaining specimens and molds for Estes in 1970 made this paper possible. Galileo Scaglia and Juan Brkljacic (Museo Municipal de Ciencias Naturales de Mar del Plata), José Bonaparte (Instituto-Fundación Miguel Lillo, Tucumán), Rosendo Pascual (Museo de La Plata), and Guillermo del Corro (Museo Argentino de Ciencias Naturales, Buenos Aires) all provided access to specimens in their collections; their courtesy is gratefully acknowledged. Miss Ana María Báez (Faculdad de Ciencias Exactas y Naturales, Buenos Aires), Sergio Archangelsky and Rosendo Pascual (Museo de La Plata) provided new data and discussion on the age and locality for the Argentine Mesozoic frogs.

Zdeněk Špinar (Charles University, Prague) kindly provided unpublished data on palaeobatrachids, and his courtesy in allowing Estes to study the large series of palaeobatrachid specimens that he has collected is also very much appreciated.

Stanley Salthe (Brooklyn College, New York) and John Bolt (University of Illinois, Chicago) offered useful comments on Lissamphibia and *Doleserpeton.* Max Hecht (Queens University, New York), who was helpful in discussion of ascaphoids, palaeobatrachids, and proanurans, also kindly provided the original photographs of *Eodiscoglossus* for use in preparing the restoration of the latter given here.

National Science Foundation Grants GB4303 and GB7176 to Estes permitted research time and travel to Europe and South America that was indispensable for the preparation of this paper.

ABBREVIATIONS. H.U.J.Z., Hebrew University, Jerusalem; M.A.C.N., Museo Argentino de Ciencias Naturales, Buenos Aires; M.C.Z., Museum of Comparative Zoology, Harvard University; M.L.P., Museo de La Plata; P.U., Princeton University; P.V.L., Instituto-Fundación Miguel Lillo, Tucumán, Laboratorio de paleontología de vertebrados.

APPENDIX 1: *VIERAELLA*

I. Postcranial skeleton

1. Vertebral column

a. The presence of nine presacral vertebrae was described by Casamiquela (1965*b*) in his study of the counterpart. Although the posterior vertebral region is poorly preserved, the difference in size between the diapophyses of the tenth vertebra and those anterior to it suggests that his interpretation is correct. The several postsacral elements figured by him (1965*b*, Figure 2) are, however, not clear on the specimen; we believe that no more is indicated than the anterior border of the postsacral transverse process.

b. Casamiquela (1965*b*) implies that Hecht (1963) believed *Vieraella* to be procoelous, but the only implication given by Hecht is that it is amphicoelous. Casamiquela (1965*b*) specifically notes that the centra of *Vieraella* show evidence of amphicoely when he states that "la casi absoluta seguridad de una cavidad anterior de los cuerpos," and later, when he refers to the fifth centrum as "un conillo de sedimento en la cara posterior, lo que hablaría precisamente de la presencia de un hueco posterior . . . ." We cannot agree that *Vieraella* is amphicoelous without question; we prefer perichordal, as do Kluge and Farris (1969). The centra are rounded ventrally, as in *Notobatrachus,* and thus appear externally to be perichordal, but we see no evidence for either anterior or posterior cavities. Only the clear dorsal and ventral surfaces can be seen in the specimen, and no cones of sediment or cavities are visible. The posterior borders of the centra appear convex, and the anterior borders concave. This configuration might lead to an interpretation as procoelous, but what in fact has happened is that the column has been dislocated and curved dorsoventrally, so that the axes of the centra are arranged in a staggered fashion, rather than along their natural axis. Each vertebra thus is displaced, so that its anterior border is morphologically *ventral* to the normal position, the posterior border morphologically *dorsal.* When the imprints on the sediment were made, the intercentral boundaries thus appear curved as a consequence.

In summary, the most that can be said is that the vertebrae of *Vieraella* appear, in external view, to have been perichordal in development. No internal structures are visible as a result of the method of preservation, and the condition of the intervertebral bodies cannot be interpreted.

2. The shoulder girdle

a. The shoulder girdle is poorly preserved. Of the bones present, the clearest imprint is of the scapula and cleithrum. The latter is not forked; the dorsal border of the scapula is well shown in dorsal view (Figure 1–1). The attachment surface for the suprascapular cartilage is clear, and a rather elongate shaft is visible. When camera lucida drawings of dorsal and ventral surfaces are superimposed, it is clear that the scapula is a rather long bone, in general appearance somewhat as in *Leiopelma,* being neither as relatively long as in higher frogs or as short as in *Ascaphus,* discoglossids, and pipids. The ventral border of the bone is difficult to interpret; some of the ventral imprint on the left side appears to represent part of the clavicle, broken and forced up on the ventral border of the scapula. If no part of the clavicle is represented here, the scapula is rather more elongate than in *Leiopelma* and is more as in advanced frogs. We believe this less likely and have restored it here (Figure 1–3) as the first of these alternatives because of the primitive features of the animal.

b. The coracoid is not well preserved, but the general outline seems to be clear. It is short, with an expanded medial border, about as in *Ascaphus* or the discoglossids, as well as many other frogs.

c. The structure of the clavicles is not easily apparent. The element labeled clavicle by both Casamiquela (1965*b*) and Reig (1961) is morphologically dorsal to the scapula and is much more probably a cleithrum, broken posteriorly and dislocated away from the scapula by crushing.

Several curved impressions occur ventrally (Figure 1–2, cl) in symmetrical positions; these probably represent parts of the clavicles. If so, these bones are arcuate as in discoglossids and many other groups of frogs, rather than straight as in ascaphids. The curved structures that we interpret as clavicles are difficult to explain in any other way, but the region is so poorly preserved that no important weight should be attached to this character.

d. The condition of the epicoracoid articulation cannot be determined with certainty in *Vieraella.* If our shoulder-girdle restoration is accurate, then closest similarity of the girdle of *Vieraella* is with Discoglossidae and Pelobatidae, both arciferal families. If the clavicle is straight, closest resemblance is to ascaphids, also arciferal. Probability thus suggests an arciferal condition, but no definite statement can be made.

3. Reig (1961) believed the radius and ulna to be separate in *Vieraella,* while Hecht (1963) interpreted it as fused and normal, and Casamiquela (1965*b*) as fused but with a patent suture. The supposed suture is, we think, only the result of crushing. Separation of the bones, either with or without suture, would be unusual but not unexpected in an early Jurassic frog, and in fact the radial and ulnar portions of some adult *Alytes* (e.g., M.C.Z. 904) are suturally separate both proximally and distally. Nevertheless, both dorsal and ventral impressions of the radioulna of *Vieraella* (especially the ventral) give the appearance of a fused bone, and we interpret it as of normal anuran type.

4. Ribs are clearly present on the third and fourth vertebrae, and clearly absent from the fifth. The presence of a rib on the second vertebra is less certain, because of poor preservation, but the outline of the transverse process is so long that a rib is probably included. We have indicated it as separate on our restoration (Figure 1–3), but its presence or absence is of no significance—it may be either present or absent in *Ascaphus* and discoglossids (Ritland, 1955*a*, Figure 5).

5. The humerus has what appears to have been a relatively large ball; it is preserved as a hollow, indicating that it was not ossified. Little development of epicondyles is visible.

6. The carpus is well preserved and is clearly shown on the latex molds, including the lateral cartilage surfaces on the fused centralia 2–3. At least seven bones are present. Centralia 2–3 appear fused, the intermedium appears to be lacking (probably fused with ulnare), and radiale and centrale 4 appear fused. No indication of carpal torsion is visible (the apparent torsion of the first finger is probably a simple dislocation). No indication of prepollex occurs on either dorsal or ventral impressions.

This is a primitive carpus, directly comparable to that of ascaphids and discoglossids, except for the lack of a prepollex. A prepollex would be expected in a primitive frog as a result of its relatively wide distribution in anurans; the absence of it is perhaps the result of postmortem loss, and no significance should be attached to this character until other specimens confirm it.

7. No clear indication of pelvic structures is present, and there is no utility in attempting identification of the few vague outlines of elements in this region.

8. Impressions of most of the left femur and a small portion of the right are present. Superimposition of camera lucida drawings of dorsal and ventral impressions shows that the bone identified as femur by Casamiquela (1965*b*) is actually tibiofibula. Figure 1–1 shows the distal end of the left femur next to the proximal end of the left tibiofibula, necessitating the change in interpretation. Concomitantly and more naturally, the paired bones identified by Casamiquela as a slender, short tibiofibula are actually tibiale and fibulare, more or less in natural position with the vague impressions of metatarsal and phalangeal elements.

II. The skull. Casamiquela (1965*b*) has rightly emphasized that breakage, crushing, and dislocation have made interpretation of the skull difficult. He notes that differential removal of sediment has included different planes of the skull on the imprint. His interpretation was perceptive and careful, and in general we agree with his over-all interpretation of the skull, but the latex molds have made it possible to delineate bone outlines more accurately and necessitated some changes.

1. Dorsal surface

a. The structures identified by Casamiquela (1965*b*) as "prevomers" are in fact nasals as Reig (1961) suggested, because the imprint of the anterior process of the vomer (or perhaps a septomaxilla) is closely associated, protruding from beneath the anterior nasal border. The natural borders of the nasals are difficult to interpret.

b. The premaxillae and nasals of Casamiquela's Figure 1 (1965*b*) are only rugose sediment, in our view, rather than bone.

c. The left maxilla is broken into several pieces, but the narial incision is visible anteriorly. Posteriorly, the limit of the maxilla is difficult to determine. On the right, the element labeled quadratojugal by Casamiquela may be the broken posterior end of the maxilla. However, because of the presence of a wider but similar element on the left, and because of what seems to be smooth bone extending forward from the squamosal-quadrate region on the right, we have interpreted this as a quadratojugal and accepted the presence of a complete maxillary arch in *Vieraella.* There are crushed bones between pterygoid and quadratojugal on the left side of the specimen that appear to represent the mandible; this, if true, tends to confirm that the latter is not being confused with quadratojugal, as noted for *Notobatrachus* in Appendix 2, III, 11.

d. The otic capsules are relatively blunt bones, not reaching far laterally. This shape is similar to that seen in ascaphids, discoglossids, and some pelobatids.

e. The frontoparietals are difficult to interpret. The imprint of the posterior half of the left bone indicates a posterolateral process extending onto the otic capsule, and a flattened dorsal table. The medial border of this fragment is straight, and it probably represents the midline. Anteriorly, two narrow bone fragments have sharply curved edges and can only be lateral borders of the frontoparietals anteriorly. We have made the simplest restoration of these elements by interpreting them as paired frontoparietals in contact on the midline posteriorly but diverging anteriorly to form a fontanelle. This differs from the condition in ascaphids and *Alytes* in lacking a posterior fontanelle. *Bombina* and *Discoglossus,* as well as many other frogs, are like *Vieraella* in this feature. Separation of the anterior and posterior portions of the frontoparietals could lead to an interpretation of separate frontal and parietal bones, but more probably this represents only breakage at a zone of weakness. *Bombina* has a median slit in this region, which certainly represents the embryological point of separation of frontals and parietals, but the lateral borders of the bone are solidly fused. It is possible that such a slit appeared in *Vieraella,* but we have not included it in the restoration.

f. The squamosal is indicated only at the junction with the quadrate. Our restoration follows that of discoglossids more than of ascaphids because no indication of a free, projecting maxillary process is preserved, as it is in both *Notobatrachus* and living ascaphids.

g. The occiput is crushed beyond interpretation, although the impression of a relatively large condyle may be seen on the dorsal right surface, near the corresponding cotyle of the atlas.

2. Ventral surface

a. Ventrally, the curve of the pterygoid is clear on the left side, but it has been dislocated posteriorly. Faint impressions of the cultriform process are present. Anteriorly, the clear impression of the choanal processes of a right vomer is present; its less-distinct medial portion and a symmetrical left vomer are also visible.

b. Ventral portions of the prearticular region of the mandible may be seen on the right, as well as the posterior curve of the pterygoid and the element here identified as a quadratojugal. The bones are too poorly preserved to suggest, as Casamiquela (1965*b*) has done, that more posterior bones other than prearticular are present (see Appendix 2, II,6). Anteriorly, the bone labeled premaxilla by Casamiquela (1965*b*, Figure 2) is perhaps a mentomeckelian, since it does not bear teeth and is rounded in ventral view.

c. The element suggested as a possible operculum and questionably labeled columella by Casamiquela (1965*b*, Figure 2) is a very indistinct impression; squared rather than slender like a columella, and posterior to the dislocated pterygoid region, this element is probably unidentifiable.

APPENDIX 2: *NOTOBATRACHUS*

The nature of four major character states in *Notobatrachus* has not been agreed upon by all workers, especially Casamiquela (1961*a*) and Hecht (1963). These are discussed in section I; some less significant disputed or undescribed character states are discussed in II and III.

I. Major character states

1. The vertebrae, described by Reig (1957) and Casamiquela (1961*a*) as amphicoelous, were believed by Hecht (1963) to be procoelous, although perhaps weakly so. Casamiquela (1965*b*) published a photograph of clearly amphicoelous vertebrae in a specimen referred to *Notobatrachus*. This was in answer to a letter from Hecht (published in part by Casamiquela, 1965*b*), who suggested that "several conditions" might have been preserved. Hecht (personal communication 1969) meant that he believed it possible that more than one taxon might be represented. Estes has examined the series of specimens of *Notobatrachus* in the Museo de La Plata and the Instituto-Fundación Miguel Lillo, as well as the type and the other figured specimen (Reig, 1957, Plates 1, 2) from the Museo Argentino de Ciencias Naturales. This suite includes 33 specimens, of which many are incomplete. Unfortunately nine catalogued specimens from the Museo Argentino and several from the Instituto-Fundación Miguel Lillo could not be found by the curators at this time. Study of the available material, however, indicates that only one type of frog is represented. Many of the specimens that are clearly parts of one individual of *Notobatrachus* (i.e., showing diagnostic elements) show undisputed evidence of perichordal vertebrae (we prefer this term to the less-meaningful "amphicoelous"; see Kluge and Farris, 1969). High-fidelity rubber molds show the detailed structure of the vertebrae; one of these molds was cut to prepare the drawing of notochordal impressions shown in Figure 1–8a. Impressions of the notochord and intervertebral cartilages are clear. No indication of asymmetry, which would suggest that the discs ever fused with either end of the centrum, could be found. Although similar perichordal vertebrae occur in *Rhinophrynus*, the *Notobatrachus* condition agrees—as far as bone allows interpretation—with *Leiopelma* and *Ascaphus* (cf. Figure 1–8b; Ritland, 1955*a*, Figure 3). No specimen indicates that anything resembling a procoelous condition existed.

2. With some hestitation, Reig (1957) suggested that the shoulder girdle of *Notabatrachus* was firmisternal, a conclusion strongly supported by Casamiquela (1961*a*, 1965*b*). Hecht (1963) found the evidence inconclusive, while Eaton (1959) compared *Notobatrachus* with ascaphids, noting that an arciferal condition was possible. This character in living frogs has been discussed extensively by Griffiths (1963, and earlier papers), who showed that definitive evidence must come from the epicoracoid cartilages, which are obviously unknown in fossils. While a definite decision cannot be made for this reason, the detailed similarity of shoulder girdles of *Notobatrachus* and *Leiopelma* give greatest weight to an arciferal condition for the fossil form (see Figure 1–7). The similarities are striking, and deserve comment here:

a. The coracoids of *Notobatrachus* are unusual in their great extent. Comparison with *Leiopelma* suggests that this is the result of medial expansion of the coracoid that limits the (usual) anterior border of the bone to an acute notch near the articulation with the scapula. The coracoids reach closer to the midline than in *Leiopelma* and apparently a greater part of the epicoracoid cartilage ossified than in the living genus. When placed in what must have been their natural pattern of articulation, the bones are well separated on the midline, however, and restoration as arciferal by comparison with *Leiopelma* seems justified. Reig (1957) lacked specimens of *Leiopelma* for his study, and *L. hochstetteri* (figured by E. M. T. Stephenson, 1952, Figure 3) is less similar to *Notobatrachus* than is *L. hamiltoni* (Figure 1–7).

b. The scapula of *Notobatrachus* was believed by Reig (1957) to be uncleft, but a number of more recently collected specimens show the scapula with great fidelity, and pars acromialis and pars glenoidalis are clearly separated by a deep cleft (Figure 1–7). This feature is discussed in greater detail in section III, 12 of this appendix.

c. The clavicle has a posterior process on its distal end that probably clasped or covered an anterior coracoid cartilage as in *Ascaphus* and *Leiopelma,* although of course this cannot be confirmed. The position of the clavicles, more or less perpendicular to the vertebral column, was a feature used by Reig (1957) as evidence for firmisterny. However, both *Leiopelma* and *Ascaphus* also have such perpendicular clavicles, and indeed this similarity, with the greater curvature of *Notobatrachus,* suggests that the Recent genera are specialized in this character state.

d. The cleithrum and suprascapula were figured by both Reig (1957) and Casamiquela (1961*a*) as separate bones. The more recently discovered material shows that, as in many other frogs, only the cleithrum is ossified. It has an anterior border that curves around to clasp the suprascapular cartilage ventrally, as in Recent forms. As in both Recent ascaphid genera and in contrast to other frog groups, the cleithrum is unforked.

e. Casamiquela (1961*a*) slightly altered the position of the coracoids from the restoration given by Reig (1957). Both of these restorations, however, separate the girdle elements more than they probably were in life. We believe that the restoration here (Figure 1–7), with the elements little separated, as in Recent frogs, is more accurate than either previous version; it is supported by P.V.L. 2194 figured by Casamiquela (1961*a*, Plate II) and kindly cast for us by José Bonaparte.

3. Hecht (1963) cited the presence of a columella (plectrum) in *Notobatrachus,* although Casamiquela (1961*a*) had believed middle-ear structures to be lacking(Reig, 1957, did not discuss ear structures). Our study confirms that of Hecht; a well-developed plectrum articulates with the otic bones in a notch on the posterolateral border of the fenestra ovalis (Figure 1–5). Moreover, we believe it probable that a faint circular or semicircular area on the fenestral border of the type specimen (and more clearly on the large specimen P.V.L. 2194, noted above, section I,2e) represents an operculum of normal anuran type (Figure 1–5). This probable operculum and the well-developed plectrum confirms the presence of a middle ear in *Notobatrachus.* This fact and the presence of a middle ear in the Jurassic discoglossid *Eodiscoglossus* (Hecht, 1970) makes it clear that no evidence supports the view of N. G. Stephenson (1951) that absence of the middle ear is primitive for ascaphids.

4. Both Casamiquela (1961*a*) and Hecht (1963) doubted the presence of mandibular teeth in *Notobatrachus*, although Reig (1957) had discussed a specimen (M.A.C.N. 17723) that he believed to demonstrate mandibular teeth. This specimen was disarticulated, and the figure of the supposed toothed dentary given by Reig is unlike that of other preserved dentaries of *Notobatrachus*, which are thin, apparently edentulous bones loosely applied to the prearticular in the characteristic anuran manner. This loose connection in combination with the presence of well-developed mentomeckelian bones, implies that the tongue-protruding mechanism was already well established in *Notobatrachus*. Functional aspects of this mechanism suggest that loss of teeth is related to tongue protrusion and that on these grounds alone the presence of dentary teeth in *Notobatrachus* is unlikely. Unfortunately the specimen cited by Reig and another similar specimen mentioned by Casamiquela (1961*a*) were missing when Estes was in Argentina in 1970.

II. Other character states have also been subject to dispute, although these are less important systematically than the above.

1. The sphenethmoids in a number of specimens have been displaced laterally by crushing. This interpretation was suggested by Reig (1957) and supported by Casamiquela (1961*a*, Plate 1B) on the basis that one of the specimens (P.V.L. 2194) showed the paired bones in natural position and fused anteriorly. Casamiquela's text gives the impression that the sphenethmoid is consistently a single ossification in *Notobatrachus*. We believe that as in ascaphids and discoglossids, the sphenethmoid begins as a paired element and that in larger specimens the two halves may fuse. The lateral portions of the anterior end of the parasphenoid serve as floor for the two halves of the bone (Figure 1–5).

2. The presence of a space between the frontoparietals anteriorly, and the striated bone surface visible between these bones on the type specimen, suggested to Reig (1957) that an internasal was present. This bone surface is with greater probability the dorsal surface of the parasphenoid crushed against the fontanelle region. For this reason we concur with Casamiquela (1961*a*) that no internasal (dermal sphenethmoid) was present (Figure 1–4).

3. The shape of the parasphenoid was interpreted as acuminate anteriorly by Reig (1957) and blunt by Casamiquela (1961*a*). The specimens used by Casamiquela to demonstrate this are actually broken anteriorly; camera lucida drawings of the type make the structure given in Figure 1–5 the correct interpretation. As Casamiquela demonstrated, material in which the thyroid processes of the hyoid did not obscure the parasphenoid posteriorly indicates that the parasphenoid corpus was more extensive than Reig was able to observe in his specimens. It is possible that this unusual shape is the result of the ventral migration of the fenestrae ovales, but it is also a possible resemblance to pipids.

4. Casamiquela (1961*a*) identified two bones with similar depressions on them as possible vomers and suggested that if the identification was correct, the bones resembled vomers of labyrinthodonts more than those of frogs. These bones (Casamiquela, 1961*a*, Figure 3) are actually ventral surfaces of the nasals, the depressions being for part of the nasal capsules. The vomer was figured correctly by Reig (1957, Figure 4), although the detailed outline given here in Figure 1–5 is more accurate.

5. Casamiquela (1961*a*) states that the *squamosal* contacts the parasphenoid and is separated from the maxilla, in contrast to the reconstruction given by Reig (1957). This is clearly a lapsus for the *pterygoid*, which contacts both the maxilla and parasphenoid, as is usual in frogs, whereas the squamosal contacts neither.

6. Because of symmetrical transverse grooves on the ventroposterior part of the prearticular in the type and P.V.L. 2196, Casamiquela (1961*a*) believed that both angular and splenial were present. In other specimens, both articulated and disarticulated, no such grooves occur. The grooves are surely the result of compression in some of the specimens, which then cracked in similar places on the lower jaw. A transverse suture would be unusual in any case, and we believe that the posterior bone is prearticular

alone and that only dentary, prearticular, and mentomeckelians are present, as in other frogs.

III. This study has also produced information about some character states that differs from all previous studies or involves features not previously described.

1. Free ribs are present on vertebrae 2–5, as noted by all previous studies. In addition, in the type specimen, the right transverse process of the sixth vertebra shows a clear suture and expanded area, although this is absent on the right. This finding confirms the suggestion of Reig (1957) that the shape of some posterior transverse processes indicated the presence of fused ribs. Ritland (1955*a*, Figure 4R) has documented a similar condition of the fifth transverse process in *Ascaphus*, and E. M. T. Stephenson (1952) cites it for *Leiopelma* as well. The *Notobatrachus* condition is, however, the most posterior indication of ribs in any frog.

2. Casamiquela (1961*a*) notes that the urostyle of *Notobatrachus* has a dorsal sulcus, rather than a dorsal keel. This is true only posteriorly; although the keel is small and is crushed in most specimens, it is still visible and rises to the height of the neural arch of the urostylar vertebra. The extensive crushing of the urostyle indicates that it was essentially hollow and probably contained (at least in part) notochordal tissue as Ritland (1955*a*, Figure 3F) has shown for *Ascaphus*. The wide, expanded shape of the urostyle agrees well with that of *Ascaphus*, less so with the more elongate, slender structure in *Leiopelma*.

3. The humeral ball in *Notobatrachus* is unossified, as shown by an unnumbered specimen from the Museo de La Plata (Figure 1–8c). In spite of the lack of ossification, it is clear that the ball was originally small and that no fossa cubitus ventralis was present (see Hecht and Estes, 1960). A similar small ball is found in pipoids, and the lack of a pronounced fossa is characteristic of palaeobatrachids (Figure 1–8d) and some discoglossids—especially a presumed discoglossid from the Cretaceous of Montana (Estes, Berberian, and Meszoely, 1969, Figure 3). (In light of the possible presence of a palaeobatrachid in this deposit, noted above, this occurrence may need restudy.) As in living ascaphids, the epicondyles of the *Notobatrachus* humerus are about equally developed and relatively large.

4. The carpus of *Notobatrachus* is remarkable. As Reig (1957) has shown, the intermedium is separate from the ulnare, a condition known otherwise only as a variation in palaeobatrachids. Reig demonstrated that at least eight bony elements were present in the carpus. More recently collected material allows additional interpretation. Reig's restoration (1957, Figure 13) was based on M.A.C.N. 17721. P.V.L. 2194 shows that the element labeled radiale by Reig is actually the fourth centrale, which is fused with the radiale but still retains clear evidence of the suture (Figure 1–6b). The radiale is therefore lost from M.A.C.N. 17721. P.V.L. 2194 also shows fused but easily distinguishable centralia 2–3, and molds of the type specimen show imprints of distal tarsals 1–2 that were not seen by Reig. The restored carpus (Figure 1–6) compares fused and unfused conditions and indicates the presence of either eleven or (if fused) nine carpals —more than is present in any other frog.

The carpus is thus most closely comparable with those of salamanders and primitive amphibians. It is, in addition, the most primitive carpus known for any lissamphibian, differing from that of the most primitive tetrapod condition only in lacking any indication of the fifth distal tarsal.

5. Reig (1957) described the teeth as having two roots and a conical crown. In P.V.L. 2194 the maxillary tooth row shows a row of tooth bases with evenly truncated tops, indicating that the teeth were probably pedicellate. No crowns seem to be preserved on any specimen seen by the senior author. The two roots mentioned by Reig are actually the two lateral portions of each pedicel that were eroded medially by tooth replacement crowns.

P.V.L. 2194 is the largest known specimen of *Notobatrachus*. Its maxillary tooth count of about 50 teeth and premaxillary count of about 15 teeth probably indicates the approximate maximum number of teeth present in any individual.

6. Reig (1957) showed long, narrow nasal processes of the premaxillae that extended posteriorly well over the nasals. Detailed study of both dorsal and ventral nasal impressions indicates that these are actually slim anterior processes of the nasals, thickened ventrally on the midline. Few of the adult specimens of *Notobatrachus* show any indication of the nasal processes of the premaxillae; usually they are broken or covered with sediment. An unnumbered immature specimen in the collection of the Museo de La Plata shows these processes clearly; they are close to the midline and extend dorsally, barely touching (or underlying) the nasals (Figure 1–4).

The premaxillae are primitive and, like those of many salamanders, have the ascending processes approximated on the midline and close to the surface, rather than deep and separated by an ethmoid or fontanelle.

7. Reig (1957) stated that the parasphenoid touched the premaxillae, as in pipids. The parasphenoid is long, like that of pipids, but the structures believed by Reig to be posteromedial extensions of the premaxillae are actually the thickened anteromedial processes of the nasals. Crushing has brought the two bones together in the type and other specimens. The anterior extent of the bone is more extensive than in other frogs, however, and is like that of pipids in this sense.

8. The palatine is absent or fused to the maxilla in *Notobatrachus.* This was already noted by Reig (1957), although he believed that the position of the bone was indicated by a line posterior to the vomers. This line was also suggested as the possible impression of a lamina orbitonasalis by Casamiquela (1961*a*). It is actually the impression of the posterior border of the nasal, crushed ventrally into the palatal region, as shown by superimposed camera lucida drawings of both dorsal and ventral surfaces in the type specimen.

9. Reig (1957) indicated that the pterygoids met the prootics but not the parasphenoid, and Casamiquela (1961*a*) pointed out the existence of a definite space between both prootic and parasphenoid, and the pterygoid. Although the actual articulation cannot be demonstrated between all three of these bones, the restoration given here (prepared by fitting together camera lucida drawings of the individual bones and checked from a number of base points) indicates that there was a blunt meeting of all three. On the dorsal surface of the parasphenoid in this region, there is a hollow space (reflected ventrally by a bulging ridge, see Figure 1–5) that probably indicates the presence of a cartilaginous basal articulation with the braincase. The medial extensions of the pterygoids of the juvenile specimen noted above (III,6) have flattened surfaces posteriorly, indicating an articulation surface with the basal process, as in some salamanders (Estes, 1965, Figure 7) and Paleozoic amphibians.

10. Casamiquela (1961*a*) has noted that the condylar plane was well posterior to the jaw articulation; he believed this to be a primitive condition for frogs. We have commented on this before and here note also that it is more likely a paedomorphic feature, as well as a variable one. Young individuals of most amphibians have a forward articulation of the jaws that grows more posterior as the animal reaches full size. This condition can be demonstrated in labyrinthodonts as well, although paedomorphosis has retained the larval structure in adults of some groups, for example, the brachyopids.

11. Both Reig (1957) and Casamiquela (1961*a*) believed that a quadratojugal was present in *Notobatrachus.* The element so identified in these studies is actually the raised lateral border of the mandible, crushed against quadrate and maxilla so as to give the impression of a complete maxillary arch. This is shown well in some specimens having the lower jaw dislocated, which permits the two areas to be studied separately. High fidelity latex molds also make it possible to trace continuity of bone (having a characteristic texture that reflects the cartilage attachment of the articular) throughout the dorsal channel of the prearticular. The original interpretation was the result of lack of ossification of the articular.

Notobatrachus is thus similar to *Leiopelma* and *Ascaphus* in the lack of a complete maxillary arch.

12. Reig (1957) stated that the scapula of *Notobatrachus* was uncleft. The new material,

as well as the availability of the latex molds, shows that the cleft is present, with a prominent separation between partes glenoidales and acromialis.

Confusion exists in the literature about this feature and its significance. Following and extending the original study of Procter (1921), Griffiths (1963) concluded that three groups of frogs could be distinguished on the basis of scapular character states: (1) the Ascaphidae and Pipidae, which have an uncleft scapula and a clavicle-scapula ratio greater than 3.0; (2) the Discoglossidae, which have a cleft scapula and a clavicle-scapula ratio greater than 3.0; (3) other frogs, which have a cleft scapula and a clavicle-scapula ratio less than 2.0.

We disagree with Griffiths' conclusion, because while *Ascaphus* has an uncleft scapula, those of *Leiopelma* and *Notobatrachus* are deeply cleft; in addition, while the scapula of *Pipa* is uncleft, those of *Xenopus* and *Hymenochirus* show either a cleft or evidence of its ontogenetic closure (especially in the latter genus, in which there is no fusion of clavicle and scapula as seen in *Xenopus*). The clavicle-scapula ratios, as Kluge and Farris (1969) correctly point out and as we have confirmed in our material, are too variable intrataxonomically to be especially useful. In any case, the method of taking measurements has not been standard for all authors, and thus published results are not necessarily comparable.

Kluge and Farris (1969) suggest that the overlap of clavicle and scapula cited by Griffiths (1963) has utility. With this we must also disagree: while overlap clearly occurs in *Ascaphus, Rhinophrynus,* and *Pipa,* no overlap is present in *Notobatrachus, Leiopelma hochstetteri,* or *L. hamiltoni,* and the condition in discoglossids is difficult to distinguish from that of pelobatids.

Our view is that the overlap of clavicle and scapula, as well as the closure of the scapular cleft, is related to the development of the extreme aquatic habit in pipids (and palaeobatrachids, which also lack a cleft) and that it is a specialization. Ritland (1955*b*) shows that muscular changes involve alteration of the acromial region in ascaphids relative to other frogs, and it is clear that no further use of these characters should be made, pending detailed study of the shoulder girdle of frogs from the functional standpoint.

REFERENCES

Beçak, W., M. Beçak, G. Schreiber, D. Lavalle, and F. Amorim. 1970. Interspecific variability of DNA content in amphibia. Experientia 26:204-206.

Bolt, J. 1969. Lissamphibian origins: possible protolissamphibian from the lower Permian of Oklahoma. Science 166:888-891.

Bonaparte, J., and G. Bossi. 1967. Sobre la presencia de dinosaurios en la Formación Pirgua del Grupo Salta y su significado cronológico. Act. Geol. Lilloana 9:25-44.

Branisa, L., and R. Hoffstetter. 1966. Nouvelle contribution a l'étude de la paléontologie du Groupe Puca (Crétacé de Bolivie). Bull. Mus. Natl. Hist. Nat., Paris 38: 301-310.

Bunge, M. 1959. Metascientific queries. C. C. Thomas Co., Springfield, Ill., 313 p.

Carroll, R. 1964. Early evolution of the dissorophid amphibians. Bull. Mus. Comp. Zool., Harvard Univ. 131:163-250.

_______. 1967. An adelogyrinid lepospondyl amphibian from the upper Carboniferous. Canadian J. Zool. 45:1-16.

Casamiquela, R. M. 1960. Datos preliminares sobre un pipoideo fósil de Patagonia. Act. y Trab. 1ro Congr. Sudam. Zool. 4:17-22.

_______. 1961*a*. Nuevos materiales de *Notobatrachus degiustoi* Reig. Rev. Mus. La Plata, Sec. Paleontol. (Nueva Ser.) 4:35-69.

_______. 1961*b*. Un pipoideo fósil de Patagonia. Rev. Mus. La Plata, Sec. Paleontol. (Nueva Ser.) 4:71-123.

_______. 1965*a*. Nuevos ejemplares de *Shelania* (Anura, Pipoidea) del Eoterciario de la Patagonia. Ameghiniana 4:41-50.

_______. 1965*b*. Nuevo material de *Vieraella herbstii* Reig. Rev. Mus. La Plata, Sec. Paleontol. (Nueva Ser.) 4:265-317.

CAZENUEVE, H. 1965. Datación de una toba de la formación Chon Aike (Jurásico de Santa Cruz, Patagonia) por el método de Potasio-Argon. Ameghiniana 4:156-158.
COLBERT, E. 1971. Tetrapods and continents. Quart. Rev. Biol. 46:250-269.
COX, C. B. 1967. Cutaneous respiration and the origin of the modern amphibia. Proc. Linn. Soc. London 178:37-47.
DARLINGTON, P. J. 1957. Zoogeography: the geographical distribution of animals. John Wiley and Sons, New York. 675 p.
_____. 1965. Biogeography of the southern end of the world. Harvard Univ. Press, Cambridge, Mass. 236 p.
DE MAR, R. 1968. The Permian labyrinthodont amphibian *Dissorophus multicinctus,* and adaptations and phylogeny of the family Dissorophidae. J. Paleontol. 42:1210-1242.
DICKINSON, W. 1971. Plate tectonics in geologic history. Science 174:107-113.
EATON, T. H., JR. 1959. The ancestry of modern amphibia: a review of the evidence. Univ. Kansas Publ. Mus. Nat. Hist. 12:155-180.
ELLIOT, D., E. COLBERT, W. BREED, J. JENSEN, AND J. POWELL. 1970. Triassic tetrapods from Antarctica: evidence for continental drift. Science 169:1197-1201.
ESTES, R. 1964. Fossil vertebrates from the late Cretaceous Lance Formation, eastern Wyoming. Univ. California Publ. Geol. Sci. 49:1-180.
_______. 1965. Fossil salamanders and salamander origins. Am. Zool. 5:319-334.
_______. 1967. Review of: Romer, A. S., Vertebrate paleontology, 3rd ed. Copeia 4:873-876.
_______. 1969. A new fossil discoglossid frog from Montana and Wyoming. Breviora, Mus. Comp. Zool., Harvard Univ. 328:1-7.
_______. 1970*a*. New fossil pelobatid frogs and a review of the genus *Eopelobates.* Bull. Mus. Comp. Zool., Harvard Univ. 139(6):293-339.
_______. 1970*b*. Origin of the Recent North American lower vertebrate fauna: an inquiry into the fossil record. Forma et Functio 4:139-163.
ESTES, R., AND J. TIHEN. 1964. Lower vertebrates from the Valentine Formation of Nebraska. Am. Midland Naturalist 72:453-472.
ESTES, R., M. HECHT, AND R. HOFFSTETTER. 1967. Paleocene amphibians from Cernay, France. Am. Mus. Novitates 2295:1-25.
ESTES, R., P. BERBERIAN, AND C. MESZOELY. 1969. Lower vertebrates from the late Cretaceous Hell Creek Formation, McCone County, Montana. Breviora, Mus. Comp. Zool., Harvard Univ. 337:1-33.
FOXON, G. 1964. Blood and respiration, p. 151-209. *In* J. Moore, [ed.], Physiology of the amphibia. Academic Press, New York.
FRIANT, M. 1960. Les batraciens anoures. Caracteres osteologiques des Discoglossidae d'Europe. Act. Zool. 41:113-139.
GANS, C., AND T. PARSONS. 1966. On the origin of the jumping mechanism in frogs. Evolution 20:92-99.
GOIN, O. B., AND C. J. GOIN. 1968. DNA and the evolution of the vertebrates. Am. Midland Naturalist 80:289-298.
GRAZZINI, C., AND M. MŁYNARSKI. 1969. Position systematique du genre *Pliobatrachus* Fejérváry. Compt. Rend. Acad. Sci. Paris 268:2399-2402.
GREGORY, J. 1950. The tetrapods of the Pennsylvanian nodules from Mazon Creek. Am. J. Sci. 248:833-873.
GRIFFITHS, I. 1956. The status of *Protobatrachus massinoti.* Nature 177:342-343.
_______. 1963. The phylogeny of the salientia. Biol. Rev. 38(2):241-292.
HAUGHTON, S. 1931. On a collection of fossil frogs from the clays at Banke. Trans. Roy. Soc. South Africa 19:233-249.
HECHT, M. K. 1959. Amphibians and reptiles. *In* P. McGrew et al., The geology and paleontology of the Elk Mountain and Tabernacle Butte Area, Wyoming. Bull. Am. Mus. Nat. Hist. 117:117-176.
_______. 1960. The history of the frogs. (Abstr.) Anat. Rec. 138:356.
_______. 1962. A reevaluation of the early history of the frogs. Part I. Systematic Zool. 11(1):39-44.
_______. 1963. A reevaluation of the early history of the frogs. Part II. Systematic Zool. 12(1):20-35.

———. 1970. The morphology of *Eodiscoglossus,* a complete Jurassic frog. Am. Mus. Novitates 2424:1-17.

HECHT, M. K., AND R. ESTES. 1960. Fossil amphibians from Quarry Nine. Postilla, Peabody Mus. Nat. Hist., Yale Univ. 46:1-19.

HECHT, M. K., AND R. HOFFSTETTER. 1962. Note preliminaire sur les amphibiens et les squamates du Landénien supérieur et du Tongrien de Belgique. Inst. Roy. Sci. Nat. Belg. 38:1-30.

HECHT, M. K., AND R. RUIBAL. 1958. Review of: Stipanicic, P. N., and O. A. Reig, 1955 (see below). Copeia 3:242.

HERBST, R. 1961. Algunos datos geológicos y estratigráficos de la zone estancia Roca Blanca y alrededores, provincia de Santa Cruz. Ameghiniana 2:55-60.

HOLMAN, J. 1963. A new rhinophrynid frog from the early Oligocene of Canada. Copeia 4:706-708.

———. 1967. Additional Miocene anurans from Florida. Quart. J. Florida Acad. Sci. 30:121-140.

———. 1968. Lower Oligocene amphibians from Saskatchewan. Quart. J. Florida Acad. Sci. 31:273-289.

HURLEY, P., AND J. RAND. 1969. Pre-drift continental nuclei. Science 164:1229-1242.

JARVIK, E. 1962. Les porolépiformes et l'origine des urodèles. Coll. Int. Cent. Nat. Rech. Sci. 104:87-101.

KEAST, A. 1971. Continental drift and the evolution of the biota on southern continents. Quart. Rev. Biol. 46:335-378.

KITCHING, J., J. COLLINSON, D. ELLIOT, AND E. COLBERT. 1971. *Lystrosaurus* zone (Triassic) fauna from Antarctica. Science 175:524-526.

KLUGE, A. 1966. A new pelobatine frog from the lower Miocene of South Dakota with a discussion of the evolution of the *Scaphiopus-Spea* complex. Los Angeles Co. Mus. Contrib. Sci. 113:1-26.

KLUGE, A. G., AND J. S. FARRIS. 1969. Quantitative phyletics and the evolution of anurans. Systematic Zool. 18(1):1-32.

KUHN, O. 1961. Die familien der rezenten und fossilen amphibien und reptilien. Meisenbach KG, Bamberg. 79 p.

———. 1962. Die vorzeitlichen Frösche und Salamander, ihre Gattungen und Familien. Jahrb. Ver. vaterl. Naturkde. Württemberg 1962:327-372.

———. 1964. *Cyrtura* Jaekel aus dem Solnhofener Schiefer ist ein Nachzugler der Temnospondyli (Amphibia, Labyrinthodontia). Neues Jahrb. Geol. Paleontol., Monatsh. 11:659-664.

LEANZA, A. 1969. Sistema de Salta, su edad, sus peces voladores, su asincronismo con el Horizonte Calcáreo-Dolomitico y con las Calizas de Miraflores y la hibridez del Sistema Subandino. Rev. Asoc. Geol. Argentina 24:393-407.

LEHMAN, J.-P. 1968. Remarques concernant la phylogénie des amphibiens, p. 307-315. *In:* T. Ørvig, [ed.], Current problems of lower vertebrate phylogeny. Proc. 4th Nobel Symp., Almqvist and Wiksell, Stockholm.

LYNCH, J. 1971. Evolutionary relationships, osteology, and zoogeography of leptodactyloid frogs. Univ. Kansas Publ. Mus. Nat. Hist. 53:1-238.

MARSH, O. 1887. American Jurassic mammals. Am. J. Sci. 33:327-348.

MCELHINNY, M., AND G. LUCK. 1970. Paleomagnetism and Gondwanaland. Science 196:830-832.

MELENDEZ, B. 1957. Leonardi's "La evolución biologica." Ediciones Fax. 450 p.

MOODIE, R. 1912. An American Jurassic frog. Am. J. Sci. 34:286-288.

———. 1914. The fossil frogs of North America. Am. J. Sci. 38:531-536.

MORENO, J. 1969. Estratigrafía y paleogeografía del cretácico superior en la cuenca del noroeste Argentino, con especial mención de los subgrupos Balbuena y Santa Barbara. Rev. Asoc. Geol. Argentina 25:9-44.

MORESCALCHI, A. 1968. Hypotheses on the phylogeny of the salientia, based on karyological data. Experientia 24:964-966.

———. 1970. Karyology and vertebrate phylogeny. Bol. Zool. 37:1-28.

NEVO, E. 1956. Fossil frogs from a lower cretaceous bed in southern Israel (central Negev). Nature 178:1191-1192.

———. 1968. Pipid frogs from the early cretaceous of Israel and pipid evolution. Bull. Mus. Comp. Zool., Harvard Univ. 136(8):255-318.

NOPCSA, F. 1930. Notes on stegocephalia and amphibia. Proc. Zool. Soc. London 2:979-995.

PARODI BUSTOS, R. 1962. Los anuros cretácicos de Puente Morales (Salta) y sus vinculaciones con *Shelania pascuali* Casamiquela (Chibut) y *E. reuningi* Haughton, de Africa del Sur. Rev. Fac. Cienc. Nat. Salta 1:83-85.

PARODI BUSTOS, R., AND J. KRAGLIEVICH. 1960. A propósito de los anuros cretácicos descubiertos en la provincia de Salta. Rev. Fac. Cienc. Nat. Salta 1:37-40.

PARODI BUSTOS, R., M. FIGUEROA CAPRINI, J. KRAGLIEVICH, AND G. DEL CORRO. 1960. Noticia preliminar acerca del yacimiento de anuros extinguidos de Puente Morales. Rev. Fac. Cienc. Nat. Salta 1:1-20.

PARSONS, T. S., AND E. E. WILLIAMS. 1962. The teeth of amphibia and their relation to amphibian phylogeny. J. Morphol. 110:375-389.

———. 1963. The relationships of the modern amphibia. Quart. Rev. Biol. 38:26-53.

PIVETEAU, J. 1937. Un amphibien du Trias Inférieur. Essai sur l'origine et l'evolution des amphibiens anoures. Ann. Paléontol. 26:135-177.

———. 1955. Anoura. Traité de Paléontol. 5:250-274.

PROCTER, J. B. 1921. On the variation of the scapula in the batrachian groups aglossa and arcifera. Proc. Zool. Soc. London 1-2:197-214.

REIG, O. A. 1957. Los anuros del Matildense. *In* P. N. Stipanicic and O. A. Reig, "El complejo porfírico de la Patagonia extraandina y su fauna de anuros." Act. Geol. Lilloana 2:231-297.

———. 1958. Proposiciones para una nueva macrosistemática de los anuros. Nota preliminar. Physis 21:109-118.

———. 1959. Primeros datos descriptivos sobre los anuros del Eocretáceo de la Provincia de Salta (Rep. Argentina). Ameghiniana 1:3-7.

———. 1960. Lineamentos generales de la historia biogeographica de los anuros. Act. y Trab. 1ro Congr. Sudam. Zool. 1:271-278.

———. 1961. Noticia sobre un nuevo anuro fósil del Jurásico de Santa Cruz (Patagonia). Ameghiniana 2:73-78.

———. 1964. El problema del origin monofilético o polifilético de los anfibios, con consideraciones sobre las relaciones entre anuros, urodelos y apodos. Ameghiniana 3:191-211.

———. 1968. Peuplement en vertébrés tetrapodes de l'Amerique du Sud. *In* C. Delamare Deboutteville and E. Rapoport, [eds.], Biologie de l'Amerique Australe. Cent. Nation. Rech. Sci. Paris 4:215-260.

———. 1970. The Proterosuchia and the early evolution of the archosaurs; an essay about the origin of a major taxon. Bull. Mus. Comp. Zool., Harvard Univ. 139:229-292.

———. 1972. *Macrogenioglottus* and the South American bufonid toads, p. 14-36. *In* W. F. Blair, [ed.], Evolution in the genus *Bufo*. Univ. Texas Press, Austin.

REIG, O. A., AND J. CEI. 1963. Elucidación morfológico-estadística de las entidades del género *Lepidobatrachus* (Anura, Ceratophrynidae). Physis 24:181-204.

RITLAND, R. M. 1955*a*. Studies on the post-cranial morphology of *Ascaphus truei*. I. Skeleton and spinal nerves. J. Morphol. 97(1):119-178.

———. 1955*b*. Studies on the post-cranial morphology of *Ascaphus truei*. II. Myology. J. Morphol. 97(2):215-282.

ROMER, A. S. 1947. Review of the Labyrinthodontia. Bull. Mus. Comp. Zool., Harvard Univ. 99:1-368.

———. 1966. Vertebrate paleontology. 3rd ed. Univ. Chicago Press, Chicago. 468 p.

SALTHE, S. N., AND N. O. KAPLAN. 1966. Immunology and rates of enzyme evolution in the amphibia in relation to the origins of certain taxa. Evolution 20:603-616.

SCHAEFFER, B. 1949. Anurans from the early tertiary of Patogonia. Bull. Am. Mus. Nat. Hist. 93:45-68.

SCHMALHAUSEN, I. I. 1958. Istoria proiskhozhdeniia amfibii. Izvest. Akad. Nauk S.S.S.R., Ser. Biol., 1958:39-58.

———. 1964. Proishoženie nazemarjh Pozvonočnyh. Izdatel'stvo Nauka, Moscow.

Trans. by Leon Kelso, 1968, as The origin of terrestrial vertebrates. Academic Press, New York. 314 p.

SEIFFERT, J. 1969*a*. Sternalelement (omosternum) eines Mitteljurassischen Anuren von SE-Aveyron/Südfrankreich. Z. System. Evolutionsforsch. 2:145-153.

_______. 1969*b*. Urodelen-Atlas aus dem obersten Bajocien von SE-Aveyron (Südfrankreich). Paläontol. Z. 43:32-36.

_______. 1972. Ein Vorläufer der Froschfamilien Palaeobatrachidae und Ranidae im Grenzbereich Jura-Kreide. Neues Jahrb. Mineral. Geol. Paleontol. 2:120-131.

SHISHKIN, M. 1970. The origin of anura and the theory of the Lissamphibia, p. 30-44. *In* Materials on the evolution of land vertebrates. [In Russian] Otdelenie Obschchei Biologii, Izdatel'stvo Nauka, Moscow.

SLABBERT, G. K., AND W. A. MAREE. 1945. The cranial morphology of the Discoglossidae and its bearing upon the phylogeny of the primitive anura. Ann. Univ. Stellenbosch 23A(2-6):91-97.

ŠPINAR, Z. V. 1966. Some further results of the study of Tertiary frogs in Czechoslovakia. Časopis mineral. geol. 11:431-440.

_______. 1967*a*. Neue Kenntnisse über den stratigraphischen Bereich der Familie Palaeobatrachidae Cope, 1965. Věst. Ústřed. ústavu geol. 42:217-218.

_______. 1967*b*. Familie Paleobatrachidae Cope, 1965, ihre taxonomische Einreihung und Bedeutung für die Phylogenie der Frösche. Věst. Ústřed. ústavu geol. 42:375-379.

_______. 1970. Předkové našich žab. (The ancestors of our frogs). Časopis Národ. Muz. 137:74-88.

_______. 1972. Tertiary frogs from central Europe. W. Junk, The Hague. 346 p.

STEPHENSON, E. M. T. 1952. The vertebral column and appendicular skeleton of *Leiopelma hochstetteri* Fitzinger. Trans. Roy. Soc. New Zealand 79(3):601-613.

STEPHENSON, N. G. 1951. On the development of the chondrocranium and visceral arches of *Leiopelma archeyi*. Trans. Zool. Soc. London 27(2):203-252.

STIPANICIC, P. N. 1957. Consideraciones sobre el denominado "complejo porfírico de la Patagonia Extraandina." *In* P. N. Stipanicic and O. A. Reig, "El complejo porfírico de la Patagonia extraandina y su fauna de anuros." Act. Geol. Lilloana 1:185-230.

STIPANICIC, P. N., AND O. REIG. 1955. Breve noticia sobre el hallazgo de anuros en el denominado "complejo porfírico de la Patagonia extraandina" con consideraciones acerca de la composición geológica del mismo. Rev. Asoc. Geol. Argentina 10:215-233.

STOLL, N., R. DOLLFUS, J. FOREST, N. RILEY, C. SARBOSKY, C. WRIGHT, AND R. MELVILLE. 1964. International code of zoological nomenclature. Intern. Trust for Zool. Nomenclature, London. 176 p.

STREMME, M. 1916. Über die durch Bandverknocherung hervorgerufene proximale Verschmelzung zweier Mittelhand oder Mittelfussknochen eines Reptiles. Arch. Biont. 4:143-144.

SZARSKI, H. 1962. The origin of amphibia. Quart. Rev. Biol. 37:189-241.

_______. 1970. Changes in the amount of DNA in cell nuclei during vertebrate evolution. Nature 226:651-652.

TAYLOR, E. H. 1941. A new anuran from the middle miocene of Nevada. Univ. Kansas Sci. Bull. 27:61-69.

THOMSON, K. 1964. The ancestry of the tetrapods. Sci. Progr. 52:244-254.

_______. 1969. The biology of the lobe-finned fishes. Biol. Rev. 44:91-154.

THOMSON, K., AND P. VAUGHN. 1968. Vertebral structure in Rhipidistia (Osteichthyes, Crossopterygii) with description of a new Permian genus. Postilla, Peabody Mus. Nat. Hist., Yale Univ. 127:1-19.

TIHEN, J. A. 1962. A review of new world fossil bufonids. Am. Midland Naturalist 68:1-50.

_______. 1965. Evolutionary trends in frogs. Am. Zool. 5:309-318.

VAUGHN, P. 1963. New information on the structure of Permian lepospondylous vertebrae—from an unusual source. Bull. So. California Acad. Sci. 62:150-158.

_______. 1965. Frog-like vertebrae from the Lower Permian of southeastern Utah. Los Angeles Co. Mus. Contrib. Sci. 87:1-18.

VIDAL, L. 1902. Nota sobre la presencia del tramo Kimeridgense en el Montsech (Lérida) y hallazgo de un batracio en sus hiladas. Mem. Roy. Acad. Cienc. Barcelona 4:263.

WAKE, D. 1966. Comparative osteology and evolution of the lungless salamanders, family Plethodontidae. Mem. So. California Acad. Sci. 4:1-111.

WATSON, D. 1941. The origin of the frogs. Trans. Roy. Soc. Edinburgh 60:195-213.

WHITING, H. P. 1961. Pelvic girdle in amphibian locomotion. *In* Vertebrate locomotion. Zool. Soc. London Symposium 5:43-57.

WIGHTMAN, W. 1950. The growth of scientific ideas. Oliver and Boyd, Ltd., Edinburgh. 495 p.

WILLIAMS, E. 1959. Gadow's arcualia and the development of tetrapod vertebrae. Quart. Rev. Biol. 34:1-31.

ZANGERL, R., AND R. DENISON. 1950. Discovery of early cretaceous mammals and frogs in Texas. Science 112:61.

2

BONES, FROGS, AND EVOLUTION

Linda Trueb

INTRODUCTION

In this age of quantitative phyletics, of character weighting, statistical analysis of characters, and computer-generated phylogenies, anuran systematists speak a new and sophisticated language. Amidst this semantic and methodological maze, one tends to lose sight of the sometimes prosaic foundation of systematics, the nature of the characters themselves. In many cases, and I believe anuran osteology is one such example, systematists have utilized characters without fully understanding their variation, significance, and distribution; without this information, attempts to designate primitive and derived states border on folly or, at best, educated guesswork. Moreover, the same characters—those involving the vertebral column and pectoral girdle—have been used time after time to virtual exclusion of many other useful osteological characters.

The value of osteological criteria in diagnosing phylogenetic relationships has been a matter of some dispute. Yet there seems to be no real alternative to their use, because the major categories of frogs, as currently recognized, are distinguished in large part on osteological characters. Furthermore, osteological features are virtually the only characters by which fossil material can be related to modern forms.

There is a severe lag between information and methodology. We have developed promising and useful methods of predicting evolutionary relationships but we have not carefully surveyed anurans with regard to osteological criteria, nor have we adequately synthesized the information which is presently available. The purpose of this paper is to provide at least an initial amendment of these circumstances. I have attempted to survey osteological variation among anurans and suggest the presence and direction of evolutionary trends where they are apparent. Where appropriate, synonymous names of bones are provided to eliminate confusion. The sometimes labored morphological descriptions are designed to facilitate precision and, it is hoped, encourage the reader and researcher to relate elements of architectural units, such as the skull or vertebral column, to one another in order to discover interrelationships and unit function. Literature resumés summarize significant research and provide a basic bibliography for the individual interested in more detailed description.

One of the most commonly encountered criticisms of a paper of this type centers on the ill-defined usage of such terms as "primitive," "derived," "advanced," "generalized," and "specialized." The most troublesome of these concepts is that of primitiveness. Countless definitions have been proposed to define "primitive" or the "primitive state of a character"; among the most lucid of the recent contributions is that of Kluge and Farris (1969), which draws heavily upon W. H. Wagner (1961). Kluge and Farris (1969: 5) state:

> To infer the primitive states of characters, we rely on available fossil material and on the criteria for primitiveness established by Wagner . . . . In order of reliability these criteria are:
> (1) The primitive state of a character for a particular group is likely to be present in many representatives of closely related groups.
> (2) A primitive state is more likely to be widespread within a group than is any one advanced state.
> (3) The primitive state is likely to be associated with states of other characters known from other evidence to be primitive.

The fossil record of anurans is so poor that for most characters, it is of little use in determining primitive character states; thus, we are left little choice but to reason inductively on the basis of these hypothetical criteria for primitiveness. The potential fallibility of the reasoning employed is obvious. The first and second criteria are based on an a priori assumption of the phylogeny of a group and do not account for the occurence of parallelism or convergence. The reliability of the third criterion is contingent upon that of the first two. These three criteria are employed throughout this paper in estimating primitive character states; however, the conclusions must always be qualified by the inherent limitations of the logic.

The term "derived" is used to denote a character that represents a deviation or change in a character state from the presumed primitive or ancestral state. "Advanced" refers to an animal's general morphological and evolutionary status that is characterized by a number of derived characters. The terms "generalized" and "specialized" are not to be misconstrued to connote "primitive" and "derived," respectively; primitive characters may be either generalized or specialized, as may be derived character states. "Generalized" is used to describe a character state or an organism which does not seem to be specifically adapted to a given function or environmental condition. "Specialized" describes a character state or an organism which is, in one or more ways, specifically adapted to a particular function or set of environmental factors to the partial or total exclusion of other functions or environmental conditions.

In the discussion section of this paper, considerable use has been made of the terms "archaic," "transitional," and "advanced" or "modern" with respect to the evolutionary status of extant anurans. The application of these terms is a convenient way of designating broad groups of anurans having combinations of characters that seem to indicate their archaic, transitional, and relatively modern or recent derivation. The usage of these terms and the taxonomic arrangement of the modern anurans follow closely that proposed by Lynch (this volume). The ascaphids and discoglossids (Superfamily Ascaphoidea) and the pipids and rhinophrynids (Superfamily Pipoidea) constitute archaic frogs. Transitional anurans include pelobatids (including pelodytids), myobatrachids (Superfamily Pelobatoidea), and sooglossids (part Superfamily Ranoidea). The remaining families are considered to be modern or advanced anurans. These are the leptodactylids, dendrobatids, bufonids, brachycephalids, rhinodermatids, pseudids, centrolenids, and hylids (Superfamily Bufonoidea) and the ranids, hyperoliids, rhacophorids, and microhylids (part Superfamily Ranoidea).

Functional osteology is a virtually untouched approach to anuran evolution, and much more descriptive morphological work remains to be done. Families, genera, and species need to be surveyed osteologically in the manner that Lynch (1969) has done for the Leptodactylidae. The paper that follows is only an initial attempt at synthesis. There doubtless are errors, both of fact and omission. However, it is hoped that this work will be repeatedly supplemented and amended and will prove to be of use in morphological and systematic research.

BONE TISSUE

Most anatomy and histology textbooks define two types of bone formation. These are endochondral or cartilage replacement bone and membrane or dermal bone, respectively. According to this classification, endochondral bones are always preformed in cartilage, which is laid down early in development. Dermal bones, in contrast, are not preformed but are laid down in membranous connective tissue by a process quite distinct from that of cartilage bone formation. Endochondral bone tacitly has been considered to be much less subject to evolutionary alteration than dermal elements and always to remain morphologically distinct from membrane bone.

Study of anuran osteology strongly suggests that at least in these lower vertebrates, bone formation is far more labile than has previously been thought. The basic developmental distinction between cartilage and dermal bone prevails. However, once bone is deposited, there are no morphological differences between the two types, and secondary restructuring can result in the coparticipation of dermal and endochondral elements to form a single structure. The dermal sphenethmoid of some hylids, the cranial structure of pipids, and the pectoral-girdle structure of *Brachycephalus* (Brachycephalidae) afford adequate examples of this phenomenon. It is also evident that in some cases, new centers of ossification can be generated as they are needed. In the case of dermal bone, this apparently can occur anywhere there is membranous connective tissue; thus, we have the occurrence of epipubic and Nobelian (postpubic) bones in *Ascaphus*, the presence of presacral vertebral shields in *Brachycephalus* and some dendrobatids and leptodactylids, and the addition of new dermal elements in the cranial structure of casque-headed hylids. This facility is more common in dermally derived elements, but it is not strictly confined to them. Cartilage is frequently laid down in myocommatous areas (e.g., vertebrae and hyoid), and apparently any such area has the capacity for cartilage formation. The occurrence of inscriptional ribs in *Leiopelma* (Ascaphidae) is one example. Once cartilage is present, it can be wholly or partially replaced by bone. Thus, in some anurans the sternum, omosternum, and suprascapula are partially or completely ossified, and the hyoid may have ossified parahyoid elements.

Another phenomenon associated, and often confused, with ossification is calcification. Calcium salts are often deposited in a cartilaginous matrix, such as the sternum, hyoid, suprascapula, and neurocranium. On gross inspection, calcification can sometimes be confused with ossification. Close examination usually reveals that calcium salts are irregularly deposited in a pattern quite distinct from normal structure of bone. Calcification, especially of the neurocranium, is usually associated with advanced age after the skeletal components have reached their maximum states of development.

In a recent paper, Moss (1968) suggested that all vertebrate dermal and dental skeletal tissues are homologous components of a unitary integumental skeleton. The author proposed that all dermally derived elements are formed by coparticipation of epidermal cells with subjacent mesodermal cells. The type of skeletogenous tissue produced by this coparticipation is controlled by the activity of the basal cell layer of the epidermis and involves inductive interactions between epidermal and dermal components. Moss outlined three categories of tissue types. "Structural epidermal derivatives" are produced by the activity of epidermal cells alone, although the author pointed out that such derivatives are thought to be produced by an inductive interchange between the dermis and epidermis. Keratinous claws, larval beaks, and teeth are included in this category. The second category includes "structural epidermal-ectomesenchymal derivatives" in which the dermal and epidermal integumental layers contribute to tissue formation. This type of formation is characteristic of teeth. The third type of formation, "structural mesodermal derivatives," is characterized by the total absence of any visible tissue formation by the epidermis. Included in this category are dermal bones, osteoderms and, according to Moss, calcified plates. I would take exception to the latter; such calcification frequently occurs in areas not directly adjacent to the integument.

There obviously is much to be learned about the formation of different types of bone and their relationships to one another. The fact that all bone has a connective tissue precursor and that mature bones of both endochondral and membranous derivations are identical in structure and capable of fusion leads one to believe that there is a continuous spectrum of bone tissue types.

HETEROCHRONY

As defined by E. M. T. Stephenson (1960), heterochrony denotes a deviation from a typical ancestral sequence in the formation of certain organs. Fundamentally, heterochrony or neoteny is concerned more with the rate of development of the gonads than with other parts of the body; we are specifically interested in the relationship of sexual development to the progress of ossification. E. M. T. Stephenson (1960) and N. G. Stephenson (1965) suggested that some kinds of osteological differences evident among groups of closely related species are the result of neotenic or heterochronous changes. Thus the developmental rate of ossification may be slower than that of the gonads, or conversely, sexual maturity may develop precociously with respect to the progress of ossification. Regardless of which way the situation is viewed, the net result is that a sexually mature specimen of one species, such as *Leiopelma archeyi,* may be osteologically similar to an immature form of a second species, such as *Leiopelma hamiltoni* (E. M. T. Stephenson, 1960).

My own observations on hylids suggest that heterochrony is operational in maintenance of osteological differences between the sexes of a single species. Among tree frogs, for example, the males tend to be smaller than the females; consequently, males are less extensively ossified than females. Frequently this results in obvious differences in the size and shape of dermal roofing bones. Thus, a sexually mature male frog may resemble an immature female more closely than it might a breeding female. This phenomenon poses potential problems in osteological studies and emphasizes the importance of noting such variation where it occurs and specifying the age and sex of individuals being described.

As a group, anurans show evidence of neotenic arrest of ossification in comparison with primitive amphibians. For example, in most frogs the neurocranium is incompletely ossified, and the quadrate remains cartilaginous, as does the hyoid apparatus. Among most arciferal genera, the epicoracoid cartilages and the omosterna and sterna, when present, do not ossify. The pubis remains cartilaginous. In comparison with primitive amphibians, the number and extent of dermal roofing bones are reduced. Thus, heterochronous arrest of bony development seems to be a basic pattern of specialization that was established early in the evolution of anurans and that continues to be operational in the evolution of species.

CRANIAL OSTEOLOGY

Literature Resumé

The first significant contribution to our knowledge of anuran cranial anatomy was W. K. Parker's (1881) monograph on the structure and development of the batrachian skull. This work includes a description of the development of the skull in *Rana,* descriptions of the skulls of adults of 69 taxa, and illustrations of the skulls and hyoid apparati of many of these taxa. Although this volume is still frequently referred to, there are major drawbacks to its use. The morphologist is likely to be confused by the outmoded classification and use of many generic names no longer accepted, whereas the systematist will find it difficult to interpret much of the terminology and some of the descriptive material in view of more recent theories of homology and development. Many of the drawings are inaccurate and vague in their detail. Despite these disadvantages, this monograph is unique in its detailed inclusion of a wide variety of anurans.

Gaupp, in Ecker's and Wiedersheim's *Anatomie des Frosches* (1896), offers the classical description of anuran anatomy, based on *Rana temporaria.* This work includes detailed treatment of osteological features, including the osteocranium and chondrocranium. The illustrations, based on dissection and graphical reconstructions, are among the work's major advantages. (The figures are frequently encountered in later publications by other authors.) The utility of Gaupp's work is diminished somewhat by the outdated terminology and the unfortunate fact that the study is based upon a relatively advanced anuran.

Another major contribution to anuran cranial osteology is the material on *Rana fusca* presented by de Beer (1937) in his monograph on the development of the vertebrate skull. The scope of this study, together with the author's frequent digressions into discussions of homologies and his concern with the nature of cartilage and bone, make this a standard reference work.

Parker, Gaupp, and de Beer constitute the three most comprehensive works on anuran cranial osteology to date. The remainder of the published information can be broadly segregated into one of two types: (1) those anatomically oriented and concerned primarily with internal and developmental features, and (2) those including osteological data as part of larger systematic studies. Papers in the first category are usually extremely detailed in their descriptions of internal cranial anatomy and are based on studies of microtomized material. The external cranial architecture is seldom described or considered, and presumably because the authors are anatomists rather than systematists, there is virtually no attempt at synthesis or interpretation of evolutionary trends or adaptive significance of the structural variation noted. In contrast, the osteological data in most systematic studies tend to deal only with external aspects of the skull, describing characters that are useful in specific or generic recognition. Frequently such descriptions are naïve and superficial and result in misinterpretation and emphasis being placed on characters of questionable taxonomic or phylogenetic value. Perhaps the most unfortunate aspect of these divergent approaches to cranial osteological studies is that rarely are both kinds of information available for a given taxon, and even more infrequently have both methods been employed in a comprehensive investigation of the cranial osteology of a species or group of species. Consequently, for many species (principally Old World species) there are significant sources of information available on internal cranial anatomy, whereas in other, more broadly studied taxa there are accounts of external osteological features. It is extremely difficult, time-consuming, and usually impossible to extrapolate facts of internal structure from generalized, external descriptions, or vice versa. Hence, much of the information on anuran cranial osteology is not comparable and thwarts attempts at synthesis.

The following is a list of significant and recent sources of osteological information on the anuran crania. In most cases, references to older works or detailed anatomical studies of a part of the cranium are cited in these papers. General reference works are not included. Papers that treat particular structures in detail (e.g., the ear) are cited elsewhere in appropriate discussions.

Family Ascaphidae: *Ascaphus:* de Villiers, 1934*a;* Pusey, 1943; van Eeden, 1951. *Leiopelma:* E. M. T. Stephenson, 1951, 1955, 1960; N. G. Stephenson, 1951; W. H. Wagner, 1934.

Family Discoglossidae: Slabbert and Maree, 1945. *Alytes:* Maree, 1945; Ramaswami, 1942; van Seters, 1922. *Bombina:* Ramaswami, 1942; Slabbert, 1945. *Discoglossus:* van Zyl, 1949.

Family Rhinophrynidae: *Rhinophrynus:* Kellogg, 1932; Walker, 1938.

Family Pipidae: Nevo, 1969. *Xenopus:* Bernasconi, 1951; Kalin and Bernasconi, 1949; Kotthaus, 1933; Paterson, 1939; Sedra and Michael, 1957. *Hymenochirus:* Paterson, 1945. *Hemipipa:* Paterson, 1955.

Family Pelobatidae: Kluge, 1966; Ramaswami, 1935*a*; Zweifel, 1956. *Pelobates:* Başoğlu and Zaloğlu, 1964. *Megalophrys:* Kruijtzer, 1931; Ramaswami, 1943.

Family Myobatrachidae: Lynch, 1969; N. G. Stephenson, 1965. *Heleophryne:* C. A. du Toit, 1930, 1931, 1934*a*; C. A. du Toit and Schoonees, 1930; Westhuizen, 1961. *Crinia:* C. A. du Toit, 1934*b*.

Family Leptodactylidae: Lynch, 1969. *Syrrhophus:* Baldauf and Tanzer, 1965. *Leptodactylus:* Heyer, 1969. *Physalaemus:* Ramaswami, 1932*a*. *Caudiverbera:* Reig, 1960; Reinbach, 1939. *Lepidobatrachus:* Reig, 1961. *Elosia:* van Eeden, 1943.

Family Brachycephalidae: *Brachycephalus:* McDiarmid, 1969; McLachlan, 1943.

Family Bufonidae: Ramaswami, 1936. *Atelopus:* McDiarmid, 1969. *Bufo:* Baldauf, 1955, 1957, 1958, 1959; Sanders, 1953; Schoonees, 1930; Sedra, 1949; Tihen, 1962. *Rhamphophryne:* Trueb, 1971. *Dendrophryniscus:* McDiarmid, 1969. *Melanophryniscus:* Badenhorst, 1945; McDiarmid, 1969. *Oreophrynella:* McDiarmid, 1969.

Family Centrolenidae: *Centrolenella:* Eaton, 1958.

Family Hylidae: *Acris:* Chantell, 1968*a*; Gaudin, 1969. *Limnaoedus:* Chantell, 1968*a*; Gaudin, 1969. *Pseudacris:* Chantell, 1968*b*; Gaudin, 1969; Stokely and List, 1954. *Smilisca:* Duellman and Trueb, 1966; Trueb, 1968. *Hyla:* Gaudin, 1969; Gillies and Peabody, 1917; León, 1969; Trueb, 1966, 1968, 1970*a*. *Triprion:* Peters, 1955; Trueb, 1970*a*. *Aparasphenodon:* Trueb, 1970*a*. *Pternohyla:* Trueb, 1970*a*. *Trachycephalus:* Trueb, 1970*a*. *Osteocephalus:* Trueb, 1970*a*. *Corythomantis:* Trueb, 1970*a*. *Argenteohyla:* Trueb, 1970*b*.

Family Ranidae: Laurent, 1940, 1941*a*, 1941*b*, 1943*a*, 1944; Ramaswami, 1934, 1935*b*; Liem, 1969 (rhacophorids and hyperoliids). *Rana:* Al-Hussaini, 1941; Bhatia and Prashad, 1918; C. A. du Toit, 1933; Feinsmith, 1962; Procter, 1919; Pusey, 1938. *Ceratobatrachus:* Laurent, 1943*b*. *Ooedozyga:* Laurent, 1943*b*. *Mantidactylus:* Laurent, 1943*c*. *Arthroleptella:* de Villiers, 1929*a*. *Hemisus:* de Villiers, 1931*a*. *Cacosternum:* de Villiers, 1931*b*. *Arthroleptides:* C. A. du Toit, 1938. *Petro-*

pedetes: C. A. du Toit, 1943. *Phrynobatrachus:* G. P. du Toit, 1933. *Hyperolius:* G. P. du Toit and de Villiers, 1932. *Schoutedenella:* Maas, 1945. *Trichobatrachus:* Laurent, 1942*a*.

Family Microhylidae: Carvalho, 1954; H. W. Parker, 1934. *Kaloula:* Boring and Liu, 1937-1938; Ramaswami, 1936. *Phrynomerus:* de Villiers, 1930. *Breviceps:* de Villiers, 1931*c*, 1933; Laurent, 1942*b*. *Anhydrophryne:* de Villiers, 1931*d*. *Probreviceps:* de Villiers, 1933. *Rhombophryne:* de Villiers, 1934*b*. *Spelaeophryne:* de Vos, 1935. *Austrochaperina:* Fry, 1912. *Elachistocleis:* Pentz, 1943. *Glyphoglossus:* Ramaswami, 1932*b*. *Phrynella:* Ramaswami, 1936. *Microhyla:* Roux, 1944.

Dermal Roofing Bones

Frontoparietal. The frontoparietals are usually paired, dermal roofing bones overlying the posterior half or third of the neurocranium (Figure 2–1). They commonly overlap the posterior portion of the sphenethmoid anteriorly, lie dorsal to the prootic with which they articulate by means of a ventral flange, the lamina perpendicularis. Posteriorly, the frontoparietals overlie the prootic and exoccipital; the frontoparietals are sometimes separated from the latter bones by connective tissue (e.g., in hylids) or fused (e.g., in several bufonids and a few leptodactylids).

There has been some dispute concerning the origin of the frontoparietal in anurans. Anatomists, observing a single center of ossification for each frontoparietal, claimed that the bone was homologous with the frontal of primitive amphibians and that the parietal had been lost; others, observing two centers of ossification that later fused to form one bone on each side of the skull, held that the frontoparietal is compound in origin, representing fusion of both the frontal and parietal of primitive amphibians. The confusion seems to have resulted from the discrepancies noted in the number of ossification centers. Thus, among the Ranidae, Leptodactylidae, and Bufonidae, one or two centers occur, depending on the species, whereas species studied in the Discoglossidae, Hylidae, and Pelobatidae are uniformly characterized by single centers of ossification. The Pipidae are unique; most species exhibit a single center of ossification, but one species, *Hemipipa carvalhoi*, has only a single, median center of ossification to form one frontoparietal (Figure 2–2b; Ramaswami, 1956). Griffiths (1954*a*) presented convincing arguments in favor of the dual origin of the frontoparietal, proposing that there has been a trend toward forward movement of the parietal ossification center to a stationary frontal ossification center. The culmination of this trend is evidenced by the consolidation of the two centers to form one center of ossification for each frontoparietal. By projection, this same line of reasoning accounts for the single center of ossification for both frontoparietals in *Hemipipa*. It is reasonable to assume that once single frontoparietal ossification centers are established, these in turn could migrate medially and fuse. Given that this is the evolutionary sequence which has occurred, then retention of two separate ossification centers must be regarded as a more primitive trait than the more advanced condition of a single center for each frontoparietal and the specialization of a single center of ossification for both frontoparietals.

In the majority of anurans the frontoparietals are paired elements (Figure 2–2). Notable exceptions occur in the families Rhinophrynidae, Pipidae, and Brachycephalidae, in which the frontoparietals of the adults are completely fused and lacking any indication of a median suture. The fusion of frontoparietal elements to each other and surrounding elements (exoccipital, prootic, sphenethmoid, and nasals) should be regarded as a specialization; with the single exception of *Hemipipa*, all of these bones arise independently of one another and fuse only late in development.

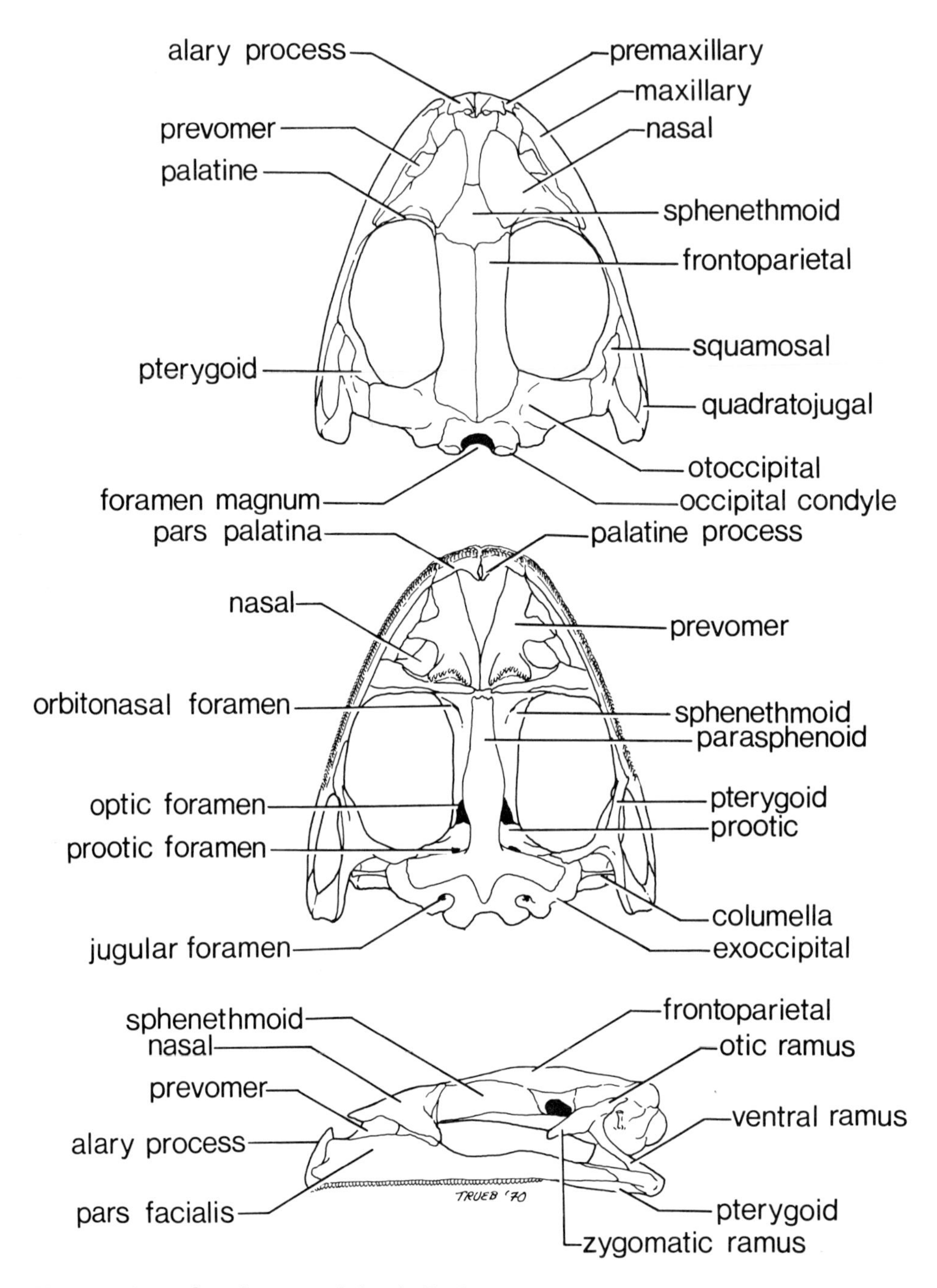

Figure 2–1. Outline drawing of the skull of *Leptodactylus bolivianus: top,* dorsal view; *middle,* ventral view; *bottom,* lateral view.

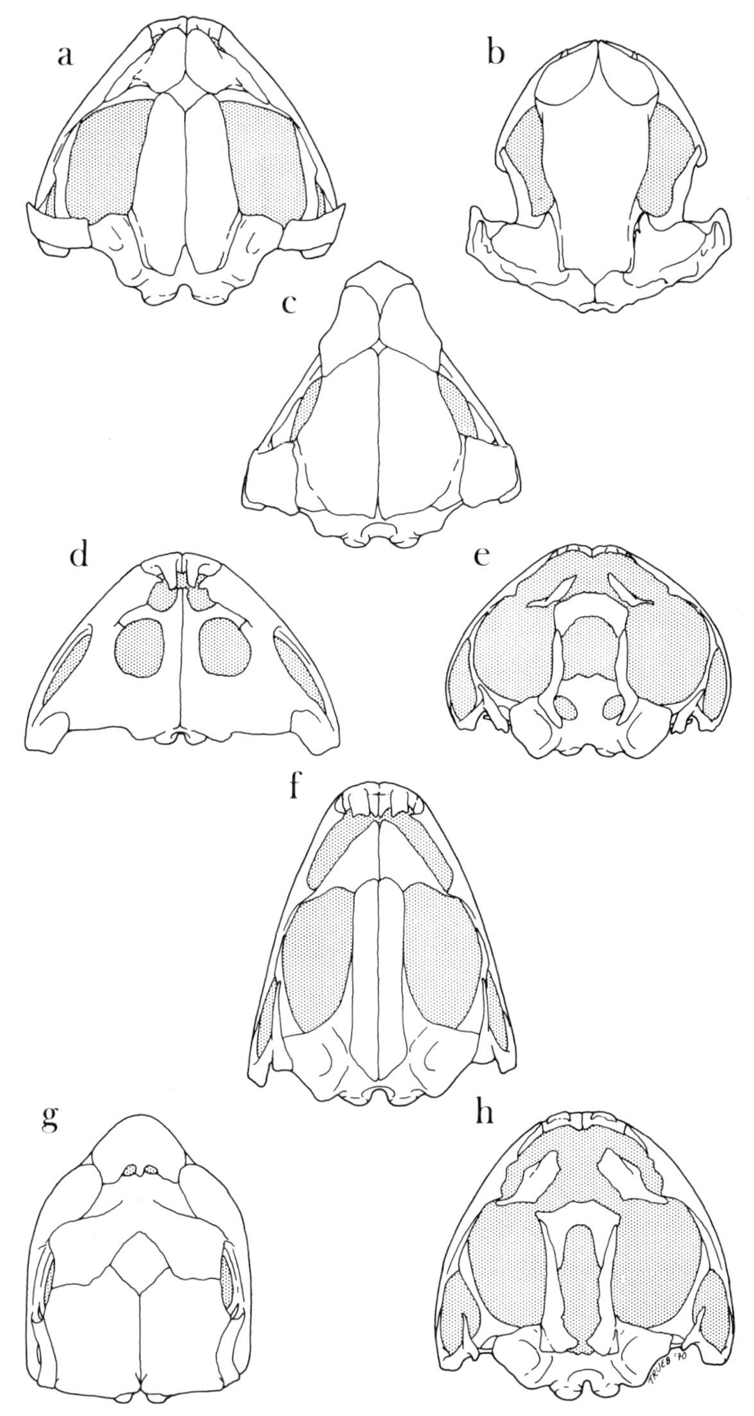

Figure 2–2. Diagrammatic dorsal views of skulls of anurans: a, *Bufo spinulosus* (Bufonidae); b, *Hemipipa carvalhoi* (Pipidae); c, *Rhamphophryne acrolopha* (Bufonidae); d, *Lepidobatrachus asper* (Leptodactylidae); e, *Centrolenella fleischmanni* (Centrolenidae); f, *Rana pipiens* (Ranidae); g, *Triprion petasatus* (Hylidae); h, *Heleophryne natalensis* (Myobatrachidae). *Black outlines,* bones; no ventral elements are shown. Open areas, such as fontanelles and orbits, are indicated by *stippled pattern.*

Other variation of the frontoparietals involves the extent to which the bones are marginally ossified. Minimally (Figure 2–2e), the elements are present as exceedingly slender longitudinal strips of ossification overlying the dorsolateral areas of the neurocranium and exposing the frontoparietal fontanelle medially (e.g., *Ascaphus* [Ascaphidae] and *Centrolenella* [Centrolenidae]); I know of no case in which the frontoparietals are absent. Extensive ossification of the frontoparietals occurs in several families (e.g., Bufonidae, Hylidae, Leptodactylidae, Ranidae, Brachycephalidae). In their maximal state of development (Figure 2–2c,d,g), the anterior margins of the frontoparietals articulate with the posterior margins of the nasals, and the medial margins articulate throughout the entire lengths of the bones; thus, the sphenethmoid and frontoparietal fontanelle are completely obscured beneath the frontoparietals. Frequently, lateral extension of the bones results in the presence of a slightly upturned supraorbital shelf or flange. Posterolaterally, the frontoparietals produce shelves which extend over, and completely roof, the lateralmost portions of the prootics; these posterolateral extensions form the posterior margins of the orbits and usually articulate with the dorsomedial margins of the squamosals laterally (Figure 2–2g). In several casque-headed species, such as *Triprion* (Hylidae), the posterior margins of the frontoparietals terminate in an upturned flange to which dorsal musculature is attached.

Between these developmental extremes, nearly every intermediate degree of frontoparietal ossification is represented in anurans. The amount of ossification is to some extent dependent on the size of the frog; thus, smaller individuals are apt to have less extensively ossified skulls than larger individuals. This should probably be regarded as an example of heterochronous change as discussed by N. G. Stephenson (1965). Disregarding size, the extent of frontoparietal ossification is obviously correlated with the habits of frogs. Arboreal species of the Centrolenidae, Hylidae, and Ranidae tend to have small frontoparietals, whereas terrestrial species (e.g., bufonids and dendrobatids) have moderately ossified to well-ossified frontoparietals. Burrowing and phragmotic anurans are characterized by extensive development, as are individuals that are aquatic (pipids).

Parenthetically, I would like to point out that the extent of ossification of the frontoparietal is one of the most reliable indices to the over-all ossification of the cranium. A small, poorly developed frontoparietal is normally indicative of minimal ossification throughout the entire cranium, whereas if the frontoparietal is well developed, one can anticipate a corresponding degree of ossification elsewhere in the cranium. Secondly, some caution should be exercised in describing the extent of medial frontoparietal ossification in the terms "frontoparietal fontanelle present, absent, small, large," etc.; such statements are misleading and imply that the fontanelle is absent or varies significantly in size and shape. The fontanelle is invariably present; it is a dorsal, ovoid opening in the neurocranial roof which varies slightly in size and location. The fontanelle is dorsally exposed to varying degrees, depending upon the location of the anteromedial and medial borders of the frontoparietals; thus, complete medial articulation of the frontoparietals obscures the fontanelle, which lies ventral to the bones.

Nasal. The nasals (Figure 2–1) are paired, dermal roofing bones overlying the olfactory region of the skull. I know of no instance in which the nasals are absent, although they may vary greatly in size and shape (Figure 2–2), or fuse to form a single element. Functionally, the nasals form a roof overlying the cartilaginous framework of the nasal capsules. The framework supporting the three main cavities, or sacs, of the olfactory apparatus consists of a cartilaginous roof, medial septum, floor, the septomaxillary, anterior and lateral cartilages, and several internal cartilages associated with the septomaxillary. These structures will be dealt with in more detail in another section of this paper. The medial septum and the dorsal roofing and ventral flooring cartilage are continuous with the cartilage in which the bone of the sphenethmoid is deposited. Thus the primary function of the nasals is to provide a protective bony covering for the cartilaginous structure of the nasal capsules anterior and lateral to the ossified part of the sphenethmoid. Secondarily, the posterolateral borders of the nasals may form a bony anterior orbital margin, and the extreme posterolateral corner (maxillary process) may be extended to articulate with the preorbital process of the pars facialis of the maxillary, thereby completing the anterior margin of the orbit and providing a strut that braces the maxillary arch against the roof of the skull.

There seems to be no doubt that the nasal is homologous with the bone of the same name of primitive amphibia. Each nasal forms from a single center of ossification. The extent of ossification is an especially labile character and varies in much the same way that the size of the frontoparietal does; that is, particularly small frogs or arboreal anurans tend to have smaller nasals, whereas larger species, or terrestrial, burrowing, or phragmotic frogs have larger nasals. In *Rhinophrynus* and the pipids, the nasals are moderate in size and uniquely characterized by their fusion with the frontoparietals and sphenethmoid. In a minimal state of development (Figure 2–2c), the nasal is present as a slender strip of bone overlying the nasal capsule on each side; the longitudinal axes of the bones frequently lie parallel to the maxillary so that the anterior ends of the bones lie near one another at the midline of the skull, but the posterior ends are widely separated and lie in the region of the anterodorsal corner of the orbit. Increased ossification produces marginal expansion. Maximally (Figure 2–2g), the nasals articulate with one another medially and overlie the anterior part of the sphenethmoid posteriorly; in such cases, the posterior margins of the nasals frequently articulate with the anterior margins of the frontoparietals. The lateral margin of the nasal may extend ventrolaterally so that a complete articulation with the pars facialis of the maxillary is effected (Figure 2–3a,f,h).

Internasal. This is a single, dermal element described and illustrated by Trueb (1970*a*). It is known to occur only in a single species, *Pternohyla fodiens* (Hylidae), and apparently has no homologue in the primitive amphibia. The narrow bone lies in the midline of the skull dorsal to the internasal septum. The anterior tip lies between the alary processes of the premaxillaries, and the posterior end articulates with the anteromedial tips of the nasals. The internasal is of membrane origin and forms late in development, probably as a specialized cranial reinforcement in the skull of the burrowing and phragmotic species. The term "internasal" should not be confused with the same name applied to the septomaxillary by Gaupp (1896) and others in some of the older literature.

Dermal Sphenethmoid. A second, single, dermal element apparently confined to the casque-headed hylids is the dermal sphenethmoid (Figure 2–2g), described and illustrated by Trueb (1966, 1970*a*). This bone is usually diamond-shaped; it articulates with the posteromedial margins of the nasals anteriorly and the anteromedial borders of the frontoparietals posteriorly. In the adult frog this element is completely and indistinguishably fused with the underlying sphenethmoid. It was named as a separate bone because during development it forms from a separate, membranous site of origin dorsal to the cartilaginous sphenethmoid. The dermal sphenethmoid ossifies as a flat plate between the nasals and frontoparietals and is only fused to the sphenethmoid below by secondary resorption and bone restructuring late in development. This element has no known homologue among primitive amphibians and, like the internasal, should probably be regarded as a specialization for cranial reinforcement and protection among many of the casque-headed, co-ossified hylids.

Maxillary Arch

Premaxillary. These paired, dermal elements at the anterior end of the skull form the anteriormost segments of each maxillary arch (Figure 2–1). The base of the premaxillary consists of a pars dentalis that may or may not bear teeth; the pars dentalis of the premaxillary articulates with the corresponding dental ledge of the maxillary forming the basic structural component of the arch. Posteriorly, the premaxillary bears a lingual ledge, the pars palatina. The palatal ledge is usually narrowest laterally; medially it expands onto a projecting palatine process. The premaxillary bears a dorsal projection known most commonly as the alary process, or infrequently as the squame. The premaxillaries are always separated from one another and from the maxillaries by narrow areas of dense connective tissue. The points of articulation involve the partes dentales and palatinae and the palatine processes.

Aside from completing and uniting the two maxillary arches, the premaxillaries apparently are of some importance in forming an anterior abutment for the nasal cartilages. The paired superior and inferior prenasal cartilages are rod-like structures which extend anteriorly from the main body of the nasal cartilages and abut against the posterior surfaces of the alary processes. A third cartilage, the crista subnasalis, lies along the dorsal surface of the pars palatina of the premaxillary and maxillary, separating these bones from the overlying nasal cartilages.

There is more variation in anuran premaxillaries than has been noted in the literature. Among primitive anurans (ascaphids and pipids), the premaxillaries are broad structures; the proportions of the maxillary to premaxillary length are reminiscent of those of salamanders. However, among most other anurans the premaxillaries are quite narrow. The pars palatina may be expanded into a broad shelf or so shallow as to be barely noticeable. It may be expanded at its lateral extreme but diminished markedly in width medially; lateral expansion is always associated with a similar expansion of the pars palatina of the maxillary and presumably related to the establishment of a strong articulation at this point. The nature of the palatine process is a useful specific character. I know of no case in which the processes are absent, but they may be so poorly developed that they are scarcely detectable (for example, *Triprion* [Hylidae]; Figure 2–5e); at the other extreme they may be long and acuminate, projecting freely into the oral cavity ventral to the prevomers and floor of the nasal capsules (for example, *Hemipipa* [Pipidae]; Figure 2–5b). The processes may articulate with one another or not. The rationale for these kinds of variation in the posterior components of the premaxillaries is obscure at present. Presumably, enlargement is related to strengthening of the upper jaw in large, terrestrial anurans, or in some frogs otherwise lacking a complete maxillary arch (for example, *Hemipipa*). In the majority of anurans, the pars and processus palatina are only moderately developed.

A second major source of premaxillary variation involves the structure and orientation of the alary processes. The processes may be exceptionally short (Figure 2–3d)—equal to or less than the height of the pars dentalis—or long (Figure 2–3b)—five times the height of the pars dentalis. When the skull is viewed in its lateral aspect, the alary processes of some frogs slope anteriorly (Figure 2–3g). The usual degree of anterior slope varies around 80 degrees with the horizontal plane of the skull. In extreme cases, for instance in some bufonids and hylids, the processes are displaced anteriorly at angles of 10 degrees to 20 degrees. The alary

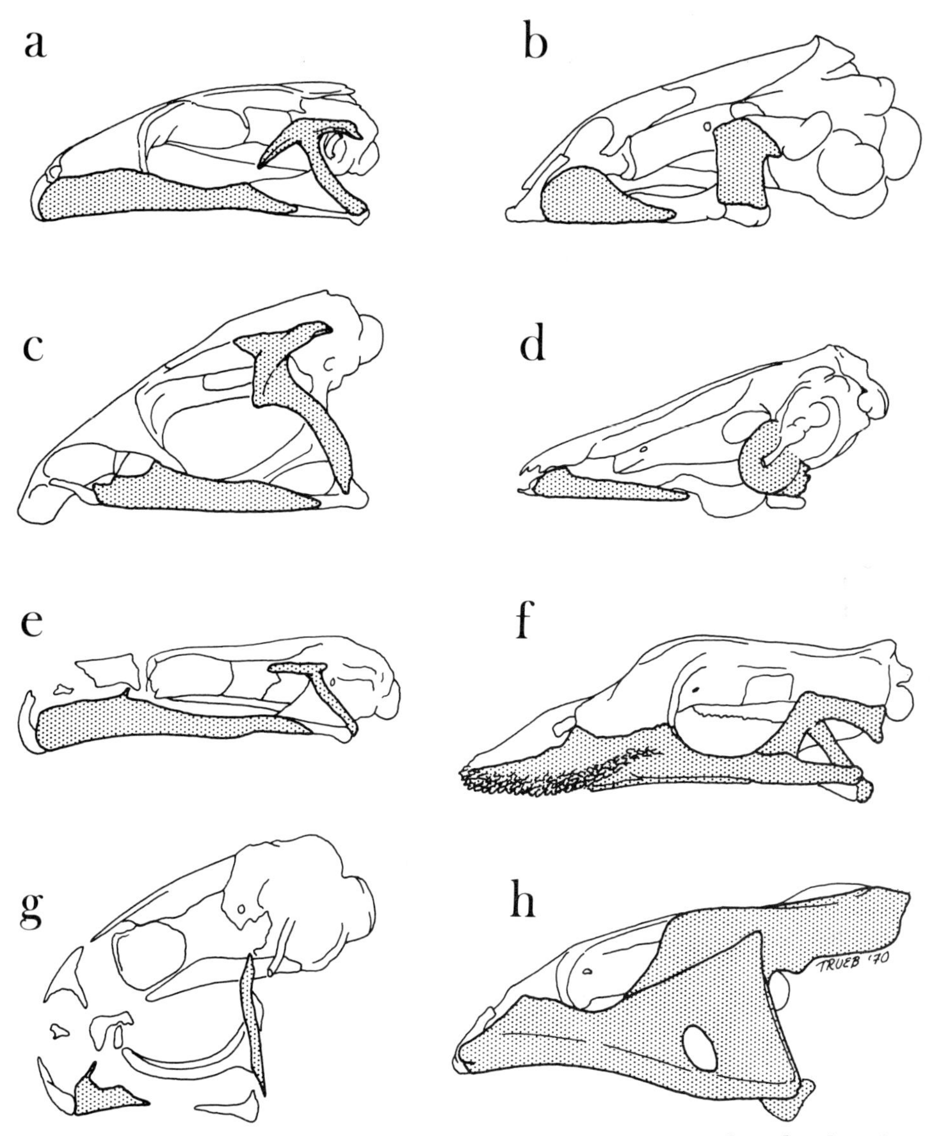

Figure 2–3. Diagrammatic lateral views of skulls of anurans: a, *Smilisca baudinii* (Hylidae); b, *Rhinophrynus dorsalis* (Rhinophrynidae); c, *Rhamphophryne acrolopha* (Bufonidae); d, *Hemipipa carvalhoi* (Pipidae); e, *Heleophryne natalensis* (Myobatrachidae); f, *Triprion petasatus* (Hylidae); g, *Notaden nichollsi* (Leptodactylidae); h, *Hemiphractus panamensis* (Hylidae). *Stippled pattern,* squamosal and maxillary bones.

processes of most anurans are nearly vertical or inclined posteriorly at angles usually no greater than 135 degrees (Figure 2–3a). The structure of the process is almost uniformly that of a thin bony shaft, which is convex anteriorly (or ventrally) and concave posteriorly (or dorsally). There is one significant exception to this trend. The genus *Plectrohyla* (Hylidae) is characterized by a bifurcate alary process. The posterior surface of the process bears a posterodorsally directed projection at approximately its mid-height. This projection serves to brace the anterior end of the nasal capsule under which the projection lies.

Prenasal. This single, dermal bone is known to occur only in two casque-headed hylid genera (*Triprion* and *Aparasphenodon*); it was described and illustrated by Trueb (1970*a*). The bone lies anterior to the premaxillaries (Figures 2–2g, 2–3f); it articulates posterolaterally with the maxillaries, posterodorsally with the nasals, and posteroventrally with the pars dentalis of the premaxillary. When present, this bone forms an integral part of the maxillary arch, performing much the same function as the premaxillaries which it covers. It should be noted that in those species having a prenasal, the alary processes of the premaxillaries are inclined anteriorly at an angle of between 10 degrees and 20 degrees, and the nasal capsules have extended forward to occupy an anterior position within the prenasal.

Maxillary. The maxillaries (Figure 2–1) are the largest components of the maxillary arches. The bones are dermal, and like the premaxillaries, each element consists of a pars dentalis which may or may not bear teeth and a lingual ledge, the pars palatina. Dorsolaterally, the maxillary bears a facial flange, known as the pars facialis. The par facialis usually has a preorbital process and less frequently a postorbital process. At most, the maxillary articulates at five separate points with the remainder of the skull. In an anterior to posterior sequence, they are (1) the pars dentalis and pars palatina at the lateral edge of the premaxillary and, where it occurs, the prenasal; (2) the preorbital process at the maxillary process of the nasal; (3) the pars dentalis and occasionally the pars palatina at the anterolateral edge of the anterior pterygoid ramus; (4) the postorbital process with the zygomatic process of the squamosal; and (5) the posterior end of the maxillary with the quadratojugal. Externally, there appears to be a sixth point of articulation between the maxillary and the distal end of the palatine bone. The palatine invests the cartilaginous planum antorbitale; its distal end lies in connective tissue in the internal angle formed between the pars palatina and the pars facialis, but it is not closely associated with either maxillary component.

The maxillary is closely applied to a series of cranial cartilages that lie along its inner surface. In the nasal region it invests the lateral aspect of the nasal floor (solum nasi) from which the crista subnasalis diverges. The crista is a rod of cartilage that lies in the angle between the pars facialis and pars palatina and extends a short distance posteriorly. In the posterior part of the nasal region, the position anteriorly occupied by the crista subnasalis is filled by a rod of cartilage known as the anterior maxillary process. Suborbitally, the cartilaginous rod is known as the posterior maxillary process; this process diverges medially from the maxillary region of the orbit with the pterygoid. There are variations on this pattern. However, it should be noted that there is a good deal of support for the upper jaw aside from obvious, external articulations. These articulations are mainly a function of the presence or absence of bones or parts of bones associated with the maxillary. Thus the maxillary always articulates with the premaxillary and pterygoid because these bones are invariably present. Articulation with the squamosal and nasal is dependent upon the degree of development of the zygomatic and maxillary processes, respectively, and articulation with the quadratojugal depends upon the presence and degree of development of this element.

Variation in the maxillary is principally a function of the development of the pars facialis and its associated parts—the preorbital and postorbital processes (Figure 2–3). In the majority of anurans, the pars facialis bears a preorbital process and is moderately developed anterior to the orbit; the flange is usually reduced in height suborbitally and nonexistent posterior to the orbit. In certain terrestrial or burrowing genera (of the families Bufonidae, Ranidae, Hylidae, Leptodactylidae) with heavily ossified skulls, the facial flange is better developed and tends to establish strong articulations with the nasals anteriorly and squamosals posteriorly. The microhylids are uniquely characterized by an anteromedial extension of the facial flange; in some members of this family, the premaxillaries are nearly covered by an overgrowth of the pars facialis of each maxillary (Figure 2–5h). In casque-headed, phragmotic species (of the family Hylidae), the basic structure of the maxillary is frequently modified by intensive ossification of its lateral surface to produce a projecting shelf or flange.

Quadratojugal. This small, dermal element completing the posterior sequence of the maxillary arch is also known as the quadratomaxillary (Figure 2–1). The quadratojugal is highly variable in its occurrence and is frequently lost or reduced in smaller frogs or those in which ossification is reduced. Reduction always proceeds in an anterior-to-posterior sequence. Thus, the first sign of reduction is the loss of the quadratojugal-maxillary articulation (Figure 2–3g); further reduction is reflected in progressive abbreviation of quadratojugal length. Since the ossification of the quadratojugal frequently invades the quadrate cartilage posteriorly, the quadratojugal, apparently absent externally, may be represented by vestigial ossification.

Dentition. The pedicellate nature of modern amphibian teeth has been described by Parsons and Williams (1962). In primitive anurans, teeth are present on the maxillary and premaxillary of the maxillary arch; they are secondarily lost in bufonids, *Rhinophrynus*, some dendrobatids and microhylids, and pipids. Lynch (1969) has described the differentiation of fang-like teeth in a few leptodactylid genera and one genus of the Myobatrachidae. Fang-like teeth characterize the hemiphractine hylids but not *Ceratobatrachus* (Ranidae), which is otherwise superficially similar to *Hemiphractus* (Hylidae). The differentiation of these teeth seems to be associated with the dietary habits of these anurans, which frequently prey upon small vertebrates.

Dermal Ornamentation

Dermal ornamentation was widespread in primitive Amphibia and was expressed by the rugosity and pitting of the dermal roofing bones. The majority of modern anurans have unornamented skulls, but some have developed different kinds of dermal ornamentation as a secondary modification or specialization to ornament or to strengthen the skull and aid in the prevention of desiccation. Cranial ornamentation occurs in representatives of the following families: Brachycephalidae, Bufonidae, Hylidae, Leptodactylidae, Myobatrachidae, Pelobatidae, and Ranidae. It involves modification of the following bones: frontoparietal, nasal, premaxillary, maxillary, prenasal, internasal, dermal sphenethmoid, and squamosal. Ornamentation can be broadly categorized into three types: exostosis, casquing, and co-ossification.

Exostosis. This is a morphological condition wherein additional membrane bone is laid down on dermal cranial elements to form ridges, crests, spines, etc. Occasionally, exostosis produces simple bony spines in the skull (e.g., *Anotheca* [Hylidae]), but usually it results in a simple reticulate pattern of bone deposition on the surface of dermal bones following metamorphosis. Proliferation of bone in certain areas and subsequent secondary reorganization result in such familiar structures as the cranial ridges or crests and carotid canals of bufonids. Additional bone deposition and subsequent modification result in more intricate patterns of dermal sculpturing. Thus, the complicated systems of spines and radiating ridges characteristic of *Triprion, Trachycephalus,* and *Aparasphenodon* (Hylidae) are refinements of the generalized reticulate pattern first established in the ontogenetic development of the species.

Casquing. Casquing involves the extension or proliferation of dermal bones to form a "casqued," or helmeted, appearance (e.g., *Ceratobatrachus* [Ranidae], *Hemiphractus* [Hylidae]). This is expressed in the occurrence of such bones as the dermal sphenethmoid, internasal, and prenasal, the formation of labial flanges on the maxillaries, and the marginal extension of nasals, frontoparietals, and squamosals. I know of no species in which casquing occurs without exostosis; therefore, I believe that all casque-headed species have some characteristic pattern of dermal ornamentation or sculpturing.

Co-ossification. Co-ossification, which is always accompanied by some degree of exostosis and casquing, involves the integration of skin overlying dermal bones into the bone structure so that the skin is fused to the bone below. Co-ossification is prevalent among arid to semiarid lowland forms and certain canopy-dwelling arboreal species.

These three types of dermal modifications seem to form a natural evolutionary sequence of specialization wherein exostosis is the most widespread and generalized, and co-ossification is the most highly specialized. The appearance of these dermal modifications in primitive tetrapods has led many authors to comment on the primitive nature of the modifications and to speculate on their occurrence among advanced anuran families. Recalling that much anuran tissue is skeletogenous and that bone formation is highly labile, one should view the varieties of dermal modifications that occur as a potential possessed by all anurans and exploited by those in which it is advantageous.

Palatal Bones

Parasphenoid. This dermal bone, which is also called the parabasal and synpterygoid, invests the neurocranium ventrally. It forms a bony bridge from the posterior, bony portion of the sphenethmoid anteriorly to the ossified part of the prootic posteriorly (Figures 2–1, 2–4). The parasphenoid is triradiate in form; the major part of the bone is termed the cultriform process or the parasphenoid rostrum, and the posterolateral extensions are usually referred to as alae or wings. I know of no species in which the parasphenoid is absent, although in some pipids it is indistinguishably fused with the overlying prootic and sphenethmoid. Variation in the parasphenoid is slight and involves the length of the cultriform process, the presence and orientation of the alae, and the presence of odontoid-like structures ventrally.

Maximally the cultriform process of the parasphenoid extends anteriorly between the prevomers (*Rhinophrynus;* Figure 2–4e); in some leptodactylids, the anterior tip lies slightly anterior to the level of the palatines. The process terminates just posterior to the level of the palatines (or antorbital process) in the majority of anurans. A few, such as the elosiine leptodactylids, dendrobatids and pipids (Figure 2–4b), tend to have truncated cultriform processes that terminate abruptly on the posterior part of the sphenethmoid. The parasphenoids of pipids are unique in lacking alae; this may be related to the incorporation of the bone into the neurocranium. The length of the alae vary among other anurans, and the orientation may be lateral or slightly posterolateral. This variation provides useful characters at the specific and generic level but hardly seems to have any phylogenetic significance. True teeth do not occur on the parasphenoid, but the posterior part of the cultriform process may bear some minor, dermal modification in the form of a smooth median keel (Figure 2–4c), a denticulate keel, or a

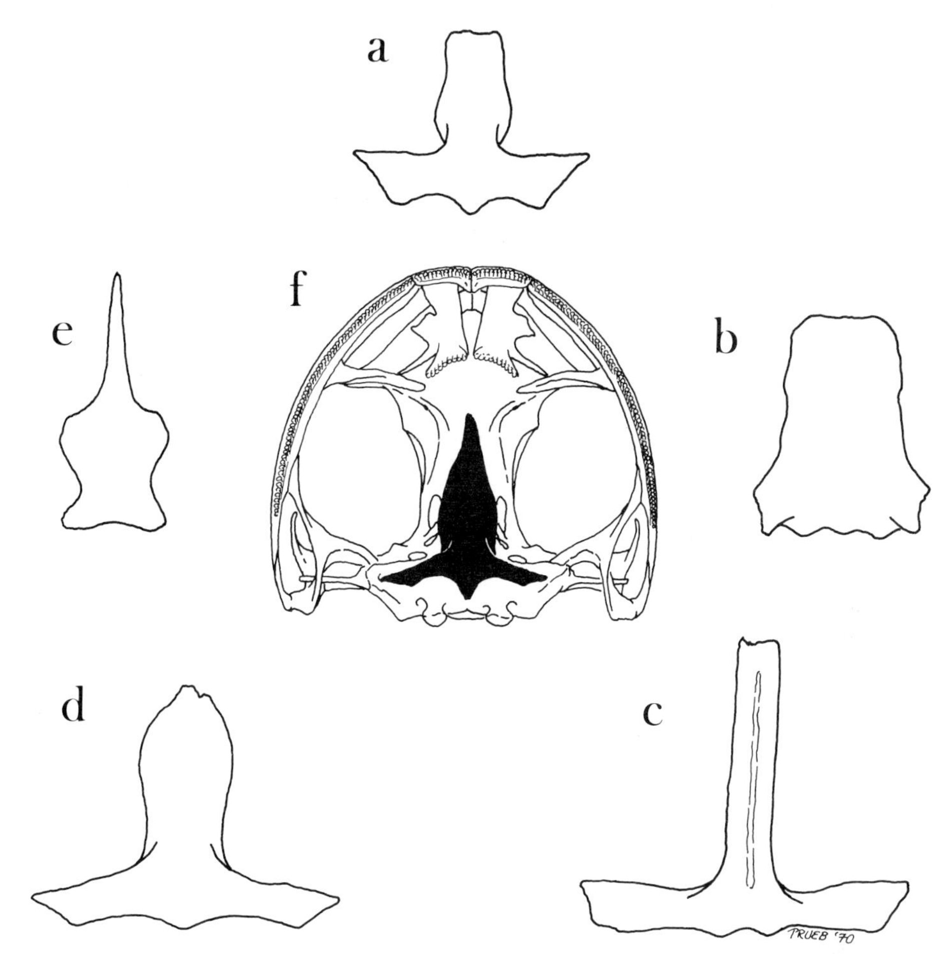

Figure 2–4. Schematic representation of variation in the anuran parasphenoid bone: a, *Kaloula pulchra* (Microhylidae); b, *Hemipipa carvalhoi* (Pipidae); c, *Pseudis paradoxa* (Pseudidae); d, *Phyllomedusa edentula* (Hylidae); e, *Rhinophrynus dorsalis* (Rhinophrynidae); f, *Osteocephalus taurinus* (Hylidae). *Solid color* in f, parasphenoid.

proliferation of tooth-like denticles. The utility of these structures is questionable; it is possible that they function with the prevomerine teeth to control prey in the oral cavity.

Prevomers. These paired, dermal bones (Figures 2–1, 2–5) are among the most variable of any of the skull. Many authors use the term "vomer" in preference to "prevomer" with reference to amphibians. Broom (1902) claimed that the mammalian vomer is homologous to the parasphenoid of amphibians and reptiles and therefore suggested that the so-called vomer of lower vertebrates would be more appropriately termed "prevomer."

The prevomers of most anurans are moderate-sized bones underlying and supporting parts of the nasal capsule and bearing true teeth. Usually the anterior ends lie in connective tissue adjacent to the palatal shelf of the premaxillary and maxillary, and the lateral wings form bony anterior, medial, and posteromedial margins of the internal naris. The dentigerous processes usually lie at a level slightly anterior to the palatines. The prevomer bears an internal, dorsal expansion which, in association with the solum nasi, supports the olfactory eminence of the principal nasal cavity. Aside from occasional disappearance, there are many specializations on the generalized pattern described above.

Minor prevomerine variation is expressed in the over-all size of the bones and the orientation of the dentigerous ridges. The latter is a useful diagnostic character at the species and genus level in which variations may be transverse, oblique, curved, or angled. Teeth are sometimes absent, and odontoids are occasionally present in the absence of true teeth.

Some pelobatids and microhylids are characterized by an apparent trend toward replacement of the palatine by parts of the prevomer. In pelobatids (Figure 2–5f), the posterolateral wing of the prevomer tends to displace the palatine (Zweifel, 1956). The palatine is not lost, but it is reduced in size and maintains an articulation with the prevomer. The situation is considerably more complex in the microhylids. There is a marked trend towards fusion of prevomerine and palatine elements (Figure 2–5g,h). There is also a tendency towards subdivision of the prevomer into discrete anterior and posterior elements; the anterior elements are associated with nasal support, whereas the posterior parts bear teeth (Figure 2–5d). The anterior elements tend to become reduced in size and occasionally disappear completely, while the posterior elements tend to lose their teeth, fuse medially to one another and laterally to the palatines. Among some microhylids it is obvious that the prevomer has fused to the palatine, forming a "vomero-palatine"; in others it seems that the palatine has been lost and only partially replaced by the prevomer. Comparative study of adults and ontogenetic material of microhylids is necessary before any significant statements can be made concerning the evolutionary trends of the palatal bones in this family.

Palatine. The palatines (Figure 2–1) are paired, dermal investing bones lying on the ventral surface of the planum antorbitale at the anteroventral margin of the orbit. The bones usually lie adjacent to the maxillary and articulate with the sphenethmoid medially. They are always edentate but may bear a ventral transverse ridge that is smooth or serrate. Frequently the palatines are reduced in length or lost (Figure 2–5b). Reduction occurs in a medial to lateral direction; that is, palatine reduction is first expressed by the loss of its articulation with the sphenethmoid. Reduction of the palatine is nearly always compensated for by increased ossification of the prevomer or ossification of the underlying cartilaginous planum antorbitale. Ossification of the latter is frequently interpreted as lateral extension of the sphenethmoid. Both the sphenethmoid and the planum are contiguous in the cartilaginous chondrocranium, but in the adult they are discrete structures.

Pterygoid

Traditionally the pterygoids have been considered to be paired, dermal palatal bones. There is no argument that the broad pterygoids of labyrinthodont amphibians are, in fact, integral parts of the palate. However, the homologues among modern anurans usually differ in appearance, function, and position from the pterygoids of primitive amphibians. The bones no longer contribute to the palate; instead, they are almost wholly involved in the jaw suspensory mechanism of the anuran skull. Accordingly, the pterygoids are here considered distinct from the palatal bones and in conjunction with the cranial suspensory mechanisms.

In the majority of modern anurans the pterygoid is a triradiate structure (Figures 2–1, 2–6), the rami being designated as anterior, medial, and posterior, respectively. The anterior ramus is the longest. It articulates with the pars dentalis of the maxillary bone anterolaterally and invests a rod of cartilage, the pterygoid

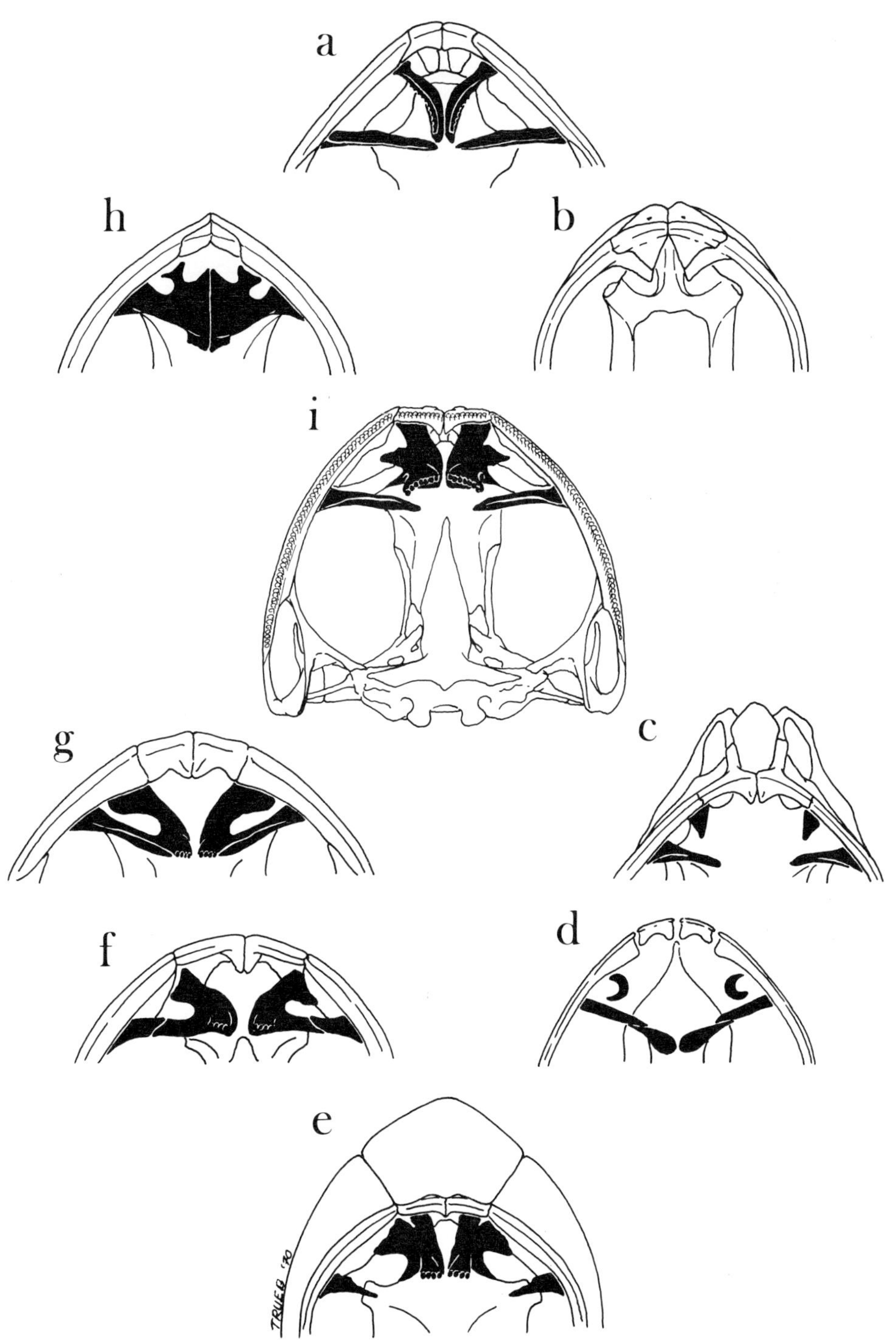

Figure 2–5. Schematic representation of variation in the anuran prevomer and palatine bones: a, *Hemiphractus panamensis* (Hylidae); b, *Hemipipa carvalhoi* (Pipidae); c, *Rhamphophryne acrolopha* (Bufonidae); d, *Stumpffia psologlossa* (Microhylidae); e, *Triprion petasatus* (Hylidae); f, *Scaphiopus bombifrons* (Pelobatidae); g, *Kaloula pulchra* (Microhylidae); h, *Asterophrys rufescens* (Microhylidae); i, *Osteocephalus taurinus* (Hylidae). *Solid color,* prevomer and palatine bones.

process, which is contiguous with the posterior maxillary process anteriorly and the pseudobasal (basal process of the Ascaphidae) and quadrate processes posteriorly. The medial ramus invests the pseudobasal or basal process; it may or may not articulate with the otic capsule. The posterior ramus invests the cartilaginous quadrate process medially and terminates posteriorly at the angle of the jaw.

There is considerable variation in the nature of the articulations of the anterior and medial rami (Figure 2–6). The anterior ramus is invariably present and bears some kind of articulation with the maxillary arch. Usually the articulation lies at the midlevel of the orbit. If the medial ramus is absent or lacks a cranial articulation, or if the skull is poorly ossified, the anterior pterygoid ramus is apt to have an extensive anterior articulation with the maxillary. Occasionally in the absence of a palatine or other ossification in the region of the planum antorbitale (e.g., some discoglossids), the anterior pterygoid ramus diverges medially from the maxillary to invest the lateral part of the planum.

The medial ramus may be present or absent, and if present, it may or may not be directly articulated with the neurocranium. Although the medial ramus may be reduced so that it does not bear a bony articulation with the otic capsule (Figure 2–6e), the ramus is usually indirectly associated with the capsule by means of a pseudobasal or basal process. (*Rhinophrynus* may prove to be an exception; see Figure 2–6a.) The medial ramus usually articulates with the anteroventral surface of the otic capsule laterally. Occasionally the articulation is expanded ventrally and, in some leptodactylids and bufonids, overlaps the ala of the parasphenoid (Figure 2–6c), or in the case of some pipids broadly underlies most of the otic capsule (Figure 2–6d).

In one discoglossid–*Barbourula* (Figure 2–6b)–and three leptodactylid species–*Cycloramphus, Hydrolaetare,* and *Zachaenus*–the pterygoid bears a peculiar ventral flange. Lynch (1969) suggested that in the leptodactylids this flange may be a specialization for muscle attachment as an adaptation for burrowing.

Suspensorium

In the adult anuran, the suspensory apparatus functions to brace and suspend the jaws against the neurocranium. The squamosal and pterygoid are the principal dermal bones involved, but their functions are of secondary importance as investing elements that protect and strengthen the basic cartilaginous system, which consists of the quadrate, pterygoid process, and the lateral wall of the neurocranium.

There are two fundamental types of suspensory apparatus among anurans, based on the nature of the articulation of the apparatus with the neurocranium. The evolution of the suspensorium is apparently correlated with the trend toward consolidation of all branches of nerves V (Trigeminal) and VII (Facial) into a single foramen, the prootic foramen. In primitive anurans, three separate foramina occur (in *Ascaphus* and salamanders), and two are present in *Leiopelma* (Ascaphidae) and the discoglossids. The primitive type, similar to that of salamanders, involves the presence of basitrabecular and basal processes; among anurans, it is found only in ascaphids. According to Pusey (1938) the basitrabecular process is produced embryologically by a lateral outgrowth from the basal plate of the skull floor. The basal process, in contrast, arises as a medial outgrowth of the quadrate. The neurocranial articulation of *Ascaphus* is thus formed by the abutment of the basal process of the quadrate against the basitrabecular process. The basal process articulates with the anterolateral floor, which is formed by the basitrabecular process, of the auditory capsule and underlies the anterior part of the ventral cartilage ledge of the auditory capsule. *Leiopelma* represents a

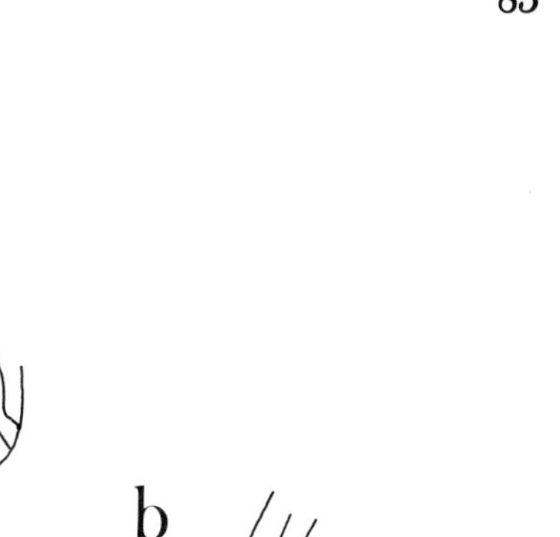

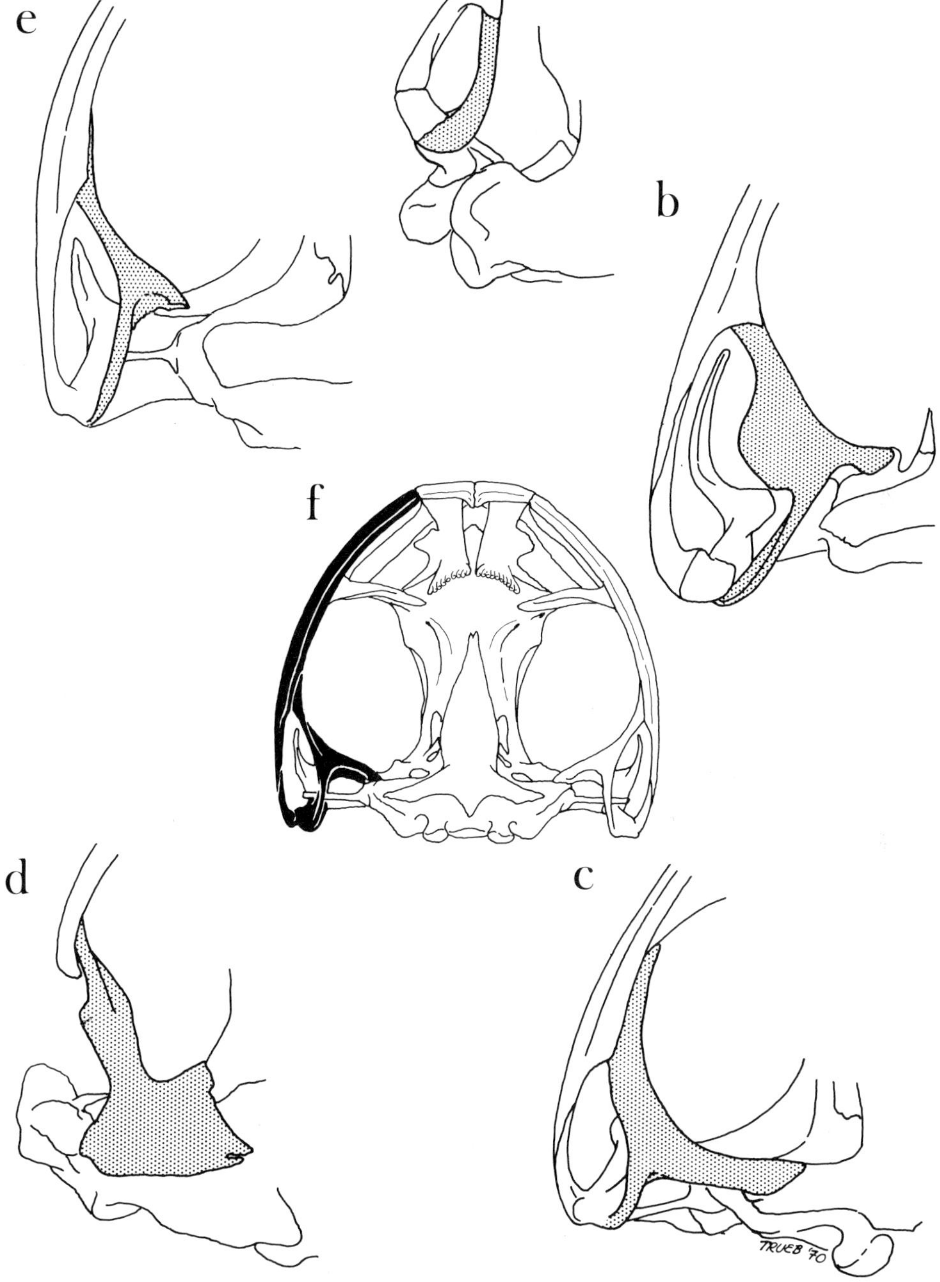

Figure 2–6. Schematic representation of variation in the anuran pterygoid bone: a, *Rhinophrynus dorsalis* (Rhinophrynidae); b, *Barbourula busangensis* (Discoglossidae); c, *Bufo arenarum* (Bufonidae); d, *Hemipipa carvalhoi* (Pipidae); e, *Aparasphenodon brunoi* (Hylidae); f, *Osteocephalus taurinus* (Hylidae). Peripheral figures show only posteroventral, right quadrant of skull; *stippled pattern*, pterygoid bones. In central figure, f, maxillary, pterygoid, and lower arm of squamosal are indicated by solid color.

slight advancement of the suspensory mechanism. The basal process of the quadrate is essentially similar to that of *Ascaphus;* however, the root of the basitrabecular process has been lost. The lateral part of the basitrabecular process underlies the cartilaginous floor of the auditory capsule and becomes fused to a posterior block of cartilage known as the postpalatine commissure. The anterior part of the basitrabecular process projects forward, lateral to the neurocranium at the level of the facial (VII) ganglion. The basal process articulates with the cartilaginous ledge of the postpalatine commissure that is formed ventral to the auditory capsule and located posteroventral to the facial ganglion and basitrabecular process.

On the basis of ontogenetic comparative anatomical evidence, Pusey (1938) hypothesized that in all anurans other than the ascaphids, the basal process of the quadrate or pteroquadrate has come to lie far anterior in the chondrocranium; it fuses with the lateral edge of the trabecula at a level just posterior to the nasal sac, thereby forming a cartilaginous bridge, the commissura quadrato-cranialis anterior. The adult pterygoid process arises from the union of the distal part of the commissura (near its junction with the quadrate process) and a small anterior projection of the commissura known as the processus quadrato-ethmoidalis. During development the medial attachment of the commissura to the trabecula is destroyed.

As a result of the loss of the basal process of the quadrate, the distal end of the basitrabecular process has secondarily fused to the quadrate process, forming what is known as the pseudobasal process of the quadrate in discoglossids and all other anurans. The pseudobasal process thus consists of the basitrabecular process and the postpalatine commissure to which the process is fused. In its primitive state the pseudobasal process incorporates the cartilaginous ledge of the commissure and is contiguous with the otic capsule. However, in most anurans the pseudobasal process has acquired a jointed articulation with the auditory capsule, thereby more closely approximating the kinetic arrangement between the basal and basitrabecular processes of ascaphids and salamanders. (Notable exceptions occur among the Bufonidae [*Bufo*], the Ranidae [*Rana grayi*], and the Leptodactylidae [*Heleophryne*].)

There is a second, immovable attachment of the suspensorium to the skull in the majority of anurans; this is the otic process (not to be confused with the "otic element" *sensu* Griffiths, 1954*b*), which attaches the quadrate bar to the otic capsule medially. The aforementioned quadrate or palatoquadrate cartilage is contiguous with the body of the quadrate cartilage that lies medial to the ventral arm of the squamosal. The larval otic process is restructured to attach the dorsomedial aspect of the quadrate bar to the lateral surface of the auditory capsule; this union is further reinforced by anterior growth and fusion of the crista parotica laterally with the muscular process of the quadrate, which is anteroventral to the union of the otic process with the quadrate (Pusey, 1938). In most anurans the quadrate remains cartilaginous, although ossification of the quadratojugal frequently invades the quadrate cartilage. True ossification of the quadrate occurs in the ascaphids, pelobatids, and *Brachycephalus,* and is probably a specialization. Fusion of all or part of the palatoquadrate suspensorium results in autosystylic jaw suspension that, so far as is known, characterizes all anurans except *Acris* (Hylidae) and *Brachycephalus* (Brachycephalidae). According to McLachlan (1943), the otic process and crista parotica are absent in both of the latter genera, and the pseudobasal process is not fused to the neurocranium; therefore, the jaw suspension should be considered autodiastylic.

Squamosal. These paired, dermal bones invest the quadrate bar laterally; they are sometimes termed paraquadrates. The squamosals are invariably present (Figures 2–1, 2–3). In lateral aspect, basically the bones are triradiate. The greatest variation occurs in the presence and nature of the anterior (zygomatic ramus) and posterior (otic ramus) arms. The zygomatic ramus may be completely absent, developed only as a knob-like tuberosity or maximally, articulate with the postorbital process of the pars facialis of the maxillary anteroventrally. Every conceivable degree of intermediate development of the zygomatic ramus occurs among anurans; development is most often positively related to the size of the skull and the extensiveness of its ossification. The posterior arm, or otic ramus, is always present and usually bears one of three relationships with the medially adjacent crista parotica. As defined by Griffiths (1954*b*), these relationships are as follows: (1) The otic ramus bears a medially expanded otic plate (*sensu* Lynch, 1969) that broadly articulates with the distal dorsal portion of the crista parotica; (2) the medial expansion of the otic ramus articulates with posterolaterally expanded frontoparietal forming a complete or partial arch (temporal arcade of Lynch, 1969) over the crista parotica; and (3) the otic ramus is small and poorly developed, and it lies laterally adjacent to the critsa parotica but does not overlap it. The first type of relationship is characteristic of many bufonids (especially *Bufo*), leptodactylids, hylids, pelobatids, and *Brachycephalus* and is associated with greatly increased cranial ossification of usually moderate-to-large terrestrial, burrowing, or phragmotic anurans. The third type is usually characteristic of smaller or arboreal species in which cranial ossification is not especially well developed; it is also typical of many microhylids and pipids.

Nasal Capsule

A generalized description of the nasal capsule is available in Trueb (1970*a*) and need not be repeated here. Among those taxa studied by microtomized sections, variation has been described in the complex cartilaginous framework of the nasal capsule. The information is so scarce and widely dispersed that it is difficult to formulate generalizations or comment on the significance of the structural variations at this time. The following trends have been observed: (1) In ascaphids, *Bombina* (Discoglossidae), and salamanders the nasal capsules tend to be depressed and laterally displaced compared with other anurans; presumably this is a primitive trait; (2) there is a tendency in some bufonids and microhylids for the crista intermedia to fuse ventrally with the solum nasi. I would agree with Baldauf (1959) that this structural modification reinforces the support of the nasal cavities but does not constitute a specialization for fossorial habits as suggested by McLachlan (1943). The septomaxillary (internasal of Gaupp) seems to be invariably present. The developmental derivation of this element is still open to question. Trueb (1970*a*) described minor variation in the septomaxillaries in hylids, as did Lynch (1969) for the Leptodactylidae. Thorough investigation of this area of the cranium would contribute significantly to our knowledge of anuran variation and undoubtedly yield interesting information concerning adaptive modifications.

Neurocranium

Sphenethmoid. This endochondral bone, which is known variously as the sphenethmoid, ethmoid, and *os en ceinture,* is invariably present as a housing around the anterior end of the brain (Figure 2–1), although it may not be ossified. The precise limits of the sphenethmoid, as of all neurocranial bones, are extremely difficult to define because all of these elements are formed in a continuous system of cartilage surrounding the central nervous system. The internasal septum of the olfactory capsule lies at the anterior extreme, and the exoccipital at the posterior extreme. Therefore, variation in these cranial elements is mainly a result of the degree of ossification, which in turn is related to the skull size, as well as the extent of ossification of protective and supporting dermal elements.

In most anurans the anterior terminus of sphenethmoid ossification lies at the posterior level of the nasals; posteriorly, ossification usually terminates around the anterior margin of the frontoparietal fontanelle. Additional ossification anteriorly—in the internasal septum—or posteriorly—around the optic foramen—should probably be viewed as a specialized feature of more heavily ossified skulls. The reduction of ossification in small frogs, in contrast, seems to be a neotenous adaptation to reduce the weight of the skull at the expense of protection for the brain. Similarly the presence of "paired" sphenethmoids in the microhylids and pipids merely reflects reduced ossification. Ontogenetically, all sphenethmoids pass through a "paired" stage. It is somewhat misleading to say that the sphenethmoid is reduced in the pipids. Although the sphenethmoid does not completely encircle the skull, it is fused with the parasphenoid ventrally; the brain is thus afforded complete protection.

Otoccipital. This term is used to designate the prootic and exoccipital, which are indistinguishably fused in the modern anurans (Figure 2–1). The prootic portion of the otoccipital lies posterior to the sphenethmoid and optic foramen; it forms the neurocranium anterior to the auditory capsules and the auditory capsules themselves. The exoccipital forms the posteriormost part of the neurocranium, including the bone around the foramen magnum and the occipital condyles. The otoccipital is subject to much the same kind of variation in ossification as is the sphenethmoid.

There is a marked evolutionary trend toward reduction of the number of cranial nerve foramina. Primitively, in *Ascaphus* there are eleven foramina for the cranial nerves. In *Leiopelma* (Ascaphidae) and the discoglossids, there are ten; in all other anurans the number varies between six and nine. Apparently this incorporation of cranial nerves into fewer foramina is responsible for the change in the suspensory mechanism noted in transitional stages in *Leiopelma* and the discoglossids. In salamanders and *Ascaphus* the branches of the trigeminal (V) nerve exit from the neurocranium by means of two separate foramina; the facial nerve (VII) has a separate foramen, and there are three separate acoustic foramina. *Leiopelma* and the discoglossids have lost the division in the trigeminal foramen, but the trigeminal and facial foramina are still separated by the prefacial commissure. In all other anura the prefacial commissure is absent; the facial and trigeminal nerves exit from the neurocranium via the prootic foramen. The majority of anurans have only two acoustic foramina. The presence of three is regarded as primitive, while the presence of only one acoustic foramen is an advanced feature.

I know of no case in which a frog lacks an auditory capsule. Frequently, the plectral apparatus is absent, but the operculum is present. It is generally agreed that the loss of the plectral apparatus is a specialization associated with aquatic and terrestrial adaptations and that the primitive anurans had fully developed auditory mechanisms. In those terrestrial frogs lacking a plectral apparatus, it has been argued (de Villiers, 1931*a*) that the presence of a large operculum is an adaptation to terrestrial life. De Villiers concluded that the operculum is phylogenetically "younger" than the plectral apparatus and that in the absence of the latter among terrestrial anurans, the operculum tends to enlarge and receive sound oscillations via the forelimbs, suprascapula, and opercular muscle.

Lynch (1969) noted an interesting variation in the occipital condyles of the otoccipital. He classified the condylar arrangement into three different types depending upon the proximity of the condyles to one another and upon the nature of the articular surfaces. This point will be discussed in more detail in the section dealing with cervical vertebrae.

Another source of variation in the neurocranium about which very little is known is the nature of the chondrocranial roof cartilages. These include the taenia tecta marginalis (forming the lateral margins of the frontoparietal fontanelle), the taenia tecta medialis, and the tectum synoticum. The taenia tecti medialis is a narrow strip of cartilage in the midline of the skull; it extends from the taenia tecta transversalis to the tectum synoticum, producing a parietal fontanelle on each side. The tectum synoticum roofs the neurocranium posterodorsally between the otoccipitals and forms the posterior margins of the parietal fontanelles. The tecti medialis and transversalis are frequently absent, and as Baldauf (1959) points out, ossification of the cartilages and synostosis with adjacent bones may obscure the presence and character of the structure.

Mandible

The mandible consists of four elements, the symphysial bone, and Meckel's cartilage, which is invested by the dermal dentary bone laterally and the dermal angular bone posteromedially. Although little attention has been directed toward the mandible, variation appears to be rather insignificant. The anterior end of Meckel's cartilage is usually ossified to produce a symphysial bone; however, this element may be absent. The symphysial bone may be synosteotically united with the dentary. Only one anuran genus, *Amphignathodon* (Hylidae), is known to possess teeth on the lower jaw. Several genera—in Leptodactylidae, Hylidae, and Ranidae—possess denticulate serrations on the dentary.

HYOLARYNGEAL SKELETON

Throughout anatomical and systematic literature on anurans there are scattered descriptions of the hyolaryngeal apparatus. These descriptions usually are limited to the structure of the hyoid plate and sometimes its associated muscles. Inclusion of the laryngeal apparatus is relatively infrequent. Although W. K. Parker (1881) illustrated and described the hyoid apparatus of many anurans, the most comprehensive and important work to date is that of Trewavas (1933) on the hyoid and larynx of the anura. In this monograph the author described the hyolaryngeal apparatus of some 60 anuran taxa and reviewed pertinent literature. It remains the classical reference work on the subject.

There is a great deal of variation in the hyoid apparatus (Figure 2–7), which is concerned principally with the presence or absence and nature of the processes of the hyoid. So far as is known, all frogs possess well-developed anterior cornua or hyale, except the pelobatids, which are characterized by more or less complete reduction of this element. It provides a useful taxonomic character at specific and generic levels; however, no real evolutionary trends are apparent. Similar degrees of variation have been noted in the posterolateral and posteromedial processes and the presence and nature of anterior hyale processes. The nature of the development of these processes is correlated with muscle attachments. Presumably, the occurrence of a parahyoid bone is primitive; it occurs in *Rhinophrynus*, ascaphids (Figure 2–7a), discoglossids (Figure 2–7c), and *Pelodytes* (Pelobatidae).

With regard to the larynx, there is variation in the presence and development of the apical and basal cartilages of the arytaenoid cartilage. It was suggested (Trewavas, 1933) that the cricoid ring of the larynx is dorsally incomplete primitively (e.g., the Pelobatidae and *Rhacophorus dennysi* of the Ranidae). The

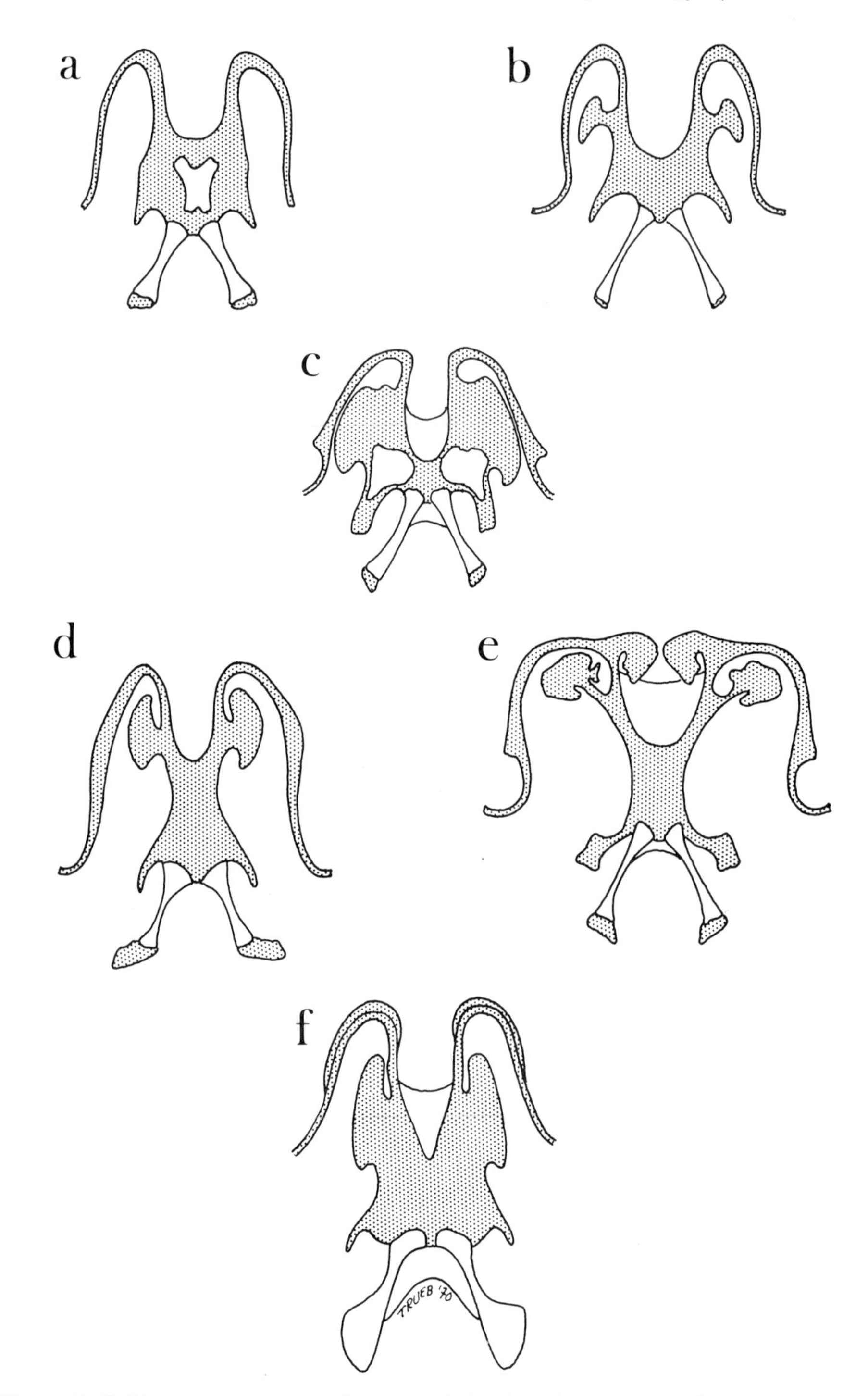

Figure 2–7. Diagrammatic ventral views of the hyoid apparatus of anurans: a, *Leiopelma hochstetteri* (Ascaphidae); b, *Leptodactylus ocellatus* (Leptodactylidae); c, *Bombina variegata* (Discoglossidae); d, *Bufo himalayanus* (Bufonidae); e, *Helioporus albopunctatus* (Myobatrachidae); f, *Kaloula pulchra* (Microhylidae). *Stippled pattern,* cartilaginous parts of the hyoid.

cricoid is complete in most anurans, although a midventral gap is characteristic of *Oreophrynella* (Bufonidae), some leptodactylids, and ranids. Paired lateral cricoid gaps occur in *Dendrobates tinctorius* (Dendrobatidae).

PECTORAL GIRDLE

Literature Resumé

Cope (1864, 1865) must be credited with initially distinguishing two different types of anuran pectoral girdles. The Arcifera were characterized by free, overlapping epicoracoids and divergent clavicles and coracoids, whereas the Raniforma had fused epicoracoids and very narrowly separated clavicles and coracoids. Boulenger (1882) subsequently used these criteria as the basis for his subdivision of the phaneroglossid anurans into two major series, the Arcifera and Firmisternia. This classification was virtually unchallenged until 1922, when Noble pointed out an apparent morphological intermediate between the arciferal and firmisternal pectoral girdles. Noble rejected the Arcifera and Firmisternia by reasoning that the firmisternal condition could have arisen independently on more than one occasion from the arciferal pectoral girdle. Later, Noble (1931) proposed another system of classification, based on vertebral column structure and thigh musculature, that superseded that of Boulenger.

Aside from the papers cited above, literature concerning the pectoral girdle is rather scarce. W. K. Parker (1886) wrote a monograph on the structure and development of the shoulder girdle and sternum in vertebrates which contains a significant amount of anuran material. The principal disservice of this work to anuran morphology is Parker's statement that the raniform (=firmisternal) girdle passes through an arciferal stage of development; this has since been shown to be incorrect. The first detailed anatomical treatment of the anuran shoulder girdle and its associated muscles is that of Gaupp (1896) describing *Rana temporaria*. In the early 1900's two papers appeared concerning the anuran pectoral girdle, one by Antony and Vallois (1914) on the significance of the ventral elements of the pectoral girdle, and a second by Braus (1919) on the breast-shoulder apparatus.

There was a significant increment of interest in the pectoral girdle during the 1920's. Procter (1921) wrote a paper on the taxonomic significance of the scapular variation, which unfortunately has been largely overlooked by subsequent authors. De Villiers began to investigate developmental and microscopic aspects of the girdle. His first paper (1922) deals with *Bombina* (Discoglossidae); subsequent papers (1924, 1929*b*) are concerned with the aglossal pipids. Fuchs (1926*a*, 1926*b*) and Roggenbau (1926*a*, 1926*b*) investigated the morphology and development of the cartilaginous elements of the pectoral girdle in *Rana fusca*. Probably the most significant papers of the period were those by Noble (1922, 1926), who pointed out the occurrence of intermediate, so-called arcifero-firmisternal pectoral girdles. Once the reliability of the pectoral girdle as a basis of major classification fell into disrepute, relatively little attention was devoted to it until the 1950's, when van Pletzen (1953) produced a paper on the morphogenesis and ontogenesis of the breast-shoulder apparatus of *Xenopus laevis*, and Griffiths (1959, 1963) redefined arcifery and firmisterny.

Structural Patterns

Arcifery. It is generally agreed that arcifery is the primitive morphological pattern of the pectoral girdle; it is also the most widespread (Table 2–1). As currently defined (Griffiths, 1963), arciferal genera are characterized by possession of posteriorly directed epicoracoid horns. These horns articulate with the sternum by means of grooves, pouches, or fossae in the dorsal surface of the sternum and provide a surface for the insertion of a pair of muscles derived from the abdominal recti. Griffiths (1963) further defined the arciferal girdle by its developmental pattern.

Table 2–1. Distribution of Pectoral Girdle Types Among Anurans

	Arciferal girdle		Firmisternal girdle				
Family	Normal	Modified	Normal	Modified	Omosternum	Sternum	Scapula
Ascaphidae	+	–	–	–	–	+	pu>3
Discoglossidae	+	–	–	–	±	+	pb>3
Rhinophrynidae	+	–	–	–	–	–	pb<2
Pipidae	–	+	–	–	–	+	pu>3
Pelobatidae	+	–	–	–	±	+	pb<2
Myobatrachidae	+	–	–	–	+	+	pb<2
Leptodactylidae	+	+	–	–	±	+	pb<2
Bufonidae	+	+	–	–	–	+	pb<2
Rhinodermatidae	–	+	–	–	+	+	pb<2
Brachycephalidae	–	+	–	–	–	–	pb<2
Dendrobatidae	–	–	+	–	+	+	pb<2
Pseudidae	+	–	–	–	+	+	pb<2
Centrolenidae	+	–	–	–	–	+	pb<2
Hylidae	+	–	–	–	±	+	pb<2
Ranidae	–	–	+	+	+	+	pb<2
Sooglossidae	–	–	–	+	+	+	pb<2
Microhylidae	–	–	+	–	±	+	pb<2

+ Presence of a structure in a family.
– Absence of a structure in a family.
± Presence and absence of a structure in a family.
pu>3 Scapula proximally uncleft, overlaid anteriorly by clavicle and with a clavicle-to-scapula ratio greater than three.
pb>3 Scapula proximally bicapitate with a clavicle-to-scapula ratio greater than three.
pb<2 Scapula proximally bicapitate with a clavicle-to-scapula ratio less than two.

The majority of arciferal genera are also characterized by fusion of the epicoracoids anteriorly in the interclavicle region. Posterior to the clavicles, the epicoracoids are usually free and overlapping. The longitudinal axis of the pectoral arch (i.e., epicoracoid cartilage) tends to be long as compared with the firmisternal girdle, and prezonal and postzonal elements tend to be less well developed than in the firmisternal pattern.

Several arciferal taxa deviate markedly from the usual structural pattern associated with arcifery. Medial fusion of the epicoracoids has simulated the firmisternal condition to varying degrees and resulted in the unfortunate appellations of "arcifero-firmisternal," "secondary," or "pseudo-firmisterny."

Firmisterny. The firmisternal pectoral girdle is derived from an arciferal pattern and distinguished from the latter by its developmental pattern (which does not recapitulate arciferal development), lack of epicoracoidal horns, and fusion of the sternum to the pectoral arch. As a general rule, firmisternous taxa are also characterized by completely fused epicoracoidal cartilages. The longitudinal axis of the pectoral arch is shorter than that of the arciferal girdle, but prezonal and postzonal elements are more elaborately developed.

A few taxa considered to be firmisternal lack epicoracoidal horns but have epicoracoid cartilages that are free and overlapping to varying degrees. These taxa have also been categorized as arcifero-firmisternal or pseudofirmisternal.

Pectoral Osteology

Procoracoid. The anterior part of the epicoracoid cartilage associated with the clavicle is usually referred to as the procoracoid, also known as the precoracoid. The procoracoid forms the anterior part of the epicoracoid cartilages (Figure 2–8). In a maximal state of development, it is associated with the scapula laterally and invested anteriorly by the clavicle. In ventral aspect, the procoracoid can be seen to lie posterior to the clavicle; it is narrowest laterally and usually widens medially to form the anterior end of the pectoral arch. The procoracoids are fused anteromedially in both arciferal and firmisternal genera.

There is a trend towards reduction of the procoracoid throughout anurans. The element is best developed among arciferal genera, particularly the ascaphids, discogolossids, and pipids (Figure 2–9g,i,j); it tends to undergo marked reduction in firmisternal genera, particularly the microhylids (Figure 2–9b). Reduction in the procoracoid occurs in a lateral to medial direction.

In ascaphids, pipids *(Pipa)*, some bufonids *(Atelopus)*, and *Brachycephalus*, there is a tendency towards ossification of the procoracoid (Figure 2–9a). In the ascaphids and pipids, only the lateral extremity of the cartilage is involved and is fused with the scapular ossification. However, in *Brachycephalus*, apparently the entire procoracoid is synosteotically united with the clavicle and scapula. I know of no other example of procoracoid ossification among anurans. The pipids exhibit one other unique procoracoidal specialization; in *Pipa* the procoracoids extend anterior to the clavicle to form the functional, zonal equivalent of the omosternum.

Epicoracoid. "Epicoracoid," as it is used here, refers to the principal cartilaginous body of the pectoral arch, posterior to the level of the clavicles (Figure 2–8). As

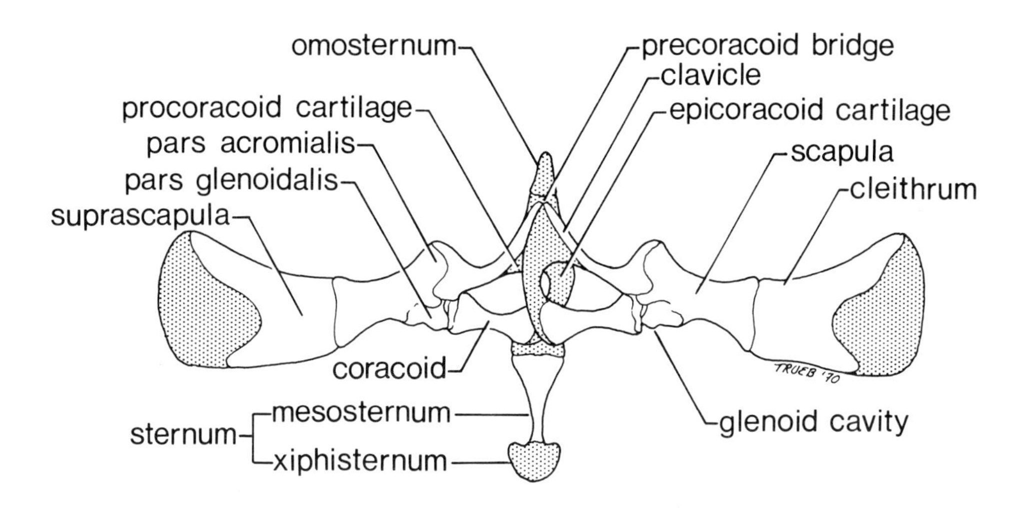

Figure 2–8. Schematic diagram of pectoral girdle of *Leptodactylus bolivianus* in ventral view. *Stippled pattern,* cartilaginous areas.

used by other authors, it may include the procoracoid or be restricted to the cartilage associated with the coracoids medially and posteriorly. Synonyms of epicoracoid are infracoracoid, coracoid cartilage, corpus sterni, entosternal, and sterni ossa media.

The basic variation in the epicoracoids involves the presence of epicoracoidal horns among arciferal groups and absence of such horns in firmisternal groups. Secondarily, there is considerable variation in the relationship of the epicoracoids to one another within each basic group. Most of the arciferal species are characterized by relatively broad epicoracoid cartilages which freely overlap one another medially, posterior to the fusion of the procoracoids at the anteromedial tip of the pectoral arch. In some bufonid genera (e.g., *Atelopus, Rhamphophryne, Oreophrynella,* and *Dendrophryniscus*) and at least one leptodactylid genus (*Sminthillus*), the procoracoid fusion is extended posteriorly to form what is known as a precoracoid bridge. Thus the two epicoracoids are fused or juxtaposed anteriorly, resulting in a so-called partially firmisternous condition. Progressive lengthening of the precoracoid bridge results in the condition characteristic of *Rhinoderma* (Rhinodermatidae; Figure 2–9e) and some *Atelopus* (Bufonidae) wherein the epicoracoids are completely fused posteriorly to the level of the sternum and a functionally firmisternous girdle is created. In *Brachycephalus* (Figure 2–9a) the epicoracoid cartilages are completely ossified; they are, however, not fused but closely juxtaposed and articulating throughout most of their lengths. The pipids seem to represent a special case of independent modification of the pectoral girdle towards a firmisternous-like condition. This family is uniquely characterized by a marked posterolateral elaboration of the epicoracoid cartilages around the posteromedial portions of the coracoid bones and by the absence of an obvious overlap of the epicoracoid cartilages medially (Figure 2–9g). Of the pipids, *Xenopus* most nearly approaches the typical arciferal condition. In this genus there is a gap between the procoracoid cartilages anteriorly. The portions of the epicoracoids lying anterior to the coracoid bones are reduced to slim cartilaginous bars that are free from one another; the posterior, expanded parts of the epicoracoids are not fused. However, they are rather firmly bound together by fibrous connective tissue. The procoracoids and epicoracoids anterior to the coracoid bones are fused in *Pipa*. The posteromedial portions of the epicoracoids, which lie ventral to the sternum, are separated from one another but united by fibrous connective tissue so that movement is all but prohibited; *Hymenochirus* represents a true, morphological firmisternous condition in which the epicoracoid cartilages are completely fused with one another and the sternum. According to de Villiers (1929*b*), there is persistent evidence of a medial line of epicoracoid fusion in the adult. This leaves little doubt that the pectoral girdle is derived from an arciferal type, despite the absence of free epicoracoidal horns. It would therefore be appropriate to refer to the pectoral-girdle condition typified by *Hymenochirus* as "pseudo-firmisternous" or as an example of "secondary firmisterny."

Epicoracoidal horns are absent in one monotypic, arciferal family, the Rhinophrynidae. The broad, freely overlapping epicoracoidal cartilages are classically arciferous, and the absence of horns is probably associated with the absence of a sternum in this family. The epicoracoidal horns are greatly reduced in a second, monotypic family, the Brachycephalidae, which also lacks a sternum (Figure 2–9a). Among the remaining arciferous families the epicoracoidal horns seem to be one of two general types. They are apparently broad, rounded, and shallow in such families as the ascaphids, pipids, discoglossids and centrolenids, which are characterized by relatively short, broad sterna, and long and acuminate in those families having more extensively developed sterna.

Among firmisternal families the epicoracoid cartilages are usually reduced in size. They are narrow, tend to be short, and are not elaborated posterolaterally around the coracoid bones (Figure 2–9d). Typically, the epicoracoids are medially fused to one another and posteriorly fused to the sternum. There is little variation in the relationship of the epicoracoids in firmisternous taxa, compared to the variation in arciferal taxa. In three species of *Rana—rugulosa, tigrina,* and *occipitalis*—an arciferal-like condition prevails (Figure 2–9f). The epicoracoid cartilages of these species are partly free and overlapping; however, the epicoracoids are fused to one another and the sternum posteromedially.

Omosternum. The omosternum is a midventral prezonal element associated with, and lying anterior to, the procoracoids and clavicles (Figure 2–8). There is considerable confusion concerning the proper name of this element and many synonyms are in use. "Omosternum" is probably the most widespread in recent literature; it was introduced by W. K. Parker (1886) to replace "episternum." The latter is frequently used interchangeably with "omosternum" in reference to the entire prezonal element, or it may refer only to the distal, expanded portion sometimes characterizing the element; in such cases, the so-called style between the episternum and procoracoids is termed the "omosternum." In this paper the prezonal element in its entirety will be termed omosternum. Other synonyms are interclavicle (in part), manubrium (in part), presternum (in part), sternum superior, prezonal sternum, and prezonal element.

The status of the omosternum in terms of its primitive or derived nature is a moot question. The element is absent among three primitive families—the ascaphids (Figure 2–9i), pipids (Figure 2–9g), and rhinophrynids. The omosternum is present but poorly developed in two (*Bombina* and *Discoglossus*) of the four discoglossid genera, and it is variably present in pelobatids (Figure 2–9h), leptodactylids, and hylids. It is present in myobatrachids, pseudids, and rhinodermatids (Figure 2–9e), and absent in centrolenids, brachycephalids (Figure 2–9a), and bufonids (Figure 2–9c) among arciferal families. The firmisternal families, Ranidae (Figure 2–9d,f), Dendrobatidae, and Sooglossidae are uniformly characterized by well-developed omosterna that are usually elaborated into a style (frequently ossified or calcified) and expanded distal portion. The microhylids are characterized by the presence of elaborate omosterna among more primitive and generalized members; consequent to the trend to reduce the anterior portion of the pectoral girdle in this family, the omosternum tends to be reduced (Figure 2–9b) or absent in many specialized members.

The distribution of omosterna among modern anurans seems to suggest that the element was absent among primitive arciferous groups in which the pectoral arch was long enough to accommodate the attachment of breast musculature. As the pectoral arch was shortened, the omosternum developed as a site of muscle attachment; thus the largest, most elaborate omosterna are found among firmisternous taxa with very short pectoral arches. In this regard, it is important to remember that the omosternum is myocommatous in origin and, hence, is an extremely labile character. Thus one can anticipate that within any large, diverse family, such as the Leptodactylidae, the omosternum is apt to be present or absent and vary considerably in its structure depending upon the configuration of the pectoral girdle and the need for prezonal muscle attachments.

Sternum. The sternum (Figure 2–8) is usually a flat, cartilaginous element associated with the posteromedial part of the pectoral girdle. It is also known as the xiphisternum, hyposternum, and sternum inferius. Occasionally the sternum is produced into a slender style and a posterior, expanded portion (Figure 2–9d,f); in this case the style is referred to as the mesosternum and the posterior part as

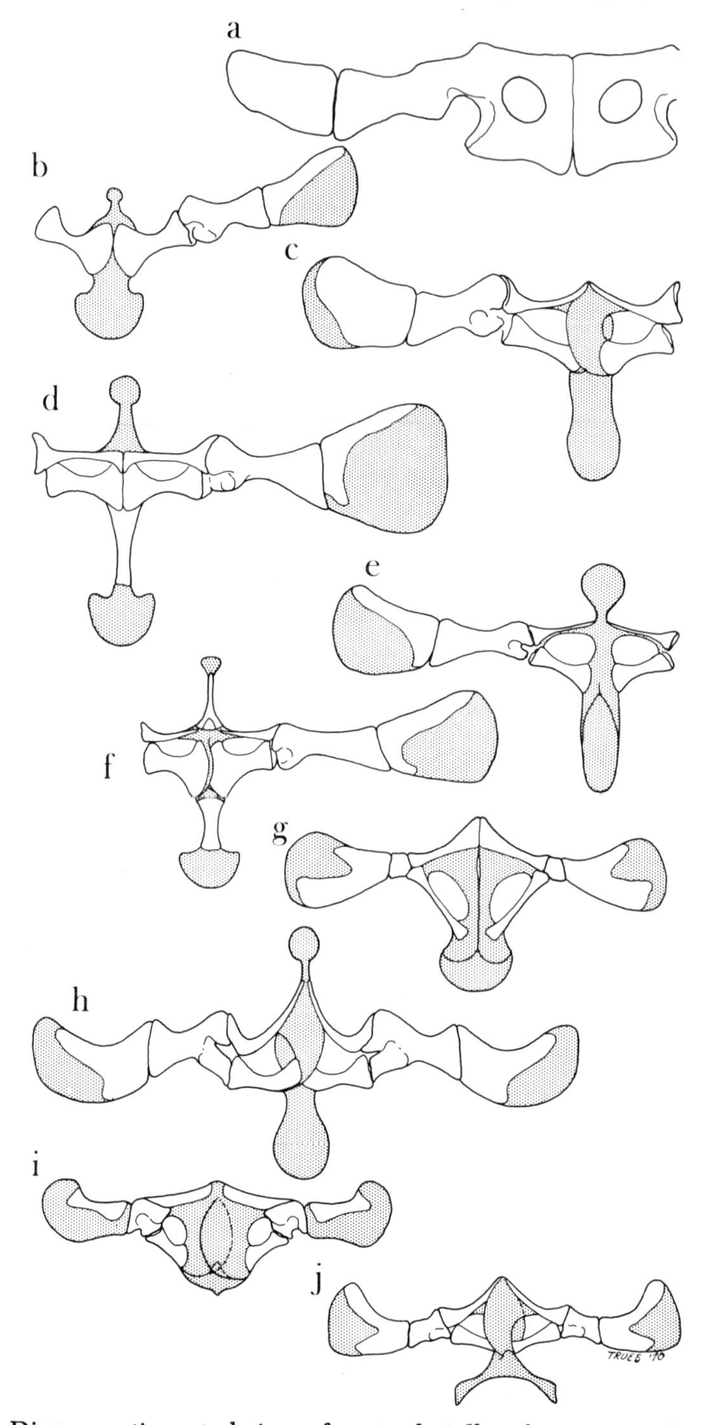

Figure 2–9. Diagrammatic ventral views of pectoral girdles of anurans: a, *Brachycephalus ephippium* (Brachycephalidae); b, *Kaloula pulchra* (Microhylidae); c, *Bufo coccifer* (Bufonidae); d, *Rana pipiens* (Ranidae); e, *Rhinoderma darwini* (Rhinodermatidae); f, *Rana rugulosa* (Ranidae); g, *Xenopus laevis* (Pipidae); h, *Scaphiopus hammondi* (Pelobatidae); i, *Ascaphus truei* (Ascaphidae); j, *Alytes obstetricans* (Discoglossidae). *Stippled pattern* cartilaginous portions of the pectoral girdle.

either the metasternum or xiphisternum. According to Griffiths (1963), sterna are of two types, depending upon their origin in arciferal or firmisternous anurans. In the latter group the sternum forms as part of the girdle and is fused to it (i.e., the epicoracoid cartilages) in the adult, whereas among arciferous frogs the sternum is formed from a discrete anlage that buds off from the coracoid primordium, migrates into the linea alba, and differentiates concurrently with the epicoracoid horns. The sterna of arciferous frogs always articulate with the epicoracoid horns dorsally and are never fused with the epicoracoid cartilages, except in *Hymenochirus* (Pipidae). Sterna are present in all but two monotypic anuran families, the Rhinophrynidae and Brachycephalidae; the sternum is also absent in one microhylid genus, *Aphantophryne*.

Variation in the sternum parallels that of the omosternum to some extent. The sternum provides area for muscle attachment; thus like the omosternum, it tends to be larger and more elaborate in those groups having shorter pectoral arches and tending towards firmisterny. The sternum is broad and shallow in the ascaphids (Figure 2–9i), pipids (Figure 2–9g), and centrolenids which lack omosterna. The discoglossid sternum is unique; it is produced into posterolaterally divergent horns (Figure 2–9j). In *Bombina* (Discoglossidae) the sternum has an anterior pair of so-called presternal pieces, or Sternalblattchen, which lie between the coracoids on the ventral aspect of the sternum. Among all other groups, except the microhylids, the sternum tends to be broad or roughly rectangular in those taxa having relatively long pectoral arches. Generally, the more reduced the pectoral arch, the more marked the tendency towards elaboration of the sternum into a proximal style and distal expanded portion that are characteristic of many ranids. Calcification or ossification may occur in the sternum; frequently, sternal styles are ossified, while the distal portions remain cartilaginous. Among some leptodactylids there is a tendency for the style to bifurcate posteriorly and for the distal cartilaginous part of the sternum to be subdivided into two discrete pieces, each associated with a ramus of the style. In microhylids the sternum never bears a bony style; furthermore, it tends to be broad and bear lateral wings in many genera.

Clavicle. The clavicles are paired, dermal investing bones associated with the procoracoids in the anterior part of the pectoral girdle (Figure 2–8). (The clavicle is also known as the furcula, acromial, and os thoracale.) Usually the clavicle invests the anteroventral margin of the procoracoid and extends over the dorsal edge of the procoracoid; laterally the clavicles articulate with the scapula. There is, however, considerable variation in the structure of this element. It is uniformly present in all groups, except the microhylids in which there is a marked tendency towards reduction and loss of this element (Figure 2–9b). Reduction usually takes place in a medial to lateral direction.

Primitively, in pipids, rhinophrynids, and ascaphids, the clavicle has a rather broad articulation with the scapula which it overlies to some extent anteriorly (Figure 2–9i,j); typically the clavicle does not overlap the scapula anteriorly. In the majority of arciferal frogs, the clavicles tend to be gently to strongly arched and separated medially (e.g., Figure 2–9h), whereas among firmisternal or "partially firmisternal" species they tend to be straight, oriented perpendicular to the longitudinal axis of the body, and in close proximity medially (e.g., Figure 2–9d). Presumably clavicle separation and curvature are primitive characteristics. Specializations noted in clavicle structure are (1) lack of articulation with the scapula due to intrusion of ossified procoracoid (Ascaphidae, *Leiopelma hochstetteri*) or reduction and modification of procoracoid (several microhylids); (2) fusion of clavicles mid-dorsally (Pipidae, *Pipa*); (3) fusion of clavicle and scapula (Pipidae, *Hymenochirus;* Bufonidae, *Atelopus, Dendrophryniscus, Melanophryniscus,* and *Oreophrynella*); (4) synostosis of clavicle, procoracoid, epicoracoid, and coracoid (Brachycephalidae, *Brachycephalus* [Figure 2-9a]); and (5) splinter-like bifurcation of lateral end of clavicle (Centrolenidae, *Centrolenella*).

Scapula. This anterolateral component of the pectoral girdle is endochondral in origin (Figure 2–8). Terms synonymous with scapula are scapula minor, collum scapulae, omoplate, and lower shoulder blade. The scapula usually articulates with the clavicle and coracoid medially, forms the anterior margin of the glenoid fossa or cavity, and articulates with the cleithrum and suprascapula.

In primitive anurans, the scapula is small—one-third or less of the length of the clavicle. It is overlain anteriorly by the clavicle and in pipids and ascaphids is proximally uncleft (Figure 9–2g,i). A slightly advanced situation is represented in *Rhinophrynus* in which the clavicle-to-scapula ratio is still greater than three, but the scapula is proximally cleft or bicapitate. The anterior head is referred to as the pars acromialis, and the posterior head as the pars glenoidalis. In all other anurans, except those in which synostosis between the scapula and other parts of the girdle has taken place, the scapula is bicapitate and considerably larger in proportion to the clavicle; the clavicle-to-scapula ratio is less than two (Table 2–1).

The scapula is always present and ossified, although its boundaries may be obscured by synostosis with peripheral bones or cartilages. Such modifications should be viewed as specializations. Fusion may involve the procoracoid (ascaphids and pipids), the clavicle (pipids), or the clavicle, procoracoid and coracoid (some bufonids and *Brachycephalus*).

Coracoid. The coracoid (Figure 2–8) is endochondrally derived and probably the least variable element in the pectoral girdle. It is known also as the clavicula, clavicula vera or posterior, and pars sternalis scapulae. The coracoid is more or less hourglass-shaped; it forms the medial border of the glenoid fossa laterally and adjoins the epicoracoid cartilages medially. Variation in the coracoid is principally a matter of the angle of the longitudinal axis of the bone, the degree of expansion of the ends of the bones, and the proximity of the medial ends of the bones to one another. Among arciferous frogs, the coracoids tend to be widely separated from one another, whereas they lie much closer together in firmisternal frogs or anurans having an arciferal girdle modified towards firmisterny. In general the coracoids tend to be obliquely oriented with respect to the longitudinal axis of the body among arciferal frogs. Firmisternal frogs, except some microhylids, are usually characterized by coracoids that are perpendicularly oriented or only at a slight angle to the midline. The shape of the coracoid seems to vary independently of the basic girdle structure. Among the pipids, for example, it is remarkably long and slender in *Xenopus,* greatly expanded medially but laterally in *Pipa,* and rather broadly expanded both medially and laterally in *Hymenochirus.* With the exception of the pipids, microhylids, and some specialized bufonids, the medial end of the coracoid tends to be as large as, or only slightly larger than, the lateral end. In *Atelopus* and *Oreophrynella* of the Bufonidae and in the microhylids, the medial part of the coracoid is much larger than the lateral end. The coracoid is fused to the scapula in *Atelopus, Dendrophryniscus,* and *Melanophryniscus* of the bufonids, and synosteotically united with the epicoracoids, procoracoids, and scapula in *Brachycephalus.*

Suprascapula. The suprascapula (Figure 2–8), also referred to as the scapula major, omolita, adscapulum, or episcapulum, basically seems to be a dorsolateral extension of the scapula. It is synchondrotically united with the scapula in the ascaphids but in all other anurans appears to be joined to the scapula by fibrous connective tissue. Primitively, in the ascaphids and pipids (frogs having small scapulae) the suprascapula is large; it is restricted in size among all other frogs. Ossification or calcification of the suprascapula occurs frequently.

Cleithrum. This dermal investing bone lies principally on the ventral (lateral) surface of the suprascapula (Figure 2–8). It may extend around the anterior margin of the suprascapula to invest a small part of the dorsal (medial) surface. In most anurans the bone is distally bifurcate, forming a ramus along the anterior edge of the suprascapula and a posterior ramus lying on the ventral (lateral) midbody of the suprascapula. There is considerable variation in the extent of ossification of the cleithrum. It is poorly developed and lacks a posterior ramus in the ascaphids. The cleithrum is reduced in centrolenids and moderately developed in most anurans; it tends to be extensively ossified in bufonids (Figure 2–9c) and discoglossids. The cleithrum and suprascapula are indistinguishably fused in *Rhinophrynus* and *Brachycephalus* (Figure 2–9a).

VERTEBRAL COLUMN

Literature Resumé

Beginning with Cope (1865) considerable emphasis was placed on the nature of the vertebral column as a major criterion in the classification of anurans. Since that time considerable discussion has been devoted to the relative reliability and merit of vertebral characters. Adequate literature reviews may be found in Noble (1922) and Griffiths (1963).

There are, perhaps, three principal contributions concerning the morphology of the anuran vertebral column and its application to anuran classification. The first of these is a paper by Nicholls (1916) in which four major types of vertebral centrum patterns—opisthocoelous, anomocoelous, procoelous, and diplasiocoelous —were used to define four phaneroglossid "tribes." Noble (1922), on the basis of Nicholls' work and his own investigations, defined four anuran suborders, combining phaneroglossids and aglossids: Opisthocoela, Anomocoela, Procoela, and Diplasiocoela. In 1931 the same author defined a fifth suborder, Amphicoela, to accommodate the Ascaphidae. Mookerjee (1931) reported on divergent developmental patterns that produce the different types of vertebral centra. Subsequently, it was shown that in at least two families—the ranids and microhylids—the nature of the vertebral centra was subject to variation, thus casting doubt on the stability and reliability of this character in higher classification. Several papers (e.g., Tihen, 1959; Madej, 1965) have reported high incidences of variation in vertebral column structure, including the numbers of presacral vertebrae. In an attempt to ameliorate the conflicting evidence, Griffiths (1963) defined three vertebral types—ectochordal, stegochordal, and holochordal—according to the formation of the centra. By this system the Rhinophrynidae and Ascaphidae are ectochordal, having, in the adult, centra ossified as cylinders enclosing a notochord; the pipids, pelobatids, and discoglossoids are stegochordal, having transversely depressed centra. All other anurans are holochordal, with cylindrical, solidly ossified centra. The system proposed by Griffiths adds yet another criterion to distinguish among primitive families and between the latter and modern families. It does not, however, aid in the classification of the remaining 12 modern frog families, nor does the author suggest the evolutionary significance or phylogenetic implications of the different centra. In the same paper, Griffiths extended the conceptual basis of Mookerjee's work on the development of the centra, drawing attention to the divergent modifications of the intervertebral cartilage in the formation of procoelous, amphicoelous, opisthocoelous, anomocoelous, and diplasiocoelous vertebral columns. Largely as a result of Mookerjee's and Griffiths' contributions, emphasis has come to be placed on the nature of the intervertebral cartilage associated with various centra types (e.g., Lynch, 1969). Unfortunately, these investigations have done little to elucidate relationships among modern anurans and strongly suggest that the vertebral column is subject to the same evolutionary lability that characterizes the remainder of the anuran skeleton.

Presacral Vertebrae

As the name implies, the term "presacral vertebrae" refers to the vertebral elements lying anterior to the sacrum. The first presacral, articulating with the occipital condyles anteriorly, is distinguished as the cervical. Vertebrae lying between the cervical and sacral vertebrae have been designated as trunk vertebrae by some authors (e.g., McDiarmid, 1969); however, I prefer to number the presacrals in an anterior to posterior sequence; the cervical is the first presacral vertebra.

Number of presacral vertebrae. The number of presacral vertebrae varies between 9 and 5 (Table 2–2). It is agreed that higher presacral counts are a primitive character, whereas fewer presacral vertebrae are associated with specialization and advancement. Reduction in the number of presacrals has been effected in two ways. Primitive anurans have presacral vertebrae incorporated into the sacral structure; this is evidenced by the presence of spinal nerve foramina in the sacral region. In specialized members of all families, except the leptodactylids, centrolenids, pseudids, hylids, and apparently the ranids, sooglossids, and microhylids, there is a tendency to reduce the number of presacral vertebrae by fusion of anterior vertebrae. This is generally manifest initially by fusion of the neural arches dorsally, while the centra remain discrete. In more specialized groups the vertebrae fuse completely. Vertebral fusion usually occurs in an anterior to posterior sequence; thus when fusion occurs it first involves the cervical and second presacral vertebrae. In the majority of frogs, fusion is limited to the first two presacrals, but the first four vertebrae are known to be fused in some *Dendrobates* (Dendrobatidae).

Cervical Vertebra. By virture of its anterior position, the cervical vertebra is structurally distinct from the other presacrals. Its anterior surface bears some kind of paired articulating surfaces known as cervical cotyles, which receive the occipital condyles of the cranium. Lynch (1969) defined three basic cotylar arrangements. Cervical cotylar Type I is characterized by cup-like cotyles displaced laterally and thus widely separated from one another (Figure 2–10l). This arrangement occurs in most advanced families (see Table 2–2) and presumably is an advanced feature. The articulating surfaces of Type II cotyles are distinctly but narrowly separated from one another (Figure 2–10k); separation may be emphasized by the presence of a deep notch between the cotyles. Type II cotyles characterize many archaic families and primitive members of more modern groups; this arrangement occurs in the Discoglossidae, Rhinophrynidae, Pelobatidae, Myobatrachidae, Leptodactylidae, Bufonidae, and Ranidae. The cotyles are confluent and represent a single articular surface in Type III (Figure 2–10j). This is only known to occur in leptodactylids (subfamily Ceratophryinae) and ascaphids. Based on the distribution of these three types of cervical cotyles, Lynch (1969) hypothesizes that Type II is primitive and that Types I and III are specialized derivations from it.

The cervical vertebra bears a pair of postzygapophyses posteriorly that artic-

Table 2–2. Distribution of Vertebral Column Characters Among Anurans

Family	Vertebral type	Number of presacral vertebrae	Inter-vertebral fusion evident	Neural arch imbricate	Neural arch non-imbricate	Cervical cotylar type	Ribs	Sacral diapophyses	Sacral-coccygeal articulation	Coccygeal transverse processes	Centrum type
Ascaphidae	Am	9	–	–	+	III	+	Expanded	Contiguous cartilage	+	e
Discoglossidae	O	8-9	–	+	–	II	+	Expanded	Bicondylar	+	s
Rhinophrynidae	O	8	+	+	–	II	–	Expanded	Bicondylar	–	e
Pipidae	O	5-8	±	+	–	I	+[a]	Expanded	Fused	±	s
Pelobatidae	A	8-9	±	+	–	II	–	Expanded	Fused or monocondylar	+	s
Myobatrachidae	A	7-8	+	+	+	I, II	–	Dilated	Bicondylar	–[b]	h
Leptodactylidae	P	8	–	+	+	I, II, III	–	Dilated or round	Bicondylar	±	h
Bufonidae	P	5-8	±	+	+	I, II	–	Expanded	Fused, monocondylar, or bicondylar	±	h
Rhinodermatidae	P	8	+	–	+	II	–	Dilated	Bicondylar	–	h
Brachycephalidae	P	7	+	+	–	I	–	Dilated	Bicondylar	–	h
Dendrobatidae	P	8	±	+	–	I	–	Dilated or round	Bicondylar	–	h
Pseudidae	P	8	–	–	+	I	–	Round	Bicondylar	–	h
Centrolenidae	P	8	–	–	+	I	–	Dilated	Bicondylar	–	h
Hylidae	P	8	–	+	+	I	–	Expanded or dilated	Bicondylar	–	h
Ranidae	P, D	8	–	+	+	I, II	–	Round	Bicondylar	–	h
Sooglossidae	P	8	–	?	?	?	–	Dilated	Contiguous cartilage	?	h
Microhylidae	P, D	8	–	+	–	I	–	Dilated	Fused or bicondylar	–	h

\+ Presence of a structure in a family.
– Absence of a structure in a family.
± Presence and absence of a structure in a family.
A Amphicoelous or procoelous with free intervertebral discs sometime during life cycle.
Am Amphicoelous with contiguous intervertebral cartilage.
D Diplasiocoelous.
O Opisthocoelous.
P Procoelous.
e Ectochordal.
h Holochordal.
s Stegochordal.
[a]Ribs present but fused to transverse processes in adult.
[b]Zygapophyseal processes occur on coccyx of *Metacrinia*.

ulate with the prezygapophyses of the second presacral. The cervical does not bear transverse processes unless fusion of the first and second presacrals has occurred.

Presacral Centra. The ventral portion of the vertebra upon which the neural arch rests is termed the centrum. Early in development and primitively, the centrum encloses the notochord. Variation in centra is expressed in two ways. According to Griffiths (1963), different developmental patterns result in one of three adult conditions. Thus the ectochordal centrum is cylindrical, hollow, and encloses a persistent remnant of the notochord; this condition is considered primitive. All traces of the notochord disappear in the stegochordal and holochordal types of centra. Both are solidly ossified, but the stegochordal centrum is dorsoventrally depressed, whereas the holochordal centrum is round in transverse section. Griffiths maintains that there are two different developmental patterns which give rise to the stegochordal centra; the centra are indistinguishable in the adults.

Alternatively, centra have been classified by the nature of their articulating surfaces. Centra having flat or very slightly concave ends are termed amphicoelous. Those that have a cup-shaped depression anteriorly and a convex surface posteriorly are procoelous. The reverse situation in which the centrum is convex anteriorly and concave posteriorly is termed opisthocoelous.

Intervertebral cartilages. A block of cartilage, which usually ossifies, lies between successive centra in the vertebral columns of all anurans. The nature of this intervertebral cartilage, its development and relationship to the centra, has been a classical criterion in distinguishing anuran suborders. Thus, if the cartilage ossifies and fuses with the anterior end of a centrum, a so-called opisthocoelous vertebra is formed; if fusion occurs with the posterior end of the centrum, a procoelous vertebra results. As originally defined, amphicoely referred to a condition in which the ossified portion of the centrum is very slightly biconcave or flat, and the intervertebral cartilage fails to ossify and is equally associated with adjacent vertebrae. In the anomocoelous vertebral column, the centra are morphologically amphicoelous; however, the intervertebral cartilage ossifies as a discrete ball between adjacent centra. The diplasiocoelous category accommodates anurans in which all of the vertebrae are procoelous, except the eighth presacral, which is biconcave and, by definition, amphicoelous.

Investigations by Mookerjee (1931) and Griffiths (1963) on the development of the vertebral column have suggested that the nature of the mature centrum is determined by the fate of the intervertebral cartilage. Early in development the intervertebral cartilage is connected both anteriorly and posteriorly to successive vertebrae. Subsequently, arcs of connective tissue may invade the cartilage. If the arc bisects the anterior portion of the intervertebral cartilage, a depression results in the end of the anteriorly adjacent centrum that accommodates the block of cartilage associated with the posterior centrum. In contrast to this opisthocoelous formation, procoely results when the connective tissue arc invades the posterior portion of the intervertebral cartilage; thus, the block of cartilage is associated with the posterior end of a centrum and articulates with the posteriorly adjacent centrum by means of a depression formed in its anterior surface. So-called anomocoely prevails when each block of intervertebral cartilage is invaded by two connective tissue arcs, one anteriorly and one posteriorly. The cartilaginous block is thus freed from both adjacent centra and articulates with each by means of depressions formed in their terminal surfaces. In the amphicoelous condition, connective tissue fails to invade the intervertebral cartilage.

On the basis of this information, it seems logical to assume that the primitive anuran vertebra is amphicoelous in character, that is, terminally flat or only slightly biconcave with uninterrupted intervertebral cartilages. Further, deviations from this pattern would seem to be dependent on the nature and number of the arcs of connective tissue which invade the cartilage. Unfortunately, the evolutionary relationships of these various modifications and their significance are unclear at present. Only the pelobatids and myobatrachids have free intervertebral bodies. It is possible that this condition developed from an ancestral "amphicoelous" type similar to the ascaphids, a type which had no free intervertebral bodies, and that after the loss of one or the other invading connective tissue arcs, opisthocoely, procoely, and diplasiocoely resulted. Opisthocoely is limited to three families—discoglossids, pipids, and rhinophrynids; on the basis of other characters these families seem to have arisen early and are markedly divergent from other anurans. Procoely, in contrast, is widespread among modern anurans, exclusive of the aforementioned families and of the many ranids and microhylids that are diplasiocoelous.

The widespread use of such terms "amphicoelous," and "anomocoelous" and their application as subordinal names are somewhat unfortunate and misleading in view of our present concept of vertebral types. For example, "amphicoelous" can logically be used to describe the centra structure of both the ascaphids (Amphicoela) and the pelobatids (Anomocoela), in the sense that opisthocoelous and procoelous define the groups of families to which they are applied. "Anomocoelous" was designed to describe the variation in vertebral patterns of the members originally assigned to the suborder; as conceived by Noble (1931) the Anomocoela included the pelobatids and pelodytids, both of which have "amphicoelous" vertebrae with free intervertebral bodies, and the sooglossids, which have uniformly procoelous vertebrae. In contrast, "diplasiocoela" denotes the formation of two types of vertebrae in reference to the presence of a single amphicoelous vertebra in an otherwise uniformly procoelous column. These terms cannot be strictly related to one another, nor are they acceptable in the original contexts of their definitions. In light of the confusion surrounding these names and their general inapplicability to anuran suborders, as those suborders are currently recognized, I suggest that their use should be restricted to the designation of the several types of vertebral columns. Moreover, it seems advisable that subordinal or superfamilial names be derived in the accepted manner from the family group names, for example, Ascaphoidea, Bufonoidea and Ranoidea.

Amended definitions of Nicholls' (1916) and Noble's (1931) terms are suggested as follows:

Amphicoelous: A vertebral column in which the vertebral centra are ectochordal, slightly biconcave or flat terminally, and the intervertebral cartilage is contiguous to, and not subdivided between, successive presacral vertebrae.

Anomocoelous: A vertebral column in which the vertebral centra are stegochordal and slightly biconcave or flat terminally. Each intervertebral cartilage is subdivided anteriorly and posteriorly, producing a free intervertebral element between adjacent centra. This element subsequently ossifies and remains free.

Opisthocoelous: A vertebral column in which the vertebral centra are ectochordal or stegochordal and each intervertebral cartilage is anteriorly subdivided, subsequently ossified, and fused to the posteriorly adjacent centrum; the mature vertebral centra is therefore convex anteriorly and concave posteriorly.

Procoelous: A vertebral column in which the vertebral centra are holochordal. Each intervertebral cartilage is posteriorly subdivided, subsequently ossified, and fused to the anteriorly adjacent centrum; the mature vertebral centra is therefore concave anteriorly and convex posteriorly.

Diplasiocoelous: A vertebral column in which the vertebral centra are holochordal. Each intervertebral cartilage is posteriorly subdivided in the procoelous manner, with the exception of the cartilage located between the eighth presacral and the sacrum; this cartilage is subdivided in the opisthocoelous manner. Thus, the first seven presacrals are concave anteriorly and convex posteriorly, whereas the eighth is biconcave.

Neural arches. The neural arches constitute the dorsal superstructure of the vertebral centra; the arches form a canal enclosing the spinal cord. Anteriorly each side of the arch bears a prezygapophysis, the articulating surface of which faces dorsomedially. Posteriorly the arch bears a pair of postzygapophyses; the articulating surfaces are oriented ventrolaterally to articulate with the prezygapophyses of the posteriorly adjacent vertebra. Dorsomedially, at the point of junction of the two sides of the neural arch, a dorsally projecting neural spine is produced. Laterally the neural arches bear diapophyses or transverse processes.

Neural arches vary in length (Figure 2–10a,b,c). Short neural arches tend to expose the centra and intervertebral structure between successive vertebrae. Such arches do not overlap and are therefore termed "nonimbricate." In contrast, the neural arches, which are more extensively developed and longer, tend to overlap and obscure the intervertebral and centrum structure ventrally. Overlapping neural arches are referred to as "imbricate." Most anurans (Table 2–2) tend to have imbricate arches, although in small, poorly ossified, or primitive members of some families, the arches are nonimbricate. The ascaphids, rhinodermatids, centrolenids, and pseudids are uniformly characterized by nonimbricate arches.

Neural spines. Neural spines are produced at the dorsomedial junction of the two halves of the neural arch. The spines tend to be absent or barely evident among nonimbricate species, and moderate to well developed in imbricate taxa (Figure 2–10d,e,f). Among those frogs having moderately large neural spines, the spines are largest on anterior presacrals and decrease in size posteriorly. The presence and development of neural spines seems to be associated with the size and degree of ossification of the species; terrestrial taxa seem to have better-developed spines than do aquatic or arboreal species. Thus, one is led to assume that there is a trend towards elaboration of neural spines in frogs whose locomotory activities or size demands more extensive muscle development and consequently more surface area for muscle insertion.

Transverse processes. The neural arches of presacral vertebrae two through nine, inclusively, bear lateral expansions known as diapophyses or, more commonly, transverse processes. Variation in the expansion (longitudinal), and respective widths (transverse), of these elements and in their orientation with respect to the longitudinal axis of the body is so great that trends are obscure. In the primitive families characterized by free ribs (ascaphids and discoglossids), the transverse processes tend to be slender and not as wide as the sacral diapophyses (Figure 2–10a); the anterior processes which articulate with ribs are narrower and more broadly expanded than the posterior processes. In the discoglossids, pipids, and rhinophrynids, there is a marked trend for the posterior four transverse processes to be inclined anteriorly at an acute angle; they are wide in the pipids but very much reduced in *Rhinophrynus*. Commencing with the pelobatids (Figure 2–10b), there is a trend towards expansion and widening of the transverse processes of the primitively rib-bearing presacral vertebrae (two, three, and four). The processes of these anterior vertebrae tend to be as wide as, or slightly wider than, the sacral diapophyses, whereas the posterior processes are subequal in length. This pattern is characteristic of many modern anurans; however, there does seem to

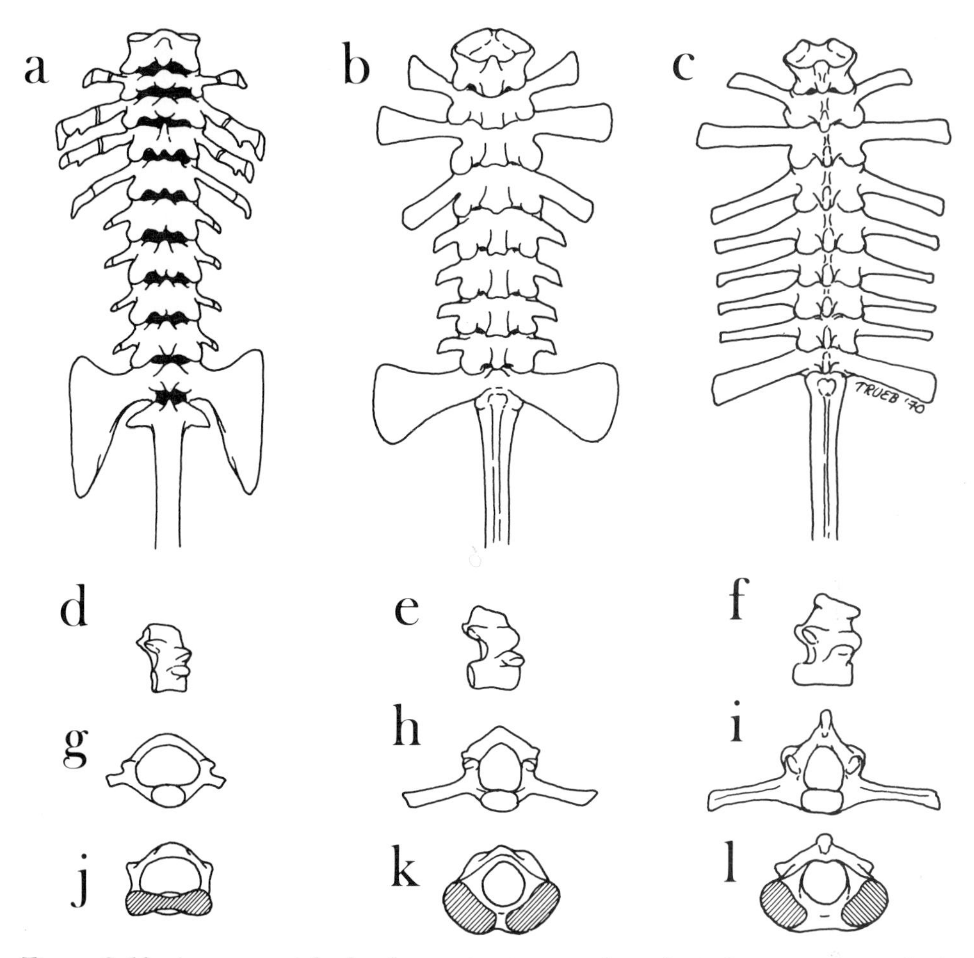

Figure 2–10. Anuran vertebral column structure: a, hypothetical primitive vertebral column having nine, nonimbricate presacral vertebrae, free ribs, expanded sacral diapophyses, and vestigial transverse processes on coccyx; b, hypothetical transitional vertebral column with eight imbricate presacrals, no free ribs, subequal transverse processes, dilated sacral diapophyses, and a monocondylar sacral-coccygeal articulation; c, hypothetical advanced vertebral column having transverse processes of nearly uniform lengths and a bicondylar sacral-coccygeal articulation; d-f, lateral views of second presacral vertebrae of each column; g-i, anterior views of second presacral vertebra of each column; j-l, anterior views of first presacral (cervical) vertebrae of each column. j, cervical cotyle Type III; k, cervical cotyle Type II; l, cervical cotyle Type I.

be a tendency associated with terrestrialism to equalize the widths of the processes (Figure 2–10c). Thus the posterior transverse processes are widened to produce vertebral column patterns characteristic of such taxa as *Atelopus* and *Bufo typhonius* (Bufonidae), *Hemiphractus* (Hylidae), and *Ceratobatrachus* (Ranidae).

Ribs. Ribs are present in only 3 of the 17 anuran families; these are the Ascaphidae, Discoglossidae, and Pipidae. Normally, three pairs of ribs that articulate with the transverse processes of the second, third, and fourth presacral vertebrae, respectively, are present (Figure 2–10a). The ascaphids occasionally have a fourth pair of ribs associated with the fifth presacral. The ribs are free in the ascaphids and discoglossids but indistinguishably fused to the transverse processes in adult pipids. The occurrence of ribs is obviously primitive. The fused condition of the ribs of pipids represents an advancement over the freely articulating ribs of the ascaphids and discoglossids.

Presacral shield. A dorsal, dermal plate of bone overlies the presacral vertebrae in a few taxa. Such shields usually have irregular, sculptured surfaces, frequently are broadly expanded, and may be composed of several separate plates, which are fused in the adult but presumably arise from separate centers of ossification. The shield is narrow, covering only the neural arches of the first four presacral vertebrae in *Lepidobatrachus asper* (Leptodactylidae). In *Ceratophrys aurita* (Leptodactylidae) the shield is broadly expanded, covering the first seven presacrals and most of their associated transverse processes, the shield is attached to the vertebral column by ligaments (Lynch, 1969). Smaller shields are present in some species of *Dendrobates* (Dendrobatidae); the association with the underlying vertebral column is undiagnosed as yet. A well-developed vertebral shield characterizes *Brachycephalus eppiphium* (Brachycephalidae); the shield covers presacral vertebrae two through seven and obscures their transverse processes. According to McDiarmid (1969), the shield is fused to some of the underlying neural spines and the distal ends of transverse processes of the third and fourth presacrals. The functional significance of these shields is not immediately evident. The only trait held in common by the species possessing the shields is terrestrialism; their sizes and the extent of ossification of the rest of the skeleton varies widely. For the present it is probably most reasonable to assume that presacral shields have been independently derived on at least three occasions and are protective modifications.

Sacral Vertebra

The sacrum is a specialized vertebra lying between the posteriormost presacral and the coccyx. It is somewhat modified from the generalized pattern of the presacral vertebrae. Like the latter, the sacrum bears a pair of prezygapophyses, which articulate with the postzygapophyses of the last presacral. Postzygapophyses are absent on the sacra of all anurans, except *Metacrinia* (Myobatrachidae). The sacral diapophyses represent modified transverse processes. They tend to be broadly expanded in some primitive families (Figure 2–10a; Table 2–2) but only moderately dilated in most (Figure 2–10b). In the majority of anurans the sacral diapophyses are only moderately expanded, a condition here designated as "dilated." Among advanced families or advanced members of some families the diapophyses are round, that is, nearly cylindrical (Figure 2–10c). The diapophyses articulate distally with the anterior ends of the ilia. Presumably the dilated, and especially the round, sacral diapophyses permit more axial flexibility than do the broadly expanded diapophyses. The more expanded diapophyses tend to be oriented at right angles to the longitudinal axis of the vertebral column, whereas those that are narrowly dilated or round tend to be inclined posteriorly.

The sacral-coccygeal articulation is subject to considerable variation. Primitively, in ascaphids, the sacrum is joined to the coccyx by cartilage in the same way that presacral vertebrae are associated with one another. Apparently this same situation prevails in the Sooglossidae; the occurrence of this primitive character in an otherwise relatively advanced family should probably be regarded as an example of paedomorphosis or neoteny. The remainder of the anurans have one of three types of sacral-coccygeal relationships. A bicondylar articulation is most widespread among primitive (other than the ascaphids) and advanced families. Secondarily, several groups have modified either the ascaphid or bicondylar plan to produce fusion of the sacrum and coccyx or formation of a monocondylar sacral-coccygeal articulation. Occasionally (in bufonids) one encounters a "weakly bicondylar" articulation, that is, an articulation intermediate between the bicondylar and monocondylar condition. The occurrence of this condition in a specialized group suggests that in these cases, monocondyly has been secondarily derived from bicondyly.

The sacrum has been modified in some genera (e.g., Bufonidae, *Oreophrynella*) by the incorporation of one or more presacral vertebrae into its structure. The presence of more than one pair of spinal nerve foramina is evidence of such incorporation and is one way in which anurans have shortened the vertebral column.

Coccyx

The coccyx or urostyle was primitively formed by the fusion of postsacral vertebral elements. There is some variation in the proportional length of the coccyx, although there is not sufficient information regarding this characteristic to extrapolate any evolutionary tendencies. There is a tendency among many anurans, especially the bufonids and ranids and the larger species of other families, for the coccyx to bear a well-developed longitudinal ridge (Figure 2–10c,d); the ridge is largest anteriorly and gradually decreases in size posteriorly. In many primitive families and primitive members of such families as the Leptodactylidae and Bufonidae (Table 2–2), the coccyx bears a vestigial pair of transverse processes anteriorly (Figure 2–10a). Such processes seem to be absent among more advanced families. The genus *Metacrinia* (Myobatrachidae) is uniquely characterized by the presence of a pair of coccygeal prezygapophyses of the sacrum anteriorly. There is minor variation in the shape of the coccyx. Usually the bone is slender, largest anteriorly at the point of its articulation with the sacrum and gradually decreasing in size posteriorly. In some taxa, notably the bufonids, the coccyx may expand laterally to a minor degree. The significance of such modifications is unknown.

PELVIS

The pelvis of primitive amphibians consisted of three paired elements: the ilium, ischium, and pubis. Among modern anurans, the girdle is primarily composed of the ilium and the ischium. The pubis is reduced to an inconspicuous, ventral cartilaginous structure which sometimes calcifies. Osteologically, there is little information available on the anuran pelvic girdle. Green (1931) studied the development of the pelvic girdle in *Rana temporaria, Xenopus laevis,* and *Bufo melanostictus,* and in the same paper, he offered a theory of pelvic mechanics. More recently, Whiting (1961) discussed the function of the pelvic girdle in amphibian locomotion, and Lynch (1962) wrote a note concerning a taxonomic character of the ilia of certain hylids. None of these papers is especially enlightening in terms of description or comparative anatomy.

Ilium. The ilia are paired, endochondral structures which articulate with the ventral surfaces of the sacral diapophyses anteriorly. Posteromedially, the ilia articulate with each other and with the ischia posteriorly (Figure 2–11). Variation in the ilium involves length of the shaft, a variety of minor modifications of shape posteriorly, and the presence or absence of protuberances. Longer shaft length is associated with saltatorial habits, whereas shorter shafts are characteristic of terrestrial or fossorial species that tend to walk rather than jump. Primitive anurans have a plain shaft that tends to be cylindrical in cross section (Figure

2–11b). Among advanced frogs the ilial shaft may bear crests or ridges (Figure 2–11c). This tendency seems to be best developed in some ranids and hemiphractine hylids. The pipids are uniquely characterized by a lateral crest (Figure 2–11d). The presence of a well-developed dorsal acetabular zone and dorsal prominences and protuberances is a specialization (Figure 2–11a). Similarily, it seems that an expanded preacetabular zone is advanced (Figure 2–11f). There is considerable variation in the angle formed by the ilia when viewed dorsally or ventrally. Ultimately this variation must relate to the relative lengths of the ilia and the sacral diapophyses; however, the functional and evolutionary significance is not clear at this time.

Ischium. The ischia are paired, posterior elements that are endochondral. There is variation in the shape of the ischium (i.e., the extent of the dorsal and posterior expansion), but so little information is available that it is useless to speculate on the significance of the differences. Moreover, investigation of the bone structure without associated study of pelvic myology would be naïve.

Pubis. The pubis is usually present as a ventral cartilaginous element between the ischium and ilium. In the majority of anurans the pubis remains unossified; however, in some it may be calcified.

Prepubic bones. Three anuran genera are known to possess a prepubic skeletal element known also as the epipubis; these are *Ascaphus* and *Leiopelma* (Ascaphidae) and *Xenopus* (Pipidae). A literature survey is presented in conjunction with a morphological study of prepubic and postpubic elements of *Ascaphus truei* by de Villiers (1934*c*), who concluded that the epipubis is in cartilaginous continuity with the pelvic girdle. The element is calcified in adult *Ascaphus* and ossified in *Leiopelma* and *Xenopus*. De Villiers suggested that the structure is homologous in the three genera possessing it and that it probably is a derivative of the linea alba. He further suggested that the epipubis may be compared with the ypsiloid apparatus of salamanders, although it is not clear whether he was proposing a functional or morphological comparison.

Postpubic bones. Postpubic, or Nobelian, bones are unique to *Ascaphus truei*. The structures lie within the so-called tail, or copulatory organ, of *Ascaphus* and are attached to the posteroventral part of the pelvic girdle. Distally the Nobelian bones are discrete, and proximally they are joined. According to de Villiers (1934*c*), the Nobelian bones are incorporated into the phallic organ and possibly act as ossa penis.

APPENDAGES

The appendages are possibly the least well known part of the anuran skeleton. The status of the long bones is relatively evident; we are concerned with the humerus and radio-ulna in the forelimb and the femur and tibia in the hind limb. It seems apparent that the proportions of the long bones vary in accordance with the locomotor and behavioral habits of the frog: Typical saltatorial anurans, such as *Rana*, are usually characterized by long limbs, whereas hopping or walking frogs, such as bufonids, tend to have proportionally shorter limbs. Although casual observation suggests that this is the case, there have been no published studies on long bone variation.

The most recent osteological discussion of the anuran hand and foot is that of Howes and Ridewood (1888), who investigated 18 species representing eight families. A review of earlier literature is included in this paper. The absence of information, the confused nomenclature, and the conflicting opinions about the

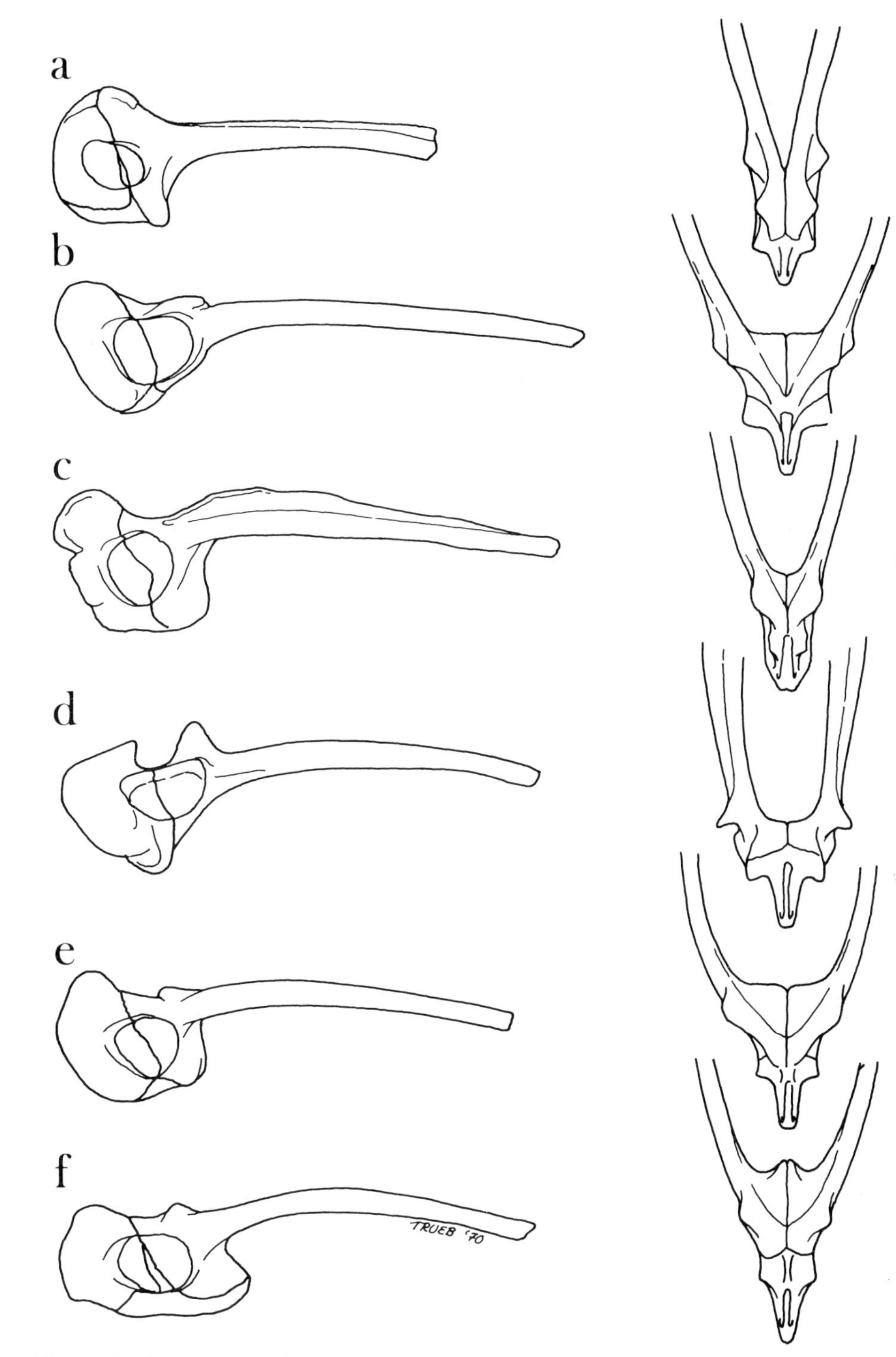

Figure 2–11. Lateral and posterodorsal views of pelvic girdles of anurans: a, *Rhinophrynus dorsalis* (Rhinophrynidae); b, *Barbourula busangensis* (Discoglossidae); c, *Hemiphractus panamensis* (Hylidae); d, *Xenopus laevis* (Pipidae); e, *Bufo arenarum* (Bufonidae); f, *Kaloula pulchra* (Microhylidae).

origins of the carpal and tarsal elements is surprising. So little is known about the osteological structure of the hand and foot that it is not possible to review variation in the way that has been done for other parts of the skeleton. The only generalized comment that can be made is that the number of carpal and tarsal elements is reduced among advanced frogs.

Forelimb. The long bones of the forelimb are constituted by the humerus proximally and the compound radius and ulna distally (Figure 2–12). The humeri of male frogs are frequently modified by the addition of large crests, in comparison to females of the same species (e.g., *Leptodactylus pentadactylus*). Presumably, increased surface area allows for the attachment of additional musculature which characterizes the forelimbs of males of many species of frogs.

The forefoot or hand of anurans is characterized by four digits that are usually composed of ten phalangeal elements in the following arrangement: 2–2–3–3. Frequently this number is reduced by loss of one or more phalanges. Among a few ranids proliferation of phalangeal elements has resulted in a phalangeal formula of 3–3–4–4. Several families—hylids, pseudids, centrolenids, ranids (in part), and microhylids (in part)—are characterized by the presence of an additional cartilaginous element, the intercalary cartilage, between the penultimate and ultimate phalanges. Among leptodactylids, bufonids, and microhylids, the terminal or ultimate phalanges are frequently modified into a variety of different shapes (Lynch, 1969; H. W. Parker, 1934; Tihen, 1965). Proximally, the phalanges articulate with metatarsal elements. The medial surface of the first, or inner, digits may bear bony excrescences in males of some taxa.

The number of carpal elements varies greatly among modern anurans, and the homologies are unresolved. Typically, there is a series of distal carpals; primitively, these were five, with one carpal opposing the metacarpal of each of the five digits. Apparently, early in amphibian evolution there was a tendency towards loss of the fifth distal carpal. Thus frogs, as an order, are usually considered to have lost this element, although Howes and Ridewood (1888) maintained that a vestige is retained in adults of pelobatids and discoglossids. The disposition of the remaining distal carpal elements depends partly upon an interpretation of prepollical structure. If, as has generally been assumed, the prepollex represents the vestigial remnant of the first digit, the cartilage with which the prepollex articulates proximally probably represents the first distal carpal, leaving carpals two, three, and four to articulate with the metacarpals of digits I, II, and III, respectively. The carpal bone articulating with the metacarpals of digit IV then could represent distal carpal five, a central carpal element, or a combination thereof. If, on the other hand, the prepollex does not represent a vestigial first digit, it would seem logical to assume that the first distal carpal has either been lost or incorporated into an adjacent structure. There seems to be no real basis to assume that all of the distal carpals have shifted position so that the first becomes associated with the metacarpal of the second digit (i.e, digit I). The second distal carpal is usually in line with the head of the metacarpal of digit II. This distal carpal is reduced in discoglossids and some leptodactylids, and while generally discrete, it may be fused with adjacent bones in *Brachycephalus,* pipids, and microhylids. The third distal carpal is reduced or displaced by the fourth in most advanced frogs; it is discrete in pipids, pelobatids, and discoglossids. The fourth distal carpal is usually enlarged and may represent the fusion of distal carpal four with central carpal three.

Among primitive amphibians there were four central carpal elements. The fate of these elements in anurans is a matter of dispute. The large bone lying adjacent to the prepollex and posterior to distal carpals two and three probably represents a fusion of central carpals one and two. The third central carpal may be the large bone lying adjacent to fused central carpals one and two and articulating with the metacarpal of digit IV. It is generally agreed that the fourth central carpal has fused with the proximal carpal element, the radiale; this bone lies adjacent to the distal head of the radius. The fused ulnare and intermedium articulate with the distal head of the ulna.

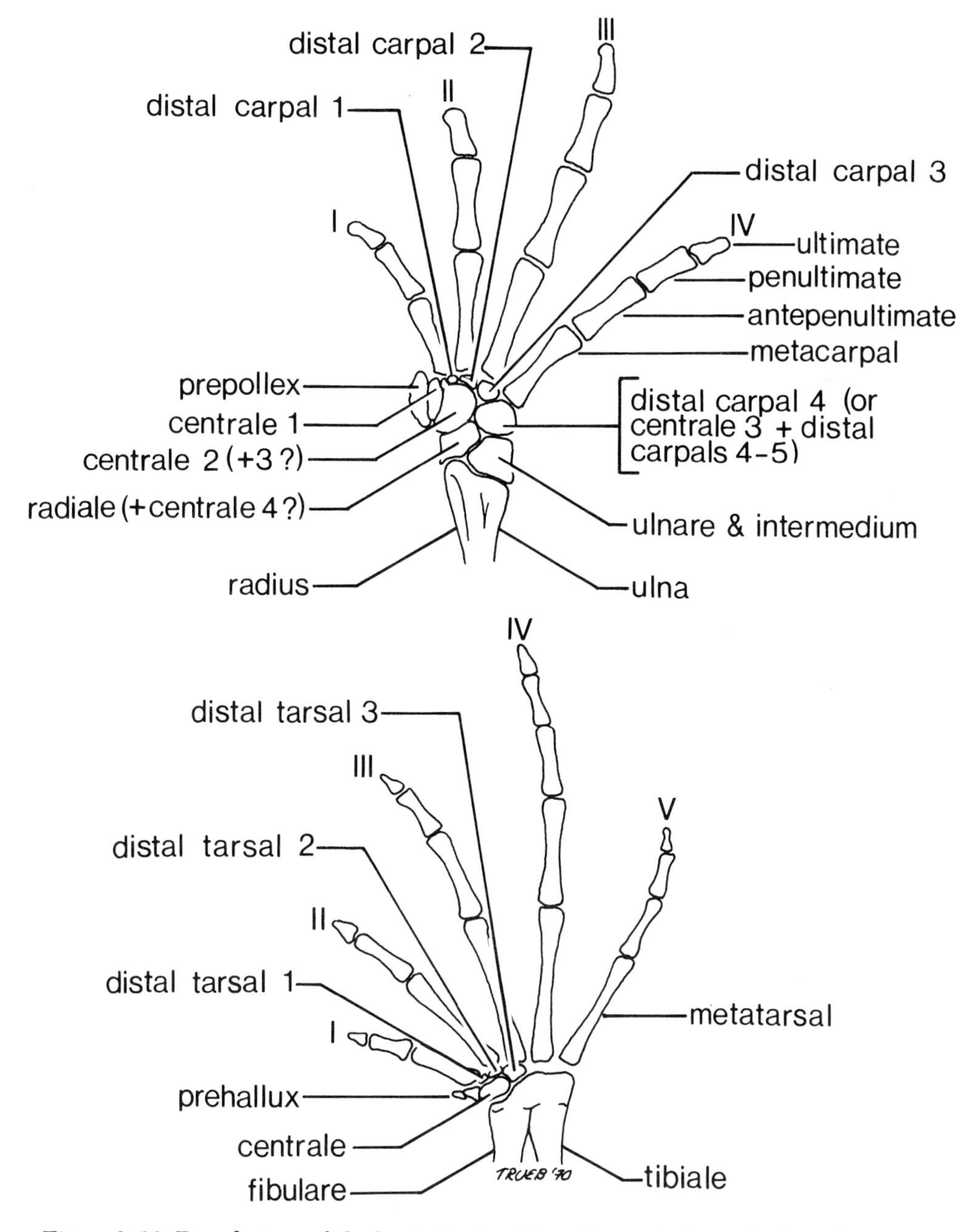

Figure 2–12. Dorsal views of the hand *(top)* and foot *(bottom)* of a male *Ascaphus truei*. Drawing adapted from that of Ritland (1955).

Hind limb. The long bones of the hind limb are constituted by the femur proximally and the compound tibia and fibula distally (Figure 2–12b). The hind foot of anurans is characterized by five digits, which are usually composed of 15 phalangeal elements in the following arrangement: 2–3–3–4–3. Frequently this number is reduced by loss of one or more phalanges. Intercalary cartilages are present in the digits of the feet of some frogs, and the terminal phalanges may be modified in the same fashion as those of the hand.

The tarsal elements have undergone greater reduction and modification than have the carpal elements. The tibiale and fibulare have become elongated and fused both proximally and distally. In some families—namely, Centrolenidae, Rhinophrynidae, and Pelobatidae (in part)—the tibiale and fibulare are fused throughout their lengths. The third proximal tarsal element, the intermedium, is fused with the head of the tibiale. Of the four original central tarsals, only one remains, articulating with the distal head of the fibulare. Conceivably this central tarsal may represent the fusion of two or more of the original series of bones. Primitive amphibians are characterized by a series of five distal tarsals, each of which articulated with metacarpals one through five. As an order, modern frogs are characterized by loss of the fourth and fifth distal tarsals. The three remaining distal tarsals are associated with the metatarsals of digits I, II, and III. In the microhylids, distal tarsals two and three tend to fuse and distal tarsal one may fuse with the central tarsal.

In addition to five digits, the hind foot of anurans bears a prehallux on its medial margin, adjacent to digit I. The prehallux consists of one or more bones articulating proximally with a central tarsal element. The prepollex tends to be better developed and to ossify earlier in males than in females. This presumably is related to its function in males of supporting nuptial pads or excrescences or, when very well developed, of forming a clasping organ. The prehallux, although present in all frogs, is best developed or hypertrophied in burrowing forms. The relationship of the prehallux and prepollex and their origins are problematic. The two organs are similar in position and structure and seem to have no homologue among primitive amphibians. It has been suggested that the prepollex represents the vestige of the first digit. If we restrict ourselves to a consideration of the forefoot, this seems to be a logical assumption. However, in the hind-foot structure the prehallux cannot be considered to be a similar derivation because all five digits have been retained and presumably the primitive amphibian ancestor bore only five digits on each foot. There obviously is considerable need for a thorough investigation of the anuran hand and foot morphology. On the basis of the present information I would suggest that the first digit of the hand has been completely lost and that the prepollex and prehallux represent corresponding organs of the forefeet and hind feet, respectively. Early in anuran evolution, these organs may have been an innovation, which has come to be functionally discrete among modern frogs.

SUMMARY OF EVOLUTIONARY PATTERNS

Cranium

Dermal Roofing Bones:

1. Dermal roofing bones tend to be the least well developed in aquatic (except among pipids), semiaquatic, and arboreal species. Roofing elements are maximally developed in anurans inhabiting arid or semiarid habitats.

2. Among larger, terrestrial anurans the frontoparietal tends to fuse with the otoccipital.

3. New centers of ossification, and consequently new dermal roofing bones (i.e., dermal sphenethmoid and internasal), tend to appear among phragmotic and fossorial species.

4. There is a tendency to consolidate multiple centers of ossification among advanced anurans.

Maxillary Arch:

1. The premaxillary tends to be reduced in length relative to the length of the maxillary.

2. The presence of pedicellate teeth on the maxillary and premaxillary is primitive. Loss of teeth or modification of the pedicellate form is a specialization.

3. Primitively, the maxillary arch is composed of three elements: the quadratojugal, maxillary, and premaxillary. Among small species and aquatic species, the quadratojugal tends to be reduced or lost. Loss of the quadratojugal tends to be compensated for by strengthening and sometimes reorientation of the pterygoid.

4. The structure of the maxillary arch is elaborated in phragmotic species and among larger terrestrial species that prey upon small vertebrates.

Dermal Ornamentation:

1. Dermal ornamentation is a secondary specialization usually found among species inhabiting arid or semiarid environments.

2. The addition of dermal bone follows an obligate sequence in modern anurans, as follows: exostosis, casquing, and co-ossification.

Palatal Bones:

1. The parasphenoid is invariably present in modern anurans and tends to be elaborated in larger species by the addition of odontoid structures.

2. The prevomer is subject to modification. It frequently is lost in smaller species whereas in other groups it has expanded to replace the palatine. Primitively, the prevomer is dentate.

3. The palatine tends towards reduction and loss.

Pterygoid:

1. Primitively, the pterygoid functions as a palatal bone. In most modern anurans its role is suspensory.

2. The medial ramus of the pterygoid tends to be reduced or lost.

3. In species which have lost the palatine, the anterior ramus of the pterygoid is extended; in those which have lost the quadratojugal, there is a tendency for the pterygoid to replace the quadratojugal functionally.

Suspensorium:

1. The basitrabecular and basal processes are lost among the majority of modern anurans in which articulation is affected by a pseudobasal process.

2. There is a tendency for the pseudobasal process to lose the bony reinforcement of the medial ramus of the pterygoid.

3. The quadrate tends to remain cartilaginous in most anurans; true ossification of this element is primitive, although secondary ossification may occur by invasion of ossification from the quadratojugal.

4. Jaw suspension is almost uniformly autosystylic. In at least two genera, the mechanism has been altered to an autodiastylic condition.

5. The squamosal tends to be reduced in all but large and extensively ossified species.

Nasal Capsule:

1. Primitively, the nasal capsules tend to be laterally displaced. Among advanced anurans there is a trend toward anteromedial migration of the capsules.

2. There is a trend toward fusion of some of the cartilaginous elements of the nasal capsule in advanced anurans.

Neurocranium:

1. The sphenethmoid is generally more extensively ossified among terrestrial and fossorial species than it is among aquatic and arboreal frogs.

2. In species that have lost the palatine, the planum antorbitale tends to ossify.

3. Cranial nerve foramina tend to be consolidated.

4. The auditory region undergoes reduction by loss of the plectral apparatus in groups that are otherwise advanced.

5. Cranial roof cartilages frequently disappear, especially the taenia tecta medialis.

Hyolaryngeal Apparatus

1. The hyoid tends to remain cartilaginous throughout life, losing ossifications characteristic of some primitive anurans.

2. There is a trend toward simplification of the hyoid plate with an attendant reduction in the size of distal processes.

3. In several anuran groups there is a tendency toward reduction of the cricoid cartilage of the larynx.

Pectoral Girdle

1. Firmisternal-like pectoral girdles have been independently derived from arciferal girdles on several occasions. The trend toward firmisterny is characterized by the changes listed below.

2. The pectoral arches tend to become shorter.

3. Procoracoid cartilages undergo reduction.

4. Epicoracoid cartilages tend to fuse in an anterior to posterior direction.

5. Clavicle is reduced.

6. Scapula is enlarged and elaborated.

7. Epicoracoid horns undergo reduction.

8. Sternum tends to become elaborated and fuse with the epicoracoid cartilages.

9. Omosternum is enlarged and elaborated.

10. Suprascapula and cleithrum tend to undergo reduction.

Vertebral Column

1. The number of presacral vertebrae tends to be reduced by one or both of two means: posterior presacral may be incorporated into the sacrum; anterior presacrals may fuse to one another.

2. Cervical cotyles tend to be displaced laterally.

3. There is an over-all trend toward ossification of the intervertebral cartilage and subsequent fusion of the ossified cartilage to one of the two adjacent vertebral centra.

4. Since transverse processes tend to become equivalent in width to one another and the sacral diapophyses, there is a trend toward reduction in the width of the anterior processes and increase in the width of the posterior processes.

5. Sacral diapophyses are reduced in length and tend to become posterolaterally oriented.

6. Primitively, intervertebral cartilage lies between the sacrum and coccyx. Bicondylar sacral-coccygeal articulations are widespread. Apparently fusion and monocondyly in advanced anurans are independently derived from a bicondylous relationship.

Pelvic Girdle

1. The ilium has tended to become longer and more elaborate by the addition of a variety of ridges, crests, and prominences.

Appendages

1. Humeri of males have tended to become larger and more elaborate than corresponding bones of females.

2. The tibiale and fibulare undergo fusion.

3. There is a tendency to reduce the number of phalanges.

4. Terminal phalanges often undergo adaptive modifications.

5. There is a trend toward elaboration of intercalary cartilages among arboreal species.

6. Distal and central carpals tend to undergo fusion.

DISCUSSION

In the preceding pages, I have attempted to present an initial assessment of anuran osteological variation and to offer suggestions about the evolutionary relationships of the morphological conditions described. Several cogent observations emerge from the data as a whole. Frogs are highly specialized organisms, as Inger (1967) and others have repeatedly pointed out. Nonetheless, within the bounds of their unique morphology, there is considerable variation. The nature of this variation suggests that osteological characters are more labile and subject to more adaptive modification than has been generally thought. This evolutionary lability seems to be discriminative, affecting some parts of the skeleton more readily than others; thus, some osteological characters appear to be more conservative than others. The basis of this differential lability becomes an issue of some importance in evaluating the potential conservatism of a given character. At least two factors are immediately apparent—the histological and developmental characters of the different types of bone tissue, and the limitations imposed on modification by the specialized morphological and functional patterns characteristic of anurans.

A brief resumé of the histology of bone tissue has been given previously. Although not all bones can be categorically stated to be strictly of endochondral or membranous origin, the basic developmental distinction prevails in the majority of cases. Dermal bone must be considered to be the older phylogenetically, but it is the younger ontogenetically and much less complicated in its development than endochondral bone. Thus, one would anticipate more variation in the deposition of dermal components, which require only a membranous connective tissue precursor, than in endochondral elements, which require a preformed cartilaginous precursor.

Among the characters used to define anurans are the following skeletal specializations listed by Romer (1966): (1) long hind limbs; (2) absence of a tail; (3) eight to ten vertebrae; (4) presence of a urostyle; (5) ribs present or, in most forms, ribs replaced by long transverse processes; (6) ilium a long rod; (7) tibia and fibula and radius and ulna fused, respectively; (8) proximal tarsals elongated and occasionally fused; and (9) pectoral girdle retaining all primitive parts. It is evident that not one of these characters, save for the pectoral girdle, involves dermal components. With respect to the pectoral girdle, it should be noted that not all frogs do retain all primitive parts; there is a tendency among the microhylids toward reduction and loss of the clavicle. (It should be noted, however, that this alteration involves a dermal element; the clavicle and cleithrum are the only two dermal bones associated with the pectoral girdle.) Thus, the above characters and the principal components of the anuran skeletal system are endochondral structures. The neurocranium, vertebral column, appendages, and pelvic girdle are examples of such structural units.

It has been shown that there are morphological variations in all endochondral structural units, but the variations are such that the integrity of the anuran skeleton remains relatively undisturbed. Should the morphology and function of any of these endochondral structural units undergo anything other than minor modification, it is highly unlikely that the resultant organism would fit the rather arbitrary set of criteria with which we characterize the Anura. In contrast to endochondral elements, dermal bones can and do undergo a great deal of modification without affecting the basic structural and functional integrity of the organism. Such changes usually involve the additional deposition or the deletion of bone and occasionally the addition or loss of a center of ossification. These alterations are almost exclusively concerned with dermal investing bones which are superficially associated with deeper-lying endochondral elements. The wide ranges of variation observed in these elements seem to represent adaptive modifications acquired by specialized anurans. Thus, while the most striking variation is associated with dermal elements, we must rely on the endochondral elements to provide us with the most conservative sets of characters upon which to base phylogenetic conclusions.

Evolutionary Trends

The statement of an evolutionary trend is a shorthand expression of the kinds and direction of morphological (or other) changes that have occurred and are occurring within a lineage of animals. If less bone and fewer bony elements occur among advanced anurans than among primitive anurans, we can state that there has been a trend toward reduction in bone. Unfortunately, such statements can rarely be made without qualification. Any attempt to formulate an evolutionary trend seems to be fraught with glaring exceptions and apparent reversals, and within any group, several evolutionary trends are likely to be operant at the same time.

Evolutionary trends represent a pattern of selective pressures acting upon a phyletic line. It is implied that any series of changes evident within a lineage represents some kind of selective or adaptive value for the organisms involved. If we review the evolutionary patterns associated with the anuran skeletal system, the trends seem to segregate into one of two general categories. The first of these comprises trends that are most often evident in the transition from archaic (i.e., the Ascaphoidea and Pipoidea) to modern anurans. These trends are restricted to changes in basic, endochondral structural units, such as the neurocranium or pectoral girdle; they tend to result in the evolution of a series of fundamental structural types frequently represented by a wide variety of adaptive types among modern anurans. The second general category includes trends which seem to be subordinate and, perhaps, more recent in origin. These are more evident among modern anurans and seem to be principally associated with the evolution of different adaptive types. Although the subordinate trends occasionally affect endochondral structures, the modifications are relatively minor; the action of secondary or subordinate trends is most evident among dermally derived structures. One can associate the primary trends with evolutionary changes that have modified the fundamental design of the frog. In general, these trends seem to have acted relatively early in the evolution of the group in order to establish efficient basic designs. In contrast, secondary trends act to modify the basic designs to particular adaptive situations. A given secondary trend may affect several basically divergent forms (as a result of primary trends) to produce similarly modified adaptive types. Thus, for example, we have the evolution of intercalary cartilages among arboreal members of both firmisternal and arciferal groups.

Primary Evolutionary Trends

Not all primary trends are archaic in nature; some, such as the trend toward reduction of the number of auditory foramina and the evolution of functionally firmisternal pectoral girdles, are evident among modern frogs. These patterns of divergence are evident at generic and familial levels and should be considered as continued evidence of specialization of the basic anuran skeletal system without particular association to adaptive types. Among primary trends there are those which were instrumental in affecting the transition from archaic anurans to intermediate forms and those which acted on intermediate types to produce the divergence characteristic of advanced anuran groups. Some aspects of the basic endochondral system of advanced anurans are still undergoing differentiation; such trends must be thought of as primary and recent.

Neurocranium. The neurocranium has undergone a generalized trend toward simplification of structure. In addition, there has been a slight, but basic, alteration in the organization of the anterior end of the skull.

The tendency to reduce the number of cranial nerve foramina is a primary evolutionary trend which shows two stages of progression. In *Ascaphus,* the branches of the trigeminal and facial nerves exit through the neurocranium by means of three foramina, whereas in *Leiopelma* only two foramina are utilized. In the pipids and all higher anurans, there is only a single prootic foramen. Apparently as a result of this consolidation of nerve foramina, the jaw suspensory apparatus undergoes a major modification that results in the establishment of a structural pattern prevalent among all anurans other than the Ascaphoidea. Primitive anurans have three auditory foramina, whereas the majority of modern (i.e., transitional and advanced) anurans have only two. There are isolated examples among advanced anurans in which only one auditory foramen exists; this presumably represents an advancement over those frogs having two foramina.

Archaic frogs (Ascaphoidea and Pipoidea) are characterized by wide premaxillaries, as compared with transitional and modern anurans. This notable difference in the size of the premaxillary relative to the maxillary is probably related to the lateral displacement of the nasal capsules in the archaic anurans. In transitional and advanced frogs the capsules are located more anteromedially, and the premaxillaries are proportionally smaller.

The pattern of simplification and reduction of the neurocranium is further evidenced among modern anurans by the failure of the quadrate to ossify, the tendency toward loss of the taenia tecta medialis, and the fusion and loss of some cartilaginous nasal capsule elements.

Vertebral Column. Although the various types of anuran vertebral columns have been described adequately, their evolutionary relationships remain enigmatic. It is generally agreed that the amphicoelous type, characteristic of the Ascaphoidea, is relatively unspecialized and primitive. As pointed out previously, all other vertebral types may be distinguished from the amphicoelous type by the ossification of the intervertebral cartilages and the nature of the connective tissue areas that invade the intervertebral cartilage. Assuming that vertebral centra types are related to one another by progression of a single evolutionary trend, it is most parsimonious to assume that the primitive amphicoelous pattern was subject to one major evolutionary innovation. This innovation resulted in the separation of the intervertebral cartilage from each of the two adjacent centra, a structural pattern which is evident among some of the transitional anurans. Among the latter (e.g., pelobatids), there is a tendency for the intervertebral disc, which is free in juveniles, subsequently to ossify and become rather firmly attached, although not fused, to the anterior centrum. This would seem to suggest that there is a definite morphological and functional value involved with complete ossification of the vertebral column and with reduction of the number of component parts. Subsequent to the separation of the intervertebral cartilage, two developmental options were available to accomplish this end. The cartilage could be attached to either the anterior or posterior centra and subsequently ossify. The next logical step involves the fusion of the two centers of ossification to form one developmental and functional unit.

I suggest the following as the evolutionary pattern of the vertebral column. From an amphicoelous ancester evolved an intermediate type having free intervertebral cartilages. This archaic stock gave rise to at least two phyletic lines. Fossil evidence suggests that the earliest of these lines is represented by the Pipoidea, which are opisthocoelous, whereas the second line has tended towards evolution of procoely. Some variations have evolved within the basically procoelous pattern. Thus, the pelobatids and myobatrachids should be thought of as having maintained a relatively primitive procoelous condition. Advanced procoely, characteristic of the sooglossids and the leptodactylids and their derivatives, probably evolved from an ancestral stock morphologically similar to that of the pelobatids and myobatrachids. As suggested previously, the diplasiocoelous vertebral column is basically procoelous in nature, save for the biconcave character of the last presacral vertebra. On the basis of this similarity, it seems reasonable to assume that diplasiocoely was evolved from a primitive procoelous condition. In the past—at the time of Noble's classification—it seemed quite clear that the ranids and microhylids could be distinguished from all other families on the basis of the diplasiocoelous nature of their vertebral columns. H. W. Parker's (1934) work on the microhylids, together with my own recent research on madagascarine ranids (in collaboration with Dr. Jean Guibé) showing that the vertebral column may be either diplasiocoelous or procoelous depending upon the species within a single genus of some Old World ranids, indicate that diplasiocoely is neither as widespread nor as constant in its appearance as has been supposed previously. The cumulative evidence strongly suggests that the time has come to reorient our thinking about the relative significance of diplasiocoely as compared with procoely and the other types of vertebral columns.

We would doubtless be much the wiser if we knew the details of the ontogenetic development of diplasiocoely compared with that of anomocoely (i.e., primitive procoely) and advanced procoely. The adaptive value, and consequent evolutionary significance, of diplasiocoely as compared with procoely should also be investigated. In the absence of such knowledge we can only speculate, a priori, on the significance of diplasiocoely. If we correctly can assume that procoely and diplasiocoely were similarly evolved from an anomocoelous or primitively procoelous ancestor, then the two types are closely related. Given the latter, perhaps we have accorded too much importance to the minor structural difference that actually distinguishes the two types of columns. If this is true, diplasiocoely should be regarded as a variation of the more widespread procoelous pattern and probably should not be used to define a major suborder of anurans. It would be more realistic to recognize all modern or advanced anurans (i.e., all leptodactyloids and derivatives, ranids, and microhylids) to be basically procoelous. The leptodactyloid frogs and their derivative (Bufonoidea) can be distinguished as strictly procoelous, with no variation in the structure of the eighth vertebra, whereas the ranids and microhylids (Ranoidea) are characterized by procoely in which the eighth presacral has one of two forms—that which has classically been associated with procoely and that which has been associated with diplasiocoely.

Clearly, this would require some redefinition of the terms mentioned previously. The definitions of amphicoelous, anomocoelous, and opisthocoelous would remain the same; however, that of procoelous should be amended as follows:

Procoelous (Bufonoidea and Ranoidea). A vertebral column in which the vertebral centra are holochordal (i.e., with no remnant of the notochord, solidly ossified, and with the centrum round in transverse section). Intervertebral cartilages anterior to the last presacral vertebra are posteriorly subdivided, subsequently ossified, and fused to the anteriorly adjacent centrum. At no time during development do free intervertebral cartilages exist. Two types of procoely are described:

1. Procoelous Type [*sensu stricto*] (sooglossids, some ranids and microhylids, and leptodactyloid frogs and their derivatives). A vertebral column in which the intervertebral cartilage between the posteriormost presacral and the sacrum is posteriorly subdivided, subsequently ossified, and fused to the centrum of the last presacral vertebra.
2. Diplasiocoelous Type (some ranids and microhylids). A vertebral column in which the intervertebral cartilage between the posteriormost presacral and the sacrum is anteriorly subdivided, subsequently ossified, and fused to the centrum of the sacral vertebra.

If we reorient our thinking along these lines, then the question of diplasiocoely versus procoely, that is, the differences in the articulation between the posterior presacral vertebra and the sacrum, is of less consequence. It becomes a character to be used at the generic or specific levels of classification. Although all modern frogs would be categorized as procoelous by this redefinition, the suborders can be distinguished by the condition of the pectoral girdle. The Bufonoidea, for example, is characterized by the presence of an arciferal pectoral girdle and a procoelous vertebral column of the "procoelous" type. The Ranoidea, in contrast, has a firmisternal pectoral girdle and a procoelous vertebral column of either the "procoelous" or "diplasiocoelous" type.

The deletion of free ribs on the anterior presacral vertebrae is a second primary trend evident in the transition from archaic to transitional anurans. Although there is no direct evidence available, I strongly suspect that ribs were never actually lost; instead, the centers of ossification fused or were consolidated with those of the abbreviated transverse processes associated with the ribs. The feasibility of this suggestion is supported by the developmental pattern observed in pipids: Free ribs are present in the young of this family; the ribs subsequently ankylose with adjacent transverse processes so that free ribs are not present in the adults. Such an ontogenetic pattern may represent a character state intermediate between that characterized by free ribs associated with short transverse processes and that characterized by the absence of free ribs but the presence of long to moderately long transverse processes on presacral vertebrae two through four.

A third primary trend acting upon the vertebral column involves the articulation of the coccyx with the sacrum. This articulation has been subject more to secondary than to primary adaptive modification. Discussion of the evolutionary trends will be found in the section dealing with secondary trends.

Pectoral Girdle. Repeated efforts have been made to utilize the pectoral girdle as a character or a basis of major anuran classification. As a consequence, a great deal of discussion and speculation has been devoted to the morphological variation of this structure. Most of this literature has been reviewed previously, and a cursory review of the literature and the morphological variation of the pectoral girdle demonstrates that the variation is not readily separable into two or three major phyletic lines, nor does it corroborate other lines of evidence upon which reasonable phylogenetic determinations have been based.

Pectoral girdle variation is a compound of archaic and recent primary trends. Among archaic frogs there is an obvious trend towards enlargement and proximal bicapitation of the scapula. The primitive scapula is small and unicapitate, whereas the scapulae of transitional and advanced anurans are large and bicapitate, except among frogs in which the articulation is obscured by fusion with adjacent elements.

The trend towards fusion of the two halves of the pectoral girdle and the establishment of partial, or full, functional firmisterny seems to be much more recent in origin and to have occurred at least seven, and probably several more, times. Referring to Table 2–1, we note that the majority of the anurans are arciferal. Those families indicated as having modified arciferal girdles have functionally firmisternal girdles, whereas those indicated as having normal firmisternal girdles have firmisternal girdles as defined by Cope, Boulenger, and Griffiths. It is my contention that no distinction should be drawn between these types of firmisterny. Instead, they all must be thought to represent a single trend among advanced frogs towards a similar alteration of pectoral girdle structure, that is, fusion of the halves of the girdle. The variety of ways in which this has been accomplished structurally is probably an indication of a number of independent origins of the condition. So-called partial firmisterny, wherein the precoracoid bridge is deepened but the girdle halves are not completely fused, should be viewed as an intermediate condition representing the evolutionary adjustment of the phyletic line from the arciferal to the firmisternal condition. The pipids furnish a fine example of an evolutionary continuum from an arciferal to a firmisternal condition. Several subordinate modifications of the pectoral girdle seem to be attendant on the shift from arcifery to firmisterny. These involve the fusion of the sternum, epicoracoid cartilages, and epicoracoid horns; reorientation of the clavicles; reduction of the procoracoid cartilages; and the elaboration of prezonal and postzonal elements. The degree of modification of these elements may be a useful index to the relative age of the group of animals when the shift from arcifery to firmisterny occurred. If this assumption is correct, the ranids and microhylids must have been among the earliest groups of animals to have achieved firmisterny, and in comparison, the dendrobatids must have acquired the condition relatively recently. The morphological conditions characterizing the functionally or partially firmisternous pipids, leptodactylids, bufonids, rhinodermatids, and brachycephalids suggest that these anurans are transitional or less advanced with respect to their acquisition of firmisterny than are the ranids, microhylids, or dendrobatids.

Two groups—the ranids and the sooglossids—are indicated as having members which have modified firmisternal girdles. This refers to a girdle pattern that in subordinate modifications is very similar to the firmisternal pattern characteristic of the ranids as a whole; however, the epicoracoid cartilages are not completely fused. It should be borne in mind that this morphological deviation is characteristic of only four species of anurans—*Rana tigrina, R. rugulosa, Sooglossus gardineri,* and *S. seychellensis.* The sooglossids, represented by only the two latter species, are considered to be a transitional group on the basis of other characters, and the ranids, as already discussed, probably represent one of the first lines to acquire firmisterny. Thus it seems logical to suspect that the sooglossids and perhaps these two species of ranids represent early phyletic offshoots of a line evolving towards typical ranid firmisterny.

Pelvic Girdle. The anuran pelvic girdle is a relatively simple structure. Two archaic trends seem evident in its structural evolution. The first involves the loss of prepubic and postpubic elements, which characterizes the ascaphids and pipids, and the second is the elongation of the ilial shaft, which characterizes most of the advanced anurans.

Appendages. The most archaic primary trend affecting the appendages seems to be involved with the relative lengthening of the hind limb elements, particularly the tibiale and fibulare—the astragalus and calcaneum. More recent trends center on the reduction and fusion of tarsal and carpal elements. Among tarsal elements there is the tendency for fusion of the tibiale and fibulare among rhinophrynids, pelobatids, and centrolenids. This should probably be considered as a logical extension of the same trend that led to the fusion of the radius and ulna in the forelimb and the tibia and fibula in the hind limb. The tendency for loss of carpal elements through reduction or fusion is most evident among advanced frogs and primarily involves deletion of distal carpal elements and consolidation of central carpals.

Secondary Evolutionary Trends

Secondary evolutionary trends are more difficult to deal with than their primary counterparts because they produce the greatest amount of evolutionary divergence and adaptive radiation. Secondary trends affect the endochondral skeleton, as well as the dermal skeleton, but their action is more obviously manifest in the latter. Furthermore, the alterations brought about on endochondral units are of relatively minor importance and subject to much more parallelism and convergence than those wrought by primary trends.

Basically, secondary trends are brought about in one of two ways. The amount of bone involved in the skeletal system may be reduced, a phenomenon usually associated with small size and arboreal and semiaquatic habits. In contrast, there may be an elaboration of skeletal components associated with an increase in size and shift toward terrestrial and fossorial habits or habitation of arid to semiarid environments.

The trend toward reduction usually involves arrest of bony development by neoteny and fusion of ossification centers, rather than actual loss of such centers. In the process of reduction, both dermal and endochondral centers are apt to be affected; one is likely to find, for example, that if dermal roofing bones are minimally developed, the sphenethmoid will be poorly ossified also. Reduction of bone is carried to such extremes in some species (e.g., small centrolenids and the leptodactylid *Notaden*) that it would seem that the lower limits of bone required for support are extremely minimal.

The trend towards elaboration usually involves the deposition of additional ossification around established centers of ossification, as well as the addition of new centers of ossification. In the process of elaboration, dermal components, which do not require a cartilaginous precursor, are affected before endochondral elements are. It has been found that elaboration in a given architectural unit, such as the skull, is usually countered by reduction elsewhere in the same unit (Trueb, 1970*a*). This strongly suggests that in contrast to the pattern observed in bone reduction, there is an upper limit to the amount of bone that is necessary for a system of support.

Cranium. In comparison with the rest of the anuran skeleton, the skull has been subject to a striking array of secondary modifications. The changes are sometimes awkward to classify, seeming to work at cross-purposes to one another adaptively. For example, ossification is usually reduced among smaller species, yet internal reduction is apt to be obscured among small, terrestrial or fossorial species in which secondary increments of dermal bone occur (*Brachycephalus* is a case in point). Thus, in making generalizations about adaptive and evolutionary trends involving the skull, a certain amount of latitude must be granted.

Size, as pointed out above, is a difficult parameter to correlate with degree of ossification. In general, it can be asserted that the smaller the frog, the less well ossified the skull. Adequate evidence is afforded by studying small members of such families as the Hylidae, Ranidae, and Leptodactylidae in comparison with larger members of the same families. The actuality of this trend is particularly obvious in comparing the relative degrees of ossification of males and females of a single species in which females are larger than males at sexual maturity. However, many small anurans have much more extensively ossified skulls than their larger relatives. These apparent exceptions occur when the small species is terrestrial or fossorial in habit, whereas the larger is arboreal or semiaquatic. Thus it is obvious that modifications successfully adapting an animal to a particular mode of life are of more significance than restrictions inherent in their size.

Despite their size, arboreal frogs (e.g., hylids, centrolenids, and some ranids) tend to be characterized by broad, depressed skulls and a generalized trend toward reduced ossification. The neurocrania are usually only moderately well ossified. The cristae paroticae of the prootic are moderately to poorly developed, sphenethmoidal ossification rarely occurs anterior to the midlevel of the nasals, and an extensive bridge of cartilage occurs between the sphenethmoid and prootic. As an apparent consequence, the ventral dermal investing bones (prevomer, palatine, and parasphenoid) are almost invariably present and moderately to well developed. There is considerable variation in dermal roofing bones, but in general their coverage is incomplete. Nasals rarely roof the entire forepart of the skull. A portion, or all, of the frontoparietal fontanelle is uncovered, and the frontoparietals afford little or no protection to the dorsum of the orbit. The squamosals usually have a poorly developed or no otic plate and the anterior arm frequently extends less than one-half the distance to the maxillary. As far as is known, all arboreal species retain maxillary and premaxillary dentition, although prevomerine dentition is absent in some. The upper jaw is moderately well ossified, although the quadratojugal is frequently absent or reduced among smaller species. The pterygoid maintains its triradiate structure; the medial articulation is weak in some species. All arboreal species retain a plectral apparatus.

In contrast to arboreal species, terrestrial anurans tend to have much less depressed skulls that are more heavily ossified and often casqued and co-ossified. Neurocranial ossification is frequently obscured dorsally by the dermal roofing bones. The nasals tend to be relatively larger and to articulate medially. Frontoparietals frequently cover the frontoparietal fontanelle and form a supraorbital flange laterally. Posterolateral development of the frontoparietals varies, but the bones commonly form a protective shelf or covering for the cristae paroticae. The squamosal usually bears an otic plate and a well-developed anterior arm. Upper jaw structure tends to be robust among terrestrial species; the quadratojugal and a well-developed pterygoid are usually retained. The premaxillaries usually bear moderate to well developed alary processes, and the partes palatinae of the premaxillary and maxillary are well developed. There is considerable variation in maxillary and premaxillary dentition in terrestrial species. Some, such as the bufonids, have lost all traces of maxillary dentition, whereas others—*Hemiphractus* (Hylidae), *Ceratobatrachus* (Ranidae), and *Ceratophrys* (Leptodactylidae)—have modified the primitive pedicellate anuran tooth to a fang-like form. By comparison with dorsal, dermal components, ventral investing bones tend to undergo reduction and modification. A well-developed parasphenoid is almost always retained, but the palatines and prevomers are frequently reduced or absent. Normally, either one or the other of the latter elements is lost, but not both. The palatine is most frequently reduced; its reduction is usually structurally compensated by ossification of the planum antorbitale or replacement by the prevomer. The plectral apparatus is often absent among terrestrial species.

Fossorial and phragmotic anurans are specialized terrestrial types. As one would expect, both of these groups display the generalized evolutionary trends characteristic of terrestrial anurans with a few peculiar modifications of their own. Fossorial anurans are particularly characterized by structurally compacted skulls that frequently have acuminate snouts. The anterior end of the skull is well developed; the sphenethmoid and internasal septum are usually well ossified, and the nasals usually form a complete roof to the anterior part of the skull. Phragmotic species are uniformly characterized by casquing, co-ossification, and a moderate to extensive development of lateral flanges along the upper jaw. In addition, the skull is completely roofed; the orbits are protected by a broad supraorbital flange, and the cranium is frequently reinforced by the development of new centers of ossification (e.g., prenasal, internasal, and dermal sphenethmoid).

Aquatic and semiaquatic species are generally characterized by fusiform heads with depressed snouts. Some of these species, for example, ranids, attain relatively large sizes; ossification is moderate, but the skulls are structurally compact. The neurocrania are almost always completely encased in bone and the olfactory capsules roofed by extensive nasals. Characteristically there is minimal distal outgrowth of roofing bones to form protective flanges or shelves. The upper jaw is usually well developed, and in species in which the quadratojugal has been lost, the pterygoid has been modified to compensate structurally.

Vertebral Column. The most obvious variation in the vertebral column involves modifications in the number of presacral vertebrae and the relative sizes of their neural arches and transverse processes, the nature of the sacral diapophyses, the characters of the sacral-coccygeal articulation, and the presence or absence of depositions of dermal bone associated with the column. The variation in these characters seems to follow patterns of secondary evolutionary trends that can be associated either with size or particular adaptive modifications.

Once free ribs disappeared in the transition from archaic to more modern frogs, a basic pattern was established wherein the transverse processes of the anterior presacral vertebrae were wider than those on the posterior presacrals. This pattern is characteristic of most modern anurans with the exception of specialized, terrestrial members of advanced families. Such species tend to be uniformly characterized by transverse processes equal in width to one another and the sacral diapophyses. An increment in the size of neural arches is clearly associated with increased size and a terrestrial mode of life. Expanded sacral diapophyses, such as those found in the pipids, seem to be associated with aquatic habits, whereas the round type, characteristic of many ranids and leptodactylids, is typical of saltatorial species. The majority of anurans have moderately dilated sacral diapophyses; this condition is the least specialized and accommodates most terrestrial and arboreal modes of locomotion.

The nature of the sacral-coccygeal articulation involves both primary and secondary trends; because most of the significant variation is secondary in nature, the evolutionary trends involved are discussed here. Morphological and paleontological evidence confirms the fact that the urostyle or coccyx represents a fusion of caudal vertebrae. If amphicoely is assumed to represent the primitive condition of the vertebral column, then it follows that separation of the sacrum and coccyx by contiguous intervertebral cartilage (as in ascaphids and sooglossids) is primitive. Furthermore, if the evolution of the articulation between the coccyx and sacrum paralleled that of the presacral vertebrae, it is logical to speculate that a monocondylar articulation evolved from the primitive acondylar condition. The presence of a monocondylar articulation, together with a coccyx bearing vestigial transverse processes among some transitional anurans, further confirms this assumption. Among advanced anurans, the most widespread type of articulation is bicondylar, which allows for vertical axial flexibility but not horizontal mobility of the coccyx. Among some aquatic anurans (pipids) and some terrestrial anurans (members of the bufonids and microhylids), the coccyx is fused to the sacrum. In a few specialized bufonid species, a monocondylar articulation prevails. Fusion seems to have occurred among species in which there is no premium placed on axial flexibility; monocondyly apparently represents a primitive character among transitional frogs and a derived condition among certain specialized, terrestrial anurans.

There is ample evidence (Table 2–2) that the number of presacral vertebrae among anurans has tended towards reduction. Experimental evidence (Madej, 1965) and many records of vertebral anomalies further suggest that the number of vertebral components is developmentally a labile character. The greatest reductions in presacral counts occur among highly specialized anurans (pipids; bufonids *Rhamphophryne, Oreophrynella,* and others) which are either aquatic or terrestrial. The deletions occur either by incorporations of posterior presacrals into the sacrum or fusion of one or more anterior presacrals late in development. The former method is associated with archaic anurans and seems to represent the primitive manner of vertebral column reduction, whereas the latter is characteristic of advanced anurans, which are typically small and terrestrial.

The development of a dermal presacral shield is an unusual phenomenon found only in the terrestrial *Brachycephalus*, some myobatrachids, and dendrobatids. This is another example of the lability of dermal bone formation and apparently develops in response to a need for added protection dorsally.

Pectoral Girdle. The pectoral girdle seems to have been subject to very few secondary modifications, and those that have occurred are of unknown significance. Among those trends which can be listed are (1) the tendency for the entire ventral portion of the girdle to ossify, which typifies some smaller, specialized bufonid derivatives (e.g., *Atelopus* and *Brachycephalus*) and (2) a trend towards reduction and loss of the clavicle in the microhylids. The adaptive significance of the primary trend towards firmisterny is in itself obscure; until this is understood, it is premature to speculate on the possible significance of these other, apparently minor, modifications.

Pelvic Girdle. Secondary modifications to the pelvic girdle include the development of flanges, crests, and protuberances of various types on the ilial shaft and probably the occasional calcification and ossification of the pubis. The significance of these modifications is uncertain. Ossification of the pubis probably strengthens the girdle. The addition of crests and protuberances provides a greater surface for muscle attachment and therefore must be related to myological differentiation and specialization.

Appendages. Like the pelvic girdle, most modifications of the long bones of the appendages involve the addition of various crests and protuberances presumably for the added attachment of muscles. There is a tendency among terrestrial species for the usual phalangeal formulae of the hand and foot to be reduced; this tendency is especially evident among specialized bufonids. The terminal phalanges of terrestrial leptodactylids and microhylids are often modified in shape. In some ranids there is an apparent multiplication of phalangeal elements, which increases the length of the digits markedly. Certain arboreal groups, namely, the centrolenids, rhacophorid ranids, hylids, and pseudids, have also increased the number of phalangeal elements by the addition of intercalary cartilages between the penultimate and terminal phalanges; the intercalary elements are ossified in the pseudids.

CONCLUSIONS

On the basis of the foregoing discussion, it should be obvious that anuran osteological characters are potentially valuable systematic tools. There is significant variation among characters at both major and lower levels of classification, and many character states seem to interrelate sequentially, forming logical and reasonably predictable evolutionary patterns. Modifications to the basic, endochondral skeleton of the frog seem to be among the oldest and most conservative evolutionary changes; these seem to have produced phylogenetic diversity characterizing currently recognized families and superfamilies. Dermal modifications and some relatively minor modifications of endochondral structures are apparently more recent. These frequently reflect adaptive radiation at generic and specific levels and, as a consequence, are highly subject to parallel and convergent development among different families. The occurrence of parallelism and convergence does not necessarily render these osteological characters useless; it does, however, indicate that we should exercise caution in their use.

It is clear that we can not rely on only one or two characters to establish major levels of classification as has been done in the past, nor can we rely on any one system of characters to predict phylogenetic relationships. Instead, we must utilize suites of individually weighted characters from a variety of disciplinary studies. Although osteology is only one of several potential sources of data, which include such disciplines as cytology, ethology, soft morphology, larval morphology, and others, it is unique and of singular importance because skeletal characters constitute our only reliable link with the fossil remains that give us our only substantial indication of the morphology and distribution of anurans in the past.

Acknowledgements

Many of the concepts discussed in this paper are the outgrowth of discussions with my colleagues. For their interest and valuable commentary I would like to thank Dr. Theodore H. Eaton, Dr. William E. Duellman, Dr. John D. Lynch, Dr. Joseph Tihen, and Mr. Stephen R. Edwards. Dr. Duellman and Dr. Eaton critically reviewed the manuscript, and the many specimens examined during the preparation of this paper were made available through the courtesy of Dr. Duellman at the Museum of Natural History, University of Kansas. Special thanks are extended to Mr. John E. Simmons, who assiduously and skillfully prepared many of the skeletal specimens. And finally, as a participant in this symposium on the Evolutionary Biology of the Anura, I am especially indebted to the Society for the Study of Amphibians and Reptiles for their generous invitation to take part in this significant and timely discussion.

REFERENCES

Al-Hussaini, A. H. 1941. The osteology of *Rana mascareniensis* Dum. et Bibr. Bull. Fac. Sci., Fuad I Univ., Cairo 24:75-89.

Antony, R., and H. Vallois. 1914. Sur la signification des éléments ventraux de la ceinture scapulaire chez les batraciens. Bibliogr. anat., Nancy 24:218-275.

Badenhorst, C. E. 1945. Die Skedelmorfologie van die Neotropiese Anure *Atelopus moreirae* de Mirando-Ribeiro. Ann. Univ. Stellenbosch 23A(2):1-17.

Baldauf, R. J. 1955. Contributions to the cranial morphology of *Bufo w. woodhousei* Girard. Texas J. Sci. 7(3):275-311.

_______. 1957. Additional studies on the cranial morphology of *Bufo w. woodhousei* Girard. Texas J. Sci. 9(1):84-88.

_______. 1958. Contributions to the cranial morphology of *Bufo valliceps* Wiegmann. Texas J. Sci. 10(2):172-186.

_______. 1959. Morphological criteria and their use in showing bufonid phylogeny. J. Morphol. 104(3):527-560.

Baldauf, R. J., and E. C. Tanzer. 1965. Contributions to the cranial morphology of the leptodactylid frog, *Syrrhophus marnocki* Cope. Texas J. Sci. 17(1):71-100.

Başoğlu, M., and Ş. Zaloğlu. 1964. Morphological and osteological studies in *Pelobates syriacus* from Izmir region, western Anatolia (Amphibia, Pelobatidae). Senck. biol. 45(3/5):233-242.

Bernasconi, A. F. 1951. Über den Ossifikationsmodus bei *Xenopus laevis* Daud. Denks. der Schweizer. Naturfor. Ges. 79(2):1-251.

Bhatia, B. L., and B. Prashad. 1918. Skull of *Rana tigrina* Daud. Proc. Zool. Soc. London 1:1-8.

Boring, A. M., and C. C. Liu. 1937-1938. Studies of the rainfrog, *Kaloula borealis*. III. An analysis of the skeletal features. Peking Nat. Hist. Bull. 12:43-46.

BOULENGER, G. A. 1882. Catalogue of the batrachia salientia s. ecaudata in the collections of the British Museum. 2nd ed. British Museum (Nat. Hist.) 503 p.

BRAUS, H. 1919. Der Brustschulterapparat der Froschlurche. Sitzungsber. d. Heidelberger Akad. d. Wissensch. 10. [not seen]

BROOM, R. 1902. On the mammalian and reptilian vomerine bones. Proc. Linn. Soc., New South Wales XXVIII (4):545.

CARVALHO, A. L. DE. 1954. A preliminary synopsis of the genera of American microhylid frogs. Occ. Pap. Mus. Zool., Univ. Michigan 555:1-19.

CHANTELL, C. J. 1968*a*. The osteology of Acris and Limnaoedus (Amphibia, Hylidae). Am. Midland Naturalist 79:169-182.

_______. 1968*b*. The osteology of Pseudacris (Amphibia, Hylidae). Am. Midland Naturalist 80:381-391.

COPE, E. D. 1864. On the limits and relations of the raniformes. Proc. Acad. Nat. Sci. Philadelphia 16:181-184.

_______. 1865. Sketch of the primary groups of batrachia salientia. Nat. Hist. Rev. 5:97-120.

DE BEER, G. R. 1937. The development of the vertebrate skull. Clarendon Press, Oxford. 552 p.

DE VILLIERS, C. G. S. 1922. Neue Beobactungen über den Bau und die Entwicklung des Brust-Schulter Apparates bei den Anuren insbesonders bei *Bombinator.* Act. Zool. Bd. III.

_______. 1924. On the anatomy of the breast-shoulder apparatus of *Xenopus.* Ann. Transvaal Mus. 10(4):197-211.

_______. 1929*a*. The development of a species of *Arthroleptella* from Johkershoek, Stellenbosch. South African J. Sci. 26:481-510.

_______. 1929*b*. The comparative anatomy of the breast-shoulder apparatus of the three aglossal anuran genera: *Xenopus, Pipa,* and *Hymenochirus.* Ann. Transvaal Mus. 13(4):37-69.

_______. 1930. On the cranial characters of the South African brevicipitid, *Phrynomerus bifasciatus.* Quart. J. Microscop. Sci. 73(292):667-705.

_______. 1931*a*. Some features of the cranial anatomy of *Hemisus marmoratus.* Anat. Anz. 71(14-16):305-331.

_______. 1931*b*. The cranial characters of the brevicipitid genus *Cacosternum* (Boulenger). Quart. J. Microscop. Sci. 74:275-302.

_______. 1931*c*. Über den Schädelbau der *Breviceps fuscus.* Anat. Anz. 72(6-9):164-178.

_______. 1931*d*. Über den Schädelbau der Brevicipitidengattung Anhydrophryne (Hewitt). Anat. Anz. 71(14-16):331-342.

_______. 1933. *Breviceps* and *Probreviceps:* a comparison of the cranial osteology of two closely related anuran genera. Anat. Anz. 75(12-14):257-265.

_______. 1934*a*. Studies of the cranial anatomy of *Ascaphus truei* Stejneger, the American "Leiopelmid." Bull. Mus. Comp. Zool., Harvard Univ. 77(1):3-37.

_______. 1934*b*. Die Schadelanatomie der *Rhombophryne testudo* Boettger in Bezug auf ihre Verwandtschaft mit den Malagasischen Brevicipitiden. Anat. Anz. 78(15-19):295-310.

_______. 1934*c*. On the morphology of the epipubis, the Nobelian bones and the phallic organ of *Ascaphus truei* Stejneger. Anat. Anz. 78:23-47.

DE VOS, C. M. 1935. *Spelaeophryne* and the bearing of its cranial anatomy on the monophyletic origin of the Ethiopian and Malagasy microhylids. Anat. Anz. 80(13-16): 241-265.

DUELLMAN, W. E., AND L. TRUEB. 1966. Neotropical hylid frogs of the genus *Smilisca.* Univ. Kansas Publ. Mus. Nat. Hist. 17(7):281-375.

DU TOIT, C. A. 1930. Die Skedelmorphologie van *Heleophryne regis.* South African J. Sci. 27:426-438.

_______. 1931. 'N Korreksie van my Verhandelung oor die Skedelmorphologie van *Heleophryne regis.* South African J. Sci. 27:408-410.

_______. 1933. Some aspects of the cranial morphology of *Rana grayi* Smith. Proc. Zool. Soc. London 715-734.

———. 1934*a*. A revision of the genus *Heleophryne*. Ann Univ. Stellenbosch 12A(2): 1-26.
———. 1934*b*. The cranial morphology of *Crinia georgiana* Tschudi. Proc. Zool. Soc. London 1-2:119-141.
———. 1938. The cranial anatomy of *Arthroleptides dutoiti* Loveridge. Anat. Anz. 86(22-24):338-411.
———. 1943. On the cranial morphology of the West African anuran *Petropedetes johnstoni* (Boulenger). South African J. Sci. 40:196-212.
DU TOIT, C. A., AND D. A. SCHOONEES. 1930. The bearing of cranial morphology upon Noble's dictum that *Heleophryne* is merely a toothed bufonid. South African J. Sci. 27:439-441.
DU TOIT, G. P. 1933. On the cranial characters of *Phrynobatrachus natalensis* (Smith). South African J. Sci. 30:394-415.
DU TOIT, G. P., AND C. G. S. DE VILLIERS. 1932. Die Skedelmorphologie van *Hyperolius horstockii* as voorbeeld van die Polypedatidae. South African J. Sci. 29:449-465.
EATON, T. H., JR. 1958. An anatomical study of a neotropical tree frog, *Centrolene prosoblepon* (Salientia, Centrolenidae). Univ. Kansas Sci. Bull. 39(10):459-472.
FEINSMITH, J. 1962. Development of the skull of *Rana sylvatica:* a morphological and histochemical investigation. Unpubl. Abstr. of paper presented at meetings of Am. Soc. Zool., December, 1962, Philadelphia.
FRY, D. B. 1912. Notes on the skull of *Austrochaperina robusta* Fry. Records Australian Mus. 8:101-106.
FUCHS, H. 1926*a*. Von der natürlichen Unterbrechung der Cartilago procoracoidea und von dem Fenster am Schultergürtel der *Rana fusca*. Anat. Anz. Bd. LXI. [not seen]
———. 1926*b*. Von der Entwickelung und vergleichend anatomischen Bedeutung des Praezonales am Brust-Schultergürtel der Amphibia anura (nach Untersuchungen am braunen Grasfrosche, *Rana fusca*). Erganzungsheft zum 61[en] Band des Anat. Anz. [not seen]
GAUDIN, A. J. 1969. A comparative study of the osteology and evolution of the holarctic tree frogs: *Hyla, Pseudacris, Acris,* and *Limnaeodus*. Ph.D. Dissertation, Univ. So. California.
GAUPP, E. 1896. A. Ecker's und R. Wiedersheim's anatomie des frosches. Vols. I, II. Friedrich Vieweg und Sohn, Braunschweig. 548 p.
GILLIES, C. D., AND E. F. PEABODY. 1917. The anatomy of *Hyla caerulea*. II. The skull. Proc. Roy. Soc. Queensland 29:117-122.
GREEN, T. L. 1931. On the pelvis of the anura: a study in adaptation and recapitulation. Proc. Zool. Soc. London Pt. 4:1259-1291.
GRIFFITHS, I. 1954*a*. On the nature of the fronto-parietal in Amphibia, Salientia. Proc. Zool. Soc. London 123:781-792.
———. 1954*b*. On the "otic element" in Amphibia, Salientia. Proc. Zool. Soc. London 124:35-50.
———. 1959. The phylogeny of *Sminthillus limbatus* and the status of the Brachycephalidae (Amphibia, Salientia). Proc. Zool. Soc. London 132:457-487.
———. 1963. The phylogeny of the salientia. Biol. Rev. 38(2):241-292.
HEYER, W. R. 1969. Biosystematic studies on the frog genus *Leptodactylus*. Ph.D. Dissertation, Univ. So. California.
HOWES, G. B., AND W. RIDEWOOD. 1888. On the carpus and tarsus of the anura. Proc. Zool. Soc. London 141-182.
INGER, R. F. 1967. The development of a phylogeny of frogs. Evolution 21(2):369-384.
KALIN, J., AND A. BERNASCONI. 1949. Über den Ossifikationsmodus bei *Xenopus laevis* Daud. Rev. Suisse Zool. T. 56.
KELLOGG, R. 1932. Mexican tailless amphibians in the United States National Museum. U.S. Natl. Mus. Bull. 160. Smithsonian Inst. Press, Washington, D.C. 224 p.
KLUGE, A. G. 1966. A new pelobatine frog from the lower miocene of South Dakota with a discussion of the evolution of the *Scaphiopus-Spea* complex. Los Angeles Co. Mus. Contrib. Sci. 113:1-26.

Kluge, A. G., and J. S. Farris. 1969. Quantitative phyletics and the evolution of anurans. Systematic Zool. 18(1):1-32.

Kotthaus, A. 1933. Entwicklung des primordial craniums von *Xenopus laevis*. Z. Wiss. Zool. 144.

Kruijtzer, E. M. 1931. Die Entwicklung des Chondrocraniums und einiger Kopfnerven von *Megalophrys montana*. Leiden: diss. 168 S. (Hollandisch).

Laurent, R. F. 1940. Contribution à l'ostéologie et à la systématique des ranides africains. Première note. Rev. Zool. Botan. africaines 34(1):74-97.

———. 1941*a*. Contribution à l'ostéologie et à la systématique des rhacophorides africains. Première note. Rev. Zool. Botan. africaines 35(1):85-111.

———. 1941*b*. Contribution à l'ostéologie et à la systématique des ranides africains. Deuxième note. Rev. Zool. Botan. africaines 34(2):192-235.

———. 1942*a*. Note sur l'ostéologie de *Trichobatrachus robustus*. Rev. Zool. Botan. africaines 36(1):56-60.

———. 1942*b*. Note sur l'ostéologie des genres *Breviceps* et *Phrynomerus* (Batraciens). Rev. Zool. Botan. africaines 35(4):417-418.

———. 1943*a*. Contribution à l'ostéologie et la systématique des rhacophorides non africains. Bull. Mus. Roy. Hist. Nat. Belg. 19(28):1-16.

———. 1943*b*. Note sur l'ostéologie de deux ranides exotiques. Bull. Mus. Roy. Hist. Nat. Belg. 19(27):1-16.

———. 1943*c*. Sur la position systématique et l'ostéologie du genre *Mantidactylus* Boulenger. Bull. Mus. Roy. Hist. Nat. Belg. 19(5):1-8.

———. 1944. Contribution à l'ostéologie systématique des Rhacophorides africains. Deuxième note. Rev. Zool. Botan. africaines 38(1-2):110-138.

León, J. R. 1969. The systematics of the frogs of the hyla rubra group in Middle America. Univ. Kansas Publ. Mus. Nat. Hist. 18(6):505-545.

Liem, S. S. 1969. The morphology, systematics, and evolution of the old world tree-frogs (Rhacophoridae and Hyperoliidae). Ph.D. Dissertation, Univ. Illinois.

Lynch, J. D. 1962. An osteological character on the ilia of *Acris crepitans* Baird, *Acris gryllus* Leconte, and *Hyla crucifer* Wied. Copeia 2:434.

———. 1969. Evolutionary relationships and osteology of the frog family Leptodactylidae. Ph.D. Dissertation, Univ. Kansas.

———. 1973. The transition from archaic to advanced frogs. This volume.

Maas, J. D. 1945. Contributions to the cranial morphology of the West African ranid *Schoutedenella muta* Witte. Ann. Univ. Stellenbosch 23A(2-6):21-42.

Madej, Z. 1965. Variations in the sacral region of the spine in *Bombina bombina* (Linnaeus, 1761) and *Bombina variegata* (Linnaeus, 1758) (Salientia, Discoglossidae). Acta Biol. Cracoviensia 8:185-197.

Maree, W. A. 1945. Contributions to the cranial morphology of the European anuran *Alytes obstetricans* (Laurenti). Ann. Univ. Stellenbosch 23A(2-6):43-89.

McDiarmid, R. W. 1969. Comparative morphology and evolution of the neotropical frog genera *Atelopus, Dendrophryniscus, Melanophryniscus, Oreophrynella,* and *Brachycephalus*. Ph.D. Dissertation, Univ. So. California.

McLachlan, P. 1943. The cranial and visceral osteology of the Neotropical anuran *Brachycephalus ephippium* Spix. South African J. Sci. 40:164-195.

Mookerjee, H. K. 1931. On the development of the vertebral column of anura. Phil. Trans. Roy. Soc. London, B, 219:165-195.

Moss, M. L. 1968. Comparative anatomy of vertebrate dermal bone and teeth. Acta Anat. 71:178-208.

Nevo, E. 1969. Pipid frogs from the early cretaceous of Israel and pipid evolution. Bull. Mus. Comp. Zool., Harvard Univ. 136(8):255-318.

Nicholls, G. C. 1916. The structure of the vertebral column in the anura phaneroglossa and its importance as a basis of classification. Proc. Linn. Soc. London, Z, 128:80-92.

Noble, G. K. 1922. The phylogeny of the salientia. I—The osteology and the thigh musculature; their bearing on classification and phylogeny. Bull. Am. Mus. Nat. Hist. 46:1-87.

_______. 1926. The pectoral girdle of brachycephalid frogs. Am. Mus. Novitates 230: 1-14.

_______. 1931. The biology of the amphibia. McGraw-Hill Book Co., New York. 577 p.

PARKER, H. W. 1934. A monograph of the frogs of the family Microhylidae. British Museum (Nat. Hist.), London. 208 p.

PARKER, W. K. 1881. On the structure and development of the skull in the batrachia. Part III. Phil. Trans. Roy. Soc. London I:1-266.

_______. 1886. A monograph on the structure and development of the shoulder-girdle and sternum in the vertebrata. Published for the Ray Society by Robert Hardwicke, 192 Piccadilly, London. 237 p.

PARSONS, T. S., AND E. E. WILLIAMS. 1962. The teeth of the amphibia and their relation to amphibian phylogeny. J. Morphol. 110:375-389.

PATERSON, N. F. 1939. The head of *Xenopus laevis*. Quart. J. Microscop. Sci. 81(322): 161-234.

_______. 1945. The skull of *Hymenochirus curtipes*. Proc. Zool. Soc. London 15(3-4): 327-354.

_______. 1955. The skull of the toad, *Hemipipa carvalhoi* Mir.-Rib. with remarks on other Pipidae. Proc. Zool. Soc. London 125(1):223-252.

PENTZ, K. 1943. The cranial morphology of the neotropical microhylid (anura) *Elachistocleis ovalis* (Schneider). South African J. Sci. 39:182-226.

PETERS, J. A. 1955. Notes on the frog genus *Diaglena* Cope. Nat. Hist. Misc., Chicago Acad. Sci. 143:1-8.

PROCTER, J. B. 1919. On the skull and affinities of *Rana subsigillata* A. Dum. Proc. Zool. Soc. London 1919:21-27.

_______. 1921. On the variation of the scapula in the batrachian groups aglossa and arcifera. Proc. Zool. Soc. London 1-2:197-214.

PUSEY, H. K. 1938. Structural changes in the anuran mandibular arch during metamorphosis, with reference to *Rana temporaria*. Quart. J. Microscop. Sci. 80:479-552.

_______. 1943. On the head of the leiopelmid frog, *Ascaphus truei*. Quart. J. Microscop. Sci. 84(II-III): 106-185.

RAMASWAMI, L. S. 1932*a*. The cranial osteology of the south Indian Engystomatidae (anura). Half-yearly J. Mysore Univ. 6(1):45-71.

_______. 1932*b*. The cranial anatomy of *Glyphoglossus molussus* (Gunther). Half-yearly J. Mysore Univ. 6(2):1-12.

_______. 1934. Contributions to our knowledge of the cranial morphology of some ranid genera of frogs. Part I. Proc. Indian Acad. Sci., B, 1:80-95.

_______. 1935*a*. The cranial morphology of some examples of pelobatidae (anura). Anat. Anz. 81(4-6):65-96.

_______. 1935*b*. Contributions to our knowledge of the cranial morphology of some ranid genera of frogs. Part II. Proc. Indian Acad. Sci., B, 2:1-20.

_______. 1936. The cranial morphology of the genera *Kaloula* Gray and *Phrynella* Boulenger (anura). Proc. Zool. Soc. London 3-4:1137-1155.

_______. 1942. The discoglossid skull. Proc. Indian Acad. Sci., B, 1:10-24.

_______. 1943. An account of the chondrocranium of *Rana afghana* and *Megalophrys*, with a description of the masticatory musculature of some tadpoles. Proc. Natl. Inst. Sci., India 9(1):43-48.

_______. 1956. "Frontoparietal" bone in anura (amphibia). Curr. Sci. 25:19-20.

REIG, O. A. 1960. Las relaciones genéricas del anuro chileno *Calyptocephalella gayi* (Dum. & Bibr.). Act. y Trab. I[ro] Congr. Sudam. Zool. 4:113-131.

_______. 1961. La anatomía del género *Lepidobatrachus* (Anura, Leptodactylidae), comparada con la de otros ceratofrinos. Act. y Trab. I[ro] Congr. Sudam. Zool. 4:133-147.

REINBACH, W. 1939. Untersuchungen über die Entwicklung des Kopfskeletts von *Calyptocephalus* Gayi (mit einem Anhang über das Os supratemporale der anuren Amphibien). Jenaische Z. für Naturwiss. 72:211-362.

RITLAND, R. M. 1955. Studies on the post-cranial morphology of *Ascaphus truei*. I. Skeleton and spinal nerves. J. Morphol. 97(1):119-178.

ROGGENBAU, C. 1926*a*. Zur frage nach einem Episternalapparat nebst einer Bemerkung über die Cartilagines epicoracoideae am Brustschultergürtel der anuren Amphibien nach Beobachtungen am braunen Grasfrosch *(Rana fusca)*. Anat. Anz. Bd. LXI. [not seen]

________. 1926*b*. Einige Beobachtungen über die Verbindungen der Cartilago praezonales mit den Procoracoidspangen und des Sternums mit den Coracoidplatten am Brustschulterapparat des braunen Grasfrosches *(Rana fusca)*. Z. für Mikroskop. Anat. Forsch. Bd. VI [not seen]

ROMER, A. S. 1966. Vertebrate paleontology. 3rd ed. Univ. Chicago Press, Chicago. 468 p.

ROUX, G. H. 1944. The cranial anatomy of *Microhyla carolinensis* (Holbrook). South African J. Med. Sci. 9:1-28.

SANDERS, O. 1953. A new species of toad with a discussion of the morphology of the bufonid skull. Herpetologica 9:25-47.

SCHOONEES, D. A. 1930. Die Skedelmorphologie van *Bufo angusticeps* (Smith). South African J. Sci. 27:456-469.

SEDRA, S. N. 1949. On the homology of certain elements in the skull of *Bufo regularis*. Proc. Zool. Soc. London 119(3):633-641.

SEDRA, S. N., AND M. I. MICHAEL. 1957. The development of the skull, visceral arches, larynx, and visceral muscles of the South African clawed toad, *Xenopus laevis* (Daudin) during the process of metamorphosis (from Stage 55 to Stage 66). Verh. Akad. Wet. Amsterdam 51(4):1-80.

SLABBERT, G. K. 1945. Contributions to the cranial morphology of the European anuran *Bombina variegata* (Linné). Ann. Univ. Stellenbosch 23A(5):67-89.

SLABBERT, G. K., AND W. A. MAREE. 1945. The cranial morphology of the Discoglossidae and its bearing upon the phylogeny of the primitive anura. Ann. Univ. Stellenbosch 23A(2-6):91-97.

STEPHENSON, E. M. T. 1951. The anatomy of the head of the New Zealand frog, *Leiopelma*. Trans. Zool. Soc. London 27(2):255-305.

________. 1955. The head of the frog *Leiopelma hamiltoni* McCulloch. Proc. Zool. Soc. London 124(4):791-801.

________. 1960. The skeletal characters of *Leiopelma hamiltoni* McCulloch, with particular reference to the effects of heterochrony on the genus. Trans. Roy. Soc. New Zealand 88(3):473-488.

STEPHENSON, N. G. 1951. On the development of the chondrocranium and visceral arches of *Leiopelma archeyi*. Trans. Zool. Soc. London 27(2):203-252.

________. 1965. Heterochronous changes among Australian leptodactylid frogs. Proc. Zool. Soc. London 144(3):339-350.

STOKELY, P. S., AND J. C. LIST. 1954. The progress of ossification in the skull of the cricketfrog *Pseudacris nigrita triseriata*. Copeia 2:211-217.

TIHEN, J. A. 1959. An interesting vertebral anomaly in a toad, *Bufo cognatus*. Herpetologica 15:29-30.

________. 1962. Osteological observations on new world *Bufo*. Am. Midland Naturalist 67(1):157-183.

________. 1965. Evolutionary trends in frogs. Am. Zool. 5:309-318.

TREWAVAS, E. 1933. The hyoid and larynx of the anura. Phil. Trans. Roy. Soc. London, B, 222(10):401-527.

TRUEB, L. 1966. Morphology and development of the skull in the frog *Hyla septentrionalis*. Copeia 3:562-573.

________. 1968. Cranial osteology of the hylid frog, Smilisca baudini. Univ. Kansas Publ. Mus. Nat. Hist. 18(2):11-35.

________. 1970*a*. Evolutionary relationships of casque-headed tree frogs with co-ossified skulls (family Hylidae). Univ. Kansas Publ. Mus. Nat. Hist. 18(7):547-716.

________. 1970*b*. The generic status of *Hyla siemersi* Mertens. Herpetologica 26(2):254-267.

________. 1971. Phylogenetic relationships of certain neotropical toads with the description of a new genus (Anura, Bufonidae). Los Angeles Co. Mus. Contrib. Sci. 216:1-40.

van Eeden, J. A. 1943. Die Skedelmorfologie van *Elosia nasus.* Licht. Tijdskr. Wetenskap. Stellenbosch 3(2):75-103.

———. 1951. The development of the chondrocranium of *Ascaphus truei* Stejneger with special reference to the relations of the palatoquadrate to the neurocranium. Act. Zool. 32:42-176.

van Pletzen, R. 1953. Ontogenesis and morphogenesis of the breast-shoulder apparatus of *Xenopus laevis.* Ann. Univ. Stellenbosch 29A(4):403-406.

van Seters, H. W. 1922. Le dévelopement du chondrocrane d'*Alytes obstetricans* avant la metamorphose. Arch. Biol. 32:373-491.

van Zyl, J. H. M. 1949. Die Beskrywende en Vergelykende Anatomie van die Skedel van *Discoglossus pictus* (Gravenhorst). Ann. Univ. Stellenbosch 26A(12):1-26.

Wagner, D. S. 1934. The structure of the inner ear in relation to the reduction of the middle ear in Leiopelmidae (Noble). Anat. Anz. 79(1-4):20-36.

Wagner, W. H., Jr. 1961. Problems in the classification of ferns, p. 841-844. *In* Recent advances in botany. Univ. Toronto Press, Toronto.

Walker, C. F. 1938. The structure and systematic relationships of the genus *Rhinophrynus.* Occ. Pap. Mus. Zool., Univ. Michigan 372:1-11.

Westhuizen, C. M. v.d. 1961. The development of the chondrocranium of *Heleophryne purcelli* Sclater with special reference to the palatoquadrate and sound-conducting apparatus. Act. Zool., Stockholm 42:1-72.

Whiting, H. P. 1961. Pelvic girdle in amphibian locomotion. *In* Vertebrate locomotion. Zool. Soc. London Symposium 5:43-57.

Zweifel, R. G. 1956. Two pelobatid frogs from the tertiary of north america and their relationships to fossil and recent forms. Am. Mus. Novitates 1762:1-49.

3

THE TRANSITION FROM ARCHAIC TO ADVANCED FROGS

John D. Lynch

INTRODUCTION

A variety of classifications of the Salientia have been proposed, but no classification has been generally accepted, although Noble's (1922, 1931) enjoyed the longest life. Anuran systematists have in general proposed schemes based largely on one character complex and have attempted to bolster their arguments with data drawn from other proposals. Classifications based on presence or absence of a tongue (Wagler, 1830; Duméril and Bibron, 1841), shape of the pupil and amplectic type (Thomas, 1854; Bruch, 1863; L'Isle, 1877), larval morphology (Orton, 1957; Hecht, 1963), pectoral girdle architecture (Cope, 1864, 1865; Boulenger, 1882; Griffiths, 1963), nature of the vertebral centra (Nicholls, 1916; Noble, 1922), and the disposition of the thigh musculature (Noble, 1922, 1931; Dunlap, 1960) are the principle classifications and while not being completely congruent are largely so. The major difficulties have been (1) incomplete knowledge of variations in families and (2) confusion of evolutionary direction of the characteristics involved. The first difficulty generally involved the toothed arcifera (Leptodactylidae) and is mitigated by a now-available monograph on the family (Lynch, 1969). The second problem has been attacked successfully by Tihen (1965) and Kluge and Farris (1969).

More satisfactory classifications than those based largely on single character complexes are those subscribing to multicharacter complexes (e.g., Tihen, 1965; Inger, 1967; and Kluge and Farris, 1969). These schemes attempted to incorporate several characteristics into the classification and to accord no excess weight to any particular characteristic a priori. In general, their proposals are less tenuous than those proposed earlier, but they do suffer from a failure to include all taxa, a failure to consider several other character complexes, and an assumption that each family now recognized is a homogeneous operational taxonomic unit (OTU).

If we are correct in assigning primitive and advanced qualities properly to the character states evidenced among the many osteological and nonosteological character complexes of frogs (see Tihen, 1965; Lynch, 1969; and Trueb, this volume, for lists), we can stratify family groups according to the number of primitive character states retained. We thus obtain an index of the relative primitiveness and, conversely, advancedness, of each family group of frogs. This exercise is instructive in that it disallows much (though not all) of the biases introduced by specialists of particular disciplines, it is multidisciplinary, and it presumably gives a more accurate representation of the primitiveness of each group. It does not provide answers about which group gave rise to which other group, except in such cases in which trends continue through several family groups. Associating family groups is done here by regarding as closely related those groups exhibiting similar patterns of primitive and derived character states.

The basis for groupings of archaic and advanced frogs stems from Reig's (1958) proposal of the Archaeobatrachia and Neobatrachia as new suborders. Reig included the Notobatrachidae, Leiopelmatidae (=Ascaphidae), Pipidae, Palaeobatrachidae, and *Eoxenopoides* in other suborders, namely, Amphicoela and Aglossa. The Archaeobatrachia included the Discoglossidae, Rhinophrynidae, and tentatively the Montsechobatrachidae, Pelobatidae, and Pelodytidae. Reig recognized twelve other families, all extant, as members of the Neobatrachia.

Reig's (1968) classification supposes either a greater phyletic distance between the Amphicoela and Archaeobatrachia than between families within these suborders or a cladistic sequence requiring hierarchical distinction for the two suborders. Evidence for a major cladistic distinction is meager, and the phyletic distance is not as great as the pipid-rhinophrynid distinction. The single basis for separation of the Amphicoela and Archaeobatrachia rests on the "amphicoelous" nature of the centrum in ascaphids and Reig's (1957) assertion that notobatrachids have "amphicoelous" centra (Hecht, 1963, disagreed and stated that the centra are procoelous; however, Estes and Reig, this volume, demonstrate "amphicoely" in the fossils).

Pipids have been considered archaic (=primitive) in osteology (subadults bear ribs, as do early pipids, see Nevo, 1968), myology, and preovulatory reproductive behavior but are recognized as being also highly specialized. Rhinophrynids share a few unique traits with pipids (namely, the form of the larvae) and have been treated as close allies of the pipids largely on those grounds. Ignoring for the moment the specializations of pipids toward aquatic existence, the pipids are archaic frogs. Tihen (1965) included in his "primitive frogs" (=archaic frogs here) the Ascaphidae, Discoglossidae, Pipidae, and Rhinophrynidae. He considered the Pelobatidae intermediate between the primitive frogs and the advanced frogs (Reig's Neobatrachia). As Tihen (1965) pointed out, his hypothetical ancestral anuran exhibits a mosaic of characteristics not duplicated among any living or fossil frog, although ascaphids and megophryine pelobatids are similar in many points. It is pertinent to emphasize at this point that discoglossids are only slightly more dissimilar. Tihen prefers to dissociate discoglossids from the mainline and retain the ascaphids somewhere near the base. The differences between ascaphids and discoglossids are few, and the difference in the nature of the centrum in these two families is greatly overstressed (refer to the family diagnoses given later).

I. HISTORICAL REVIEW OF CLASSIFICATIONS

The first serious classifications of frogs were those proposed by Cope (1864, 1865) and later modified by Boulenger (1882). These authors settled on pectoral architecture as the most reliable source of interpreting the relationships of the Anura. In the case of both authors, aglossy was considered so large a modification that the pipids (Dactylethridae and Pipidae) were isolated subordinally. The remaining frogs initially were grouped into three infraorders—the Bufoniformia, the Arcifera, and the Raniformia—although in later years these authors considered the presence or absence of maxillary arch dentition of decreasing importance.

The Bufoniformia included those species lacking teeth, lacking manubrium sterna, having dilated sacral diapophyses, and having or not having cartilaginous arches on the sternum. Included taxa (by present standards) were the Rhinophrynidae (including Hemisinae), Engystomatidae (Bufonidae in part, Rhinodermatidae, Microhylidae in large part), Brachymeridae (Microhylidae in part, Myobatrachinae in part), Bufonidae, and Dendrobatidae (Cope noted that this family possessed manubrium sterna and round sacral diapophyses).

The Arcifera included those species having teeth, the manubrium present or absent, dilated or round sacral diapophyses, and the coracoid and epicoracoid bones divergent and connected by the epicoracoidal cartilages that overlap across the chest. Included taxa were the Discoglossidae, Asterophryidae (Microhylidae in part, Pelobatidae in part, Cycloraninae in part), Scaphiopodidae (Pelodytidae, Pelobatinae, Myobatrachidae in part, Leptodactylidae in part), Hylidae, and Cystignathidae (Myobatrachidae and Leptodactylidae in large part, Pseudidae, Dendrobatidae in part).

The Raniformia included those species having teeth, cylindrical sacral diapophyses, the coracoids and epicoracoids parallel and united medially, an osseous manubrium, and usually an osseous sternal style. Only the Ranidae were included.

The subsequent modifications made on this scheme were to consider the absence of maxillary dentition of little consequence. The resulting changes were association of the Engystomatidae, Brachymeridae, and Dendrobatidae with the Ranidae as the Firmisternia. The Bufonidae were placed in the Arcifera (Boulenger, 1882). Boulenger (1882) considered the presence and absence of teeth on the maxillary arch as of familial worth, although in his later works he concluded that the characteristic was of less significance.

Nicholls (1916) divided anurans into four groups on the basis of the centra, sacro-coccygeal articulation, and presence of free ribs. The first group he called Opisthocoela—frogs with opisthocoelous centra, the sacral centrum biconvex, not fused with coccyx, free ribs. Nicholls included only the Discoglossidae. The second he termed Anomocoela—frogs with procoelous centra, the sacrum ankylosed to the coccyx or articulating with it by a single condyle, no ribs. Nicholls included only the Pelobatidae. The third he labeled Procoela—frogs with procoelous centra, the sacrum free and articulating with the coccyx by a double condyle, no ribs. He included the Bufonidae, Cystignathidae (=Leptodactylidae), and Hylidae. The fourth he called Displasiocoela (=Firmisternia)—frogs with the first seven centra procoelous, the eighth biconcave, the sacrum biconvex and articulating with the coccyx by a double condyle, no ribs. Nicholls included the Ranidae and Engystomatidae. Noble (1922) pointed out that several firmisternal genera exhibited a procoelan, not a diplasiocoelan, condition, namely, *Atelopus, Brachycephalus, Dendrobates, Geobatrachus, Hyloxalus, Phyllobates, Rhinoderma,* and *Sminthillus.* He also cited some ranids *(Arthroleptis, Cardioglossa)* as procoelan and noted some synsacral examples of *Rana.* Noble considered these and a few other ranids exhibiting nondiplasiocoelan conditions as aberrant. Noble (1922) gave less weight to dentition than had previous workers. His writings reflect that he considered the presence or absence of teeth on the maxillary arch as, at best, a "generic" character. His principle objection to the Cope-Boulenger scheme stems from his discovery of an apparent arcifero-firmisternal frog, *Sminthillus* (Noble, 1921, 1922). His conviction that all gradations between arcifery and firmisterny were evident is summarized in his paper on the girdle of brachycephalids (Noble, 1926*a*).

Noble (1922, 1931) modified Nicholls' diagnoses but retained the groupings. The Opisthocoela was further credited with arciferal or arcifero-firmisternal pectoral girdles. The Anomocoela were additionally diagnosed as arciferal. The Procoela were credited with arcifery, arcifero-firmisterny, or firmisterny. The Diplasiocoela were considered uniformly firmisternal. Noble added the Pipidae to the Opisthocoela and added the Brachycephalidae to the Procoela. In his later work

(1931), he removed *Ascaphus* and *Leiopelma* from the Discoglossidae and placed them in the Leiopelmidae as the only family of the Amphicoela (vertebrae biconcave). Noble amplified this classification with a dissertation on the thigh musculature, especially the disposition of the distal tendons of the thigh. He provided myological data to bolster the classification originally proposed by Nicholls. The Amphicoela were regarded as being the only frogs retaining tail-wagging muscles, and this suborder, Opisthocoela, and Anomocoela were separated from higher frogs by the lack of a discrete *m. sartorius.* The Procoela were distinguished from the Diplasiocoela in having the tendon of the *m. semitendinosus* inserting ventral to that of the *mm. gracilis,* instead of dorsal to it.

Noble's scheme was attacked by Parker (1934), who reported that fully one-half of the genera of microhylid frogs are procoelous. Latsky (1930), Parker (1940), and Griffiths (1959*a*) further criticized the importance of the distal thigh musculature in constructing a phylogeny of frogs. Parker's (1940) criticism is especially important, for he demonstrated a continuum of character states in the disposition of the tendons of the *m. semitendinosus* and *m. gracilis major* in myobatrachine and cycloranine frogs. In spite of the criticisms attacking the fundamental character complexes forming the basis of Noble's classification, herpetologists (e.g., Brattstrom, 1957) used the classification in almost blissful ignorance until Griffiths (1959*b*, 1963) challenged the transitional and hence invalid nature of the pectoral architecture character complex. Dunlap (1960) reinvestigated the thigh musculature and, with minor exceptions, arrived at the same conclusions that Noble had. Dunlap's study is deficient only in his small samples. More extensive studies of single muscles have cast doubt on their reliability (see Tihen, 1960, for an analysis of loss of the *m. adductor longus* in various lines of bufonids).

Orton (1953, 1957) divided the Anura into four groups distinguished exclusively on the basis of larval morphology. She used the presence or absence of mouthparts and the form of the spiracle in frogs having aquatic larvae. Type I larvae had paired, ventrolateral spiracles, had no denticles, beaks, or lips, and possessed long barbels. Two families were included—Pipidae and Rhinophrynidae. Type II larvae differed from Type I in lacking barbels and in having a single, ventral spiracle and lips but no denticles or beak. This type included only Microhylidae. Type III and Type IV larvae were diagnosed in having denticles and beaks, although Type III larvae had the denticles in multiple rows instead of single rows. Furthermore, the spiracle of Type III larvae was like that of Type II larvae—single and ventral—whereas Type IV larvae had a single, sinistral spiracle. Neither Type III nor Type IV larvae had barbels. Type III larvae were represented by two families—the Ascaphidae and Discoglossidae—and Type IV larvae included all other frogs. Orton (1957) suggested that the larval types (I, II, III, IV) might be in phylogenetic order. Hecht (1963), with no discussion, stated that the larval types were in phylogenetic order and proposed a temporal sequence for their appearance: early Jurassic for Type I, middle Jurassic for Types II and III, and very late Jurassic for Type IV. Objections to this scheme are many and from various points of view. Griffiths (1963) objected on the grounds that spiracular position was known to be polymorphic in *Pelodytes,* that *Pseudohymenochirus* has denticles, that *Lepidobatrachus* has paired spiracles, that several frogs of supposedly Type IV have nearly ventral spiracles (e.g., *Phyllomedusa*), and that some ranids have multiple rows of denticles. Griffiths and Carvalho (1965) provided additional discrediting data from a number of South American frogs relative to spiracle position (largely hylids and some microhylid species). An even more complete analysis is provided by Kluge and Farris (1969). Some of Griffiths' (1963) objections were discredited by Kluge and Farris (*Pelodytes* polymorphy, denticles of *Pseudohymenochirus,* ventral spiracle of *Phyllo-*

medusa). These authors did point out that the "spiracles" of Type I frogs were different from those of all other frogs, with the possible exception of *Lepidobatrachus*. Kluge and Farris (1969) recorded seven character states of the spiracle and an eighth for the condition seen in pipids and rhinophrynids (pseudospiracles).

Tihen (1965) argued that the larval types of Orton were not in phylogenetic sequence and asserted that Types I and II were paedogenetic, not primitive. He considered Type III to be primitive and the other types to be derived from it or one of its descendent types (IV). Starrett (this volume) has reinvestigated the bases for Orton's tadpole types and concluded, based on internal morphology of the throat and jaws, that there are no known exceptions to the four categories. She considers Type I most primitive, Type II slightly more advanced, and Types III and IV relatively similar and advanced. The four types are argued to represent four basic levels of anuran organization.

Reig (1958) proposed dividing the Anura into four suborders: Amphicoela (Leiopelmatidae, Notobatrachidae), Aglossa (Palaeobatrachidae, Pipidae, and Eoxenopoididae), Archaeobatrachia (Discoglossidae, Pelobatidae, Pelodytidae, and Rhinophrynidae), and Neobatrachia (advanced frogs; superfamily A = Procoela, superfamily B = Diplasiocoela, except including Heleophrynidae). He proposed no new justifications for the scheme but relied on diagnoses provided by earlier workers. Hecht (1963) criticized Reig's arrangement in noting that the Amphicoelan family Notobatrachidae was probably procoelous, in dismissing the Palaeobatrachidae as a wastebasket of Tertiary European frog fossils and the Eoxenopoididae as incapable of definition, and in stating his belief that microhylids were "super-primitive." Much of Hecht's justification for his classification rested on "feel."

Griffiths (1959*a*, 1959*b*, 1963) reinstated the shoulder girdle architecture as a major criterion in frog classification. The reinstatement resulted from redefinition of arcifery and firmisterny on the bases of developmental patterns and the free epicoracoidal horns. The previously enigmatic arcifero-firmisternal genera were all transferred to arciferal groups, associations predicted by Noble (1922, 1931). The major distinction between Griffiths' and Noble's arrangement was the placement of the Dendrobatidae, which Noble considered procoelan and Griffiths considered diplasiocoelan.

Griffiths (1963) did not dwell as extensively on suborders as had all previous anuran systematists, although he contended that the Anura are diphyletic. The justification for this proposal rests only on the presence of nine presacral vertebrae in Ascaphidae and eight or fewer in other frogs. Griffiths rejected the contention by several authors (Walker, 1938; Orton, 1957; Reig, 1958; Hecht, 1963) that the Rhinophrynidae are primitive frogs and followed Noble (1931) in considering *Rhinophrynus* allied to leptodactylids and bufonids. His discussion of relationships indicates the following informal groupings based on his work: Ascaphidae forms an isolated unit; discoglossids and pipids are primitive groups (they possess ribs) but are not especially closely related; the remaining families—the higher frogs—are divided into two series, one arciferal, and a second firmisternal; pelobatids are intermediate between primitive frogs (i.e., Ascaphidae, Pipidae, and Discoglossidae) and advanced frogs (i.e., Rhinophrynidae, Leptodactylidae, Hylidae, Bufonidae, Atelopodidae, Ranidae, Rhacophoridae, Microhylidae, Phrynomeridae, and Sooglossidae); rhinophrynids are isolated; leptodactylids and hylids form a unit allied to a bufonid-atelopodid unit; ranids, including dendrobatids, are closely allied to rhacophorids and form one of three units of firmisternal frogs (the others are a microhylid-phrynomerid unit and a unit for the zoogeographically isolated sooglossids of the Seychelles).

The preceding studies were principally concerned with distinguishing groups and illustrating differences and distinctness of suprafamilial units. Three other studies, discussed below, are chiefly concerned with illustrating similarities and continuities of relationships.

Tihen (1965) postulated two major anuran lineages. The first included the "primitive" anurans exhibiting amphicoely or opisthocoely and having free ribs at some stage, namely, Ascaphidae, Discoglossidae, Notobatrachidae, Pipidae, and Rhinophrynidae (although they lack ribs at any stage). The second major lineage included all other frogs. Tihen characterized this line as essentially procoelous and lacking ribs. He argued that the Pelobatidae was the stem family for all the advanced frogs, with the possible exception of the Microhylidae. His phylogenetic sketch differs from others in that he isolated the archaic frogs from the mainstem of anuran phylogeny; he made this adjustment because he considered the archaic frogs too specialized to have given rise to a pelobatid. Elsewhere in his discussion he seems less convinced of this argument, especially when he considers ascaphids, discoglossids, and possibly notobatrachids. As cited below (Section IV), the archaic frogs belong to two groups—one that consists of the three families cited above and a second, specialized, group that consists of pipids, rhinophrynids, and the extinct palaeobatrachids. The group comprised of the ascaphids, discoglossids, and possibly notobatrachids is not so specialized as Tihen implied and could in fact be regarded as ancestral to pelobatids, especially the megophryines. This suprafamilial grouping of archaic frogs is the same as Tihen's, with the exception of not recognizing the opisthocoelous families as a unit. The characteristics of the Discoglossidae preclude a close relationship with the pipoids but do not preclude a close relationship with ascaphids or *Notobatrachus*. In part, Tihen's proposal was anticipated by Noble (1924*a*), who considered pelobatids, as represented by megophryines, direct descendents of leiopelmids, the "primitive discoglossids." Implied in Noble's argument is that opisthocoelous anurans represent an early side branch of anuran evolution.

Inger (1967) proposed an anuran phylogeny in which characters were not weighted, except by selection. Unlike previous authors he did not utilize suprafamilial groups. His classification was based on three myological characteristics, three vertebral characteristics, the architecture of the girdle, and three larval characteristics. The phylogeny produced by this nonweighted classification was modified by "unique character states." Unlike earlier classifications (except Hecht's, 1963), Inger's arrangement separated microhylids and ranoids, with the microhylids representing one of the earliest dichotomies.

Inger's scheme was roundly criticized by Kluge and Farris (1969) on several grounds, the most significant of which were poor character selection (larval vent position, condition of *m. adductor longus,* and scoring of two tail-wagging muscles instead of one) and character weighting according to the "uniqueness" of the character. Inger also dismissed those characteristics for which evolutionary direction was not established and disagreed with the criterion for estimation of primitiveness proferred by Wagner (1961), which is that a primitive state is likely to be distributed among taxa in agreement with states of other characters known to be primitive.

The phylogenies produced by Inger (1967) and Kluge and Farris (1969) are deficient in two regards: (1) they include only part of the anuran families, and (2) they assume that each family is a homogeneous unit (this last criticism applies more to Inger's account than Kluge and Farris' but is applicable to at least part of Kluge and Farris' account). Kluge and Farris admittedly used few characteristics because their study was designed to interpret phylogeny, and their selection of frogs as the subject was a secondary consideration. They allowed the constraints imposed by Inger (1967)—the form of families recognized and the characters used—to limit the value of the phylogeny produced. By a more careful coding of character states they eliminated some of the problems in Inger's classification but were forced to reduce the number of character complexes analyzed from ten to six and then subdivided the six into eleven. The strength of their method (i.e., quantitative phyletics) enabled them to provide an index of phyletic distances between groups. Their classification bears similarities to those argued by previous authors: they consider (1) ascaphids most primitive and more or less allied with discoglossids, pipids, and rhinophrynids; (2) pelobatids intermediate between the complex of primitive frogs and the advanced families; (3) hylids, leptodactylids, bufonids, and atelopodids as closely allied (the Procoela); and (4) ranids, rhacophorids, and microhylids as the most specialized families. They departed from some earlier proposals when they separated the discoglossids from the other "opisthocoela" and argued a close relationship between pipids and rhinophrynids.

In the following discussions I have departed from some current usages of family group names. I use leptodactyloid frogs to include all frogs presently placed in the Leptodactylidae and collectively refer to the Old World leptodactyloids (Cycloraninae, Heleophryninae, and Myobatrachinae) as the Myobatrachidae and the New World leptodactyloids (Ceratophryinae, Elosiinae, Leptodactylinae, and Telmatobiinae) as the Leptodactylidae. Unless otherwise stated the atelopodids are combined with the Bufonidae, the phrynomerids with the Microhylidae, and the hyperoliids and rhacophorids with the Ranidae.

II. CHARACTERISTICS USED IN ANURAN CLASSIFICATION

As discussed above, most attempts at anuran classification have partially failed because too few characteristics were employed in deducing relationships. A kindred symptom was the drive to distinguish differences among groups as opposed to demonstrating continuities. I have utilized the character complexes employed by earlier authors on the subject, as well as a few others that I found especially germane in my study of leptodactyloid frogs (Lynch, 1969). Several character complexes employed by previous authors have been modified or in some cases discarded. As is true for most groups, our knowledge of intragroup variations is not complete. Scarce hints at the procoelous nature of many firmisternal, presumed diplasiocoelous frogs preceded the work demonstrating the widespread nature of procoely in these families (Parker, 1934; Liem, 1969). As more characteristics are studied they should provide substantial grounds for testing the classification proposed here (Section IV) on the bases of thirty-eight characteristics of frogs. These thirty-eight characteristics and character complexes are discussed below.

Each of the thirty-eight characteristics employed here is divided into primitive and derived character states. Determination of which character state is primitive is a major factor in the use of such characteristics. The derived states of some characteristics are apparent advancements and of others are apparent specializations. Separation of these two categories is essential because they are not of equal value in projecting a classification and phylogeny.

I have divided the thirty-eight characteristics into two categories according to the level of confidence one has in considering the primitive to derived states to be in correct evolutionary sequence. A third category includes those characteristics in which the trend is primitive to derived and also specialized. The three categories of characteristics are here referred to as first-degree characteristics, second-degree characteristics, and third-degree characteristics.

A) *First-Degree Characteristics:* These are the characteristics for which the inference of evolutionary sequence (primitive to derived) has the highest degree of confidence. Primitive character states are those which are represented in most or all amphibian groups (Urodela, Gymnophiona, and extinct orders) and in anurans. Other states of these characteristics found in frogs are considered derived. The primitive state in the subset (Anura) is more likely to be widespread among representatives of closely related groups of the set (Amphibia) than is a derived state in the subset (Anura). First-degree characteristics do not include those unique to the subset (Anura). The following characteristics discussed in this section are considered first-degree characteristics: 1, 4, 5, 8, 11, 12, 18, 21, 22, 26, 30, and 36.

B) *Second-Degree Characteristics:* These are the characteristics in which the distributions of the character states (primitive, derived) among the Anura are closely correlated with the distributions of first-degree characteristics among the Anura; second-degree characteristics differ from first-degree characteristics in usually having all character states restricted to the Anura. The following are considered second-degree characteristics: 6, 7, 9, 10, 15, 19, 23, 24, 29, 31, 32, 33, 34, and 35. Characteristics 27 and 28 are also included here, but considerable debate about the use of these characteristics and which character states are primitive obtains (see my discussion of these characteristics and Starrett's paper in this volume).

C) *Third-Degree Characteristics:* These include characteristics for which the inferred evolutionary sequence is obtained in the same way as for first-degree characteristics but for which the sequence is *primitive to specialized.* In general, such derived character states are representative of one or a few groups. The following are considered third-degree characteristics: 2, 3, 13, 14, 16, 17, 20, 25, 37, and 38.

1. Number of presacral vertebrae. The apparent reduction in number of presacral vertebrae in pipids was noted early by anuran systematists. Use of this characteristic stems largely from the work of Nicholls (1916) and the subsequent work of Noble (1922). However, these authors distinguished synostosis as a minor modification and did not think simple synostotic reduction to as few as five presacral elements especially significant. Griffiths (1963) stressed the number of presacral vertebrae as a major character in anuran classification. He regarded the higher number primitive and a decrease in number derived. Griffiths separated ascaphids from other frogs because the Ascaphidae have nine presacral vertebrae while other frogs have eight or fewer. Kluge and Farris (1969) published a table of vertebral counts giving ranges of seven to eight for most families and eight to nine for ascaphids, discoglossids, and pelobatids.

As noted by Kluge and Farris (1969), eight presacral vertebrae in ascaphids is an anomaly or due to synostosis, as is seven in discoglossids and pelobatids. However, they regard seven presacral vertebrae to be nonanomalous and not due to synostosis in pipids, bufonids, atelopodids, leptodactylids, hylids, ranids, rhacophorids, and microhylids. Furthermore, they contend that six presacral vertebrae in pipids and atelopodids is not anomalous whereas only five presacral vertebrae in pipids is due to synostosis, but they assert that five presacral vertebrae in *Oreophrynella* is due to reduction.

My study of anuran vertebral columns and the data provided by Noble (1922, 1931) and Tihen (1960) yields results at major odds with the data provided by Kluge and Farris (1969). Nine presacral vertebrae occur normally only in ascaphids. The following families have eight presacral vertebrae, although either the cervical may be synostotically fused with the second or the eighth may be fused with the sacrum: Discoglossidae, Rhinophrynidae, Pelobatidae, Pelodytidae, Myobatrachidae, Leptodactylidae, Rhinodermatidae, Pseudidae, Hylidae, Centrolenidae, Dendrobatidae, Sooglossidae, Ranidae, and Microhylidae. Pipids and bufonids represent departures. The range of variation in pipids (Figure 3–1) is from five (cervical fused to second vertebra, seventh and eighth fused to sacrum) to eight (all free). In bufonids the same range is seen with *Oreophrynella,* which has five presacrals, and most toads, which have eight. Incorporation of presacral vertebrae into the sacrum is seen in *Oreophrynella.* True reduction to seven vertebrae by deletion through the sacrum is known for two genera of toads (Tihen, 1960, 1965). The character states and coding are as follows: nine presacrals 0, eight presacrals 2, seven presacrals 4.

2. Fusion of cervical and second vertebrae. Tihen (1960) considered this character state derived from the nonfused character state in bufonids. Taylor (1941) and Zweifel (1956) cited the fusion of the two elements in pelodytids as a support for family distinction. Noble (1931), among others, cited the fusion of these two elements as a characteristic of the Palaeobatrachidae. Some of the counts of seven presacral vertebrae cited by Kluge and Farris (1969) are a result of fusion of the cervical and second vertebrae. The fusion normally occurs in the pipids *Hymenochirus, Pipa,* and *Pseudohymenochirus;* in the myobatrachids, *Heleophryne,* and most cycloranines; in the pelodytids and rhinodermatids; in several bufonid genera, including all "atelopodids" except *Brachycephalus;* and in the ranoid, *Hemisus.* A free cervical is scored as 0, the fusion as 2.

3. Fusion of sacrum and coccyx. Freedom of the two elements is considered primitive (score 0) and fusion derived (score 2). The fusion of the sacrum and coccyx is an anomaly in most frogs (e.g., *Bombina;* Madej, 1965) but was regarded as characteristic of pipids and the Holarctic pelobatids (*Pelobates, Scaphiopus,* and *Spea*) by Nicholls (1916) and Noble (1922). Noble and most subsequent herpetologists recognized the nondiagnostic quality of the characteristic among pelobatids. A few paleontologists, however, regard the fusion of the two elements as diagnostic of pelobatids. The fusion of the elements in the Pelobatinae is known only among fossils of Miocene or younger age (Estes, 1970) and is variable among recent *Pelobates* (Noble, 1924*a*).

Fusion of the sacrum and coccyx occurs in two genera of brevicipitine microhylids (Parker, 1934). Tihen (1960) reported fusion in some specialized African bufonids and fusion also occurs in some neotropical bufonids (Trueb, 1971).

Fusion of these elements precludes observations of the nature of the sacrococcygeal articulation (character 35), although the Oligocene *Macropelobates* and the aberrant *Pelobates* have a monocondylar articulation, as do megophryine pelobatids.

4. Shape of the vertebral centra. Nicholls (1916) stimulated research on the nature of the centra in frogs by pointing out that procoelous centra were found in arciferal frogs, whereas diplasiocoelous vertebral columns were found only in firmisternal families. Noble (1922, 1931) based much of his classification on the nature of the vertebral centra and noted a general correspondence of the distribution of the character states among frog families with thigh musculature, presence of ribs in adults and young, and girdle architecture. Noble recognized five suborders of frogs, characterized in large part by the form of the vertebral column.

Parker (1934) provided the first serious objection to the use of the diplasiocoelous character state of vertebral centra when he pointed out that approximately one-half of the firmisternal and presumed diplasiocoelous microhylid genera were uniformly procoelous. Nicholls (1916) and Noble (1922) reported

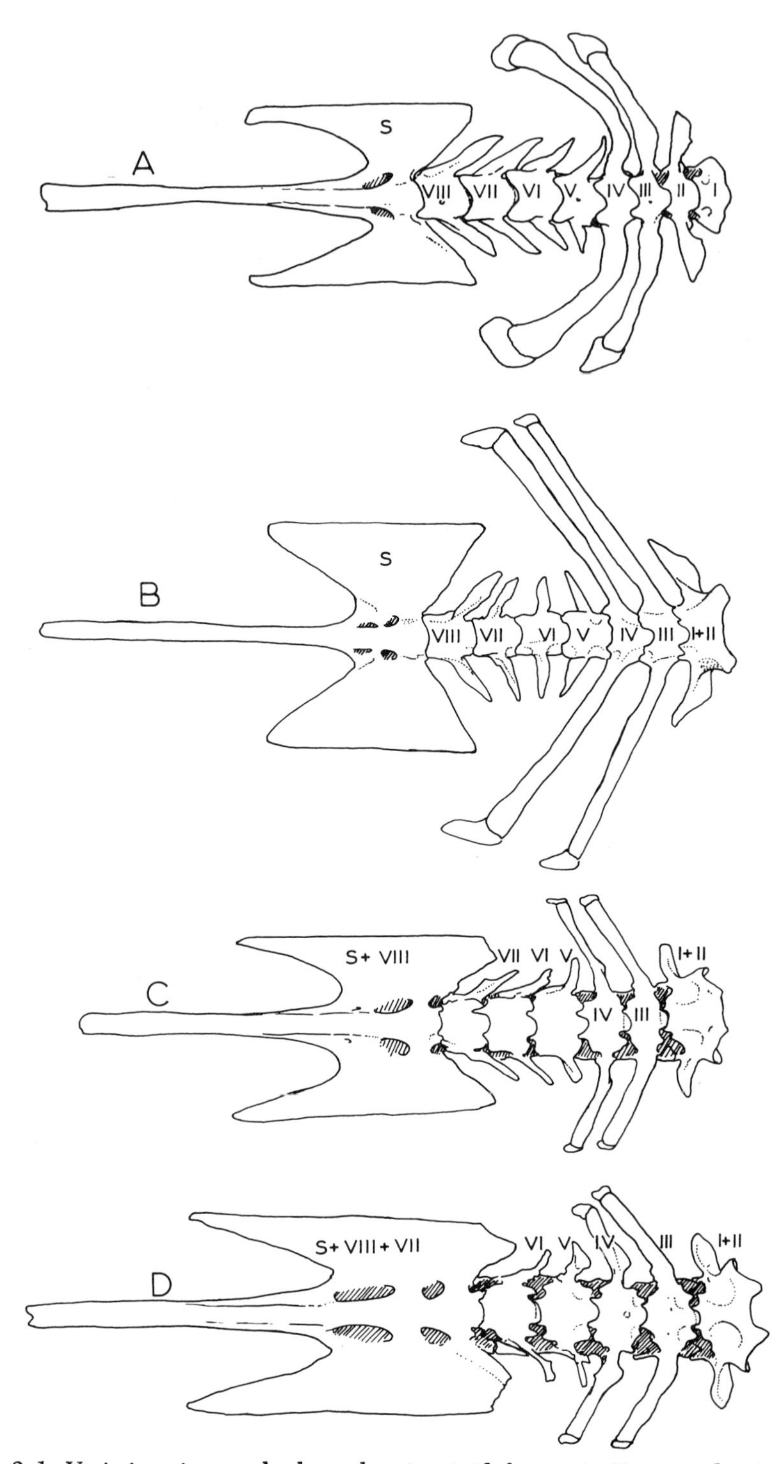

Figure 3–1. Variations in vertebral number in pipid frogs: A, *Xenopus laevis*, JDL-S 228; B, *Pipa pipa*, JDL-S 1309; C and D, *Hymenochirus boettgeri*, JDL-S 799 and 790, respectively. S, sacrum; roman numerals, vertebrae numbers.

procoelous individuals of some ranids and rhacophorids but regarded the condition as anomalous. Laurent (1940, 1941, 1942, 1943*a*, 1943*b*) reported apparent consistency of procoely in a few ranids and several rhacophorids. Liem (1969) found that 37 of 54 species of rhacophorids were procoelous and that the remainder, as well as all hyperoliids and most ranids, were uniformly diplasiocoelous.

Griffiths (1959*b*, 1963) discarded in large measure the use of amphicoelous, opisthocoelous, procoelous, and displasiocoelous vertebral columns in his classification, although he summarized developmental data for some groups. He concluded that only the opisthocoelous character state could be regarded as significant, because the other character states were poorly defined or of variable occurrence. Griffiths reoriented the attention on the vertebral columns to the shape of the transversely sectioned centrum. He distinguished three types—ectochordal, holochordal, and stegochordal—and considered ectochordy to be primitive. Inger (1967) and Kluge and Farris (1969) devoted considerable attention to Griffiths' new character. Inger disagreed with Griffiths and contended that holochordy is primitive. Kluge and Farris argued in favor of Griffiths' position and rejected Inger's arguments, but they did not accord as much weight to the characteristic as had Griffiths and Inger.

Tihen (1965) thought that the evolutionary trends evident in the nature of anuran vertebral centra pointed to the following sequence: a first stage of free intervertebral bodies (approximately equivalent to "amphicoely"); a second stage in which three patterns are evident—(a) opisthocoelous (intervertebral body fused to anterior end of centrum), (b) procoelous (intervertebral body fused to posterior end of centrum), and (c) amphicoelous (notochord persistent, not divided by connective tissue arcs). Tihen regarded the "anomocoelous" condition (procoelous with the sacrum and coccyx fused) and diplasiocoelous condition (procoelous, except for a biconcave eighth vertebra and biconvex sacral centrum) as minor deviations of the procoelous condition. Kluge and Farris (1969) agreed with Tihen's analysis. The "amphicoelous" condition (notochordal tissue not invaded by connective tissue arcs) is not equivalent to the character state in which the intervertebral bodies are free but is precedent in ontogeny. Griffiths (1959*a*) considered the trait paedomorphic in sooglossids, which reflects his opinion that sooglossids are very advanced frogs.

The nature of the centrum is here coded as follows: notochord persistent, not invaded by connective tissue arcs, 0; intervertebral bodies free in adults, 0; intervertebral bodies fused to anterior end of centrum (opisthocoely), 2; and intervertebral bodies fused to posterior end of centrum (procoely, but including "anomocoely" and "diplasiocoely"), 2. The coding of this character is not as judicious as it might be. The procoelous character state includes two groups, one of which includes pelobatines, pelodytids, and many myobatrachid frogs in which the intervertebral bodies are observably free in immature specimens (Boulenger, 1897) but fused in adults. The other group includes the remaining procoelous (and diplasiocoelous) anurans in which the intervertebral body is not a discrete element in immature individuals.

Griffiths (1959*a*, 1963) regarded dendrobatids as ranoids because they exhibit firmisternal pectoral architecture and because he thought they have a ranoid pattern in the disposition of the distal tendons of the thigh. Noble (1922, 1926*a*, 1931) and Lynch (1969) consider dendrobatids derivatives of elosiine leptodactylids and regard the pectoral girdle and thigh musculature character states exhibited by dendrobatids convergent with the states seen in ranoids. Griffiths argued that the synsacry—fusion of eight presacral and sacrum—in *Dendrobates* prevented consideration of the nature of the vertebral centra in dendrobatids when assessing their relationships (i.e., bufonoid or ranoid, approximately equivalent to pre-Parker Procoela or Diplasiocoela). Griffiths, however, either ignored or was not aware of the procoelous condition exhibited by most dendrobatids (Noble, 1922, 1931; Lynch, personal observation).

5. Atlantal intercotylar distance. The atlantal cotyles are narrowly separated in most of the genera of archaic frogs and the transitional groups but are widely separated in the majority of advanced frogs, with the exception of bufonids, rhinodermatids, and some leptodactylids. The separation of the cotyles reflects the increase in the intercondylar distance, although the intercotylar space is more readily measured than is the intercondylar distance. Direction of change is presumed to be from closely juxtaposed cotyles (0) to widely separated cotyles (2). There is considerable intergeneric variation in this character among leptodactyloids, but the character states appear to be diagnostic at the subfamily or tribe level in all families of frogs.

6. Dilation of the sacral diapophyses. Early classifications utilized the dilation or nondilation of the sacral diapophyses as a major character. Noble (1922) largely dismissed the character as being of major significance, although he frequently cited it as an indicator of relationships. Parker (1934) argued that broadly dilated sacral diapophyses were primitive and rounded, or nondilated, diapophyses were derived. He based his argument on the assumption that early frogs were swimmers or ambulatory, rather than saltating, and the shape of the diapophyses were functionally integrated with the type of movement. Tihen (1965) regarded the evolutionary trend of the shape of the diapophyses to be from broadly dilated to round. Broadly dilated sacral diapophyses characterize the four archaic families, the three transitional families, and only a few advanced families (e.g., Bufonidae, Rhinodermatidae, Sooglossidae, Microhylidae, Hylidae, and Centrolenidae). Within the leptodactyloids (Myobatrachidae and Leptodactylidae) there is a transition from broadly dilated diapophyses through moderate dilation (myobatrachids and primitive leptodactylids) to round sacral diapophyses (most leptodactylids). Some advanced frogs (all Microhylidae, Sooglossidae, and the questionable ranids of the subfamily Scaphiophryninae) have broadly dilated sacral diapophyses.

7. Presence of transverse processes on the coccyx. The coccyx of anurans is the result of fusion of the caudal vertebrae. A few anurans exhibit discrete ossification centers in the development of the coccyx (Griffiths, 1963). The presence of transverse processes suggests that the frogs exhibiting the processes are at an earlier stage of evolutionary development, not having reached the stage where the processes are lost in the process of simplifying the postsacral elements of the vertebral column. It is obvious that processes may be redeveloped to serve as specialized muscle attachments. However, until examples of redevelopment are advanced, the presence of transverse processes is coded as primitive (0) and the absence as derived (2).

8. Presence of ribs. The presence of free ribs in some frogs and the absence of free ribs in most frogs, coupled with the presence of free ribs in the majority of the remaining extinct and extant Amphibia, have led herpetologists to conclude that free ribs are primitive and their absence derived. Noble (1924*b*) pointed out that the ribs were free in young pipids, and Nevo (1968) recorded free ribs in early Cretaceous pipids. This character has been divided into two character states: (a) ribs free in adults and subadults or free in subadults, and (b) ribs not free at any stage in the development of the individual (Noble, 1931; Inger, 1967). Kluge and Farris (1969) scored all frog families, except the Ascaphidae,

Discoglossidae, and Rhinophrynidae, as having ribs absent in adults but present in subadults; their character-state distribution is reminiscent of the proposal by Féjérvary (1918) that all frogs have ribs, but he regarded the transverse processes as ribs. They assumed that the ossification center at the end of the transverse process was a rib and noted the absence of the center in *Rhinophrynus*. However, it is debatable that the center they observed is a rib remnant. To follow their argument is to argue that ascaphids, discoglossids, and pipids have a second "rib" at the distal end of the first (Figure 3–2). The "center" observed by Kluge and Farris cannot be categorically pronounced a rib without considerable research on the subject. My observations suggest that the "center of ossification" is merely a cartilaginous cap at the end of the transverse process or rib, a cap that picks up alizarin stain as do most calcified cartilages.

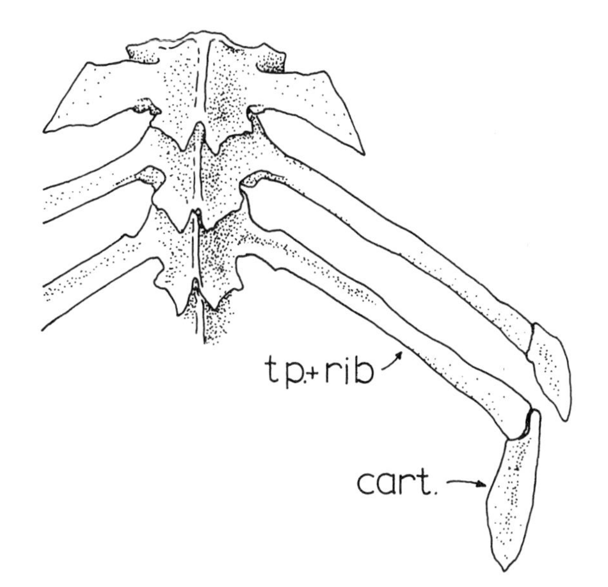

Figure 3–2. Anterior vertebrae of *Pipa pipa*. Tp.+rib, transverse process + rib; cart., cartilaginous cap.

I have coded the presence of discrete ribs in adults and subadults as primitive (0). I have subdivided the character state described as ribs not discrete in adults and subadults into two character states: (a) the transverse processes of the anterior presacral vertebrae are elongate, or (b) the transverse processes of the anterior presacral vertebrae are not elongate, that is, approximately as long as the sacral diapophyses. Character state (a) is coded as 2, character state (b) as 4. Long transverse processes of the anterior presacral vertebrae are found in the transitional groups, except Myobatrachinae, whereas nonelongate processes occur in rhinophrynids and all advanced frogs, except for some bufonids. The second character shift (state *a* to state *b*) is a second-degree characteristic.

9. Length of transverse processes of posterior presacral vertebrae. Hecht (1960) considered short transverse processes of the posterior presacral vertebrae characteristic of primitive frogs. The transverse processes are short and directed very strongly anteriorly in all archaic frogs and many transitional frogs, especially pelobatids and pelodytids. Moderate length transverse processes are found in a broad spectrum of frogs as are long (as long as the sacral diapophyses) transverse processes. My data do not suggest that the latter two character states are discrete from one another, but shorter transverse processes are roughly correlated with more primitive groups. Short processes are considered primitive (0), and long processes derived (2).

10. Degree of imbricateness of neural arch. Griffiths (1963) and Tihen (1965) regarded imbricate neural arches (spinal canal completely roofed) as primitive (0) and exposure of the nerve canal as derived (2). In large measure this character reflects degree of ossification, but it is not entirely dependent on that factor. *Notaden,* an Australian myobatrachid, exhibits extensive reduction of bone and has imbricate neural arches. I agree with Tihen that the shift from state 1 (imbricate) to state 2 (nonimbricate) is not a quantum jump but a gradual one, and hence the coding of this trend is somewhat debatable. The character is retained because selecting some characters that reflect evolutionary trends will point up relationships rather than enhance the distinctions between groups.

11. Pectoral girdle architecture. Inger (1967), like Noble (1922, 1926*a*, 1931), used a very loose definition of arcifery and firmisterny and failed to recognize the importance of Griffiths' (1963) analysis of the architecture of the girdle. Griffiths redefined the two terms in order not to depend solely on the degree of overlap of the epicoracoidal horns, and he pointed out that the two girdle types may have quite distinct developmental patterns. Arcifery (*sensu* Griffiths) is coded as primitive (0), and firmisterny as derived (2). The pseudofirmisterny exhibited by a number of bufonids and leptodactyloids is not coded as distinct from arcifery, except in the case of apparent true firmisterny in dendrobatids.

12. Prezonal elements. The presence or absence of the omosternum has been used in defining family groups of frogs from the earliest serious attempts at producing classifications (Cope, 1864). The presence of prezonal elements is coded as primitive (0), and the absence as derived (2). Osseous prezonal elements, such as those seen in many ranids, are not coded as a separate character state. Few groups universally lack prezonal elements, although the Centrolenidae are characterized by the absence of prezonal elements and very few bufonids (e.g., *Bufo haematiticus* and *Nectophrynoides*) possess them. I have not coded the presence of an osseous omosternum as a separate character state, although to do so would increase the apparent advancedness of many ranoids because the osseous element is apparently derived from the cartilaginous character state. Osseous prezonal elements are found in very few nonranoids (e.g., the leptodactyline *Lithodytes*), but they are found in all members of the ranoid subfamilies—Arthroleptinae, Astylosterninae, Cacosterninae, Hemisinae, Hyperoliinae, Mantellinae, Petropedetinae, Raninae (including the Cornuferinae of Noble, 1931, *fide* Inger, 1966), and Rhacophorinae—except the Scaphiophryninae, which may be more appropriately regarded as a microhylid group, and the Sooglossidae.

13. Postzonal elements. *Brachycephalus* (not assigned to a family here, although usually considered an atelopodid-bufonid), Hemisus (Hemisinae, Ranidae), and *Rhinophrynus* (Rhinophrynidae) are peculiar among frogs because they lack the sternum. The absence of the sternum is coded as derived (2), and the presence as primitive (0). The postzonal elements are degraded but not lost in many ranoids (Noble, 1924*a*). I have not coded as separate character states the nature of the sternum—whether it is cartilaginous or has an osseous element (sternal style or metasternal element). The osseous sternal element is found in some pelobatids, one subfamily of leptodactylids, and several ranoid groups (Cacosterninae, Mantellinae, Petropedetinae, Raninae, and Rhacophorinae).

14. Zonal elements. The reduction and loss of anterior portions of the zonal region of the pectoral girdle in many microhylid frogs has been used in deducing relationships and defining genera in the family (Carvalho, 1954; Parker, 1934). Most authors have recognized that the reduction from osseous clavicles to cartilaginous precoracoidal elements to loss of precoracoidal elements is a trend that has been repeated many times within the family (Parker, 1934; Griffiths, 1963). The reduction is also seen in the ranid subfamily Cacosterninae. The presence of osseous elements is considered primitive (0), the reduction to cartilaginous elements derived (2), and loss of the elements more derived (4).

15. Suture between clavicle and scapula. In an effort to quantify the degree of uncleftness of the scapula and clavicle-to-scapula length ratio characters used by Griffiths (1963), Kluge and Farris (1969) utilized the degree of anterior overlap of the clavicle on the scapula. They coded overlap as primitive and nonoverlap as derived. The two character states are coded as 0 and 2, respectively. The overlapping clavicle and scapula character state is found in ascaphids, discoglossids, palaeobatrachids, pipids, and rhinophrynids, but not in other frogs.

16. Phalangeal formulae. The widespread phalangeal formulae among anurans are 2–2–3–3 and 2–2–3–4–3, which are considered the primitive state (0). Reduction in the formulae is rare among frogs and characterizes only very small groups (e.g., Rhinophrynidae). Increase in the phalangeal formulae is more common and characterizes some large groups (e.g., Hylidae). Hylid, centrolenid, and pseudid frogs are the only arciferal frogs that have intercalated cartilages or bones resulting in phalangeal formulae of 3–3–4–4 and 3–3–4–5–4. Pseudids are distinctive in having long intercalary bones (Savage and Carvalho, 1953), but they have the same functional relationships between the terminal and penultimate phalanges that are seen in all other frogs having intercalated elements (Lynch, unpublished data). The firmisternal frogs with intercalated elements include one microhylid (Phrynomerinae, *Phrynomerus*) and three ranoid groups (Hyperoliidae or Hyperoliinae, Mantellinae, and Rhacophorinae), all of which are moderate in numbers of species and genera. Frogs with intercalated elements are scored as derived (2).

17. Tarsal bones, astragalus, and calcaneum. All but three living genera of frogs have the two elongated tarsal bones free. *Centrolene* and *Centrolenella* (Centrolenidae) and *Pelodytes* (Pelodytidae) have the elements more or less fused (Taylor, 1941, 1951; Eaton, 1958); the fusion is considered a derived (and specialized) character state (2) because the nonfused condition is widespread among anurans and occurs before the fusion in the ontogeny of centrolenids and pelodytids.

18. Number of tarsalia. The primitive character state (0) is the freedom of tarsalia 1 and 2; tarsalia 3 and 4 are fused. In this state there are three independent bones. The derived character state (2) is freedom of tarsalia 1, with tarsalia 2, 3, and 4 fused as a single element. Three independent tarsalia is considered a primitive character state, compared to two (Howes and Ridewood, 1888). My data for this character are incomplete, and groups have been scored with very few genera examined. The significance of the characteristic is somewhat diminished because several groups have not been well investigated; however, data are available for representatives of all but a few small and rare groups (e.g., Scaphiophryninae, Sooglossidae, among others). Three elements are found in all archaic families, except Pipidae and Rhinophrynidae; they occur in some transitional families, namely, Pelodytidae and some Myobatrachidae; and they are found in only a few advanced families, namely, Rhinodermatidae and several family groups of the Ranidae (e.g., Hyperoliinae, Mantellinae, and part or all of the subfamilies Astylosterninae and Cacosterninae). This character was investigated very early (Howes and Ridewood, 1888), and with the exceptions of work by Laurent (1940, 1941, 1942, 1943*a*, 1943*b*), Liem (1969), and myself, the complex has been ignored. A thorough investigation of the tarsalia in all anurans would be a welcome and important contribution to understanding the relationships of ranoid frogs.

19. Parahyoid ossification. The parahyoid ossification occurs in all Ascaphidae, Discoglossidae, Pelodytidae, and Rhinophrynidae, and in spite of its limited occurrence, its presence is considered primitive (Trewavas, 1933; Walker, 1938). All other anuran groups lack the median element and exhibit the derived character state (2).

20. Completeness of cricoid ring. The widespread nature of a complete ring-like cricoid cartilage in frogs is considered the primitive character state (0), although some frogs (e.g., ranids) have a ventrally incomplete ring early in development (Griffiths, 1959*a*). A middorsal gap of the cricoid ring is considered derived (2) and is found only in pelobatids, pelodytids, and rhinophrynids. A ventral cricoid gap occurs in all myobatrachines and in sooglossids; this condition is also considered derived (2).

21. Maxillary arch dentition. The presence (primitive character state, 0) or absence (derived character state, 2) of teeth on the maxillary arch (maxilla and premaxilla) was considered a characteristic of major proportions in anuran classifications during the nineteenth century; the characteristic is now considered of very little worth, in spite of the significance of edentuloseness in an entire family (Bufonidae). Loss of teeth on the maxilla and premaxilla has been regarded as a generic or specific character in almost all groups, except the Bufonidae and Microhylidae.

22. Presence of skull bones. Anurans exhibit increased specialization and advancement among the Amphibia in the reduction of the number of bones of the skull. The skull elements most frequently lost are the quadratojugals, columellae, palatines, and prevomers. Few groups have added bones, but the casque-headed hylids are a notable exception (Trueb, 1970).

The palatines and prevomers tend to be lost by groups of related genera, whereas loss of the quadratojugals and columellae occurs sporadically among the genera and families of frogs. Parker (1934) and Lynch (1969) summarized the cases of loss of the columellae among frogs. Summaries of the loss of other elements are not available but are widely scattered in the literature. Loss of elements is regarded as derived. A score of 2 is provided for each bone lost.

23. Amplexus position. Although minor variations in clasping position have been reported for many species of frogs and although undue stress and significance has been accorded the postovulatory amplectic position of *Alytes* (Discoglossidae), anurans exhibit two patterns of amplexus—axillary or inguinal. Inguinal amplexus characterizes all genera of the Ascaphidae, Discoglossidae, Pipidae, Rhinophrynidae, Pelobatidae, and Pelodytidae. I (Lynch, 1969) characterized myobatrachids as having inguinal amplexus, but Ian Straughan (personal communication 1970) informs me that no member of the Cycloraninae is known to exhibit inguinal amplexus, although most literature reports are to the contrary. All members of the Myobatrachinae exhibit inguinal amplexus. Data are not available for *Heleophryne* or for the sooglossids. In contrast, all other frogs exhibit axillary amplexus, with the possible exceptions of *Breviceps* (Microhylidae) and two of the three species of *Batrachyla* (Leptodactylidae). Data are lacking for most microhylids and for some small groups of ranids.

The postovulatory cephalic amplexus of *Alytes* is stressed by authors seemingly convinced that behavioral traits are not of major systematic importance. Boulenger (1897) pointed out that preovulatory amplexus and amplexus during ovulation in *Alytes* is inguinal, as it is in the other discoglossid genera.

Inguinal amplexus is considered primitive (0), and axillary amplexus derived (2). The distribution of these character states suggests that the direction cited above for this evolutionary trend is correct. Frogs exhibiting inguinal amplexus lay fewer eggs than do frogs exhibiting axillary amplexus (except in the cases where direct development obscures this pattern) and breed in quiet water unless they have developed major specialization (intromittant organ in *Ascaphus*, terrestrial eggs in *Leiopelma* and some Australian myobatrachids). Axillary amplexus results in a shorter distance between the vents of the mating pair and presumably allows advanced frogs to breed in a greater diversity of habitats, including moving water of streams, than can frogs exhibiting inguinal amplexus. Rabb (this volume) suggested that inguinal amplexus restricts the behavioral repertoire and hence may be primitive.

24. Pupil shape. Living frogs, in direct light, exhibit four shapes of the pupil: triangular with a ventral angle, round, a horizontal slit or ellipse, and a vertical slit or ellipse. The triangular pupil occurs in three of the four discoglossid genera (*Barbourula, Bombina,* and *Discoglossus*); *Alytes* has vertical pupils. Round pupils are found in pipids and some microhylids. The horizontal slit is found in all advanced families, except four genera of hylids, six genera of leptodactylids, a variety of microhylids (many with round pupils), all but a few ranids (Astylosterninae, Hemisinae, and all but one genus of the Hyperoliinae, *Hyperolius*), and in several genera of the Myobatrachidae (all Myobatrachinae, except *Uperoleia,* and all Cycloraninae, except *Heleioporus, Mixophyes,* and *Neobatrachus*). Vertical pupils occur in ascaphids, *Alytes* (Discoglossidae); rhinophrynids; pelobatids; pelodytids; the myobatrachids *Heleophryne, Heleioporus, Mixophyes, Neobatrachus,* and *Uperoleia;* the leptodactylids *Caudiverbera, Hydrolaetare, Hylorina, Lepidobatrachus, Limnomedusa,* and *Telmatobufo;* the hylids *Agalychnis, Pachymedusa, Phyllomedusa,* and *Nyctimystes;* the microhylid *Dyscophus;* and the ranids of the subfamilies Astylosterninae, Hemisinae, and Hyperoliinae, except *Hyperolius.*

The distribution of the vertical pupil (i.e., its presence in most archaic and transitional families) suggests that this character state is primitive and the other character states derived. Most authors have presumed that the vertical pupil is derived because the character state is less common than the horizontal pupil. Vertical pupils are associated with almost exclusively nocturnal or crepuscular frogs, whereas frogs having horizontal, round, or triangular pupils exploit a variety of habitats. An argument could be advanced that the frogs with vertical pupils are specialized in habitat selection and the state of the character of pupil shape, but it is equally defensible to argue that the vertical pupil is primitive. I contend that the character-state distributions support the latter hypothesis and have coded vertical pupils as primitive (0) and the other character states as independently derived (all 2).

25. Tongue. The absence of a tongue is restricted to pipids, although other totally aquatic frogs exhibit a reduction in tongue size *(Batrachophrynus).* Loss of the tongue represents a major character modification, but it is scored as derived (2), and the presence of a tongue as primitive (0).

26. Developmental pattern. Free-living aquatic larvae represent the primitive amphibian and anuran condition. Departures from this pattern are derived. The primitive condition in anurans and amphibians is to have relatively numerous, small, pigmented eggs laid in water. Departures from this pattern are also derived. However, numerous specializations all tending toward direct development occur in many lines of anurans. I have divided this character into five character states here termed stages 1 through 5. Coding is as follows: Stages 1, 2, and 3 are coded as primitive (0), stage 4 as derived (2), and stage 5 as derived from 4 (4). Stages 1, 2, and 3 are coded equally because the frogs can be considered in characters 27 and 28 (tadpole morphology), whereas stages 4 and 5 cannot. The five stages are described below:

Stage 1. Frogs in this stage lay numerous, small (about 1 mm in diameter), pigmented eggs in water. The entire developmental sequence occurs in water.

Stage 2. Frogs exhibiting this stage lay usually a smaller number of larger eggs that are not pigmented and are laid in water. The entire development occurs in water.

Stage 3. Frogs exhibiting this stage lay relatively few, nonpigmented eggs in sites other than water—on land or on vegetation. The eggs hatch outside of the aqueous environment, but the tadpole then enters water, where its completes its development. Examples of frogs that exhibit this stage include *Alytes, Heleophryne,* the Batrachylini (Leptodactylidae, Telmatobiinae), most if not all phyllomedusine hylids, centrolenids, and many ranoids.

Stage 4. This stage represents a considerable specialization over stage 3. The relatively large, few in number, and nonpigmented eggs are laid in terrestrial situations, and upon hatching, the larvae complete their development in the decomposing jelly mass. The larvae are not aquatic. Examples include the grypiscine (Leptodactylidae, Telmatobiinae) genera *Cycloramphus* and *Zachaenus,* the melanobatrachine *Hoplophryne* (Microhylidae), and very few other genera.

Stage 5. This stage is complete direct development. It may take two avenues. One is represented by several species of hylids (*Gastrotheca* and allies) and some pipids (e.g., *Pipa pipa*); the eggs are large, few in number, and nonpigmented, and they develop in the skin or pouch on the back of the parent. The other is the more common case—the few, large, and nonpigmented eggs are laid in a terrestrial situation, the larvae complete their development within the egg membranes, and upon hatching, metamorphosed frogs leave the eggs. The second avenue is exhibited by the ten genera of the Eleutherodactylini (Leptodactylidae, Telmatobiinae), the *Leptodactylus marmoratus* group, some Australian myobatrachids, *Leiopelma* (Ascaphidae), many microhylids (Parker, 1934; Carvalho, 1954), sooglossids, and some ranoids (e.g., Arthroleptinae).

Ascaphids exhibit stages 1 and 5. Discoglossids exhibit stages 1, 2, and 3. Pipids exhibit stages 1, 3, and 5. Rhinophrynids, pelobatids, and pelodytids exhibit stage 1. Myobatrachids exhibit stages 1, 2, 3, and possibly 4 and 5. Leptodactylids exhibit stages 1, 3, 4, and 5. *Rhinoderma* exhibits what is probably best considered an impressive specialization of stage 3. Pseudids exhibit stage 1. Bufonids exhibit stage 1, 2 *(Atelopus),* and the peculiar intrauterine pattern of *Nectophrynoides* (Orton, 1949). Centrolenids and phyllomedusine hylids exhibit stage 3; other hylids exhibit stages 1 (some depart slightly, laying numerous, small, pigmented eggs on leaves above water, e.g., *Hyla ebraccata* among others), 2, 3 (*Gastrotheca marsupiatum* complex), and 5 (Gastrothecine pattern). Dendrobatids exhibit stage 3. Sooglossids exhibit stage 5. The patterns within the Microhylidae are only partially known. Phrynomerines and most microhylines exhibit stage 1, although some microhylines exhibit stage 5. Insofar as is known, all other microhylids exhibit stage 5, except the Melanobatrachinae, which exhibit stage 4. The enigmatic firmisternal Scaphiophryninae exhibit stage 1, 2, or 3 (data incomplete). Arthroleptine ranids exhibit stage 5, as do some members of the subfamily Cacosterninae. Other ranids, rhacophorids, or hyperoliids exhibit stages 1, 2, or 3, but most or all ranids often assigned to the Cornuferinae (e.g., *Platymantis*) exhibit direct development (stage 5).

Orton (1949) incorrectly regarded direct development as uncommon, although it is widely distributed among frog families. The large Neotropical group, including *Eleutherodactylus* and its allies, and the speciose Indo-Australian ranids (*Platymantis* and allies) and microhylids account for a relatively impressive percentage of frog species. The apparent rarity of direct development is best explained as *naïveté* of holarctic-oriented herpetologists, rather than an assessment of frogs on a world-wide, particularly tropical, basis.

27. Larval mouthparts. Orton (1953), Hecht (1963), and Starrett (this volume) considered the absence of mouthparts primitive because the absence of papillae, denticles, and a horny beak is a simpler character state than the presence of such structures and is precedent in development. Tihen (1965) and Kluge and Farris (1969) considered the absence of mouthparts as a derived, probably paedomorphic, character state. They considered Orton's Type III larvae as having the primitive character state (horny beaks present, denticles present in multiple rows, lips and papillae present); tadpole Types I, II, and IV were considered derived. Types I and II differ from III and IV in lacking papillae, denticles, and horny beaks. I differs from II in having barbels and in lacking lips. Type IV differs from Type III in having single rows of denticles instead of multiple (double or triple) rows. The tadpole types as originally defined also differed in spiracle disposition (Orton, 1953). For purposes of the present study, multiple rows of denticles on each "ridge" (row of multiple rows) is considered the primitive character state (0), and the single row of denticles is considered derived (2). Loss of the denticles and beaks is considered more derived (coded as 4).

This characteristic is discussed more fully by Starrett (this volume), but in view of our differences in coding the character states, a few comments here are germane. The developmental sequence may in fact repeat the phylogenetic sequence in the development of the character states of this characteristic. Pipoids have the most simple mouthparts (internal and superficial structures), except the presence of barbels, and are reminiscent of a stage early in ontogeny. The mouthparts of microhylids are nearly identical with those of pipoids, except for small changes in the internal support of mouthparts (addition of infralabial cartilage), whereas the mouthparts of all other frogs are considerably more complex both internally (cartilages and musculature) and superficially (beaks and denticles). The morphological similarities in mouthparts between pipoids and microhylids are reflected in the similarity of food gathering of these two groups—both are filter feeders.

The similarities in food gathering could be argued to be a reflection of morphological constraints imposed by phylogeny, or the morphological similarities could be argued to be specializations to exploit the same food source. The former argument suggests the similarity stems from relationship, while the latter suggests the similarity stems from convergence.

Testing the two alternatives is difficult. It requires that we distinguish true primitiveness in character states from paedomorphy in character states (if the phylogenetic sequence coincides with the ontogenetic sequence). The only means of distinguishing the alternatives is to compare the distributions of all characteristics relative to the distribution of a particular characteristic and then to select the most parsimonious phylogeny as being the correct one (see Section IV). To assume that the microhylids are in fact an early branch of the Anura requires many more cases of parallelism among other characteristics than result if microhylids are not considered one of the earliest anuran groups (see Section IV).

28. Spiracle number and position. Anuran larvae have one or two spiracles, except for some frogs that exhibit direct development (Orton, 1949; Kluge and Farris, 1969). Orton (1953) assumed that the paired spiracles (Type I) of pipids and rhinophrynids represented a primitive condition, as did the single midventral spiracle (Types II and III) of microhylids and ascaphoids, and that the single sinisteral spiracle (Type IV) seen in most advanced and transitional families was derived. Starrett (this volume) argues that paired spiracles are primitive because two openings occur early in the development of all frogs.

Kluge and Farris (1969) presented an analysis of spiracle position and number in frogs. They pointed out that the paired spiracles of Type I larvae (Pipidae and Rhinophrynidae) were not homologous to the spiracles of fish, salamanders, or most other anurans, and termed them pseudospiracles. The observation further strengthened the argument that pipids and rhinophrynids are more closely related to one another and less closely related to other frogs. They considered the pseudospiracular condition derived from their condition 4. They also pointed out that the Patagonian leptodactylid *Lepidobatrachus* has paired spiracles, but Starrett (this volume) associates *Lepidobatrachus* with Type IV larvae. Their analysis of the remaining character states (1-7) is complicated, and the reader should refer to their paper. It can be noted here that the presence of a median, ventral spiracle located at the posterior rim of the branchial chamber (their condition 4) is considered derived from the absence of a spiracle *(Leiopelma)*; that the sinistral spiracle is derived from a median, ventral condition; that loss of a spiracle, midventral position, and slightly sinistral of midventral position are derived from the sinistral condition; and that a posterior extension of the midventral spiracle toward the vent is derived from condition 4. I have not retained the complex coding employed by Kluge and Farris, although if data were available their coding is more judicious than mine. The character states could be coded following a condensation of Kluge and Farris' (1969) analysis as follows: single, midventral spiracle (0), single, sinistral spiracle (0), paired "pseudospiracles" (2). An alternative is to code paired spiracles as primitive (0), based on ontogenetic data, and a single spiracle as derived (2). I have followed the latter alternative and have not separated the position states of single spiracles. Starrett has changed the emphasis of this characteristic to include developmental region of the forelimbs relative to the branchial chamber.

29. Outer metatarsal tubercle. The character states are absent or present. The archaic frog families, the pelobatids, and pelodytids uniformly lack an outer metatarsal tubercle, as do many other frogs. However, an outer metatarsal tubercle occurs only among advanced frogs and a few myobatrachids (mostly Myobatrachinae). I suggest that the absence is primitive (0) to the presence (derived 2) of an outer metatarsal tubercle, although there are some examples that suggest secondary loss (*Eupsophus juninensis;* Leptodactylidae, Telmatobiinae).

The presence or absence of the outer metatarsal tubercle is not related to the degree of fossoriality exhibited by various frog species. Fossorial frogs with an outer metatarsal tubercle generally have large, spade-like outer metatarsal tubercles, as well as spade-like inner metatarsal tubercles, whereas fossorial frogs lacking an outer metatarsal tubercle have a single, large, spade-like tubercle.

30-34. Myological Characters

The following five characters are included with some reluctance because of the demonstrated variability and plasticity of myological characters (Kluge and Farris, 1969). Some of these characters are included because of historical use and because homoplasic development is less apparent in them than it is in others. Kluge and Farris rejected some characters because they exhibited near continuums and were trend characters. I agree with Tihen (1965) that such characters are often more useful in deducing phylogenies than are the more often used discontinuous-varying characters.

30. Tail muscle; *M. caudaliopuboischiotibialis.* Since Noble's (1922) work on anuran musculature, herpetologists have taken note of the supposedly homologous tail muscle of ascaphids and concluded that possession of the muscle is primitive to the absence of it. I have coded the character states in this manner. However, it should be borne in mind that the *m. caudaliopuboischiotibialis* of ascaphids may be a coccygeal head of the *m. semimembranous* (Ritland, 1955) and not homologous with the tail muscles of urodeles (Kluge and Farris, 1969).

31. *M. semitendinosus–m. sartorius.* Noble (1922) and many subsequent authors have used the nonseparation (primitive, 0) or separation (derived, 2) of the *m. sartorius* and *m. semitendinosus* as a major character in illustrating relationships among frogs. The muscles are not separated in archaic frogs, except for *Rhinophrynus,* although a tendency for partial separation is seen in *Discoglossus* and *Xenopus* (Dunlap, 1960; Noble, 1922, 1931). The muscles are united in pelobatids and pelodytids but are separated in all other families. In some myobatrachids the two muscles are not completely divided. Kluge and Farris (1969) rejected this character as noncodable because variation is not entirely discrete between the two character states (i.e., united or entirely separated). They also apparently depended on their criticism of the use of the *m. adductor longus* to apply to the muscle complex discussed here. Inger (1967) used the presence (derived) or absence (primitive) of a *m. adductor longus* as a systematic tool in his analysis, but as Kluge and Farris (1969) properly pointed out, he was apparently unaware of the recorded variation in this character. The muscle is supposedly absent in the four archaic families but is present in *Discoglossus.* The muscle was supposedly present in all other families but is lacking in some pelobatids, leptodactylids, most rhacophorids (=Rhacophorinae, Mantellinae), and a variety of bufonids (Tihen, 1965; Kluge and Farris, 1969; Liem, 1969).

32. Accessory tendon of *m. glutaeus magnus.* The coding of this character follows Dunlap (1960), who reported the absence in primitive groups and the presence in advanced groups. Data for this character are meager, and its use must be considered tenuous.

33. Accessory head of *m. adductor magnus.* The accessory head of the *m. adductor magnus* is absent in primitive frogs, present but small in transitional families, and usually present among advanced frogs. The character was coded as absent (primitive) and present (derived), although data were far from complete.

34. Insertion of *Mm. sartorio-semitendinosus* relative to *m. gracilis.* Noble (1922) divided the "advanced" frogs into two groups—the Procoela and Diplasiocoela (=Bufonoidea and Ranoidea *auct.,* respectively)—on the basis of vertebral centra and the disposition of the tendon of the *m. semitendinosus* relative to the *m. gracilis*—(a) inserting or passing ventral or (b) inserting or passing dorsal to the *m. gracilis.* State (a) characterized the Procoela and state (b) the Diplasiocoela. State (a) also characterizes the archaic frogs and pelobatoids, with the following exceptions: tendon penetrates *Mm. gracilis* in *Rhinophrynus* (Rhinophrynidae) and several Myobatrachidae, and it passes dorsal to the *Mm. gracilis* in some myobatrachids (Parker, 1940). The tendon of the *m. semitendinosus* penetrates the *m. gracilis* in some but not all species of *Physalaemus,* in all dendrobatids, and in the elosiine *Crossodactylus* (Leptodactylidae). Sooglossids and ranoids (Microhylidae and Ranidae *s.l.*) exhibit state (b). Character state (a) is coded as primitive (0), and state (b) as derived (2).

35. Sacro-coccygeal articulation. This characteristic appears to be highly dependent on the nature of the notochord-centrum complex of the vertebral column. The character states are (a) monocondylar, (b) bicondylar, and (c) fused. State (c) is certainly derived but is as equally derivable from state (a) as it is from state (b). All examples exhibiting state (a), the presumed primitive character state, are not equal; some toads (e.g., *Mertensophryne*) have apparently evolved state (a) from state (b) by deletion of the sacral vertebra through the coccyx (Tihen, 1960, 1965). States (a) and (b) are recorded as intraspecifically variable in *Pelodytes* (Boulenger, 1897; Noble, 1924*a*). This example is suspect, however, because the character was checked by dissection rather than dry skeleton or alizarin preparation. State (c) is coded as character 3. State (a) is coded as primitive (0), and (b) as derived (2).

36. Parasphenoid bone. Salamanders lack posterolateral alae on the parasphenoid bone, but frogs possess them (Figure 3–3A). The exceptions to this statement are the absence of alae in pipids and rhinophrynids. In pipids the median rami of the pterygoids contact or nearly contact the parasphenoid and thus provide a ventral cover to the proötic usually covered by the parasphenoid alae. This region is exposed in *Rhinophrynus* because the median ramus of the pterygoid is extremely small or absent. The striking absence of parasphenoid alae in these two families cannot be readily explained because of a common adaptation—pipids are totally aquatic, and *Rhinophrynus* is one of the most specialized burrowing frogs known. Palaeobatrachids lack alae on the parasphenoid (Špinar, 1966), as do the earliest ascaphoid fossils, *Vieraella* and *Notobatrachus* (Estes and Reig, this volume). This characteristic is considered a first-degree characteristic; the absence of parasphenoid alae is coded as primitive (0), and the presence as derived (2).

37. Organ of Bidder. The presence or absence of an organ of Bidder in bufonoids was utilized by Davis (1936) to remove the Leptodactylidae from Noble's (1922, 1931) Bufonidae. Griffiths (1959*b*, 1963) characterized the Atelopodidae-Bufonidae complex as uniformly possessing an organ of Bidder at some stage of development. He noted the absence of the organ in *Brachycephalus, Geobatrachus, Rhinoderma,* the composite genus *Sminthillus,* and the dendrobatids—all members of Noble's (1922, 1926*a*, 1931) Brachycephalidae—but he retained *Brachycephalus* as an atelopodid because of the squamoso-otic architecture and the extensive epicoracoidal fusion. The persistence of an organ of Bidder is a paedomorphic trait consistent with other paedomorphic features of bufonids (Griffiths, 1959*b*).

38. Larval barbels. Orton (1953) characterized Type I anuran larvae in part by the presence of barbels. The only anurans exhibiting the character state of barbels present are the pipids having aquatic larvae, *Rhinophrynus,* and at least one member of the Palaeobatrachidae (Špinar, 1966). There are differences in the number of and placement of barbels (Figure 3–4). Pipids have one pair of lateral barbels, while palaeobatrachids and rhinophrynids have more than one pair. Špinar (1966) concluded that *Palaeobatrachus grandipes* had more than one pair (his figures suggest three or more pairs) of barbels lateral to the terminal mouth.

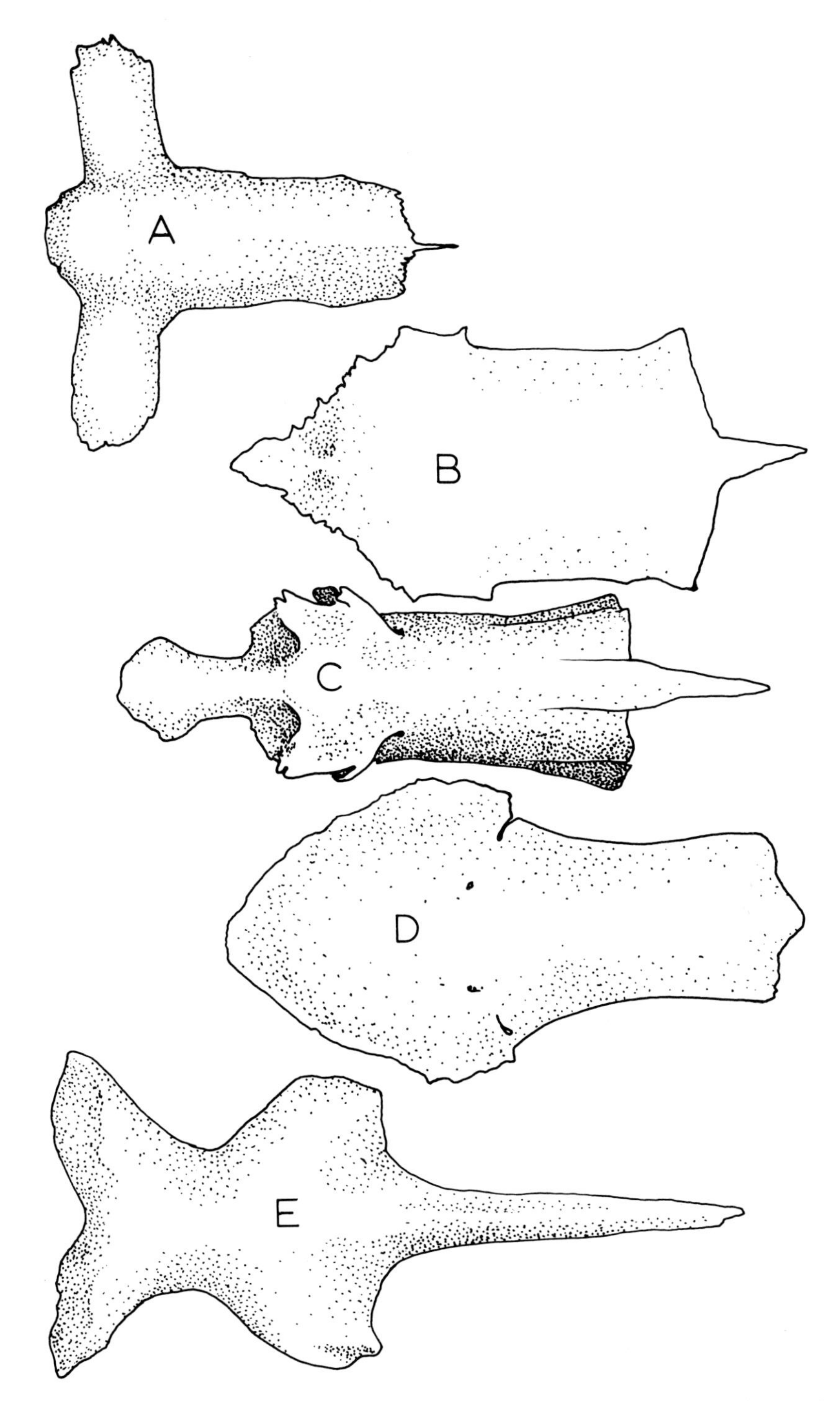

Figure 3–3. Parasphenoid bones in A, *Bombina variegata,* JDL-S 532; and pipoids B, *Pipa pipa;* C, *Xenopus laevis,* JDL-S 227, in which the parasphenoid is fused with the orbitosphenoid; D, *Hymenochirus boettgeri,* JDL-S 931; and E, *Rhinophrynus dorsalis,* JDL-S 1044.

Rhinophrynus has a ventral chin barbel, four pairs of lateral barbels, and one pair of barbels dorsal to the mouth (Orton, 1943). The restricted occurrence of the barbels suggests that the presence of barbels is derived (2) and the absence primitive (0).

III. WEIGHTING OF CHARACTERS

In recent years (since Sokal and Sneath, 1963), systematists have tended to use all characters equally and not weight some as more important than others. It is not my intent to debate the merits or drawbacks of either viewpoint, and my comments here serve only to explain my weighting of characters. When the monophyly of all derived character states can be questioned, equal weighting results in a less subjective classification than is obtained by unequal weighting. Character selection—be it by subjectively choosing more "fundamental" characters, by selecting characters at random, by virtue of noncorrelation, or as a function of one's experience with a given group—is a form of unequal weighting, as Inger (1958) observed, but it is of decreasing significance as more characters are utilized. Depending on how one groups characteristics, the classification proposed here (Section IV) is more or less heavily weighted. I have used only three larval traits, but I have used 24 osteological characteristics. The preponderance of osteological traits is defensible in that, in general, those are the only sort of traits one can use in comparing living groups to the true phylogeny (i.e., the fossil record).

Kluge and Farris (1969) employed Farris' (1966) concept of "conservatism of characters by constancy" in assigning variable weighting to their eleven characters. The merits of such a scheme are great, but theory is quite different from practice, as Kluge and Farris (1969) acknowledged. The variation within and among family groups for the 38 characters employed here is not fully known. My subjective judgments of relationships between groups is more dependent on the less variable characters than it is on the more variable ones (Section IV). Quantification of within-OTU and between-OTU variation is not deemed practical at this time; hence, the technique employed here is not termed quantitative phyletics.

Inger (1967) utilized a seemingly subjective "uniqueness" index to assess variable weighting of characters (Kluge and Farris, 1969). He also appears to apply Wilson's (1965) "uniqueness" concept to derived states within the group, whereas Wilson intended the concept for primitive character states. The technique employed here requires the following assumption: *The primitive character state arose once in the taxon at hand (Anura), whereas derived states may have arisen repeatedly among the divisions of the Anura.*

Assignment of direction of change (evolutionary trend) tends to make a strictly phenetic classification more phylogenetic. It requires some subjective judgment by the author and is not as simple as Kluge and Farris (1969) imply. They utilized much of the theory developed by Sporne (1954) and Wagner (1961) and summarized the methods of inference of primitive character states and their reliability (in order, 1–4), as follows:

1. The primitive state may be inferred from conditions seen in fossil forms.
2. Primitive state of a character for a particular group is likely to be widespread among representatives of closely related groups.
3. Primitive states are more likely to be widespread within a group than is any one advanced state.
4. Primitive states are likely to be associated with states of other characters known from other evidence to be primitive.

Individually, each of the criteria is hazardous, but collectively they allow a

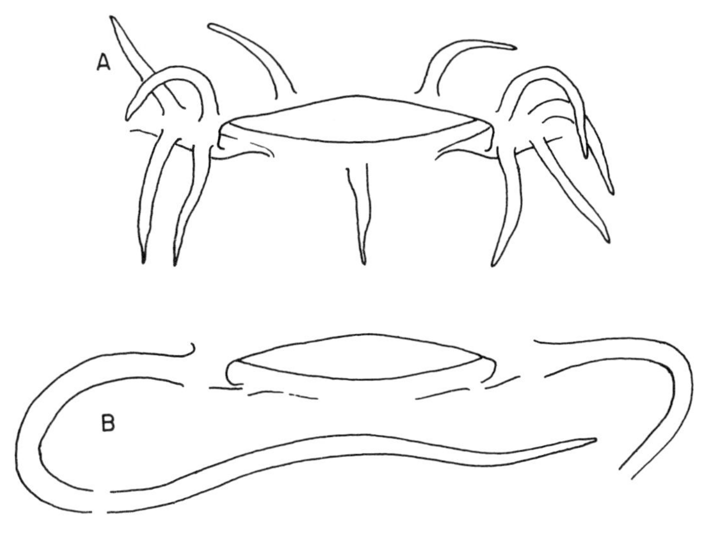

Figure 3–4. Mouths of pipoid larvae. A, *Rhinophrynus* (after Orton, 1943); and B, *Xenopus* (after Orton, 1953).

reasonably repeatable estimate. Criterion 2 is somewhat hazardous in that "closely related groups" is undefined and is best restricted to all Amphibia, rather than used within the Anura. Criterion 3 is likewise hazardous. Using that criterion alone, one could conclude that the presence of wings is primitive in insects or that choriovitelline and chorioallantoic placentae are primitive in mammals.

Kluge and Farris (1969) unduly criticized Inger's (1967) rejection of anuran fossils as instructive elements for determining primitive and advanced character states. Inger is correct in rejecting existing fossils as not especially meaningful and in asserting that we have no "truly primitive" anuran fossils. Fossil frogs are clearly related to existing groups or are so poorly preserved as to be of no use.

In projecting Figure 3–5, a graphic comparison of the relative primitiveness of each anuran group (43 family groups), all 38 characteristics (Section II) were used without regard for the distinctions between first-degree, second-degree, and third-degree characteristics. The variability within groups was partially compensated for by using intermediate values. In projecting a dendrogram (Figure 3–6), characteristics exhibiting little infrafamilial variation were used before those exhibiting more infrafamilial variation. Third-degree characteristics (specializations) were used; if they are excluded from the "degree of primitiveness" exercise (Figure 3–5), 23 groups become less advanced by values of 1 (1 microhylid group), 2 (11 groups), 3 (4 groups), 4 (4 groups, including pipids, pelobatids, and pelodytids), 5 or 6 (2 microhylid groups), or 8 (rhinophrynids).

IV. THE CLASSIFICATION OF ANURANS

The characteristics used herein to project a familial and suprafamilial classification of the Anura are discussed in Section II, and the raw data in the form of diagnoses for each of the seventeen extant families recognized here are summarized in sections V through VII. Two separate approaches are made with the same set of data. The first involves an inspection of the "degree of primitiveness" of each group, and the second is to construct a cladistic diagram for anurans based on the available data and the assumptions that evolution is irreversible and that the cladogram having the fewest evolutionary steps has a higher probability of being correct than one having more evolutionary steps.

Degree of Primitiveness

According equal weight to each of the 38 characteristics (except in those cases where two derived states were arranged in series), each of 43 extant family groups was scored for those 38 characteristics for which data were available. The total scores for each family-group varies from 10 (Ascaphidae) to a range of 53–59 (Melanobatrachinae, Microhylidae). A taxon with all character states derived would have a value of 82. Sums provide an index of the maximum "primitiveness" of each group. When data for a particular character are not known, the sum indicates maximum (not necessarily true) primitiveness. I have in those cases recorded a maximum and minimum range for the primitiveness of the group for which data are not available; this is the reason that some groups are represented

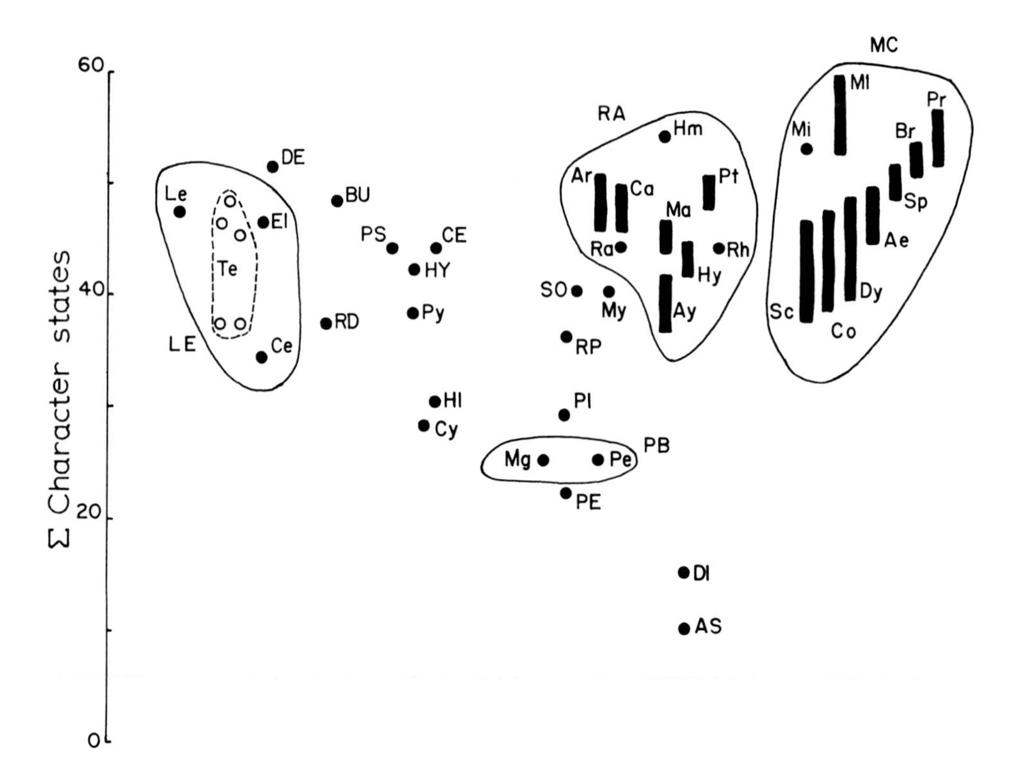

Figure 3–5. Relative primitiveness and advancedness of 44 family groups (family, subfamily, or tribe) of frogs. Lower values reflect greater primitiveness. Ar, Arthroleptinae; AS, Ascaphidae; Ae, Asterophryinae; Ay, Astylosterninae; Br, Brevicipitinae; BU, Bufonidae; Ca, Cacosterninae; CE, Centrolenidae; Ce, Ceratophryinae; Co, Cophylinae; Cy, Cycloraninae; DE, Dendrobatidae; DI, Discoglossidae; Dy, Dyscophinae; El, Elosiinae; Hl, Heleophryninae; Hm, Hemisinae; HY, Hylidae, except Phyllomedusinae; Hy, Hyperoliinae; LE, Leptodactylidae; Le, Leptodactylinae; Ma, Mantellinae; Mg, Megophryinae; Ml, Melanobatrachinae; MC, Microhylidae; Mi, Microhylinae; My, Myobatrachinae; PB, Pelobatidae; Pe, Pelobatinae; PE, Pelodytidae; Pt, Petropedetinae; Pr, Phrynomerinae; Py, Phyllomedusinae; Pl, Pipidae; PS, Pseudidae; RA, Ranidae; Ra, Raninae; Rh, Rhacophorinae; RD, Rhinodermatidae; RP, Rhinophrynidae; Sc, Scaphiophryninae; SO, Sooglossidae; Sp, Sphenophryninae; Te, Telmatobiinae. *Open circles* refer to the five tribes of the subfamily.

on Figure 3–5 by a spot (all data available) or a bar (some data lacking). Some values are probably too high and others too low, but the magnitude of error is slight.

Superficial examination of Figure 3–5 reveals that three levels of organization are represented among living frogs. The lowest level includes the Ascaphidae and Discoglossidae; the intermediate level includes the pelobatids, pelodytids, pipids, and two subfamilies of myobatrachids (Heleophryninae and Cycloraninae); and the highest level of organization includes the broadest range of variation and the remaining groups of frogs, with the least specialized members being the Ceratophryinae (Leptodactylidae) and Phyllomedusinae (Hylidae). Included toward the lower end of the range is the archaic family Rhinophrynidae. The most specialized frogs include the dendrobatids, some ranids (Hemisinae), and some microhylid subfamilies. The data summary in Figure 3–5 is correctly regarded as an evolutionary grade of the anurans, and the subdivisions evident (three levels of organization) are of relatively little interest, except in providing an index of relative phenetic distance between entities. Without considering the cladistics involved, a serious classification could not be advanced from Figure 3–5 alone. Equal values obtain, for example, in two units (*a* and *b*) exhibiting the following character state distributions: *a* with primitive character states for characteristics 1-19 and derived states for characteristics 20-38, and *b* with the reverse pattern; both taxa would have indices of primitiveness of 38 but would be quite unlike one another.

The characteristics showing less infrafamilial (family-group) variation are better for deducing phylogenetic arrangements than are those with more infrafamilial variation (Farris, 1966; Kluge and Farris, 1969), because the former are less likely to exhibit multiple character shifts than are the latter. Those characteristics illustrating much infrafamilial variation are, in general, those in which gradual changes prevail, instead of discrete steps. The characteristics discussed in Section II that exhibit high degrees of infrafamilial variation include the following: 10, 12, 18, 21, 22, 24, 26, 29, and 32.

Cladistic Arrangement

No single characteristic among the 38 can be demonstrated to have character states immune from reversibility; neither can a serious argument be advanced that the shift from state *a* to state *b* for character *j* occurred but once in the phylogeny of the Anura. The assumptions of irreversibility and single (or few) origin of a derived state are inherent in the Rule of Parsimony. Assuming that evolution is highly parsimonious carries with it the assumption that convergence, homoplasy, and parallelism are highly rare. This assumption and admission ignore the variation in the breadth of the major adaptive zone entered by the group in question. Groups evolving in narrow adaptive zones exhibit higher frequencies of homoplasy than groups evolving in broader adaptive zones (Bock, 1963). The adaptive zone need not be necessarily an environmental segment; it may be the morphological and functional constraints assumed by the earliest members of the group and transmitted to the descendants (in the present case, the elongated pelvic apparatus and associated adaptations). Systematists desiring to infer the phyletic history of a group must make subjective judgments; these judgments take the form of making principal use of subjectively determined fundamental characters as the chief source for deducing a phylogeny. The subjectivity is only slightly masked when one calculates the variance of a character and its character states within and among groups. One result depends on agreement of other specialists in the group that the proper "fundamental" characters were selected, and the other result depends on mathematics (a lower or higher ratio) to be inspected by other specialists, who then must agree that the ratios were of the best order of magnitude and that the most proper "fundamental" characters were accorded appropriate weight. The advantage of the second over the first is that the second is repeatable and perhaps slightly less subjective than the first. Both methods have the same aim.

The indices of relative primitiveness (Figure 3–5) are of importance in providing evidence that suborders are not realistic entities within so small a group as the Anura. Suborders and infraorders are of value in subdividing large groups but serve more to confuse than to clarify the relationships of the members of a comparatively small group in a narrow adaptive zone. It is more instructive to speak of levels of organization or superfamilies than to burden a group of perhaps 2000 species with two (Cope, 1865) to five (Noble, 1931) suborders. Reig's (1958) arrangement is less objectionable in that he, like others, recognized the absurdity of separating bufonoids and ranoids as suborders; he combined them into the Neobatrachia, with superfamilies A and B. Reig's classification bears some similarity to my three levels of organization. However, he subdivided the archaic frogs into three suborders (Amphicoela, Aglossa, and Archaeobatrachia), the last of which included discoglossids, rhinophrynids, pelobatids, and pelodytids (part of my "transitional families"), and divided the myobatrachids among the superfamilies of the Neobatrachia, with cycloranines and myobatrachines as bufonoids in the family Leptodactylidae, and *Heleophryne* as a monotypic family (Heleophrynidae) of the ranoid superfamily.

In constructing a cladogram of the Anura (Figure 3–6), the following assumptions were made: *The taxon with the largest number of primitive character states is the most primitive group.* Hence the Ascaphidae, which has 34 (35 if fossils are considered, characteristic 36) characteristics exhibiting primitive character states, is the most primitive family of living anurans—a conclusion shared with most other workers, except those who stress the larval characteristics (see below). *Subsequent dichotomies were arranged so that the dendrogram produced the fewest number of character shifts.* Those characteristics with the least infrafamilial variation were preferred when otherwise equivalent choices were available. Thus, characteristics having high degrees of infrafamilial variation are characterized as having many shifts (10–20) from state *a* to state *b*, whereas those having low degrees of infrafamilial variation are characterized as having few shifts (1–5) from state *a* to state *b*.

The relationships of the Anura are indicated on the dendrogram. The ascaphids and discoglossids are considered early, independent offshoots that are followed by a third early offshoot, giving rise to three families—the extinct palaeobatrachids, pipids, and rhinophrynids. The five families mentioned above are separated from the remaining families by a larger hiatus than separates any of the five. This hiatus is the basis for dividing the Anura into two series, the Archaic frogs and the Advanced frogs. The pelodytids and pelobatids are independent branches, and following these is a major dichotomy (in terms of the diversity of the branches, not in terms of so many character shifts) into one series that has myobatrachines and sooglossids as the most primitive members and ranids *(sensu latu)* and microhylids *(sensu latu)* as the more advanced members and a second series that has the African heleophrynines and cycloranines as the most primitive members and leptodactylids, bufonids, and hylids *(sensu latu)* as more advanced members. To judge the distance between groups on Figure 3–6, the number of changes in the internodal spaces provide an index. The long gap between the node for the origin of the pelobatids and the node for the origin of the lines leading to ranoids and bufonoids results from the constraints imposed by two dimen-

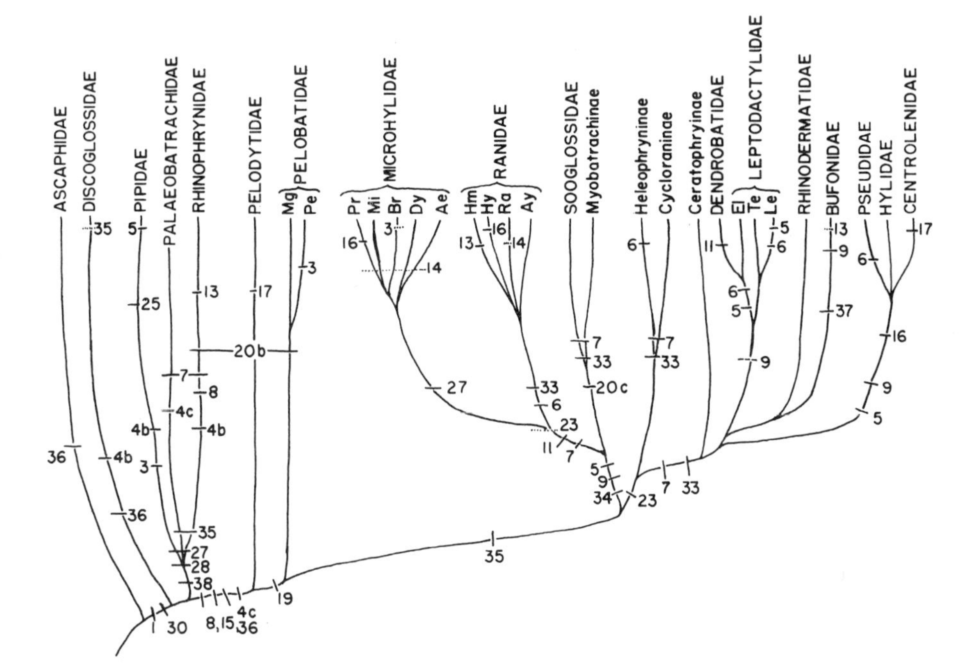

Figure 3–6. Dendrogram illustrating relationships of frogs as proposed here. *Numbers* refer to character state shifts for numbered characteristics (see Section II of this paper); *solid lines* cutting lineages mean uniform shift in character states; *dotted lines* cutting lineages mean the states are variable in derived groups. Not all 38 characteristics are included on the figure.

sions, not because the pelobatids are considered phylogenetically distant from the other groups (Figure 3–7).

Discussion

As used here, the term "archaic frogs" denotes an anomalous grouping perhaps best thought of as a paraphyletic grade-clade. The two units within the complex are distinct from one another on many characteristics. The ascaphid-discoglossid group is certainly a natural unit (paraphyletic), and the closeness of relationship is best conveyed by the use of a superfamily, the Ascaphoidea. The other archaic families pose more of a difficulty because each has specialized forms, as well as advanced, but they probably represent a monophyletic grouping. Kluge and Farris (1969) united them as an ancient unit and apparently were more impressed by the few similarities than the many differences, which is evident in their character selection. This criticism is not especially meaningful, however, since their intent was not an anuran phylogeny but the development of a method. The similarities between pipids and rhinophrynids involve characters 4, 6, 8, 9, 10, 14, 15, 17, 23, 26, 27, 28, 29, 30, 33, 36, and 38. Character states that are uncommon to frogs on the whole but that are shared by these two families include characters 4 (shared with discoglossids), 28 (not shared; see discussion of character), 36 (not shared), and 38 (not shared). The two families exhibit different character states for 11 characters; in 6 characters, pipids are variable between two character states, one of which is shared by *Rhinophrynus*, as well as other frogs. Perhaps also significant, rhinophrynids, pelobatids, and pelodytids share a unique character state

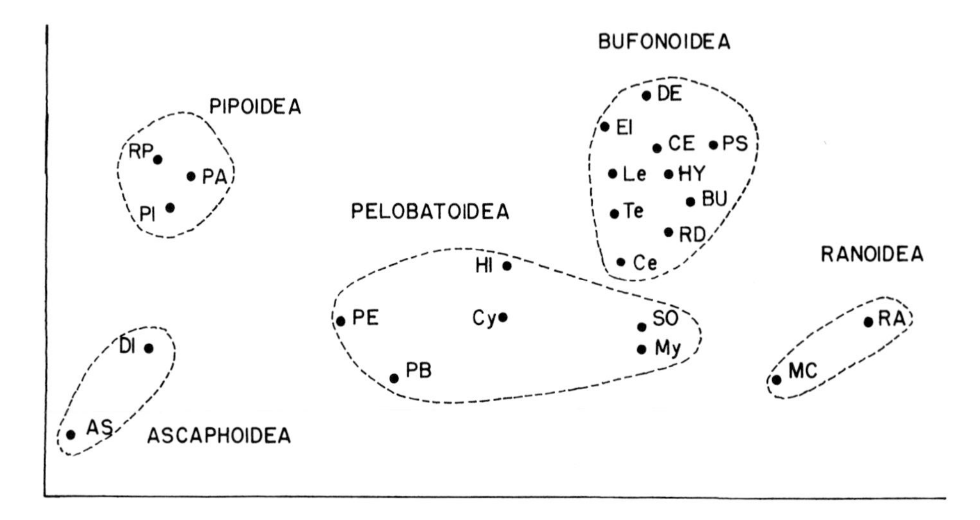

Figure 3–7. Suprafamilial relationships of frogs. For explanation of abbreviations, see Figure 3–5, with the addition of the Paleobatrachidae. Distance along the axis approximates character state shifts on the mainline of anuran evolution, whereas distance along the abscissa approximates character state shifts on divergent lines of anurans from the mainline.

of character 20. *Rhinophrynus* has similar hyolaryngeal morphology to *Pelodytes* (parahyoid ossification and reduced hyalia), which need not be necessarily pronounced as convergent as Walker (1938) did; the similarity could be equally due to parallelism, which would imply a closer relationship, or homoplasy. The aglossy of pipids precludes any comparisons of pipid hyolaryngeal anatomy with those of rhinophrynids and pelodytids. The evidence suggesting an origin of a pipid-rhinophrynid complex from the Ascaphoidea rests largely on shared characters (see above), especially the presence of free ribs in young and fossil pipids, the presence of opisthocoelous centra, and the presence of a parahyoid ossification in rhinophrynids; however, there are other characters shared among ascaphoids, pipoids, and pelobatoids. Opisthocoely could have been accomplished independently by pipids and rhinophrynids and need not have been due to an origin from a discoglossoid. The free intervertebral discs of pelobatids would seem to have equal potential for opisthocoely as they do for procoely.

The rhinophrynids and pipids are, by virtue of each group's possession of a variety of primitive character states that they do not share, considered primitive frogs that have independently specialized into very different habitats. Pipids have specialized in an aquatic mode of existence, masking much of their morphology under the specializations acquired in adapting to that narrow zone; rhinophrynids are equally specialized into the narrow fossorial zone and share many similarities with other burrowers (Australian myobatrachines, African hemisine ranids, and some microhylids). The specializations have obscured much of their relationship to one another. Like any other kind of frog, they both exhibit a mosaic of primitive and advanced character states. Including the pipids and rhinophrynids in a common superfamily, Pipoidea, implies a closer relationship than may in fact be the case, but separating them into two superfamilies is somewhat less documented and implies a greater dissimilarity than is the case. Because earlier students of this problem considered very few characteristics, the pipids and rhinophrynids were considered very primitive. If two of eleven characters are the presence of free ribs at some stage and the presence of parahyoid ossification, the two families are scored as possessing one primitive and one advanced character state each, while all advanced frogs are scored as having two advanced character states. The use of few characters, no matter how judicious their coding, accords undue emphasis on the primitive aspects of the pipid-rhinophrynid mosaics. Pipids and rhinophrynids share more character states with each other and with palaeobatrachids than any one of these share with any other anuran group. This similarity suggests a relationship that is best indicated by inclusion in the Pipoidea. The differences between pipoids and other frogs support this level of distinction and not more. Some character states shared by pelobatoid and some pipoid frogs may reflect traits of a common ancestor that have been independently lost by some pipoid groups. For example, the absence of a parahyoid ossification in pipids cannot be considered a compelling reason for dissociating them from rhinophrynids. The presence of a middorsal gap in the cricoid cartilage of pelobatids, pelodytids, and rhinophrynids is probably significant, but comparison with pipids is impossible because of the aglossal specializations.

Tihen (1965) suggested that the families discussed above represented a phyletic line divergent from a common ancestor with the pelobatoid (and advanced) frogs. My proposal differs only in that the Ascaphoidea are considered to include Tihen's "common ancestor." All data available point to frogs of a pelobatoid organization (level 2, Figure 3–5) being ancestral to all frogs of the third level of organization (Figure 3–5), with the possible exception of the pipoids.

The remaining frogs could be included in a grouping as the "advanced frogs" or could be broken into two units, one of which I am calling the transitional families and a second, the advanced families (=Neobatrachia of Reig). If the use of suborders (groupings of superfamilies) can be defended for the Anura, it is to use one suborder for the archaic frogs (Ascaphoidea and Pipoidea) and a second for the remaining frogs (transitional and advanced families).

The separation of the pelobatids and pelodytids is tenuous because few characters are involved. Although the two are combined as a single family by some authorities, they are maintained as family-groups (subfamilies). The distinctions between pelobatids (and pelodytids) and the Old World leptodactyloids are not as great as are implied by the inclusion of the Old World leptodactyloids in the bufonoid series and the exclusion of the pelobatoids. The cycloranines and heleophrynines are more similar to pelobatoids than are the myobatrachines. The myobatrachines exhibit a complex of character states similar to those of the ranids and microhylids, whereas the cycloranines (and heleophrynines) have relatively few distinctions from the New World leptodactyloids and their presumed derivatives (bufonids, rhinodermatids, centrolenids, hylids, pseudids, and dendrobatids). My proposal is that the myobatrachid stock now represented by the Myobatrachinae and the Sooglossidae gave rise to ranids and microhylids, while the myobatrachid stock now represented by the Cycloraninae and Heleophryninae gave rise to the Leptodactylidae, which in turn gave rise to the New World frog families and to the near cosmopolitan Bufonidae and Hylidae. Cladistically, the Myobatrachinae should be included with the ranoids (Microhylidae and Ranidae), and the Cycloraninae and Heleophryninae should be included with the bufonoids. The present arrangement—inclusion of the myobatrachids in the Leptodactylidae—isolates part of a grade as a family. An alternative is to separate the leptodactyloids into two families (Myobatrachidae and Leptodactylidae). A third alternative is to use three families—one for myobatrachines, a second for the cycloranines and heleophrynines, and the third for the New World leptodactylids. A fourth alternative would be the inclusion of the cycloranines and heleophrynines in the Leptodactylidae and the restriction of Myobatrachidae to the present

Myobatrachinae. I have considerable sympathy for the last suggestion; however, the function of this paper is not to propose new and novel arrangements of families, but to attack a broader picture, that is, the transition. In that vein, I will use the Myobatrachidae to include the Myobatrachinae, Cycloraninae, and Heleophryninae, although I am cognizant that this action only reduces the charge that this unit is a grade (because the inclusion of the leptodactylids and myobatrachids as a single family is a larger, and less desirable, grade). The Pelobatoidea is used here as a paraphyletic grouping that includes four families—Pelobatidae, Pelodytidae, Myobatrachidae, and Sooglossidae. The placement of the Sooglossidae here deserves further comment inasmuch as current literature (Griffiths, 1959*a*, 1963) places the family as a ranoid offshoot. Griffiths (1959*a*, 1963) placed the group as ranoid because the girdle has some resemblance to the firmisternal type and the thigh musculature (the disposition of the distal tendons, characteristic 34) was ranoid. However, as Griffiths admitted, there is some doubt that the girdle is firmisternal. The thigh musculature characteristics are not unique to the Microhylidae and Ranidae but are duplicated by several myobatrachines (Parker, 1940). The cricoid cartilage of sooglossids is similar to that of ranid embryos, as well as to the condition seen in myobatrachines. The absence of the cartilago apicalis in sooglossids and myobatrachines and the presence of the element in most ranids and some microhylids (Trewavas, 1933) suggest a closer relationship between myobatrachines and sooglossids than has been previously proposed. The osteological similarities between the two groups are extensive, but shared character states are duplicated in both microhylid and ranid groups. The separation of myobatrachines and sooglossids will be treated elsewhere (in preparation).

The separation of the "advanced frogs" (=Reig's Neobatrachia) into two units (one clustered about bufonids and leptodactylids and the other clustered about ranids and microhylids) has been done by virtually every student of anuran classification since the mid-1800's, with the exception of those authors who isolated the microhylids on the basis of the larval morphology. The excess weighting of larval characteristics has been criticized by Griffiths (1963), Griffiths and Carvalho (1965), Tihen (1965), and Kluge and Farris (1969) and is further criticized below. Starrett (this volume) provides additional support for classifications based on larval morphology, and her proposal is at major odds with mine. The relationships of the microhylids (and phrynomerines) with ranids, while perhaps ancient, are evident in the sharing of character states by the two groups. The differences between the two families are mitigated when one examines the Scaphiophryninae, a group tossed between ranids and microhylids by various authors. Parker (1934) considered the Scaphiophryninae ranids because Angel (1931) reported Type IV larvae for the group and because the sphenethmoid is entire in these frogs, in contrast to the divided sphenethmoid in microhylids. However, scaphiophrynines have dilated sacral diapophyses, palatal folds, and a hyolaryngeal apparatus similar to that seen in at least some microhylines (Trewavas, 1933). Persons considering the scaphiophrynines to be microhylids must presume that some microhylids do not have Type II larvae or that Angel's (1931) report is in error.

The separation of bufonoids and ranoids rests on three characteristics and their character states (Table 3–1). The dendrobatids are the most frequently displaced group—they are grouped with bufonoids on characteristics 4 and 34 but are grouped with ranoids on characteristic 11 (girdle type). The significance of the diplasiocoelous condition can be seriously challenged on several points, the most obvious of which is the nonsystematic distribution of the diplasiocoelous character state among microhylid and ranid genera. The monophyly of firmisternal architecture of the pectoral girdle is likewise questionable. A generic analysis of leptodactyloid frogs (Lynch, 1969) compelled me to derive dendrobatids from elosiine leptodactylids, which have little in common with microhylids or ranids. Noble (1922) used the disposition of the distal tendons of the thigh musculature to separate bufonoids and ranoids. Three character states are evident (Figure 3–8): the bufonoid pattern, the intermediate pattern, and the ranoid pattern. The bufonoid pattern has the tendon of the *m. semitendinosus* passing ventral to the *mm. gracilis* (exhibited by bufonids; centrolenids; hylids; leptodactylids, except for some species of two genera; pseudids; rhinodermatids; all archaic frogs, except as noted in the discussion of characteristic 34; cycloranine and heleophrynine myobatrachids; pelobatids; and pelodytids). The intermediate pattern, which Griffiths (1959*b*, 1963) put with the ranoid pattern, has the tendon of the *m. semitendinosus* piercing the *mm. gracilis* complex (i.e., passing between the *m. gracilis major* and *m. gracilis minor*) and is exhibited by dendrobatids, *Crossodactylus* (elosiine leptodactylids), some species of *Physalaemus* (leptodactyline leptodactylids), and by some genera of the Myobatrachinae. The ranoid pattern has the tendon of the *m. semitendinosus* passing dorsal to the *mm. gracilis* and is exhibited by sooglossids, some genera of myobatrachines, microhylids, and ranids.

Table 3–1. Characteristics Used to Separate the Bufonidea and Ranoidea

Trait	Bufonoidea[a]	Ranoidea[b]
Vertebral centra (4)	Procoelous	Diplasiocoelous[c]
Girdle architecture (11)	Arciferal, except for Dendrobatidae	Firmisternal
Insertion of *m. semitend.* relative to *mm. grac.* (34)	Ventral[d]	Dorsal

[a]Leptodactylidae, Rhinodermatidae, Bufonidae, Dendrobatidae, Hylidae, Pseudidae, and Centrolenidae.

[b]Microhylidae and Ranidae (Hyperoliidae and Rhacophoridae).

[c]Nearly one-half of the microhylid genera are procoelous; about one-half of the Rhacophorinae are procoelous, as are some or all of the Arthroleptinae and Astylosterninae.

[d]The tendon of the *m. semitendinosus* passes between the *m. gracilis major* and *m. gracilis minor* in dendrobatids and some leptodactylids, for instance, *Crossodactylus* (Elosiinae) and some species of *Physalaemus* (Leptodactylinae).

If the three patterns represent steps in an evolutionary sequence, the ranoids are probable descendants of myobatrachid or leptodactylid frogs. This idea does not conflict with any other set of data, except that involving larval mouthparts; in that area of research, different points of view and different weighting prevail. The available data do not support the idea that the neobatrachians should be divided into two suborders, and a case could be advanced against their subdivision into superfamilies. The eight families involved do tend to group into two complexes, one comprised of microhylids and ranids, and the other comprised of bufonids, hylids, and four neotropical groups. These two complexes are here referred to as superfamilies, the Ranoidea and Bufonoidea, respectively.

An Alternative Phylogeny

Orton (1953, 1957), Hecht (1963), and Starrett (this volume) have used larval characteristics at the expense of other characteristics in deducing a phylogeny. Their views differ from mine in that they have used only larval data to di-

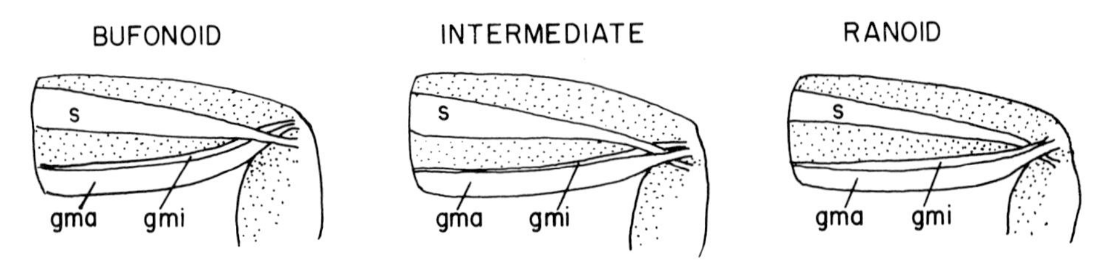

Figure 3–8. Diagrammatic representation of three character states in the disposition of the distal tendons of the thigh: s, *sartorius-semitendinosus* complex; gma and gmi, *gracilis major* and *gracilis minor*, respectively.

vide the anurans into four major groups (suborders in Starrett's arrangement), each of which is characterized by a distinctive larval type. Accepting their basic argument—namely, that Type I larvae (pipoids) are representative of the most ancient line of anurans, that Type II larvae (microhylids) are representative of an only slightly less ancient line of anurans, and that Type III larvae (ascaphoids) represent a major advancement over Types I and II and have given rise to frogs having Type IV larvae (pelobatoids, bufonoids, and ranoids, excluding the microhylids)—their dendrogram (Figure 3–9) may be compared with mine (Figure 3–6). Accepting their arrangement requires that my coding of characteristic 27 be in error (reversed evolutionary sequence) and requires a number of "extra" evolutionary steps (if the evolutionary directions coded into the character states of characteristics 1–26, 28–38 are accepted). The "extra" evolutionary steps are most apparent and most numerous for the Microhylidae (the group most displaced).

This alternative phylogeny could of course be correct. As I stated earlier (sections III, IV), I regard multiple character-state shifts possible and probable for all 38 characteristics, bar none. The arrangement based exclusively on larval characteristics (Figure 3–9) simply requires more character-state shifts (higher degree of homoplasy) than does my arrangement (Figure 3–6). If, however, we presume evolution to be highly parsimonious, the alternative phylogeny must be considered more tenuous than that diagrammed in Figure 3–6.

V. THE ARCHAIC FROGS

The characteristics of each family (Ascaphidae, Discoglossidae, Pipidae, Palaeobatrachidae, and Rhinophrynidae) are enumerated below. The 38 characteristics used in stratifying groups (see Section II) are listed for each family, their variabilities noted, and following those characteristics, miscellaneous, generally unique traits of individual families are given, where appropriate.

Ascaphidae

(1) nine (rarely, eight) presacral vertebrae; (2) atlas not fused to second vertebra; (3) sacrum not fused to coccyx; (4) notochord persistent, centra "amphicoelous"; (5) atlantal cotyles juxtaposed; (6) sacral diapophyses broadly dilated; (7) coccyx usually bearing transverse processes; (8) ribs free in subadults and adults; (9) transverse processes of posterior presacral vertebrae short or obsolete; (10) neural arches imbricate; (11) pectoral girdle arciferal; (12) prezonal elements present; (13) postzonal elements present; (14) zonal elements bony; (15) scapula overlaid anteriorly by clavicle; (16) phalangeal formulae 2–2–3–3, 2–2–3–4–3; (17) tarsal bones free; (18) three tarsalia; (19) parahyoid

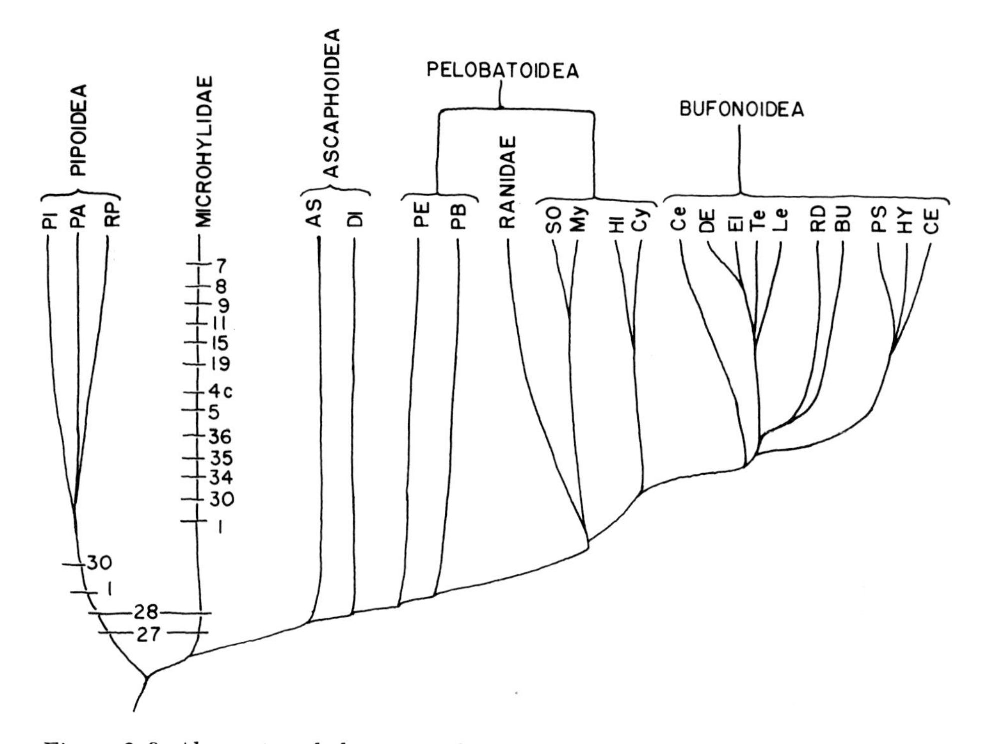

Figure 3–9. Alternative phylogeny to that in Figure 3–6. Basic divisions presented by workers on larvae are accepted. *Numbers and dashes* refer to the extra steps required beyond those steps diagrammed in Figure 3–6. Distributions here of character states of characteristics 27 and 28 suggest that they are miscoded in Section II of this paper; if so, each saves one step for this phylogeny and adds one for that in Figure 3–6.

ossification present; (20) cricoid cartilage complete; (21) maxilla dentate; (22) skull bones present, except for loss of columella; quadratojugal lost in *Ascaphus,* not in *Leiopelma;* (23) inguinal amplexus; (24) pupil vertical; (25) tongue present; (26) larvae aquatic in *Ascaphus,* but *Leiopelma* exhibits stage V of direct development; (27) denticles and beak present in *Ascaphus,* denticles in multiple rows; (28) spiracle single, median, ventral in *Ascaphus;* (29) outer metatarsal tubercle absent; (30) *m. caudaliopuboischiotibialis* present; (31) *m. sartorius* not discrete from *m. semitendinosus;* (32) *m. glutaeus magnus* lacking accessory tendon; (33) *m. adductor magnus* lacking accessory head; (34) tendon of *m. semitendinosus* inserts ventral to *mm. gracilis;* (35) sacro-coccygeal articulation is monocondylar; (36) parasphenoid of living genera having alae, Jurassic fossils lack parasphenoid alae; (37) organ of Bidder absent; (38) larvae lacking barbels.

Discoglossidae

(1) eight (rarely, nine) presacral vertebrae; (2) atlas not fused to second vertebra; (3) sacrum not fused to coccyx; (4) vertebral centra opisthocoelous; (5) atlantal cotyles juxtaposed; (6) sacral diapophyses broadly dilated; (7) coccyx usually bearing transverse processes; (8) ribs free in subadults and adults, although generally ankylosed to transverse processes in *Bombina;* (9) transverse processes of posterior presacral vertebrae short or obsolete; (10) neural arches imbricate; (11) pectoral girdle arciferal; (12) prezonal element present in *Dis-*

coglossus, absent in others; (13) postzonal elements present; (14) zonal elements bony; (15) scapula overlaid anteriorly by clavicle; (16) phalangeal formulae 2–2–3–3, 2–2–3–4–3; (17) tarsal bones free; (18) three tarsalia; (19) parahyoid ossification present; (20) cricoid cartilage complete; (21) maxilla dentate; (22) skull bones present, except for loss of columella in *Bombina;* (23) inguinal amplexus; (24) pupil vertical in *Alytes,* triangular in *Barbourula, Bombina,* and *Discoglossus;* (25) tongue present; (26) larvae aquatic, *Barbourula* in stage II, *Alytes* in stage III; (27) denticles and beak present in larvae of known genera (larvae of *Barbourula* not known), denticles in multiple rows; (28) spiracle single, median, ventral; (29) outer metatarsal tubercle absent; (30) *m. caudaliopuboischiotibialis* absent; (31) *m. sartorius* not discrete from *m. semitendinosus;* (32) *m. glutaeus magnus* lacking accessory tendon in *Alytes* and *Discoglossus,* tendon fleshy in *Barbourula* and *Bombina;* (33) *m. adductor magnus* lacking accessory head; (34) sacro-coccygeal articulation monocondylar in *Barbourula* and *Bombina,* bicondylar in *Alytes* and *Discoglossus;* (36) parasphenoid with lateral alae; (37) organ of Bidder lacking; (38) larvae lacking barbels. The sternum of discoglossids is triradiate.

Palaeobatrachidae (data incomplete)

(1) seven or eight presacral vertebrae; (2) atlas fused to second vertebra; (3) sacrum not fused to coccyx; (4) centra procoelous, stegochordal; (5) atlantal cotyles juxtaposed; (6) sacral diapophyses dilated; (7) no transverse processes on coccyx; (8) ribs free in subadults, ankylosed to transverse processes in adults; (9) transverse processes of posterior presacral vertebrae short; (10) neural arches imbricate; (11) pectoral girdle arciferal; (14) zonal elements bony; (15) scapula overlaid anteriorly by clavicle; (16) phalangeal formulae 2–2–3–3, 2–2–3–4–4; (17) tarsal bones free; (21) maxillary arch dentate; (22) quadratojugals lost; (26) larvae aquatic; (27) labial cartilages described for *Palaeobatrachus grandipes,* which has a terminal mouth; (35) sacro-coccygeal articulation bicondylar; (36) parasphenoid lacking lateral alae; (38) larvae with at least three pairs of lateral barbels.

Pipidae

(1) eight presacral vertebrae (not seven or eight as stated by Kluge and Farris, 1969); (2) atlas fused to second vertebrae in *Hymenochirus, ?Pseudohymenochirus,* and *Pipa,* but separate in *Xenopus;* (3) sacrum fused to coccyx; (4) vertebral centra opisthocoelous; (5) atlantal cotyles juxtaposed in *Xenopus,* separate in other genera; (6) sacral diapophyses broadly dilated; (7) transverse processes apparently present on coccyx, fused to sacral diapophyses as a web of bone; (8) ribs free in subadults, ankylosed to transverse processes in adults; free in adults of early Cretaceous pipids; (9) transverse processes of posterior presacral vertebrae short; (10) neural arches imbricate; (11) pectoral girdle firmisternal *(Hymenochirus)* or arciferal *(Xenopus);* (12) prezonal elements absent; (13) postzonal elements present; (14) zonal elements bony; (15) scapula overlaid anteriorly by clavicle; (16) phalangeal formulae 2–2–3–3, 2–2–3–4–3; (17) tarsal bones free; (18) two tarsalia; (19) parahyoid ossification absent; (20) cricoid cartilage complete; (21) maxilla dentate in *Hymenochirus, Pseudohymenochirus, Xenopus,* and *Pipa parva,* edentate in other *Pipa;* (22) quadratojugal lost, prevomer lost in some, columella lost in some (possibly all); (23) inguinal amplexus; (24) pupil round; (25) tongue absent; (26) larvae aquatic, except for some species of *Pipa;* (27) denticles, beaks, lips, and suprarostral cartilage absent; (28) pseudospiracles present, paired,

ventrolateral; (29) outer metatarsal tubercle absent; (30) *m. caudaliopuboischiotibialis* absent; (31) *m. sartorius* not discrete from *m. semitendinosus;* (32) accessory tendon of *m. glutaeus magnus* present; (33) no accessory head of *m. adductor magnus;* (34) insertion of *m. semitendinosus* penetrates *mm. gracilis;* (35) sacro-coccygeal articulation obscured by fusion of sacrum and coccyx, monocondylar in early Cretaceous fossils; (36) parasphenoid lacking lateral alae; (37) organ of Bidder lacking; (38) larvae with a single pair of lateral barbels. Pipids have a single median opening to the eustachian tubes.

Rhinophrynidae

(1) eight presacral vertebrae; (2) atlas not fused with second vertebra; (3) sacrum and coccyx not fused; (4) intervertebral bodies largely cartilaginous, adhering to anterior cotyle, which produces opisthocoelous condition; (5) atlantal cotyles juxtaposed; (6) sacral diapophyses broadly dilated; (7) coccyx usually lacking transverse processes; (8) ribs absent in adults and subadults, transverse processes of vertebrae two, three, and four not elongated; (9) transverse processes of posterior presacral vertebrae short; (10) neural arches imbricate; (11) pectoral girdle arciferal; (12) prezonal elements absent; (13) postzonal elements absent; (14) zonal elements bony; (15) scapula overlaid anteriorly by clavicle; (16) phalangeal formulae 2–2–3–3, 1–2–3–4–3; (17) tarsal bones free; (18) number of tarsalia not known; (19) parahyoid ossification present; (20) cricoid cartilage incomplete dorsally; (21) maxilla edentate; (22) palatines lost or fused with prevomers, columella lost; (23) inguinal amplexus; (24) pupil vertical; (25) tongue present; (26) larvae aquatic; (27) denticles, beaks, lips, and suprarostral cartilage absent; (28) pseudospiracles present, paired, ventrolateral; (29) outer metatarsal tubercle absent; (30) *m. caudaliopuboischiotibialis* absent; (31) *m. satorius* largely distinct from *m. semitendinosus;* (32) no accessory tendon of the *m. glutaeus magnus;* (33) no accessory head of *m. adductor magnus;* (34) insertion of *m. semitendinosus* penetrates *mm. gracilis;* (35) sacro-coccygeal articulation bicondylar; (36) parasphenoid lacks lateral alae; (37) organ of Bidder absent; (38) larvae with one pair of dorsal barbels, four pair of lateral barbels, and a single chin barbel.

VI. TRANSITIONAL FAMILIES

Two complexes of frogs are considered sufficiently intermediate between the Archaic frogs and advanced frogs to warrant tagging them "transitional." Noble (1922, 1931) and most subsequent authors recognized the intermediate or transitional nature of the pelobatoids (Pelobatinae, Megophryinae, and Pelodytidae), and it has been only in recent years that any stress was applied to recognizing the significance of the extraholarctic members of the family. For the purposes of this paper, the genus *Pelodytes* is considered the sole living representative of the Pelodytidae.

The other complex included here is the Old World leptodactyloids and one derivative family, the Sooglossidae. The Old World leptodactyloids are currently grouped into three subfamilies (Heleophryninae, Cycloraninae, and Myobatrachinae) distributed in southern Africa and in the Papuan region. The myobatrachines differ in many respects from the cycloranines and heleophrynines and share few character states. Sooglossids are tenuously separated from myobatrachines.

Pelobatidae (Megophryinae, Pelobatinae)

(1) eight presacral vertebrae; (2) atlas rarely fused, except as an anomaly, to sec-

ond vertebra; (3) sacrum fused to coccyx in pelobatines, rarely so in megophryines; (4) vertebral centra procoelous in pelobatines, notochord persistent in megophryines; intervertebral discs free in subadult pelobatids; (5) atlantal cotyles juxtaposed; (6) sacral diapophyses broadly dilated; (7) coccyx usually bearing transverse processes, which may be incorporated into a bony web between coccyx and sacral diapophyses; (8) ribs not apparent in adults or subadults, transverse processes of anterior presacral vertebrae elongate; (9) transverse processes of posterior presacral vertebrae short or obsolete, except in *Megophrys;* (10) neural arches imbricate; (11) pectoral girdle arciferal; (12) prezonal element present; (13) postzonal element present, sometimes ossified; (14) zonal elements bony; (15) scapula not overlaid anteriorly by clavicle; (16) phalangeal formulae 2–2–3–3, 2–2–3–4–3; (17) tarsal bones free; (18) two tarsalia; (19) parahyoid ossification absent; (20) cricoid cartilage incomplete dorsally; (21) maxilla dentate; (22) quadratojugal lost in *Scaphiopus* and in some megophryine genera; (23) inguinal amplexus; (24) pupil vertical; (25) tongue present; (26) larvae aquatic; (27) denticles and beak present, denticles in single rows in pelobatines and most megophryines, but some megophryines have developed a funnel mouth and lost denticles and beaks; (28) spiracle single, sinistral; (29) outer metatarsal tubercle absent; (30) *m. caudaliopuboischiotibialis* absent; (31) *m. sartorius* not discrete from *m. semitendinosus;* (32) *m. glutaeus magnus* bearing accessory tendon; (33) no accessory head of *m. adductor magnus;* (34) insertion of *m. semitendinosus* ventral to *m. gracilis major;* (35) sacro-coccygeal articulation monocondylar, although sometimes obscured by synsacry; (36) parasphenoid with lateral alae; (37) organ of Bidder lacking; (38) larvae without barbels.

Pelodytidae

(1) eight presacral vertebrae; (2) atlas normally fused to second vertebra; (3) sacrum not fused to coccyx; (4) vertebral centra procoelous, intervertebral discs free in subadults; (5) atlantal cotyles juxtaposed; (6) sacral diapophyses broadly dilated; (7) coccyx bearing transverse processes; (8) ribs not apparent in adults or subadults, transverse processes of anterior presacral vertebrae elongate; (9) transverse processes of posterior presacral vertebrae short; (10) neural arches imbricate; (11) pectoral girdle arciferal; (12) prezonal element present; (13) postzonal elements present; (14) zonal elements bony; (15) scapula partially overlaid anteriorly by clavicle; (16) phalangeal formulae 2–2–3–3, 2–2–3–4–3; (17) tarsal bones fused; (18) three tarsalia; (19) parahyoid ossification present; (20) cricoid cartilage incomplete dorsally; (21) maxilla dentate; (22) all skull bones present; (23) inguinal amplexus; (24) pupil vertical; (25) tongue present; (26) larvae aquatic; (27) denticles and beak present, denticles in single rows; (28) spiracles single, sinistral; (29) outer metatarsal tubercle absent; (30) *m. caudaliopuboischiotibialis* absent; (31) *m. sartorius* not discrete from *m. semitendinosus;* (32) *m. glutaeus magnus* bearing accessory tendon; (33) no accessory head of *m. adductor magnus;* (34) insertion of *m. semitendinosus* ventral to *m. gracilis;* (35) sacrococcygeal articulation is monocondylar; (36) parasphenoid with lateral alae; (37) organ of Bidder lacking; (38) larvae lacking barbels.

Myobatrachidae (Cycloraninae, Heleophryninae, Myobatrachinae)

(1) eight presacral vertebrae; (2) atlas fused to second vertebra in *Heleophryne,* all cycloranines, except *Cyclorana, Lechriodus,* and *Mixophyes,* free in three genera of cycloranines and in all myobatrachines; (3) sacrum not fused with coccyx; (4) notochord persistent and intervertebral bodies free in subadults, tend to be fused to posterior end of centrum in adults, but *Cyclorana, Lechriodus,*

and *Mixophyes* may not have free intervertebral discs as subadults; (5) atlantal cotyles juxtaposed in Cycloraninae and Heleophryninae, widely separated in Myobatrachinae; (6) sacral diapophyses rounded in *Heleophryne,* dilated in other genera; (7) transverse processes on coccyx of some genera in each subfamily; (8) ribs not apparent in adults or subadults, transverse processes of anterior presacral vertebrae elongate in Cycloraninae and Heleophryninae, not elongate in Myobatrachinae; (9) transverse processes of posterior presacral vertebrae short in most genera, long in some cycloranines and some myobatrachines; (10) neural arches imbricate in most cycloranines and heleophrynines, generally not in myobatrachines; (11) pectoral girdle arciferal; (12) prezonal element present in most genera, absent in one cycloranine and one myobatrachine; (13) postzonal elements present, cartilaginous; (14) zonal elements bony; (15) scapula not overlaid anteriorly by clavicle; (16) phalangeal formulae 2–2–3–3, 2–2–3–4–3, except in some specimens of *Limnodynastes peronii* that have 1–2–3–3; (17) tarsal bones free; (18) two tarsalia; (19) parahyoid ossification absent; (20) cricoid cartilage incomplete ventrally in myobatrachines, complete in other subfamilies; (21) maxilla dentate in most, edentate in *Notaden* of Cycloraninae and in most genera of Myobatrachinae; (22) *Heleophryne* and the cycloranines have lost no skull bones, *Glauertia, Metacrinia, Myobatrachus, Pseudophryne,* and *Uperoleia* have lost the prevomers, and *Pseudophryne* has also lost the columellae; (23) inguinal amplexus; (24) pupils vertical in one myobatrachine—*Uperoleia*—in three cycloranines—*Heleioporus, Mixophyes,* and *Neobatrachus*—and in *Heleophryne;* (25) tongue present; (26) aquatic larvae in *Heleophryne,* all but two cycloranines—*Kyarranus* and *Philoria*—and in some myobatrachines—*Crinia* (part), *Pseudophryne* (part), *Glauertia,* and *Uperoleia;* (27) denticles in single rows and beak present; (28) spiracle single, sinistral; (29) outer metatarsal tubercle absent in heleophrynines, absent in all but two species of cycloranines—*Adelotus brevis* and *Limnodynastes tasmaniensis*—and present in all myobatrachines, except six species of *Crinia* and the genus *Taudactylus;* (30) *m. caudaliopuboischiotibialis* absent; (31) *m. sartorius* largely distinct from *m. semitendinosus;* (32) *m. glutaeus magnus* bearing accessory tendon; (33) accessory head of *m. adductor magnus* present but small; (34) tendon of the *m. semitendinosus* inserts ventral to the *mm. gracilis* in heleophrynines and most cycloranines, but it penetrates the *gracilis* complex in a few specialized cycloranines and in some myobatrachines and passes dorsal to the *mm. gracilis* in the remaining myobatrachines; (35) sacrococcygeal articulation bicondylar; (36) parasphenoid with lateral alae; (37) organ of Bidder lacking; (38) larvae lacking barbels.

Sooglossidae

(1) eight presacral vertebrae; (2) atlas not fused to second vertebra; (3) sacrum not fused to coccyx; (4) notochord persistent; (5) atlantal cotyles widely separated; (6) sacral diapophyses dilated; (7) small transverse processes on coccyx; (8) ribs absent in adults, transverse processes of anterior presacral vertebrae not elongated; (9) transverse processes of posterior presacral vertebrae somewhat shortened; (10) neural arches not imbricate; (11) pectoral girdle arciferal or "firmisternal with secondary modifications toward arcifery" (Griffiths, 1959*a*); (12) prezonal element present; (13) postzonal element present; (14) zonal elements bony; (15) scapula not overlaid anteriorly by clavicle; (16) phalangeal formulae 2–2–3–3, 2–2–3–4–3; (17) tarsal bones free; (18) unknown; (19) parahyoid ossification absent; (20) cricoid cartilage incomplete ventrally; (21) maxillary arch dentate; (22) columellae absent; (23) unknown; (24) pupil horizontal; (25) tongue present; (26) development direct; (27–28) not applicable;

(29) outer metatarsal tubercle present; (30) *m. caudaliopuboischiotibialis* absent; (31) *m. sartorius* discrete from *m. semitendinosus;* (32) unknown; (33) unknown; (34) tendon of *m. semitendinosus* passes dorsal to *mm. gracilis;* (35) sacro-coccygeal articulation monocondylar; (36) parasphenoid with lateral alae; (37) organ of Bidder absent; (38) not applicable.

VII. THE ADVANCED FROGS

The following families are recognized and diagnosed below: Leptodactylidae, which is restricted to neotropical genera; Bufonidae, including the Atelopodidae but not *Brachycephalus,* a genus that is not assigned to a family here; Rhinodermatidae, which is comprised of *Rhinoderma* only; Pseudidae; Hylidae; Centrolenidae; Dendrobatidae; Ranidae, including Hyperoliidae and Rhacophoridae; and Microhylidae, including *Phrynomerus.*

Leptodactylidae

(1) eight presacral vertebrae; (2) atlas not fused to second vertebra; (3) sacrum not fused with coccyx; (4) vertebral centra procoelous; (5) atlantal cotyles juxtaposed in Ceratophryinae, Telmatobiini, and Odontophrynini, widely separated in Elosiinae, Leptodactylinae, Batrachylini, Grypiscini, and Eleutherodactylini; (6) sacral diapophyses weakly dilated in Ceratophryinae, Telmatobiini, Odontophrynini, and Batrachylini, round in other tribes and subfamilies; (7) transverse processes on coccyx rare, common in some Telmatobiini; (8) ribs absent in adults and subadults, transverse processes of anterior trunk vertebrae elongate in Ceratophryinae and some genera of Telmatobiinae, Telmatobiini, and Odontophrynini; (9) transverse processes of posterior presacral vertebrae shortened in Ceratophryinae, Odontophrynini, and some Telmatobiini, long in other groups; (10) neural arches imbricate in Ceratophryinae, Odontophrynini, Telmatobiini, and some Leptodactylinae, nonimbricate in Batrachylini, Eleutherodactylini, Grypiscini, Elosiinae, and most Leptodactylinae; (11) pectoral girdle arciferal; (12) prezonal element absent in *Lepidobatrachus* (Ceratophryinae), Odontophrynini, and a few species of *Eleutherodactylus;* (13) postzonal elements present, osseous in Leptodactylinae; (14) zonal elements bony; (15) scapula not overlaid anteriorly by clavicle, (16) phalangeal formulae 2–2–3–3, 2–2–3–4–3, except in *Euparkerella,* which has a 2–2–3–2 formula for the hand; (17) tarsal bones free; (18) two tarsalia; (19) parahyoid ossification absent; (20) cricoid ring normally complete, rarely incomplete ventrally; (21) maxilla dentate in nearly all species, edentate in *Batrachophrynus,* some *Physalaemus, Sminthillus,* and some *Telmatobius;* (22) quadratojugals lost in *Batrachyla, Crossodactylus,* and *Hylorina* (Telmatobiines) and *Pleurodema* and *Pseudopaludicola* (leptodactylines); the columellae are lost in *Crossodactylodes, Euparkerella, Holoaden, Paratelmatobius,* some *Eleutherodactylus,* some *Eupsophus,* several *Telmatobius,* and *Telmatobufo;* (23) axillary amplexus, with possible exception of two of the three species of *Batrachyla;* (24) pupil horizontal in most genera, vertical in *Lepidobatrachus* (Ceratophryinae), *Caudiverbera, Hylorina, Telmatobufo* (Telmatobiini, Telmatobiinae), and *Limnomedusa, Hydrolaetare* (Leptodactylinae); (25) tongue present, reduced in size and adherent in *Batrachophrynus;* (26) larvae aquatic, except Grypiscine genera in stage IV of direct development, Eleutherodactyline genera and *Adenomera* (Leptodactylinae) in stage V, Batrachyline genera in stage III, and several groups in stage II; (27) denticles and beak present in aquatic larvae, except in genus *Lepidobatrachus;* (28) spiracle single, sinistral, except in *Lepidobatrachus,* which has paired, ventrolateral spiracles; (29) outer

metatarsal tubercle absent in Ceratophryinae and about one-half of the genera of Telmatobiini, present in others; (30) *m. caudaliopuboischiotibialis* absent; (31) *m. sartorius* discrete from *m. semitendinosus;* (32) accessory tendon of *m. glutaeus magnus* present; (33) accessory head of *m. adductor magnus* present; (34) insertion of tendon of *m. semitendinosus* passes ventral to *mm. gracilis* in all but *Crossodactylus* and some species of *Physalaemus* in which the tendon passes between the *m. gracilis major* and *m. gracilis minor;* (35) sacro-coccygeal articulation bicondylar; (36) parasphenoid with lateral alae; (37) organ of Bidder absent; (38) larvae without barbels.

Bufonidae

(1) seven or eight presacral vertebrae; (2) atlas fused to second vertebra in *Atelopus, Cacophryne, Dendrophryniscus, Melanophryniscus, Oreophrynella, Pelophryne,* and *Rhamphophryne;* (3) sacrum normally fused to coccyx in *Laurentophryne, Nectophryne, Pelophryne,* most *Rhamphophryne,* and *Wolterstorffina;* (4) vertebral centra procoelous; (5) atlantal cotyles juxtaposed; (6) sacral diapophyses dilated; (7) transverse processes on coccyx occurs very rarely, possibly abnormally; (8) ribs absent in adults and subadults, transverse processes of anterior presacral vertebrae elongate in a few genera including part of *Bufo;* (10) neural arches imbricate in most, if not all, genera; (11) pectoral girdle arciferal in sense of Griffiths, some pseudofirmisternous, e.g., *Atelopus* and *Cacophryne;* (12) prezonal element absent, except in *Nectophrynoides, Bufo haematiticus* complex; (13) postzonal element present, cartilaginous; (14) zonal elements bony; (15) scapula not overlaid anteriorly by clavicle; (16) phalangeal formulae 2–2–3–3, 2–2–3–4–3, some reduction in *Pelophryne, Crepidophryne, Didynamipus,* and possibly others; (17) tarsal bones free; (18) two tarsalia; (19) parahyoid ossification absent; (20) cricoid ring complete; (21) maxilla edentate; (22) palatine lost in *Didynamipus, Nectophryne,* and *Pelophryne,* quadratojugals are lost or nearly so in *Ansonia, Laurentophryne, Nectrophryne, Pelophryne,* and *Wolterstorffina,* columellae are lost in some *Bufo, Atelopus,* and four African genera; (23) axillary amplexus; (24) pupil horizontal; (25) tongue present; (26) larvae aquatic, except in *Nectophrynoides,* which is ovoviviparous, several genera have aquatic tadpoles but are in stage II of direct development, and some known in stages III through V; (27) denticles and beak present, denticles in single rows; (28) spiracle single, sinistral; (29) outer metatarsal tubercle present; (30) *m. caudaliopuboischichiotibialis* absent; (31) *m. sartorius* discrete from *m. semitendinosus;* (32) accessory tendon of *m. glutaeus magnus* present; (33) accessory head of *m. adductor magnus* present; (34) insertion of *m. semitendinosus* passes ventral to *mm. gracilis;* (35) sacro-coccygeal articulation bicondylar, except in those cases where the joint is obscured by synsacry and in those cases where a presacral vertebra has been deleted through the sacrum (*Mertensophryne* and some species of *Rhamphophryne*); (36) parasphenoid with lateral alae; (37) organ of Bidder present at some stage of development; (38) larvae lacking barbels.

The genus *Brachycephalus* has been included in the Atelopodidae and hence in the Bufonidae (if the two are combined); McDiarmid (unpublished manuscript) isolated the genus familially. *Brachycephalus* differs from bufonids in the following ways: atlantal cotyles widely separated, sternum absent, and organ of Bidder absent. In addition, the genus has several peculiarities that are tentatively considered specializations.

Rhinodermatidae

(1) eight presacral vertebrae; (2) atlas fused with second vertebra; (3) sacrum

not fused to coccyx; (4) vertebral centra procoelous; (5) atlantal cotyles juxtaposed; (6) sacral diapophyses broadly dilated; (7) no transverse processes on coccyx; (8) ribs absent in adults and subadults; transverse processes of anterior trunk vertebrae not elongate; (9) transverse processes of posterior presacral vertebrae not shortened; (10) neural arches imbricate; (11) pectoral girdle arciferal, "pseudofirmisternal"; (12) prezonal element present; (13) postzonal element present; (14) zonal elements bony; (15) scapula not overlaid anteriorly by clavicle; (16) phalangeal formulae 2–2–3–3, 2–2–3–4–3; (17) tarsal bones free; (18) three tarsalia; (19) parahyoid ossification absent; (22) palatines, prevomers lost, quadratojugal greatly reduced; (23) axillary amplexus; (24) pupil horizontal; (25) tongue present; (26) larvae not aquatic, eggs are large, laid in terrestrial site, and development occurs in vocal sac of male; (27) denticles and beak present, denticles in single rows; (28) spiracle single, sinistral; (29) outer metatarsal tubercle lacking; (30) *m. caudaliopuboischiotibialis* absent; (31) *m. sartorius* discrete from *m. semitendinosus;* (32) unknown; (33) unknown; (34) insertion of *m. semitendinosus* is ventral to the *mm. gracilis;* (35) sacro-coccygeal articulation bicondylar; (36) parasphenoid with lateral alae; (37) organ of Bidder absent; (38) larvae without barbels.

Pseudidae

(1) eight presacral vertebrae; (2) atlas not fused to second vertebra; (3) sacrum not fused to coccyx; (4) vertebral centra procoelous; (5) atlantal cotyles juxtaposed; (6) sacral diapophyses cylindrical; (7) no transverse processes on coccyx; (8) ribs absent in adults and subadults, transverse processes of anterior trunk vertebrae not elongate; (9) transverse processes of posterior presacral vertebrae long; (10) neural arches not imbricate; (11) pectoral girdle arciferal; (12) prezonal element present; (13) postzonal elements present; (14) zonal elements bony; (15) scapula not overlaid anteriorly by clavicle; (16) phalangeal formulae 3–3–4–4, 3–3–4–5–4; (17) tarsal bones free; (18) two tarsalia; (19) parahyoid ossification lacking; (20) cricoid ring complete; (21) maxilla dentate; (22) no skull bones lost; (23) axillary amplexus; (24) pupil horizontal; (25) tongue present; (26) larvae aquatic; (27) denticles and beak present, denticles in single rows; (28) spiracle single, sinistral; (29) outer metatarsal tubercle present; (30) *m. caudaliopuboischiotibialis* absent; (31) *m. sartorius* discrete from *m. semitendinosus;* (32) accessory tendon on *m. glutaeus magnus;* (33) *m. adductor magnus* with accessory head; (34) tendon of *m. semitendinosus* passes ventral to *mm. gracilis;* (35) sacro-coccygeal articulation bicondylar; (36) parasphenoid with lateral alae; (37) organ of Bidder absent; (38) larvae lack barbels. The intercalated element is long and osseous.

Hylidae

(1) eight presacral vertebrae; (2) atlas not fused to second vertebra; (3) sacrum not fused with coccyx; (4) vertebral centra procoelous; (5) atlantal cotyles widely separated; (6) sacral diapophyses dilated in most genera, cylindrical in *Acris* and a few neotropical hylids; (7) no transverse processes on coccyx; (8) ribs absent in adults and subadults, transverse processes of anterior trunk vertebrae not elongated; (9) transverse processes of posterior presacral vertebrae long; (10) neural arches imbricate in most, if not all, phyllomedusines, nonimbricate in most hylids; (11) pectoral girdle arciferal; (12) prezonal element present, except in *Allophryne* and a few other genera; (13) postzonal elements present; (14) zonal elements bony; (15) scapula not overlaid anteriorly by clavicle; (16) phalangeal formulae 3–3–4–4, 3–3–4–5–4; (17) tarsal bones free; (18) two

tarsalia; (19) parahyoid ossification absent; (20) cricoid ring complete; (21) maxilla dentate, except in *Allophryne;* (22) quadratojugals are frequently lost, only a few genera are so characterized, however; (23) axillary amplexus; (24) pupils horizontal, except in phyllomedusines and *Nyctimystes,* which have vertical pupils; (25) tongue present; (26) larvae aquatic, except hemiphractines and several *Gastrotheca* and allied genera that exhibit specialized direct development in the tissue of the back or in pouches, some genera that are in stage III of direct development (e.g., Phyllomedusinae), and at least some New Guinean *Hyla* (Tyler, 1968) that exhibit stage II; (27) denticles and beak present in most aquatic larvae, and if present, denticles in single rows, but a number of species lack denticles (e.g., *Hyla microcephala* group), and loss of the beak is less common; (28) spiracle single, usually sinistral (but see Kluge and Farris, 1969; Griffiths and Carvalho, 1965) but highly variable between near-ventral to sinistral; (29) outer metatarsal tubercle present in many species, absent in many species; (30) *m. caudaliopuboischiotibialis* absent; (31) *m. sartorius* discrete from *m. semitendinosus;* (32) no accessory tendon of *m. glutaeus magnus* (among the few studied); (33) accessory head of *m. adductor magnus* present; (34) insertion of *m. semitendinosus* is ventral to *mm. gracilis;* (35) sacro-coccygeal articulation is bicondylar; (36) parasphenoid with lateral alae; (37) organ of Bidder absent; (38) larvae without barbels. The intercalated element of hylids is short and cartilaginous.

Centrolenidae

(1) eight presacral vertebrae; (2) cervical not fused to second vertebra; (3) sacrum not fused to coccyx; (4) vertebral centra procoelous; (5) atlantal cotyles widely separated; (6) sacral diapophyses dilated; (7) no transverse processes on coccyx; (8) ribs absent in adults and subadults, transverse processes of anterior presacral vertebrae not elongated; (9) transverse processes of posterior presacral vertebrae long; (10) neural arches nonimbricate; (11) pectoral girdle arciferal; (12) prezonal elements absent; (13) postzonal elements present; (14) zonal elements bony; (15) scapula not overlaid anteriorly by clavicle; (16) phalangeal formulae 3–3–4–4, 3–3–4–5–4; (17) tarsal bones fused; (18) two tarsalia; (19) parahyoid ossification absent; (20) cricoid ring complete; (21) maxilla dentate; (22) no skull bones consistently lost; (23) axillary amplexus; (24) pupil horizontal; (25) tongue present; (26) larvae aquatic, all in stage III of direct development; (27) denticles and beak present, denticles in single rows; (28) spiracle single, sinistral; (29) outer metatarsal tubercle absent; (30) *m. caudaliopuboischiotibialis* absent; (31) *m. sartorius* discrete from *m. semitendinosus;* (32) no accessory tendon of *m. glutaeus magnus;* (33) *m. adductor magnus* with accessory head; (34) insertion of *m. semitendinosus* is ventral to *mm. gracilis;* (35) sacrococcygeal articulation is bicondylar; (36) parasphenoid with lateral alae; (37) organ of Bidder absent; (38) larvae lacking barbels. The intercalated element is short and cartilaginous.

Dendrobatidae

(1) eight presacral vertebrae; (2) cervical not fused to second vetrebra; (3) sacrum not fused to coccyx; (4) vertebral centra procoelous (the nature of the centra of the posterior presacral vertebrae in *Dendrobates* and some *Phyllobates* is obliterated by synsacry involving the first presacral and sacral); (5) atlantal cotyles widely separated; (6) sacral diapophyses cylindrical; (7) transverse processes usually present on coccyx; (8) ribs absent in adults and subadults, transverse processes of anterior presacral vertebrae not elongated; (9) transverse

processes of posterior presacral vertebrae long; (10) neural arches nonimbricate; (11) pectoral girdle firmisternal; (12) prezonal element present; (13) postzonal element present; (14) zonal elements bony; (15) scapula not overlaid anteriorly by clavicle; (16) phalangeal formulae 2–2–3–3, 2–2–3–4–3; (17) tarsal bones free; (18) two tarsalia; (19) parahyoid ossification lacking; (20) cricoid ring complete; (21) maxilla dentate or not; (22) palatines absent, prevomers sometimes absent; (23) axillary amplexus; (24) pupil horizontal; (25) tongue present; (26) larvae aquatic, although carried on back of parent during early development, most if not all species in stage III of direct development; (27) denticles and beak normally present, except in funnel-mouthed species, denticles in single rows; (28) spiracle single, sinistral; (29) outer metatarsal tubercle present; (30) *m. caudalio-puboischiotibialis* absent; (31) *m. sartorius* discrete from *m. semitendinosus;* (32) accessory tendon of *m. glutaeus magnus* present; (33) *m. adductor magnus* bearing accesory head; (34) insertion of *m. semitendinosus* pierces *mm. gracilis;* (35) sacro-coccygeal articulation is bicondylar; (36) parasphenoid with lateral alae; (37) organ of Bidder absent; (38) larvae without barbels.

Ranidae

(1) eight presacral vertebrae; (2) cervical not fused to second vertebra, except in *Hemisus;* (3) sacrum not fused to coccyx; (4) vertebral centra procoelous or diplasiocoelous in which vertebra VIII is biconcave and the sacrum is biconvex; (5) atlantal cotyles widely separated; (6) sacral diapophyses cylindrical, except in Scaphiophryninae, which have dilated sacral diapophyses (Parker, 1934; Parker considered them ranids and not microhylids because they possess entire, instead of divided, sphenethmoids and because they have Type IV larvae, but he cited their incipient palatal folds and a microhyline hyolaryngeal apparatus); (7) no transverse processes on coccyx; (8) ribs absent in adults and subadults, transverse processes of anterior presacral vertebrae not elongated; (9) transverse processes of posterior presacral vertebrae long; (10) neural arches imbricate in few genera (e.g., Astylosterninae), nonimbricate in most genera; (11) pectoral girdle firmisternal; (12) prezonal elements present, omosternum usually ossified; (13) postzonal elements present, except in *Hemisus,* ossified style present in Cacosterninae, Mantellinae, Petropedetinae, Raninae, and Rhacophorinae, absent in Arthroleptinae, Astylosterninae, Hyperoliinae, and Scaphiophryninae; (14) zonal elements ossified in most subfamilies, cartilaginous or lost in Cacosterninae and Scaphiophryninae; (15) scapula not overlaid anteriorly by clavicle; (16) phalangeal formulae 2–2–3–3, 2–2–3–4–3 in all but Hyperoliinae, Mantellinae, and Rhacophorinae, which exhibit 3–3–4–4, 3–3–4–5–4; (17) tarsal bones free; (18) two or three tarsalia; (19) parahyoid ossification absent; (20) cricoid ring complete; (21) maxillary arch edentate in Hemisinae, Mantellinae, Scaphiophryninae, and some Arthroleptinae; (22) bone loss is relatively uncommon; columellae and quadratojugal lost in Hemisinae (my data are very incomplete); (23) axillary amplexus; (24) pupil vertical slit in Astylosterninae, Hemisinae, and some Hyperoliinae, horizontal in all others; (25) tongue present; (26) larvae aquatic in Astylosterninae, some Cacosterninae, Hemisinae, Hyperoliinae, ?Mantellinae, Petropedetinae, most Raninae, Rhacophorinae, and Scaphiophryninae, direct development (stage V) exhibited by Arthroleptinae, some Cacosterninae, and several Raninae, and stages II and III of direct development common especially in the Rhacophorinae; (27) denticles and beak present, denticles in single rows; (28) spiracle single, sinistral; (29) outer metatarsal tubercle absent in Astylosterninae, Cacosterninae, Hemisinae, most Hyperoliinae, Rhacophorinae, and many Arthroleptinae and Raninae, present in Petropedetinae and Scaphio-

phryninae, as well as many Arthroleptinae and Raninae; (30) *m. caudaliopuboischiotibialis* absent; (31) *m. sartorius* discrete from *m. semitendinosus;* (32) accessory tendon of *m. glutaeus* absent in many genera, present in few (in my small sample); (33) accessory head of *m. adductor magnus* present; (34) insertion of *m. semitendinosus* dorsal to *mm. gracilis;* (35) sacro-coccygeal articulation is bicondylar; (36) parasphenoid with lateral alae; (37) organ of Bidder absent; (38) larvae without barbels. The ranids with intercalated elements have short and cartilaginous elements.

Microhylidae

(1) eight presacral vertebrae; (2) cervical not fused to second vertebra; (3) sacrum not fused to coccyx, except in brevicipitine genera *Breviceps* and *Probreviceps;* (4) vertebral centra procoelous in Cophylinae and Sphenophryninae, eighth vertebra biconcave, sacrum biconvex in most Asterophryninae, Brevicipitinae, Dyscophinae, Melanobatrachinae, Microhylinae, and Phrynomerinae; (5) atlantal cotyles widely separated; (6) sacral diapophyses broadly dilated; (7) no transverse processes on coccyx; (8) ribs absent in adults and subadults, transverse processes of anterior presacral vertebrae not elongated; (9) transverse processes of posterior presacral vertebrae long; (10) neural arches imbricate in some genera, nonimbricate in others; (11) pectoral girdle firmisternal; (12) prezonal elements absent in Asterophryinae, Microhylinae, Phrynomerinae, and Sphenophryninae, and in some Dyscophinae and some Melanobatrachinae, retained in Brevicipitinae and Cophylinae; (13) postzonal elements present; (14) anterior zonal elements lost in Phrynomerinae, cartilaginous or lost or retained bony zonal elements in some genera, including Asterophryninae, Melanobatrachinae, Microhylinae, and Sphenophryninae, cartilaginous or reduced in Cophylinae and Dyscophinae, bony and present in Brevicipitinae; (15) clavicle, if present, not overlaying scapula; (16) phalangeal formulae 2–2–3–3, 2–2–3–4–3 in all genera, except *Phrynomerus,* which has 3–3–4–4, 3–3–4–5–4, and Melanobatrachines, which have a reduced formula; (17) tarsal bones free; (18) two tarsalia; (19) parahyoid ossification absent; (20) cricoid ring complete; (21) maxilla dentate in Dyscophinae and some Cophylinae, edentate in all others; (22) columella lost in Melanobatrachinae (reduction and loss of palatine is a common trait in many genera and suprageneric groups, see Parker, 1934); (23) axilliary amplexus, although at least one report of inguinal amplexus in *Breviceps* is available; (24) pupil horizontal or round, except in dyscophines where it is vertical; (25) tongue present; (26) larvae aquatic in many Microhylinae and Phrynomerinae, but other genera, including some microhylines, exhibit stages IV and V of direct development; (27) denticles, beak, and suprarostral cartilage absent; (28) spiracle single, median in all but one microhyline, where it is posterodorsolateral; (29) outer metatarsal tubercle absent in Asterophryinae, Cophylinae, Phrynomerinae, Sphenophryninae, some Melanobatrachinae, some Dyscophinae, and some Microhylinae, but present in all Brevicipitinae; (30) *m. caudaliopuboischiotibialis* absent; (31) *m. sartorius* discrete from *m. semitendinosus;* (32) accessory tendon on *m. glutaeus magnus* in at least Brevicipitinae, Melanobatrachinae, and Microhylinae; (33) no accessory head on *m. adductor magnus* in at least three genera of Microhylinae; (34) tendon of *m. semitendinosus* passes dorsal to *mm. gracilis;* (35) sacro-coccygeal articulation bicondylar, except in brevicipitines having sacro-coccygeal fusion; (36) parasphenoid with lateral alae; (37) organ of Bidder absent; (38) larvae lacking barbels. The intercalated elements in phrynomerines are short and cartilaginous.

The present recognition or nonrecognition of several traditional family groups of advanced frogs requires some explanation. I have departed from the tradition of recognizing the Atelopodidae because the unit is not defensible, except on the grounds of "pseudofirmisterny." The recognition of a bufonoid group for that reason requires that we accord equal recognition to the Asiatic genus *Cacophryne,* the African *Didynamipus,* and the American *Atelopus, Brachycephalus, Dendrophryniscus, Melanophryniscus,* and *Oreophrynella.* Parker (1931) presented reasonable arguments for deriving *Didynamipus* from the highly specialized (Tihen, 1960) *Nectophryne,* and Noble (1926*b*) did the same for deriving all of the American genera from *Bufo* but deferred from stating so because he was convinced (1926*b*) that *Bufo* was a recent immigrant to South America. He postulated that the immediate ancestors were extinct ("lost").

I have accorded familial rank to *Rhinoderma* as a monotypic family, Rhinodermatidae. *Geobatrachus* has been associated with *Rhinoderma* since Noble's (1922) work but is poorly known anatomically and may be a procoelous microhylid. *Rhinoderma* is usually accorded subfamily rank, but its relationships are not well represented as either a dendrobatid (Cei, 1962) or a leptodactylid (Griffiths, 1963). In many respects *Rhinoderma* agrees with the atelopodine grade of bufonids, in others it is like leptodactylids, and in a few respects is unique. As a family unit it is either to be regarded as a distinctive leptodactylid subfamily or as a monogeneric family (Lynch, 1969).

Pseudids and centrolenids are either accorded family rank or considered subfamilies of the Hylidae (Burger, 1954; Griffiths, 1963). Pending a better understanding of the relationships of the four genera and the interrelationships of the Neotropical hylids, the centrolenids and pseudids should be considered distinct families. Familial separation for these groups is as tenuous as the case involving the ranoid families (see below).

Griffiths (1959*b*, 1963) considered *Colostethus (Prostherapis), Dendrobates,* and *Phyllobates* to constitute a subfamily (Dendrobatinae) of the Ranidae. His data were in part erroneous. He demonstrated that dendrobatids are truly firmisternal and argued that they have a ranoid pattern in the disposition of the thigh muscles. He contended that the synsacral condition in *Dendrobates* precluded use of the vertebral centra in assessing relationships and apparently ignored the data of uniform procoely in *Colostethus* and *Phyllobates* (Noble, 1922), although he asserted that arthroleptine ranids are frequently procoelous, lack an osseous sternal element (like dendrobatids and unlike many ranids), and have similar reproductive patterns. Noble (1922, 1926*a*) and Lynch (1969) presented at least an equally compelling argument for considering dendrobatids firmisternal derivatives of elosiine leptodactylids. Ultimately, Griffiths' argument rests on his heavily weighted pectoral architecture character.

Following students of African frogs, I have considered *Phrynomerus* a subfamily of microhylids. Parker (1931, 1934) recognized the relationships of the genus but apparently considered the presence of intercalary cartilages as worthy of familial recognition.

Three ranoid family groups are often accorded family rank; three subfamilies possess intercalated cartilages (Hyperoliinae, Mantillinae, and Rhacophorinae), whereas the other seven lack the cartilages. Laurent (1951) divided the ranoids into two families on the basis of the presence or absence of an osseous sternal style (metasternum). The subfamilies with the style are more advanced with respect to that character than are those lacking the style, but Laurent's arrangement ignores many character complexes that vary differently among the subfamilies (Liem, 1969). Liem (1969) recognized three families: one for Laurent's hyperoliinae (Hyperoliidae), one for the Mantellinae and Rhacophorinae (Rhacophoridae), and a third (Ranidae) for Laurent's remaining subfamilies (Arthroleptinae, Astlylosterninae, Cacosterninae, Cornuferinae, Hemisinae, Petropedetinae, Raninae, and Scaphiophryninae). The inclusion of the Madagascaran scaphiophrynids is probably erroneous but is continued here. The Cornuferinae is probably not separable from the Raninae (Inger, 1966) and is here combined with that subfamily. Use of the subfamily name cannot be continued inasmuch as the type-species of *Cornufer* is an *Eleutherodactylus* (Zweifel, 1966). Rather than seek the next available family group name or coin another, I suggest that the ranoids should be seriously reviewed to assess the reasonability of Noble's (1931) subfamilial arrangement. Inger (1954, 1966) has studied the relationships of some of the genera in each subfamily found in the Western Indo-Australian archipelago, and his analysis does not inspire confidence in Noble's separation of the Cornuferinae and Raninae.

Liem's (1969) analysis appears to have more strength than those proposed by Noble (1931) or Laurent (1951 and other papers), but is subject to some criticism in that his considerations apply only to those groups having intercalated cartilages and does not address itself to the larger question of intra-ranoid relationships. Pending such an analysis, the status of the Hyperoliidae and Rhacophoridae should be held in abeyance and a single family, Ranidae, recognized. Griffiths (1963) combined the ranids with intercalated cartilages as a single family, Rhacophoridae, presumably following Noble's (1931) recognition of the group as a family.

REFERENCES

ANGEL, F. 1931. Contribution à'étude de la Faune de Madagascar, Reptilia et Batrachia. Faune Colloq. Franç., Paris 4:495-597.

BOCK, W. J. 1963. Evolution and phylogeny in morphologically uniform groups. Am. Naturalist 97:265-285.

BOULENGER, G. A. 1882. Catalogue of the batrachia salientia s. ecaudata in the collections of the British Museum. 2nd ed. British Museum (Nat. Hist.) 503 p.

———. 1897. The tailless batrachians of Europe. Vol. 1. The Ray Society, London. 210 p.

BRATTSTROM, B. H. 1957. The phylogeny of the salientia based upon skeletal morphology. Systematic Zool. 6(2):70-74.

BRUCH, C. 1863. Neue Beobachtungen zur Naturgeschichte der einheimischen Batrachier und Bericht über das. Brutjahr. Würzb. Nat. Z. 4:92-151.

BURGER, W. L. 1954. Two family-groups of neotropical frogs. Herpetologica 10:194-196.

CARVALHO, A. L. DE. 1954. A preliminary synopsis of the genera of American microhylid frogs. Occ. Pap. Mus. Zool., Univ. Michigan 555:1-19.

CEI, J. M. 1962. Batracios de Chile. Ed. Univ. Chile, Santiago. 128+cviii p.

COPE, E. D. 1864. On the limits and relations of the raniformes. Proc. Acad. Nat. Sci. Philadelphia 16:181-184.

———. 1865. Sketch of the primary groups of batrachia salientia. Nat. Hist. Rev. 5:97-120.

DAVIS, D. D. 1936. The distribution of Bidder's organ in the Bufonidae. Zool. Ser. Field Mus. Nat. Hist. 20:115-125.

DUMÉRIL, A. M. C., AND G. BIBRON. 1841. Erpétologie générale. Vol. 8. Librairie encyclopédique de Roret, Paris. 792 p.

DUNLAP, D. G. 1960. The comparative myology of the pelvic appendage in the salientia. J. Morphol. 106:1-76.

EATON, T. H., JR. 1958. An anatomical study of a neotropical tree frog, *Centrolene prosoblepon* (Salientia: Centrolenidae). Univ. Kansas Sci. Bull. 39(10):459-472.

ESTES, R. 1970. New fossil pelobatid frogs and a review of the genus *Eopelobates*. Bull. Mus. Comp. Zool., Harvard Univ. 139(6):293-349.

FARRIS, J. S. 1966. Estimation of conservatism of characters by constancy within biological populations. Evolution 20:587-591.
FÉJÉRVARY, A. M. 1918. Über die rüdimentären Rippen der anuren Batrachier. Verh. Zool.-Botan. Ges. Wien 1918: 114-128.
GRIFFITHS, I. 1959*a*. The phylogenetic status of the sooglossinae. Ann. Mag. Nat. Hist., Ser. 13, 2:626-640.
_______. 1959*b*. The phylogeny of *Sminthillus limbatus* and the status of the Brachycephalidae (Amphibia, Salientia). Proc. Zool. Soc. London 132:457-487.
_______. 1963. The phylogeny of the salientia. Biol. Rev. 38(2):241-292.
GRIFFITHS, I., AND A. L. DE CARVALHO. 1965. On the validity of employing larval characters as major phyletic indices in amphibia, salientia. Rev. Brasiliera Biol. 25:113-121.
HECHT, M. K. 1960. A new frog from an eocene oil-well core in Nevada. Am. Mus. Novitates 2006:1-14.
_______. 1963. A reevaluation of the early history of the frogs. Part II. Systematic Zool. 12(1):20-35.
HOWES, G. B., AND W. RIDEWOOD. 1888. On the carpus and tarsus of the anura. Proc. Zool. Soc. London 1888:141-182.
INGER, R. F. 1954. Systematics and zoogeography of Philippine amphibia. Fieldiana: Zool. 33:181-531.
_______. 1958. Comments on the definition of genera. Evolution 12:370-384.
_______. 1966. The systematics and zoogeography of the amphibia of Borneo. Fieldiana: Zool. 52:1-402.
_______. 1967. The development of a phylogeny of frogs. Evolution 21(2):369-384.
KLUGE, A. G., AND J. S. FARRIS. 1969. Quantitative phyletics and the evolution of anurans. Systematic Zool. 18(1):1-32.
LATSKY, L. 1930. Die Sistematiese Posiesie van *Heleophryne* met Betrekling tot die Klassifikasie van Noble (1922). South African J. Sci. 27:442-445.
LAURENT, R. F. 1940. Contribution à l'ostéologie et à la systématique des ranides africains. Première note. Rev. Zool. Botan. africaines 34(1):74-97.
_______. 1941. Contribution à l'ostéologie et à la systématique des rhacophorides africains. Première note. Rev. Zool. Botan. africaines 35(1):85-111.
_______. 1942. Note sur l'ostéologie de *Trichobatrachus robustus.* Rev. Zool. Botan. africaines 36(1):56-60.
_______. 1943*a*. Sur la position systématique et l'ostéologie du genre *Mantidactylus* Boulenger. Bull. Mus. Roy. Hist. Nat. Belg. 19(5):1-8.
_______. 1943*b*. Contribution à l'ostéologie et à la systématique des rhacophorides non africains. Bull. Mus. Roy. Hist. Nat. Belg. 19(28):1-16.
_______. 1951. Sur la nécessité de supprimer la famille des Rhacophoridae mais de créer celle des Hyperoliidae. Rev. Zool. Botan. africaines 45(1):116-122.
LIEM, S. S. 1969. The morphology, systematics, and evolution of the old world treefrogs (Rhacophoridae and Hyperoliidae). Ph.D. Dissertation, Univ. Illinois.
L'ISLE, A. DE. 1877. Sur un genre nouveau de batraciens bufoniformes. J. Zool. 6:472-478.
LYNCH, J. D. 1969. Evolutionary relationships and osteology of the frog family Leptodactylidae. Ph.D. Dissertation, Univ. Kansas.
MADEJ, Z. 1965. Variations in the sacral region of the spine in *Bombina bombina* (Linnaeus, 1761) and *Bombina variegata* (Linnaeus, 1758) (Salientia, Discoglossidae). Acta Biol. Cracoviensia 8:185-197.
NEVO, E. 1968. Pipid frogs from the early cretaceous of Israel and pipid evolution. Bull. Mus. Comp. Zool., Harvard Univ. 136(8):225-318.
NICHOLLS, G. C. 1916. The structure of the vertebral column in anura phaneroglossa and its importance as a basis of classification. Proc. Linn. Soc. London, Z, 128:80-92.
NOBLE, G. K. 1921. Five new species of salientia from South America. Am. Mus. Novitates 29:1-7.
_______. 1922. The phylogeny of the salientia. I—the osteology and the thigh musculature; their bearing on classification and phylogeny. Bull. Am. Mus. Nat. Hist. 46:1-87.

———. 1924*a*. A new spadefoot toad from the oligocene of Mongolia with a summary of the evolution of the Pelobatidae. Am. Mus. Novitates 132:1-15.

———. 1924*b*. Contributions to the herpetology of the Belgian Congo based on the collection of the American Museum Congo Expedition, 1909-1915. Bull. Am. Mus. Nat. Hist. 49:147-347.

———. 1926*a*. The pectoral girdle of the brachycephalid frogs. Am. Mus. Novitates 230:1-14.

———. 1926*b*. An analysis of the remarkable cases of distribution among the amphibia, with descriptions of new genera. Am. Mus. Novitates 212:1-24.

———. 1931. The biology of the amphibia. McGraw-Hill Book Co., New York. 577 p.

Orton, G. L. 1943. The tadpole of *Rhinophrynus dorsalis*. Occ. Pap. Mus. Zool., Univ. Michigan 472:1-7.

———. 1949. Larval development of *Nectophrynoides tornieri* (Roux), with comments on direct development in frogs. Ann. Carnegie Mus. 31:257-276.

———. 1953. The systematics of vertebrate larvae. Systematic Zool. 2(2):63-75.

———. 1957. The bearing of larval evolution on some problems in frog classification. Systematic Zool. 6(2):79-86.

Parker, H. W. 1931. Parallel modifications in the skeleton of the amphibia salientia. Arch. Zool. Italiano 16:1239-1248.

———. 1934. A monograph of the frogs of the family Microhylidae. British Museum (Nat. Hist.), London. 208 p.

———. 1940. The Australasian frogs of the family Leptodactylidae. Novitates Zool. 42(1):1-106.

Reig, O. A. 1957. Los anuros del Matildense. *In* P. N. Stipanicic and O. A. Reig, "El complejo porfírico de la Patagonia extraandina y su fauna de anuros." Act. Geol. Lilloana 1:231-297.

———. 1958. Proposiciones para una nueva macrosistemática de los anuros. Nota preliminar. Physis. 21:109-118.

Ritland, R. M. 1955. Studies on the post-cranial morphology of *Ascaphus truei*. II. Myology. J. Morphol. 97(2):215-282.

Savage, J. M., and A. L. de Carvalho. 1953. The family position of neotropical frogs currently referred to the genus *Pseudis*. Zoologica 38:193-200.

Sokal, R. R., and P. A. Sneath. 1963. Principles of numerical taxonomy. W. H. Freeman and Co., San Francisco. 359 p.

Špinar, Z. V. 1966. Some further results of the study of Tertiary frogs in Czechoslovakia. Časopis mineral. geol. 11:431-440.

Sporne, K. R. 1954. Statistics and the evolution of dicotyledons. Evolution 8:55-64.

Taylor, E. H. 1941. A new anuran from the middle miocene of Nevada. Univ. Kansas Sci. Bull. 27:61-69.

———. 1951. Two new genera and a new family of tropical frogs. Proc. Biol. Soc. Washington 64:33-37.

Thomas, A. 1854. Note sur la génération du Pélodyte ponctué, avec quelques observations sur les batraciens anoures en général. Ann. Sci. Nat., Ser. 4, 1:290-293.

Tihen, J. A. 1960. Two new genera of african bufonids, with remarks on the phylogeny of related genera. Copeia 3:225-233.

———. 1965. Evolutionary trends in frogs. Am. Zool. 5:309-318.

Trewavas, E. 1933. The hyoid and larynx of the anura. Phil. Trans. Roy. Soc. London, B, 222(10):401-527.

Trueb, L. 1970. Evolutionary relationships of casque-headed tree frogs with co-ossified skulls (family Hylidae). Univ. Kansas Publ. Mus. Nat. Hist. 18(7):547-716.

———. 1971. Phylogenetic relationships of certain neotropical toads with the description of a new genus (Anura, Bufonidae). Los Angeles Co. Mus. Contrib. Sci. 216: 1-40.

Tyler, M. J. 1968. Papuan hylid frogs of the genus *Hyla*. Zool. Verh. Leiden 96:1-203.

Wagler, J. G. 1830. Naturliches System der Amphibien, mit vorangehender Classification der Säugethiere und Vögel. Stuttgart und Tübingen, Munich. 354 p.

Wagner, W. H., Jr. 1961. Problems in the classification of ferns, p. 841-844. *In* Recent advances in botany. Univ. Toronto Press, Toronto.

WALKER, C. F. 1938. The structure and systematic relationships of the genus *Rhinophrynus.* Occ. Pap. Mus. Zool., Univ. Michigan 372:1-11.

WILSON, E. O. 1965. A consistency test for phylogenies based on contemporaneous species. Systematic Zool. 14:214-220.

ZWEIFEL, R. G. 1956. Two pelobatid frogs from the tertiary of north america and their relationships to fossil and recent forms. Am. Mus. Novitates 1762:1-49.

———. 1966. *Cornufer unicolor* Tschudi 1838 (Amphibia, Salientia); request for suppression under the plenary powers. Bull. Zool. Nomencl. 23:167-168.

4

BIOCHEMICAL TECHNIQUES AND PROBLEMS IN ANURAN EVOLUTION

Sheldon I. Guttman

INTRODUCTION

Recent findings in molecular biology indicate that protein structure results from the indirect translation of the genetic message encoded in part of a DNA strand. The polypeptide chain produced in this manner can be considered the primary phenotype of an individual gene (Guttman, 1972*a*). Various interactions and combinations of these polypeptide chains, at particular time intervals, yield the secondary and tertiary phenotypes of the organism and result in its morphological and physiological attributes. The more similar the genetic constitution of two organisms, the more closely they are related. From this basic observation, it follows that the degree of similarity between the proteins of organisms is directly proportional to their degree of relationship (Sibley, 1964).

One can determine genetic relationship directly, by comparing proteins, or indirectly, by studying molecules whose formation is dependent upon enzymes. Since enzymes are proteins and enzyme systems are probably highly specific, indolealkylamines, pteridines and their precursors, and metabolites may be considered genetic traits and used with proteins for determining genetic similarity.

Biochemical data from a variety of organisms have been used successfully to determine evolutionary relationships. Proteins of the *Drosophila virilis* species group were examined electrophoretically by Hubby and Throckmorton (1965). The biochemical phylogeny derived from their data closely paralleled a cytological phylogeny. It was possible to reconstruct, in part, some of the gene pools that existed early in the history of the species group. Butler and Leone (1967) immunologically compared 66 species of beetles from 24 families; several unsuspected relationships were indicated. Starch-gel electrophoresis and agar-gel precipitin reactions showed that a revision of the taxonomy of the Hominoidea is required (Goodman, 1964).

BIOCHEMICAL TECHNIQUES

Chromatographic, electrophoretic, and serological techniques are the most frequently used methods for biochemical comparisons of organisms.

A. Chromatography

Chromatography may be defined as a technique in which the components of a mixture are caused to migrate at different rates through an apparatus that involves equilibration of solutes between a stationary phase and a moving phase

(Cantarow and Schepartz, 1957). A small amount of sample is applied near one end of a strip of filter paper, the end is dipped into an appropriate solvent mixture, and the solvent is permitted to travel through the paper by capillarity. At the end of the desired period of time, the position of the solvent "front" is marked and the paper is dried. Spraying the paper with the proper reagent then reveals the various constituents of the mixture as separate spots. The ratio of the distance traveled by any particular substance to that covered by the solvent is known as the ratio of the "fronts," or R_f, and is usually constant for each substance in a particular solvent. Usually, this simple, or "one-dimensional," chromatogram is unable to separate some of the constituents of a complex mixture. In such a case, a sample is placed in one corner of a square sheet of paper that is chromatographed in one direction with one solvent system, then turned and run with another solvent system in a direction at right angles to the first, producing a "two-dimensional" chromatogram. This technique has been utilized for separation of integumental pteridines and biogenic amines following the extraction of these compounds in alcohol or acetone.

B. Electrophoresis

Protein mixtures are fractionated by electrophoresis due to the differential migation of the components in an electric field (Bier, 1959). The rate of migration, and therefore the distance migrated, depends upon two opposing forces: the net charge of the protein molecule and the resistance of the medium through which the protein migrates. Net charge is determined by the dissociation of acidic and basic groups and the possible binding of buffer ions to which the proteins are exposed. When the buffer composition is held constant, the average number of charges on particular proteins is fixed; therefore, the electrical force applied to these components, under a given potential gradient, is also fixed. The mobility of the protein through the medium, then, depends upon the two opposing forces: the electrical force and the force of frictional retardation.

Two basic types of electrophoresis are utilized for protein separations: zone electrophoresis *on* buffer impregnated solids and *in* concentrated gels. With the former method, the proteins migrate over the surface of the solid, which may be filter paper, cellulose acetate, or other material. However, many serum proteins of vastly different molecular size have essentially equal mobilities with this technique because the greater frictional retardation of the large proteins often is compensated for by their greater net charge (Smithies, 1959). In concentrated gels, such as polyacrylamide and starch, the larger molecules are retarded by the small pores in the gel; therefore, fewer proteins have equal mobilities. In addition, with electrophoresis on buffer impregnated solids, the protein molecules sometimes adsorb to the solid; this does not occur with the latter technique. Therefore, when separations are made in concentrated gels, they are more complete. For example, Brown (1964) separated the serum proteins of *Bufo woodhousei* into eight fractions using paper electrophoresis, whereas Guttman (unpublished data) resolved sixteen fractions in plasma from the same species by starch-gel electrophoresis.

Various histochemical methods are available for localizing proteins on an electropherogram. Techniques are known for proteins in general, and specific stains detect lipoproteins, glycoproteins, proteins containing copper or iron, enzymes and hemoglobin, either free or in combination with haptoglobins (Shaw and Koen, 1968; Smithies, 1959).

When techniques for electrophoresis in concentrated gels were first introduced, the large number of protein zones observed made it necessary to correlate these results with those obtained previously by electrophoresis on buffer-impregnated solids. Correlation was obtained by inserting a section of a buffer-impregnated electropherogram into a gel slab, adjacent to a sample of whole serum, and then subjecting both samples to electrophoresis. Proteins common to the section and the whole serum were compared. Results indicated that the separation procedure in concentrated gels differed from that on buffer-impregnated solids (Smithies, 1959). For example, a protein that migrated in the α_1-globulin region on filter paper, migrated as a prealbumin in starch-gels. Some proteins that were slower than the major β-globulin component in gels migrated faster, as α_2-globulins, on filter paper. These findings suggested that if the two processes were combined and run at right angles to each other, a more complete separation would result than if either process were used alone. Poulik and Smithies (1958) discuss the two-dimensional procedure in detail.

C. Serology

The proteins to be investigated, or antigens, are injected into unrelated species and stimulate the synthesis of specific serum globulins, or antibodies, which are complementary to the introduced antigens. When the serum from an immunized animal, or antiserum, is exposed to the original antigen, an antigen-antibody reaction occurs, and a precipitate is produced (Hirschfeld, 1960).

The antibodies and antigens are usually allowed to diffuse through a stabilizing medium, such as an agar gel; this process is called double diffusion. When parts of an immunologically homogeneous antigen-antibody system meet in optimal proportions, one or more precipitation lines appear in the medium. The number of precipitation lines ideally yields information concerning the number of antigens in the extract against which there are antibodies in the antiserum. It appears that each immunoprecipitate acts as a selective barrier; the antigens and antibodies that produced it cannot pass, while immunologically unrelated molecules pass unhindered. Each immunoprecipitate's location depends upon the concentration and diffusibility of the components. Therefore, it is probable that different complexes appear as distinct precipitin lines. The Ouchterlony technique (Ouchterlony, 1948) is the most widely adopted double-diffusion method; however, it has its limitations because a single precipitation line may be formed from several antigen-antibody complexes. Additionally, it is extremely difficult to delimit the actual number of lines present when a complex mixture, such as serum, is analyzed (Poulik, 1960).

The development of immunoelectrophoresis resulted when zone electrophoresis and double-diffusion antigen-antibody reactions were combined. This method takes advantage of the benefits of both techniques and eliminates some of the drawbacks of each. The samples are first subjected to electrophoresis in a supporting medium; following this, the separated components are allowed to diffuse from the supporting medium into a layer of agar. Antiserum is applied into a well located parallel to the axis of electrophoretic migration of the antigens and is permitted to diffuse perpendicularly into the same agar layer. A precipitation line results whenever homologous antigens and antibodies meet in optimal proportions.

Perhaps the most sophisticated serological procedure is quantitative microcomplement fixation (MC′F). This technique, developed by Wasserman and Levine (1961), has several advantages not available with other serological methods. First, because it is a microtechnique, 100 to 1000 times less antigen and antiserum are required than for more conventional procedures (Salthe and Kaplan, 1966). Second, MC′F is sensitive to very small modifications in antigen structure and can distinguish between human hemoglobins A, S, and C, which differ by only two amino-acid residues out of a total of 574 (Reichlin et al., 1964, 1966).

Diluent, antibody, complement, and varying concentrations of antigen are pipetted into a series of tubes. These mixtures are incubated for approximately 18 hours at 2° C to 4° C, and then sensitized erythrocytes are added directly to the reaction mixtures (Wasserman and Levine, 1961). After allowing hemolysis to proceed for 60 mintues, the reaction mixtures are immersed in an ice bath in order to discontinue hemolysis and are then centrifuged to sediment-unlysed erythrocytes. The supernatant is analyzed for hemoglobin in a spectrophotometer. Cross reactions are used to generate an index of dissimilarity (I.D.), the factor by which the antiserum concentration must be raised so that a particular antigen yields a MC′F reaction equal to that produced by the homologous antigen (Sarich and Wilson, 1966).

PHYSIOLOGICAL AND DEVELOPMENTAL FACTORS THAT INFLUENCE PATTERNS

Prior to establishing evolutionary relationships based on biochemical analyses, physiological and developmental variation must be considered. Qualitative variation of anuran patterns caused by changes in physiological condition has not been reported. However, quantitative fluctuations are common; these occur due to disease (Bucovaz and Kaplan, 1957; Guttman, unpublished data), to starvation (Brown, 1964), and with the initiation of breeding condition (Cei and Bertini, 1961).

Sexual differences have also been found to contribute to biochemical variation. Females of the genus *Rana* possess an esterase not found in the males (Augustinsson, 1959). Chen (1967) studied eleven species of amphibians, and he noted that one, *Bombina variegata,* appeared to show a sexual difference. In a more refined analysis of the serum of this frog, Huchon et al. (1968) conclusively demonstrated, using two-dimensional electrophoresis and immunoelectrophoresis, that one of the β-lipoproteins is present only in the female (Figures 4–1 and 4–2). Rossi (1960) noted that between thirty and sixty days postmating, there are differences in the protein fractions of male and female *Bufo bufo.*

Parental influences on lactate dehydrogenase (LDH) in early development of *Rana* were studied by Wright and Moyer (1966, 1968). They found that the maternal pattern persists in the embryo until heartbeat stage, at which time paternal and hybrid bands first appear. The maternal contribution to the LDH remains high and persists until at least the eleventh day after feeding begins.

Investigations of larval and metamorphosed anurans indicate that unique patterns are present at each developmental stage. Frieden (1961) and Bennett and Frieden (1962) found that young tadpoles have a low plasma albumin concentration and that as the tadpole matures, albumin concentration increases, reaching a maximum immediately following metamorphosis. The percentage of albumin in the plasma of *Xenopus laevis,* however, continues to rise after the tadpoles metamorphose (Frieden, 1961). Frieden also noted that hemoglobin is different in tadpoles and metamorphosed frogs. In *Rana catesbeiana,* there is an increase in both the number of molecular species of serum proteins and the total quantity of proteins as tadpoles develop legs and metamorphose (Richmond, 1968). Bagnara and Obika (1965) found that some pteridines (hynobius blue, sepiapterin, and drosopterin) are present in larval and newly metamorphosed *Bufo punctatus* but not in the adult. In contrast, ranachrome-3 is present in adult *Rana* and *Hyla cinerea* but is absent in the larvae. Sepiapterin is present in all larval hylids and ranids but persists only in adults of *Hyla arenicolor* and *Rana sylvatica.*

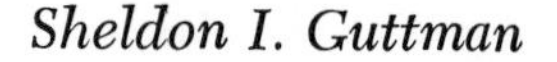

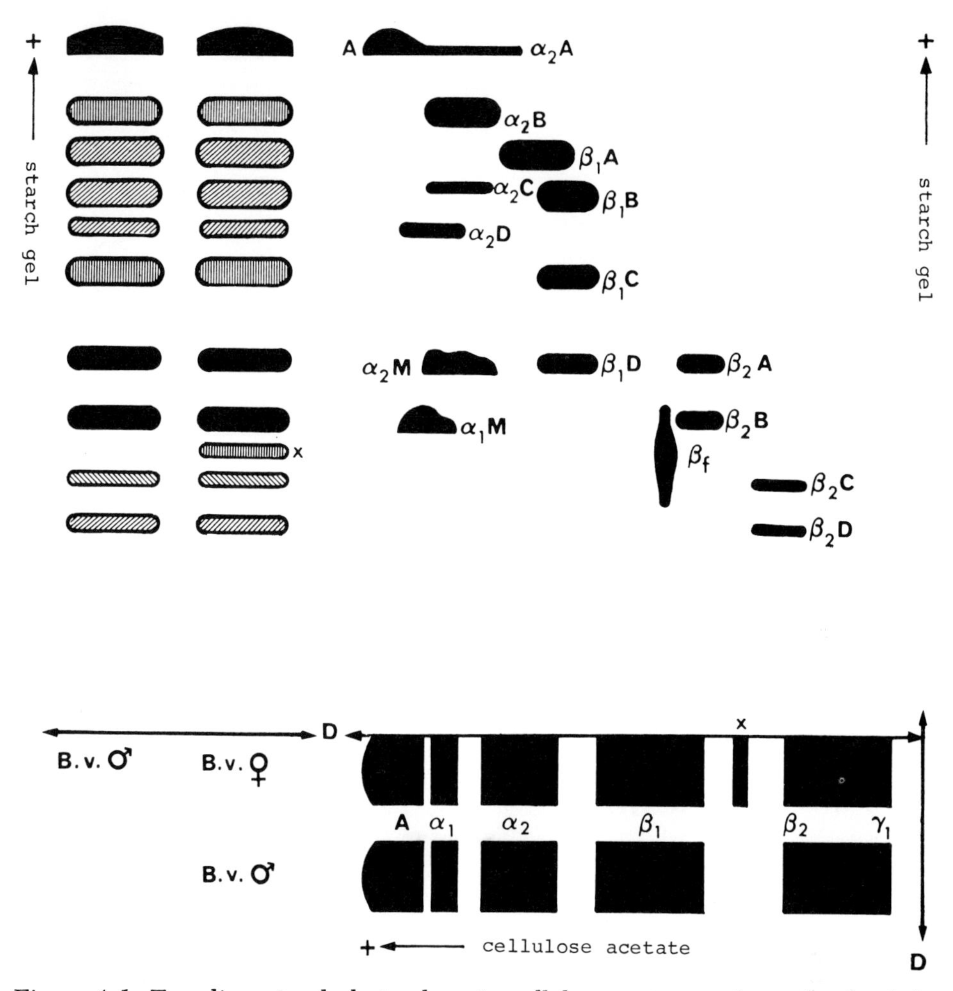

Figure 4–1. Two-dimensional electrophoresis, cellulose acetate, and starch-gel, of the adult serum from the two sexes of *Bombina variegata*. The x and the β_f signs indicate the female-specific fraction. Reprinted, by permission of Compt. Rend. Acad. Sci. Paris, from Huchon, Chalumeau-Le Foulgoc, and Gallien (1968).

Polymorphism

Each organism in a population is morphologically, physiologically, and biochemically unique (Mayr, 1963; Williams, 1956). Prior to the development and use of current sophisticated procedures, the biochemical individuality of anurans went largely unnoticed.

Since the utilization of techniques with increased resolution commenced, many anurans with variant molecules have been discovered. Hemoglobin polymorphism has been noted in *Bufo rangeri* and *B. regularis* (Guttman, 1967), the *B. americanus* species group (Guttman, 1969), *B. arenarum* and *B. spinulosus* (Brown and Guttman, 1970), and eleven other species of *Bufo* (Guttman, 1972*b*). Toad transferrins have been found to be even more variable than hemoglobins (Brown and Guttman, 1970; Dessauer et al., 1962; Guttman, 1967, 1969, 1972*b*). In fact, as many as eight transferrin and five albumin alleles are present in single populations of *B. americanus* (Figures 4–3, 4–4). *B. americanus* esterases are also

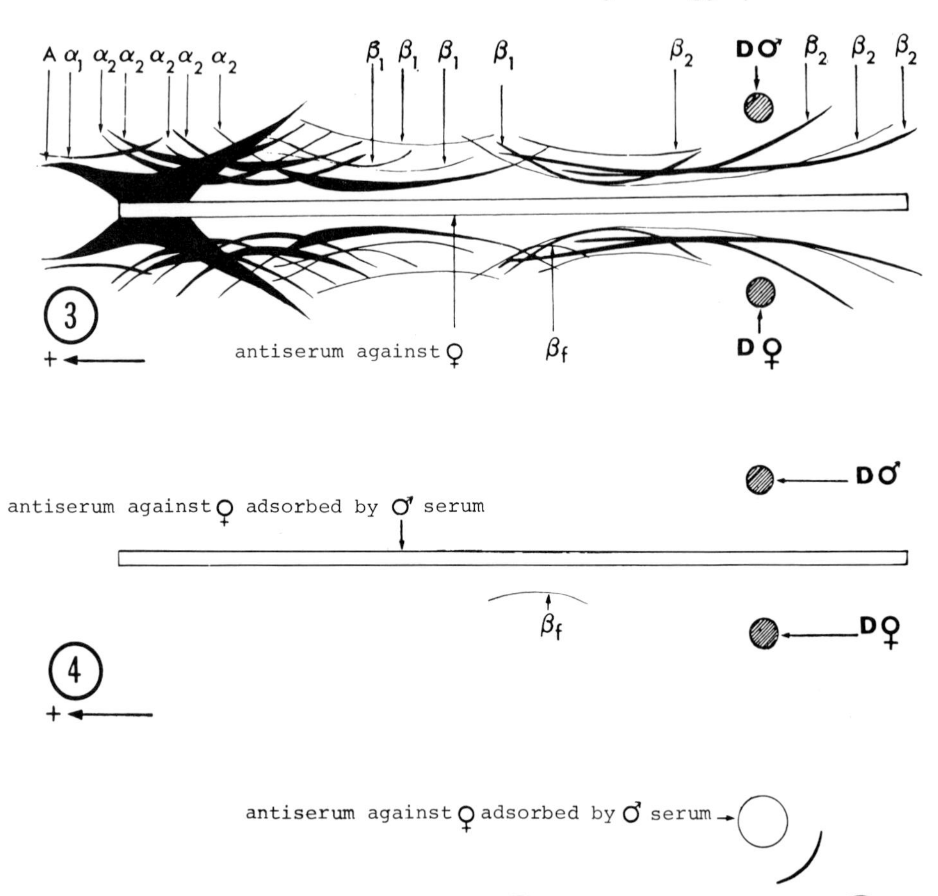

Figure 4–2. Immunological analysis: 3, comparative immunoelectrophoretic analysis of the serum of a male and female *Bombina* against antiserum to female *Bombina* serum; 4 and 5, comparative analysis of the serum of a male and female *Bombina* against antiserum to female *Bombina* serum that was previously allowed to react with the serum of a male *Bombina*. 4 was by immunoelectrophoresis, and 5 was by double-diffusion. Reprinted, by permission of Compt. Rend. Acad. Sci. Paris, from Huchon, Chalumeau-Le Foulgoc, and Gallien (1968).

polymorphic (Figure 4–5). Electrophoretic patterns of whole plasma varied in single populations of *B. americanus* (Balsano et al., 1965), *B. valliceps* and *B. woodhousei* (Fox et al., 1961). Polymorphism was exhibited by the transferrins, nicotinamide adenine dinucleotide linked malate dehydrogenase, three liver esterase loci (Dessauer and Nevo, 1969), and heart lactate dehydrogenase of *Acris crepitans* (Salthe and Nevo, 1969). Electrophoretic patterns of lens and muscle homogenates from *Bufo americanus, B. fowleri, Rana clamitans,* and *R. pipiens* also showed intrapopulation variation (Schmiel and Guttman, unpublished data). Ralin et al. (unpublished data) noted polymorphism in malate dehydrogenase and indophenol oxidase in populations of *Hyla chrysoscelis* and *H. versicolor.* Variation in electrophoretic patterns of *B. fowleri* and *B. valliceps* parotoid venom was found by Wittliff (1964); in addition, he noted individual variation in ninhydrin reactions on chromatograms (Wittliff, 1962).

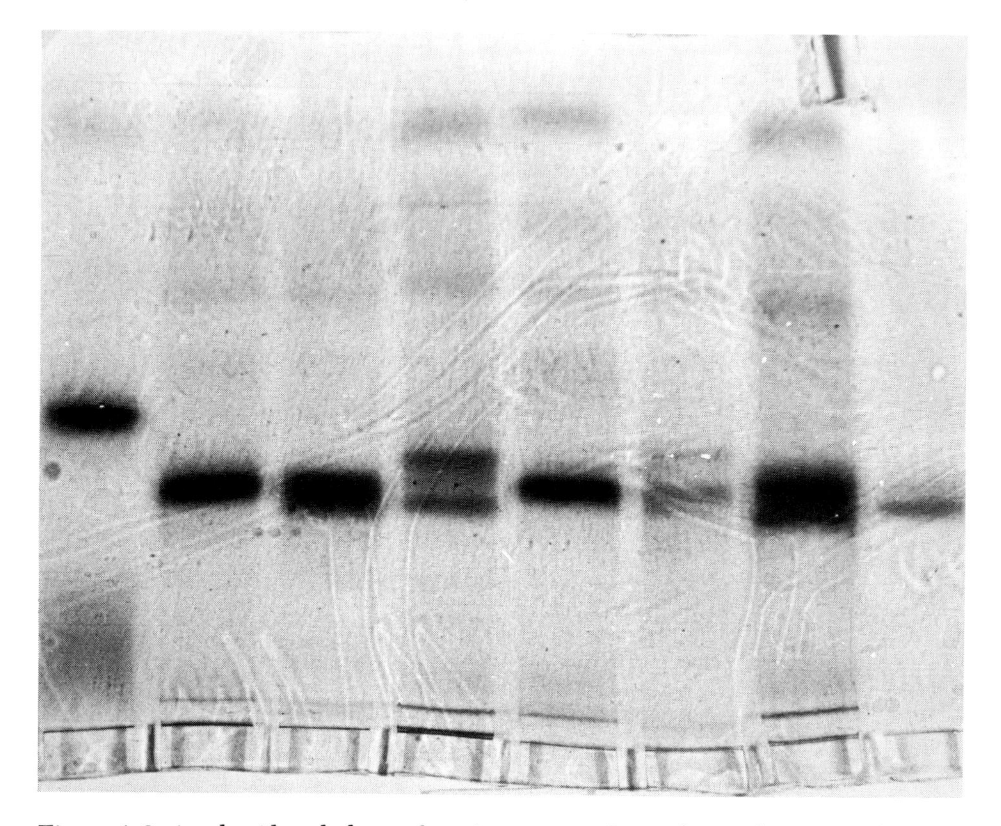

Figure 4–3. Acrylamide-gel electrophoretic patterns of transferrins from a single population of *Bufo americanus*. Vertical acrylamide-gel electrophoresis was performed using a Tris-glycine (0.10 M Tris, 0.05 M glycine) buffer, pH 8.9. Samples were prepared by the rivanol precipitation technique (Guttman, 1967). A 7.5 per cent gel was used, electrophoresis was performed at 400 volts for 2.5 hours, and the gel was stained in amido black. The intensely stained bands are the transferrins. Samples were, from left to right, human Tf C and seven *B. americanus*. The anode is at the top; the origin at the bottom.

Interpopulation Variation

Mayr (1963) considers the variation of geographically isolated populations to be a universal phenomenon. Just as they may differ in morphological attributes, one would expect them to differ in the frequency of occurrence of particular proteins or to be characterized by specific proteins (Dessauer and Fox, 1964). Since biochemical individuality exists, as a consequence of sexual reproduction no two groups of organisms can be identical. Every local population is under strong and continual selection pressure for maximal fitness in its particular habitat, and therefore, if the populations exist in different environments, minor differences may become accentuated (Mayr, 1963).

Balsano et al. (1965) found distinct frequencies of four of nine proteins in adjacent *Bufo americanus* populations. Chromatograms of *Bufo* skin secretions indicate population specificity (Cei et al., 1968, 1972; Hunsaker et al., 1961; Low, 1968, 1972; Porter, 1964, 1968). Each subspecies of *Leptodactylus* examined by Erspamer et al. (1964) exhibited a particular amine spectrum. Geographic variation of heart lactate dehydrogenase was noted in *Rana pipiens* (Salthe, 1969) and *Acris crepitans* (Salthe and Nevo, 1969). Geographic divergence of *Acris crepi-*

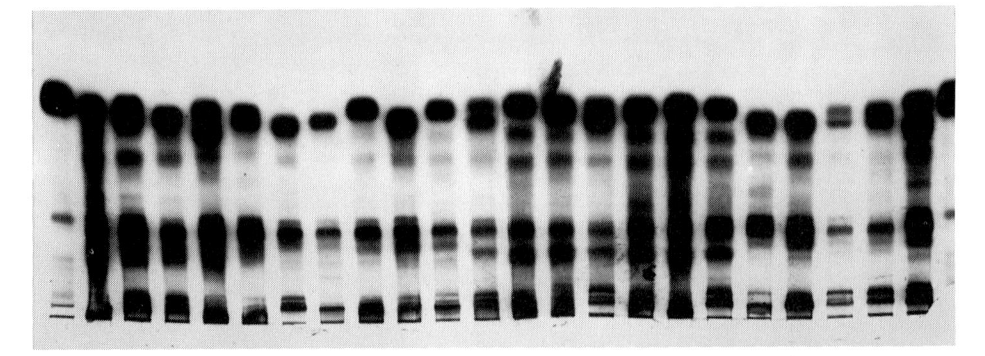

Figure 4–4. Acrylamide-gel plasma electrophoretic patterns from a single population of *Bufo americanus.* Vertical acrylamide-gel electrophoresis was performed, using a Tris-glycine (0.10 M Tris, 0.05 M glycine) buffer, pH 8.9. A 10 per cent gel was used, electrophoresis was performed at 350 volts for 3 hours, and the gel was stained in amido black. The intensely stained, rapidly-migrating components at the top of the figure are the albumins. Samples were, from left to right, human plasma, twenty-two *B. americanus,* and human plasma. The anode is at the top; the origin at the bottom.

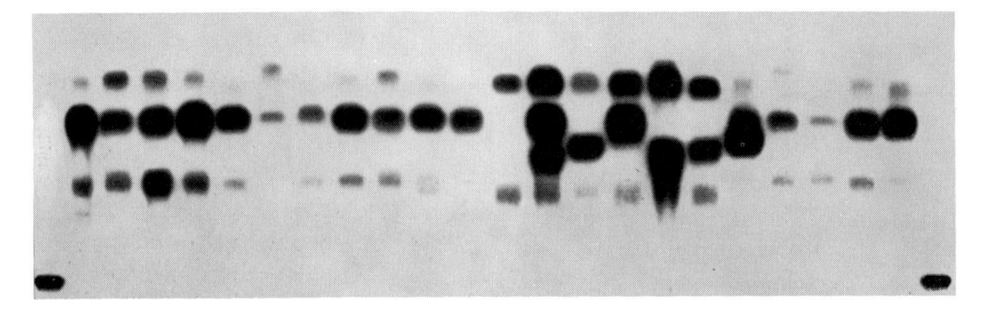

Figure 4–5. Acrylamide-gel electrophoretic patterns of plasma esterases from a single population of *Bufo americanus.* Vertical acrylamide-gel electrophoresis was performed as for Figure 4–4. Gels were sliced and stained for esterase activity as described by Shaw and Koen (1968). Samples were the same individuals used in Figure 4–4. The anode is at the top; the origin at the bottom.

tans into Plains, Delta, and Appalachian groups was indicated by the presence of specfic globin (Figure 4–6), transferrin (Figure 4–7), and liver esterase variants (Dessauer and Nevo, 1969).

The amount of interbreeding that is occurring and has occurred may be determined by comparisons of protein frequencies (gene frequencies) in adjacent populations. Upon examination of sufficient populations, the direction of gene flow and evolutionary pathways may be elucidated. For example, Abramoff et al. (1964) obtained an approximation of the route taken by *Bufo americanus* in colonizing the islands in the upper Lake Michigan area. Their electrophoretic and serological data indicate that populations inhabiting the islands are serologically similar to each other and are more closely related to the populations of the eastern mainland than to those of the western mainland. Guttman (1967) found, by examining the transferrins and hemoglobins of thirteen populations of the widespread African toad *Bufo regularis,* that this species may be composed of two forms that meet and exchange genes in Rhodesia and Kenya (Figure 4–8).

Apparently, *Acris crepitans* has differentiated into two groups: a Plains group, adapted to the grassland biome, and an Appalachian group, adapted to the deciduous forest biome (Dessauer and Nevo, 1969). Each group is characterized by specific proteins. A Delta group appears to represent a clinal bridge between the other two groups. The biochemical evidence indicates gene flow among the three groups. This occurs chiefly in the southern portion of the range, which indicates a long continuum there and a more recent colonization in the northern segment of the range following glacial recession.

Interspecific Comparisons

Although the previous discussions stressed the large amount of individual, population, and interpopulation molecular variability, conspecific individuals usually share more molecules than heterospecific individuals. Species specific patterns were reported for species in the following genera of anurans: *Acris, Alytes, Bombina, Bufo, Hyla, Leptodactylus, Melanophryniscus, Microhyla, Pleuroderma, Pseudacris, Rana, Scaphiopus,* and *Xenopus* (Barrio, 1964; Bertini and Cei, 1959,

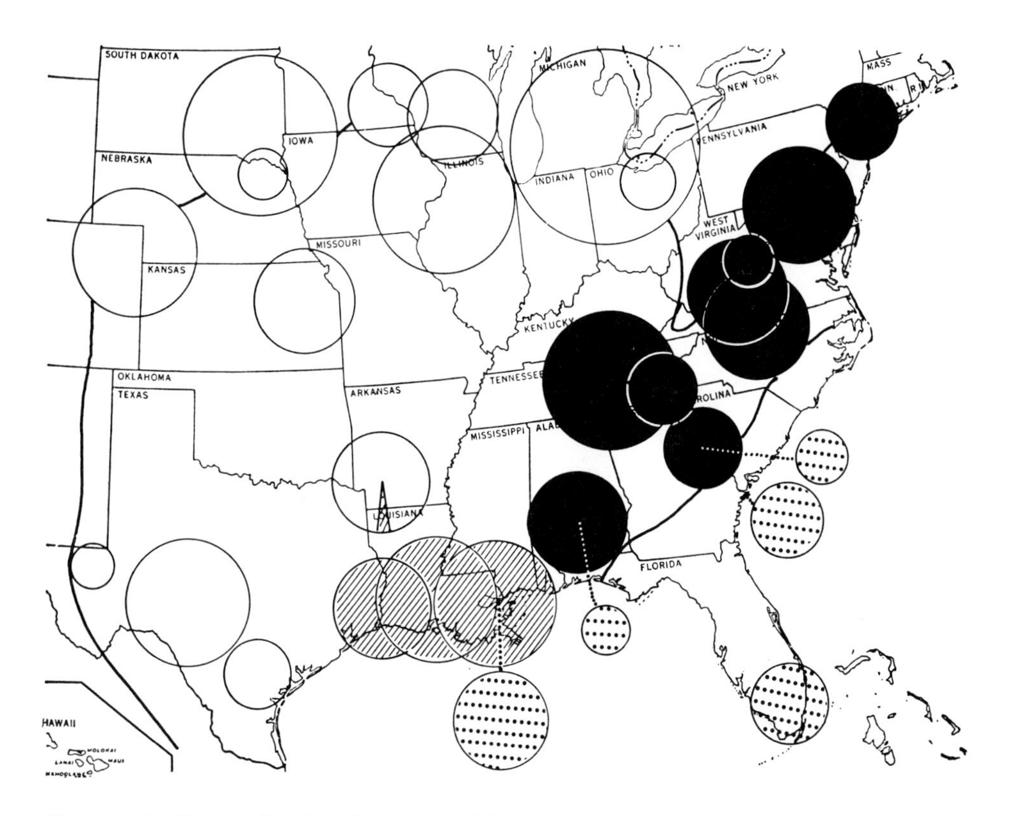

Figure 4–6. Geographic distribution of globin polypeptides among populations of *Acris*. Pie graphs are centered over collecting sites, except for certain populations of *A. gryllus*. Total area of a circle is proportional to the sample size; areas of sectors are in proportion to the frequency of different globin polypeptides in the sample. *White circles* of sectors indicate the presence of polypeptides 1, 2, 3, 4, and 5; *black* indicates the presence of 1_s, 2_s, and 3_s variants; *lines* show the occurrence of variant 5_s; *dots, A. gryllus* pattern. Reprinted, by permission of Plenum Publishing Corp., from H. C. Dessauer and E. Nevo, Geographic variation of blood and liver proteins in cricket frogs, Biochemical Genetics 3(1969).

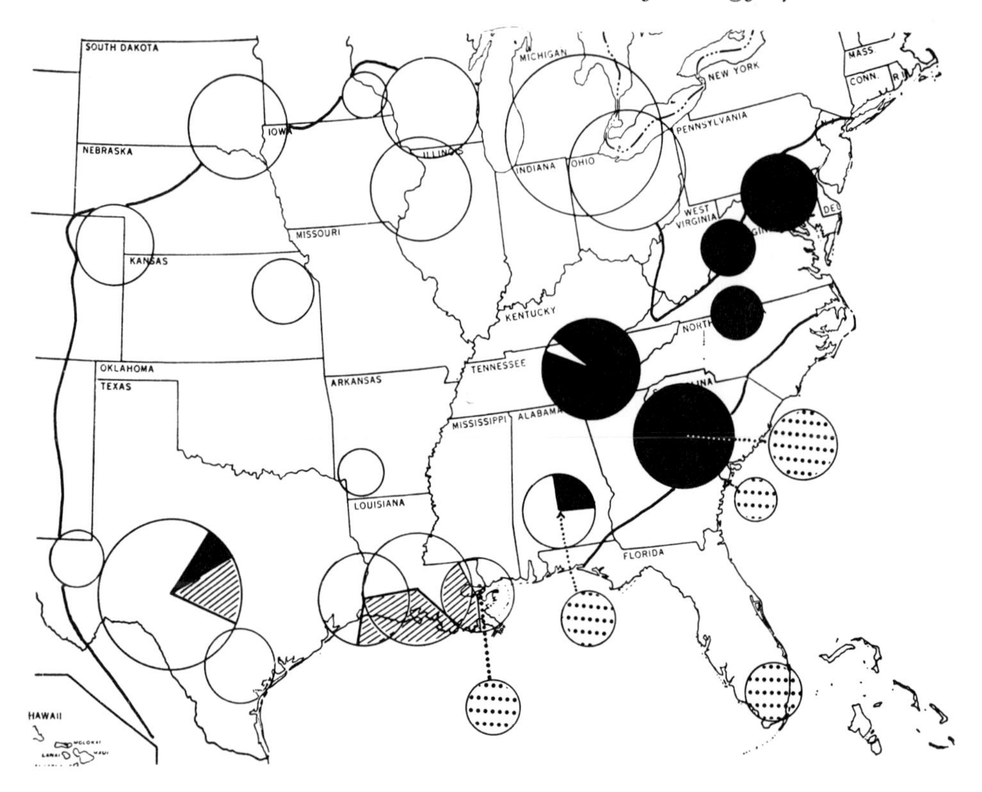

Figure 4-7. Geographic distribution of transferrins among populations of *Acris* (see legend of Figure 4–6). *White circles* represent the frequency of Tr W; *black circles,* $TrS_1 + TrS_2$; *lines,* $TrF_1 + TrF_2$; *dotted circles,* $TrGr_1 + TrGr_2$. Reprinted, by permission of Plenum Publishing Corp., from H. C. Dessauer and E. Nevo, Geographic variation of blood and liver proteins in cricket frogs, Biochemical Genetics 3(1969).

1961; Cei and Bertini, 1961; Cei et al., 1968, 1972; Chen, 1967; Dessauer and Fox, 1964; Dessauer and Nevo, 1969; Erspamer et al., 1964; Fox et al., 1961; Hebard, 1964; Hunsaker et al., 1961; Low, 1972; Porter, 1964; Porter and Porter, 1967; Ralin et al., (unpublished data); Schmiel and Guttman, (unpublished data); Wittliff, 1962).

Interspecific Hybridization

The proteins and biogenic amines of isolated conspecific populations and heterospecific populations are different. Biochemical studies of hybrids can provide evidence regarding the mechanisms of inheritance of these molecules, the amount of gene flow occurring between populations, and possibly, the role of hybridization in species formation.

Recent investigations indicate that the inheritance of particular anuran blood proteins is under simple genetic control. Codominant, multiple alleles existing at a single locus were suggested as the genetic mechanism for the inheritance of transferrin and hemoglobin differences found in *Bufo* (Fox et al., 1961; Guttman, 1969, 1972*b*). Comprehension of this pattern of inheritance led to a clarification of the hybridization and introgression between *Bufo regularis* and *B. rangeri* in South Africa (Figure 4–8; Guttman, 1967). Electrophoretic analysis of proteins

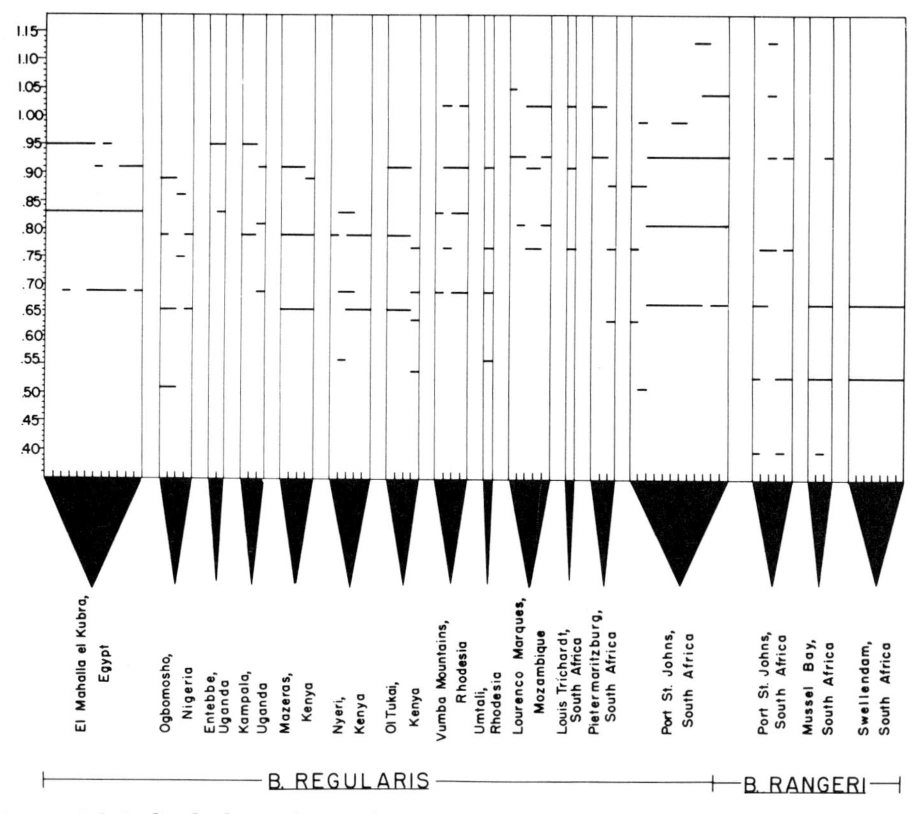

Figure 4–8. Individual transferrin phenotypes of *Bufo regularis* and *B. rangeri. Vertical axis,* relative mobility; *small spaces between vertical lines* along the horizontal axis, the number of individuals sampled from each population. Reprinted, by permission of Pergamon Press, from S. I. Guttman, Transferrin and hemoglobin polymorphism, hybridization and introgression in two African toads, *Bufo regularis* and *Bufo rangeri,* Comparative Biochemistry and Physiology, 23 (1967).

also showed that gene exchange occurs between some members of the *Bufo americanus* species group in the United States (Guttman, 1969) and between *B. arenarum* and *B. spinulosus* in Argentina (Figure 4–9; Brown and Guttman, 1970), but there is no exchange of genes between *Acris crepitans* and *A. gryllus* (Dessauer and Nevo, 1969).

Hybrids typically have two hemoglobin or transferrin components, one from each parental species. However, hybrid effects may occur; in some crosses, new, hybrid bands are noted (Guttman, 1972*b*), while in others, there is a suppression of parental types (Brown and Guttman, 1970; Guttman, 1972*b*). Chromatographic patterns of venoms from hybrid *Bufo fowleri* × *B. valliceps* resembled *B. fowleri,* while electrophoretic patterns of five of seven hybrids were identical to those of *B. valliceps* (Wittliff, 1964). The other two hybrids were unlike either parent (Figure 4–10).

Although this paper deals with the utilization of biochemical techniques for elucidating anuran evolutionary patterns, an excellent study by Uzzell and Goldblatt (1967) of the pattern within the *Ambystoma jeffersonianum* complex must be mentioned. They found that the diploid species *A. jeffersonianum* and *A. laterale* were each homozygous for a different protein component (Table 4–1;

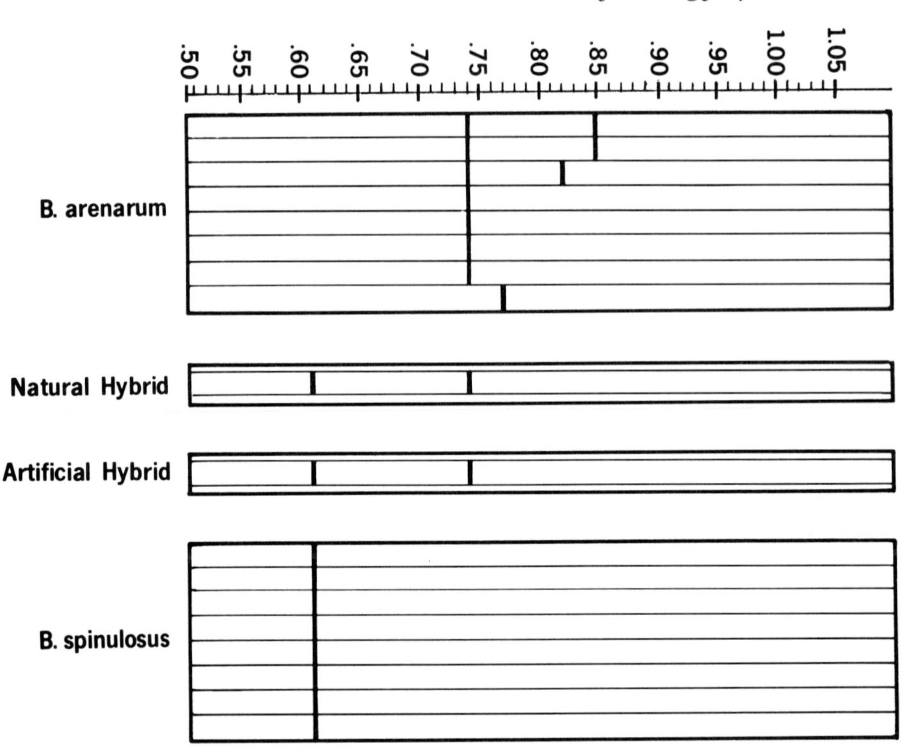

Figure 4–9. Transferrin phenotypes of *Bufo arenarum,* a natural hybrid *B. arenarum* × *B. spinulosus,* an artificial hybrid *B. arenarum* × *B. spinulosus,* and *B. spinulosus.* *Vertical axis,* relative mobility. Individual phenotypes are represented horizontally. Reprinted, by permission of American Midland Naturalist, from L. E. Brown and S. I. Guttman, Natural hybridization between the toads *Bufo arenarum* and *Bufo spinulosus* in Argentina, American Midland Naturalist, 83 (1970).

Figure 4–11). However, the triploid species *A. platineum* and *A. tremblayi* each showed one of these bands in the homozygous condition, and, they believe, the second component was represented only by a single gene. From these patterns and other criteria, Uzzell and Goldblatt reasoned that the triploid *A. platineum* has two sets of *A. jeffersonianum* chromosomes and one of the *A. laterale* chromosomes, whereas *A. tremblayi* possesses two sets of *A. laterale* chromosomes, and one set of *A. jeffersonianum* chromosomes.

Higher Taxa

Prior to discussing the relationships of higher taxa determined by biochemical analysis, mention should be made of the perils that might befall a researcher working solely with biochemical data. The dangers inherent in proposing taxonomic changes based only on chemical data are many. As an extreme example, it is conceivable that such a worker might place the genus *Bufo* in the plant order Ranales because of the common occurrence of bufotenine (Alston and Turner, 1963). Similarly, pufferfish and *Taricha* might be united due to the similarity of tetrodotoxin and tarichatoxin (Brown and Mosher, 1963). It appears that Bagnara and Obika (1965) fell into this trap. They noted that *Bufo* and *Scaphiopus* possess bufochrome, 2-amino–4-hydroxypteridine (AHP), and riboflavin in the same relative concentrations and, therefore, may be more closely related

Table 4–1. R_f Values for Two Serum Proteins of Salamanders of the *Ambystoma jeffersonianum Complex*

Species	0.67 protein		0.77 protein	
	(mean)	*(range)*	*(mean)*	*(range)*
A. jeffersonianum				
3 ♂♂	0.66	0.66		
5 ♀♀	0.66	0.64–0.69		
All individuals	0.66	0.64–0.69		
A. platineum				
10 ♀♀	0.67	0.64–0.72	0.78	0.76–0.80
A. tremblayi				
6 ♀♀	0.68	0.65–0.70	0.76	0.74–0.78
A. laterale				
6 ♂♂			0.77	0.76–0.78
12 ♀♀			0.78	0.75–0.81
All individuals			0.77	0.74–0.81
All individuals of complex	0.67	0.64–0.72	0.77	0.74–0.81

Source: Uzzell and Goldblatt (1967). Reproduced by courtesy of the Allen Press, Inc.

than present taxonomy indicates. In the same vein, since *Xenopus* skin contains bufochrome, AHP, and xanthopterin and since all three are also present in toads, they state that *Xenopus* might be closely related to bufonids. However, due to the fact that insects and fish also possess all these pteridines, their taxonomic speculations appear to be of little value.

Errors in determining relationships might also be caused by the presence of intraspecific variation. Figure 4–12 illustrates drawings of the total plasma patterns of individuals of certain *Bufo americanus* species. Individuals A and B are *B. americanus;* A from Minnesota, and B from Maryland. Patterns C and D are from *B. hemiophrys* from the same population in Minnesota. If one calculates percentage of affinity by the formula:

$$\text{percentage of affinity} = (\text{bands in common for individuals A + B} / \text{total bands in A + B}) \times 100,$$

then the affinity between individuals A and B is 69 per cent. However, if one calculates the affinity between individual A and individuals C or D, the affinities are 90 per cent and 93 per cent, respectively. The relationship between C and D is only 81 per cent. Thus, we have the unusual situation in which *B. americanus* A is more similar to either of the two *B. hemiophrys* than they are to each other. If one was unaware of the considerable degree of protein variation in this species group, he might consider revising the taxonomy of the group.

When electrophoresis is performed in concentrated gels, the resulting detailed separations appear to obscure broad patterns, and therefore, this technique appears to be of limited use in discerning relationships between families or higher taxa (Dessauer and Fox, 1964). In contrast, paper electrophoresis allows these patterns to be observed; thus, when higher taxa are compared, it is usually a better technique than the high resolution methods. Homologous proteins found in some genera may be present within a limited electrophoretic mobility range, pat-

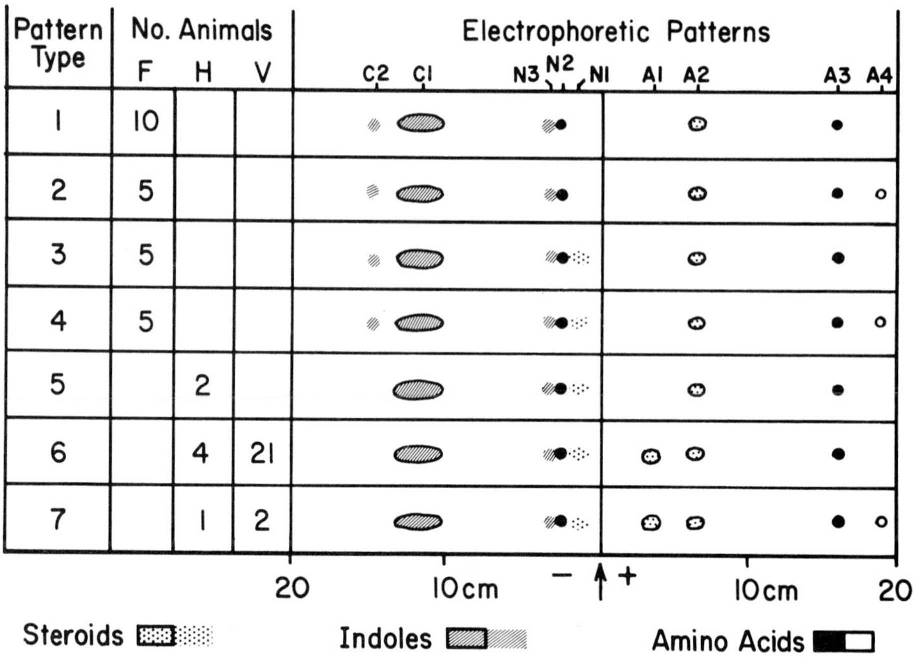

Pattern Type	No. Animals F	H	V
1	10		
2	5		
3	5		
4	5		
5		2	
6		4	21
7		1	2

Electrophoretic Patterns: C2 C1 N3 N2 N1 A1 A2 A3 A4

20 10cm − ↑ + 10cm 20

Steroids Indoles Amino Acids

Figure 4–10. A summary of the varieties of electrophoretic patterns and their distributions in parotoid secretions from F, *Bufo fowleri;* V, *B. valliceps;* and H, their natural hybrids. Types of shading indicate the chemical class of fractions separated. The type of migration: C, cathodal; A, anodal; and N, neutral. Steroid and indole spots which are shaded and outlined depict intense staining; those not outlined depict slight staining. Dark amino acid spots indicate intense staining; light spots indicate slight staining. Reprinted, with permission, from James L. Wittliff—"Venom Constituents of *Bufo valliceps,* and Their Natural Hybrids Analyzed by Electrophoresis and Chromatography" in TAXONOMIC BIOCHEMISTRY AND SEROLOGY, edited by Charles A. Leone, Copyright © 1964 The Ronald Press Company, New York.

terns from some families may lack specific components, and total protein patterns of some orders may have unique mobilities.

All genera of anurans, except *Bufo,* can be distinguished from salamanders by the presence of transferrins that migrate slowly and that are of low concentration (Dessauer et al., 1962). When the mobilities of hemoglobins from animals of distantly related families were compared, phylogenetic trends were suggested (Dessauer et al., 1957). Dessauer and Fox (1956) were able to construct a tentative key to the orders of amphibians and reptiles based on plasma patterns; they used such features as movement toward the cathode, anode, or both, total migration, and the presence or absence of certain fractions.

Chromatographic analysis of venom has been useful in determining evolutionary trends within the genus *Bufo* (Low, 1972). The primitive venom type contains 5-OH tryptamine but lacks other indoles, such as bufotenine, that require further metabolic pathways. This type is represented by several Old World and New World species (Figure 4–13). Radiations from this stock involved acquiring bufotenine and/or bufoviridine and other indoles. Cei et al. (1968, 1972) chromatographed skin extracts from 62 species and subspecies of *Bufo* and were able to derive lines of evolution essentially similar to those of Low (1972). Pteridine analysis showed that all bufonids possess bufochromes, adult ranids have ranachrome-3, bufonids do not have ranachrome-3, and ranid skins do not yield bufochrome (Bagnara and Obika, 1965).

Immunological techniques are also valuable for establishing degrees of evolutionary relationship. The analysis of degree of antigenic correspondence, among a variety of protein homologues, is an excellent method for studying the effects of long-range evolutionary trends on the proteins (Goodman, 1964). Unfortunately, few workers interested in anuran evolution have explored the use of this technique.

Cei (1964) used simple mixtures of serum and antiserum to determine the relationship of four families of South American frogs. Neotropical ranids were immunologically different from the Leptodactylidae, Hylidae, and Bufonidae. *Phyllomedusa* showed high immunological affinities to the Bufonidae and Leptodactylidae. Cei (1965) determined, by precipitin tests, that the Ceratophryidae should be kept as a separate family from the Leptodactylidae.

Perhaps the most ambitious biochemical study of higher taxa made thus far was the attempt by Salthe and Kaplan (1966) to immunologically determine the time of origin of various amphibian families. Their conclusions were that the frogs and salamanders diverged during the Permian, that the major radiation of the

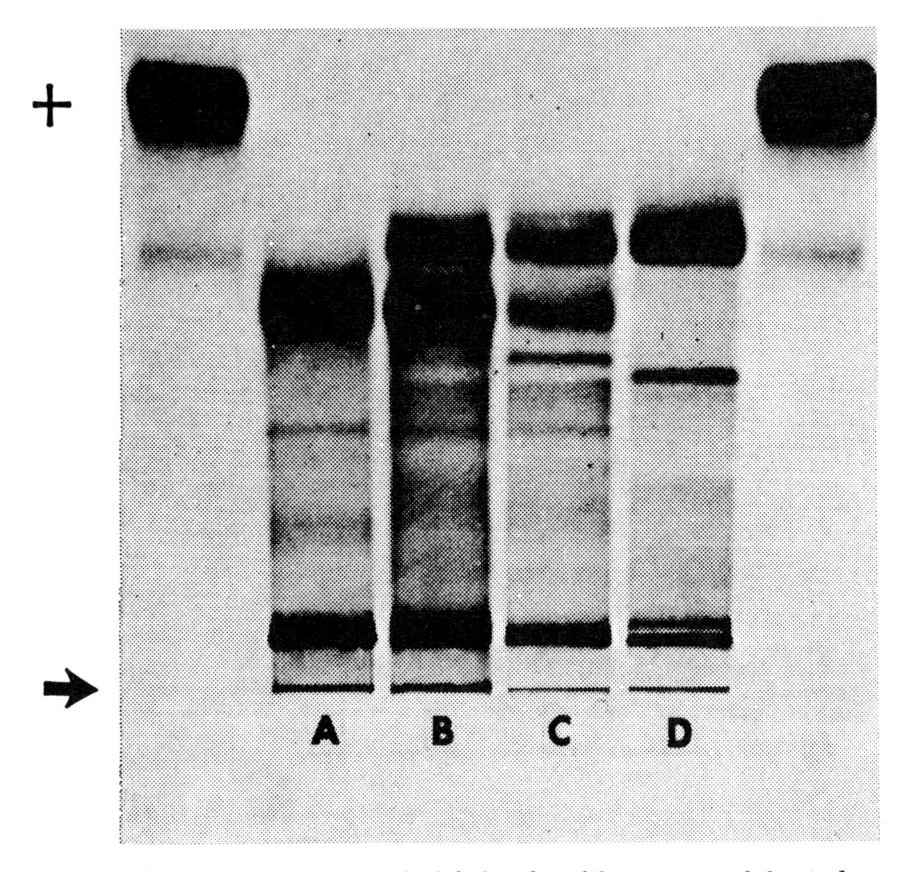

Figure 4–11. Serum protein patterns of adult females of four species of the *Ambystoma jeffersonianum* complex: A, *A. jeffersonianum*, Delaware County, Ohio; B, *A. platineum*, Boone County, Indiana; C, *A. tremblayi*, Washtenaw County, Michigan; D, *A. laterale*, Cook County, Illinois. Bovine serum albumin used for control. Reprinted, by permission of Allen Press, Inc., from Uzzell and Goldblatt (1967).

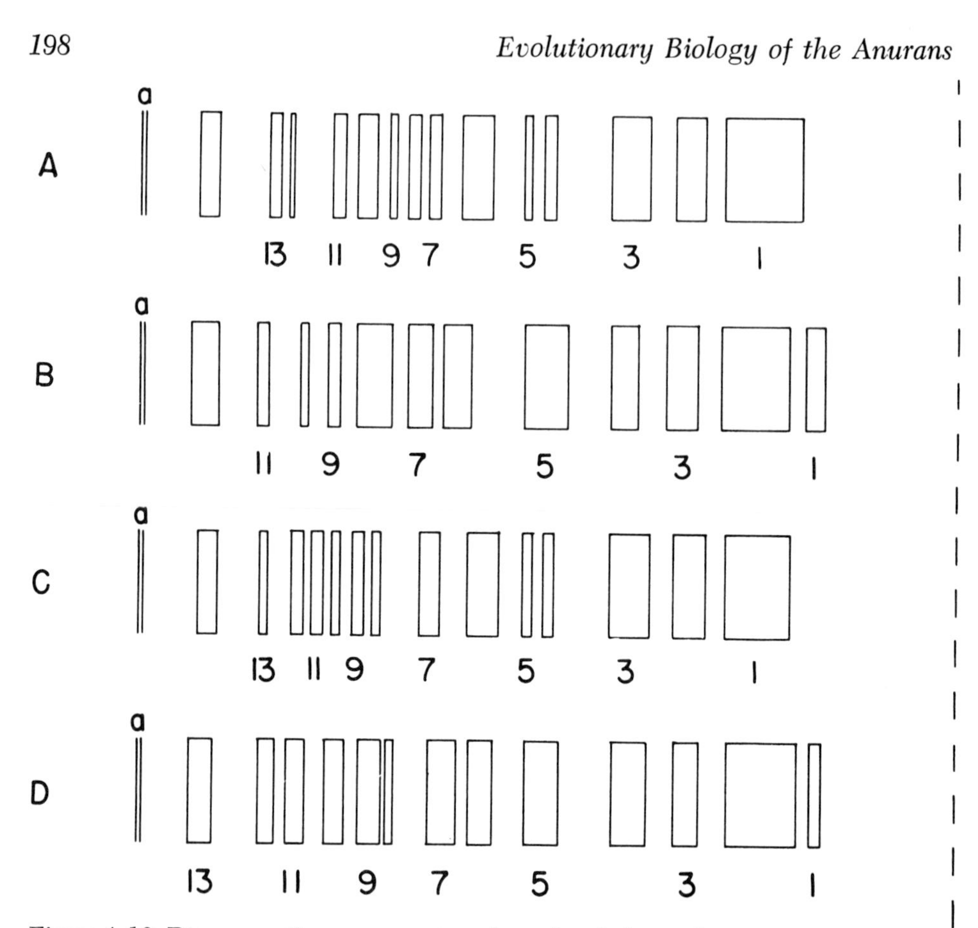

Figure 4–12. Diagrammatic representation of starch-gel electrophoretic patterns of *Bufo* plasma proteins. Vertical starch-gel electrophoresis was performed using a sodium borate (0.06 M sodium hydroxide, 0.33 M boric acid) electrode buffer, pH 8.0 and a sodium borate (0.01 M sodium hydroxide, 0.03 M boric acid) gel buffer, pH 8.0. A 13.5 per cent gel was used, electrophoresis was performed at 150 volts for twenty hours, and the gel was stained in amido black. See text for identification of individual patterns. a, the origin; the anode is to the right. *Dashed line,* mobility of a control human prealbumin. Capital letters denote individuals from populations described in text.

urodeles occurred between early Triassic and mid-Jurassic, that the amphiumid-plethodontid line diverged from other salamanders during the Triassic, with the former line splitting during the Cretaceous, and that the rhacophorids diverged from a ranid stock sometime in the Cretaceous.

PERSPECTIVES FOR THE FUTURE

As documented in the preceding sections, biochemical techniques have been of great value in studying anurans. They may be used to investigate developmental changes, as well as population genetics, speciation, and higher order systematics. It is the purpose of this paper to present the biochemical methods available and their use with the hope that they will be employed to good advantage in future studies; the potential contribution of these techniques for the elucidation of evolutionary relationships is considerable.

New, nonmorphological techniques are uncovering numerous examples of speciation. Identification of cryptic species of *Hyla* (Ralin et al., unpublished data) and *Pleuroderma* (Barrio, 1964) were aided by electrophoretic methods; Manwell and Baker (1963) described a sibling species of sea cucumber by using hemoglobin and enzyme differences. A great deal of biochemical analysis remains to be done in this area.

Biochemical trends in anuran evolution remain to be documented. Other than the detailed intrageneric studies by Cei et al. (1968, 1972) and Low (1972) on the toad genus *Bufo,* biochemical researchers have investigated only a few species within a genus, or a single species from each of several anuran genera or families. Intensive analyses similar to those just cited must be made for many genera before any trends can be observed. In addition, techniques must be standardized so that results obtained in different laboratories can be used to complement one another. Results from paper electrophoresis cannot be compared to those from starch-gel or acrylamide-gel electrophoresis; chromatograms run in one solvent in one laboratory cannot be compared to those run in another solvent in a different laboratory, and so on. This standardization will demand coordination among individuals interested in anuran evolution in order to realize a common goal.

Current knowledge of anuran phylogenetic relationships can be clarified by the use of sophisticated biochemical procedures such as quantitative microcomplement fixation. MC′F has been employed, on a small scale, to elucidate relationships among some colubrid snakes (George and Dessauer, 1970). As part of a proposed International Biology Program subprogram, MC′F will be used to generate indices of dissimilarity for the neotropical leptodactylids and their derivatives.

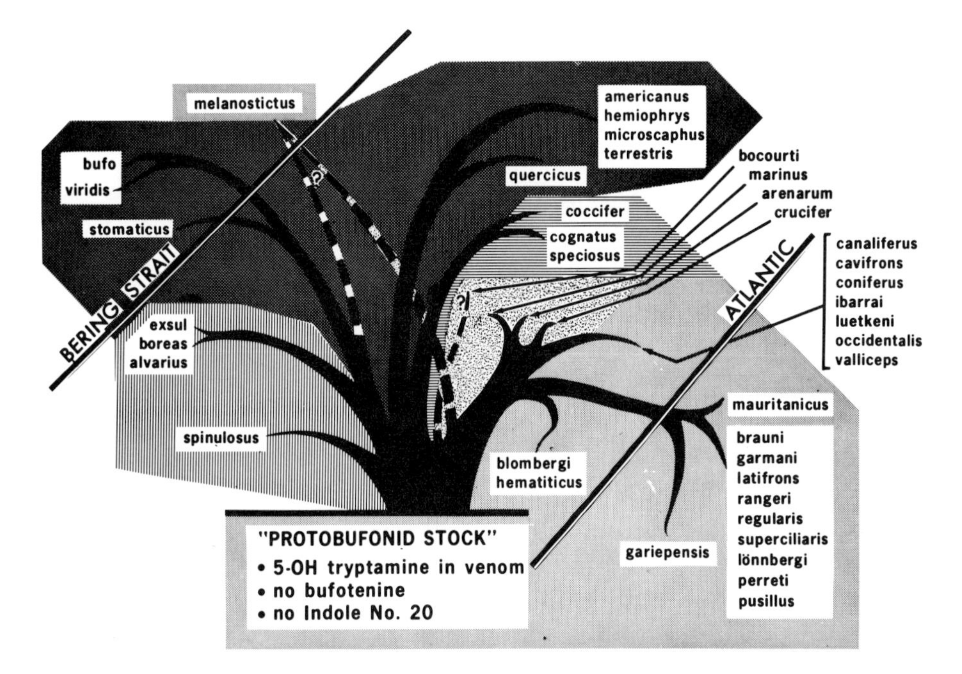

Figure 4–13. Diagrammatic representation of one possible scheme of bufonid evolution, with pertinent biochemical data superimposed: *Light stippling,* venom containing 5-OH tryptamine; *granular stippling,* venom containing 5-OH tryptamine and bufotenine; *dark stippling,* venom containing 5-OH tryptamine, bufotenine, and Indole No. 20; *horizontal lines,* venom containing 5-OH tryptamine and Indole No. 20; *vertical lines,* venom containing bufotenine and Indole No. 20. Diagrammatic representation from Low (1967).

The use of biochemical comparisons of molecules to determine degrees of evolutionary relationship is in its infancy. Crick (1958) noted that a great deal of evolutionary information is contained in the amino acid sequences of organisms. Once the technique is perfected, Sibley (1964) feels that the ultimate method of the biochemical evolutionist will involve comparisons between complete amino acid sequences of homologous proteins. Until such a time, relatively simple chromatographic, electrophoretic, and immunological methods will continue to furnish information of considerable evolutionary significance.

This work was supported in part by National Science Foundation Grant GB23601, in part by American Philosophical Society Penrose Fund Grant No. 5153, and in part by a Grant-in-Aid of Research from The Society of Sigma Xi.

REFERENCES

Abramoff, P., R. Darnell, and J. Balsano. 1964 Serological relations of toad populations of the Lake Michigan area, p. 515-525. *In* C. A. Leone, [ed.], Taxonomic biochemistry and serology. Ronald Press, New York.

Alston, R. E., and B. L. Turner. 1963. Biochemical systematics. Prentice-Hall, New York. 404 p.

Augustinsson, K. 1959. Electrophoresis studies on blood plasma esterases. II. Avian, reptilian, amphibian and piscine plasmata. Acta Chem. Scand. 13:1081-1096.

Bagnara, J. T., and M. Obika. 1965. Comparative aspects of integumental pteridine distribution among amphibians. Comp. Biochem. Physiol. 15:33-49.

Balsano, J. S., P. Abramoff, and R. M. Darnell. 1965. Serological variation in the toad populations of adjacent ponds. Am. Zool. 5:56.

Barrio, A. 1964. Especies cripticas del genero *Pleuroderma* que conviven en una misma area, identificados por el canto nupcial (Anura, Leptodactylidae). Physis 24:471-489.

Bennett, T. P., and E. Frieden. 1962. Metamorphosis and biochemical adaptation in amphibia, p. 483-556. *In* M. Florkin and H. S. Mason, [ed.], Comparative biochemistry IV. Academic Press, New York.

Bertini, F., and J. M. Cei. 1959. Electroferogrames de proteines seriques en el genero *Bufo*. Acta I^ro^ Congr. Sudam. Zool. 4:161.

———. 1961. Seroprotein patterns in the *Bufo marinus* complex. Herpetologica 17: 231-238.

Bier, M. 1959. Electrophoresis. Academic Press, New York. 563 p.

Brown, L. E. 1964. An electrophoretic study of variation in the blood proteins of the toads, *Bufo americanus* and *Bufo woodhousei*. Systematic Zool. 13:92-95.

Brown, L. E., and S. I. Guttman. 1970. Natural hybridization between the toads *Bufo arenarum* and *Bufo spinulosus* in Argentina. Am. Midland Naturalist 83:160-166.

Brown, M. S., and H. S. Mosher. 1963. Tarichatoxin: isolation and purification. Science 140:295-296.

Bucovaz, E., and H. M. Kaplan. 1957. Electrophoretic study of serum proteins in frog red leg disease. Am. J. Physiol. 191:428-430.

Butler, J. E., and C. A. Leone. 1967. Immunotaxonomic investigations of the coleoptera. Systematic Zool. 16:56-63.

Cantarow, A., and B. Schepartz. 1957. Biochemistry. W. B. Saunders, Philadelphia. 867 p.

Cei, J. M. 1964. Some precipitin tests and preliminary remarks on the systematic relationships of four South American families of frogs. Serol. Mus. Bull., Rutgers Univ. 30:4-6.

———. 1965. The relationships of some ceratophrynid and leptodactylid genera as indicated by precipitin tests. Herpetologica 20:217-224.

Cei, J. M., and F. Bertini. 1961. Serum proteins in allopatric and sympatric populations of *Leptodactylus ocellatus* and *L. chaquensi*. Copeia 3:336-340.

CEI, J. M., V. ERSPAMER, AND M. ROSEGHINI. 1967. Taxonomic and evolutionary significance of biogenic amines and polypeptides occurring in amphibian skin. I. Neotropical leptodactylid frogs. Systematic Zool. 16:328-342.

———. 1968. Taxonomic and evolutionary significance of biogenic amines and polypeptides in amphibian skin. II. Toads of the genera *Bufo* and *Melanophryniscus*. Systematic Zool. 17:232-245.

———. 1972. Biogenic amines, p. 233-243. *In* W. F. Blair, [ed.], Evolution in the genus *Bufo*. Univ. Texas Press, Austin.

CHEN, P. S. 1967. Separation of serum proteins in different amphibian species by polyacrylamide gel electrophoresis. Experientia 23:1-8.

CRICK, F. H. C. 1958. On protein synthesis. *In* The biological replication of marcomolecules. Soc. Exptl. Biol. Med. Symposium 12:138-163.

DESSAUER, H. C., AND W. FOX. 1956. Characteristic electrophoretic patterns of orders of amphibia and reptilia. Science 124:225-226.

———. 1964. Electrophoresis in taxonomic studies illustrated by analyses of blood proteins, p. 625-647. *In* C. A. Leone, [ed.], Taxonomic biochemistry and serology. Ronald Press, New York.

———, AND Q. L. HARTWIG. 1962. Comparative study of transferrins of amphibia and reptilia using starch-gel electrophoresis and autoradiography. Comp. Biochem. Physiol. 5:17-29.

DESSAUER, H. C., W. FOX, AND J. R. RAMIREZ. 1957. Preliminary attempt to correlate paper electrophoretic migration of hemoglobins with phylogeny in amphibia and reptilia. Arch. Biochem. Biophys. 71:11-16.

DESSAUER, H. C., AND E. NEVO. 1969. Geographic variation of blood and liver proteins in cricket frogs. Biochem. Genet. 3:171-188.

ERSPAMER, V., M. ROSEGHINI, AND J. M. CEI. 1964. Indole-, imidazole-, and phenyl-alkylamines in the skin of thirteen *Leptodactylus* species. Biochem. Pharmacol. 13:1083-1093.

FOX, W., H. C. DESSAUER, AND L. T. MAUMUS. 1961. Electrophoretic studies of blood proteins of two species of toads and their natural hybrid. Comp. Biochem. Physiol. 3:52-63.

FRIEDEN, E. 1961. Biochemical adaptation and anuran metamorphosis. Am. Zool. 1:115-149.

GEORGE, D. W., AND H. C. DESSAUER. 1970. Immunological correspondence of transferrins and the relationships of colubrid snakes. Comp. Biochem. Physiol. 33:617-627.

GOODMAN, M. 1964. Problems of primate systematics attacked by the serological study of proteins, p. 467-486. *In* C. A. Leone, [ed.], Taxonomic biochemistry and serology. Ronald Press, New York.

GUTTMAN, S. I. 1967. Transferrin and hemoglobin polymorphism, hybridization and introgression in two African toads, *Bufo regularis* and *Bufo rangeri*. Comp. Biochem. Physiol. 23:871-877.

———. 1969. Blood protein variation in the *Bufo americanus* species group of toads. Copeia 2:243-249.

———. 1972*a*. Biochemistry of the blood. *In* Symposium on methods of studying evolution of amphibians. Held in Bogota, Colombia, in 1968. In press.

———. 1972*b*. Blood proteins, p. 265-278. *In* W. F. Blair, [ed.], Evolution in the genus *Bufo*. Univ. Texas Press, Austin.

HEBARD, W. B. 1964. Serum-protein electrophoretic patterns of the amphibia, p. 649-657. *In* C. A. Leone, [ed.], Taxonomic biochemistry and serology. Ronald Press, New York.

HIRSCHFELD, J. 1960. Immunoelectrophoresis—procedure and application to the study of group-specific variations in sera. Science Tools 7:18-25.

HUBBY, J. L., AND L. H. THROCKMORTON. 1965. Protein differences in *Drosophila*. II. Comparative species genetics and evolutionary problems. Genetics 52:203-215.

HUCHON, D., M. TH. CHALUMEAU-LE FOULGOC, AND L. GALLIEN. 1968. Mise en évidence, au niveau des protéines sériques de l'adulte chez *Bombina variegata* L.

(Amphibien, Anoure), d'une protéine spécifique du sexe femelle. Compt. Rend. Acad. Sci. Paris 266:399-402.

Hunsaker, D., R. E. Alston, W. F. Blair, and B. L. Turner. 1961. A comparison of the ninhydrin positive and phenolic substances of parotoid gland secretions of certain *Bufo* species and their hybrids. Evolution 15:352-359.

Low, B. S. 1968. Venom polymorphism in *Bufo regularis.* Comp. Biochem. Physiol. 26:247-257.

_______. 1972. Evidence from parotoid gland secretions, p. 244-264. *In* W. F. Blair, [ed.], Evolution in the genus *Bufo.* Univ. Texas Press, Austin.

Manwell, C., and C. M. A. Baker. 1963. A sibling species of sea cucumber discovered by starch gel electrophoresis. Comp. Biochem. Physiol. 10:39-53.

Mayr, E. 1963. Animal species and evolution. Belknap Press, Cambridge, Mass. 797 p.

Ouchterlony, O. 1948. *In vitro* method for testing the toxin-producing capacity of diptheria bacteria. Acta Pathol. et Microbiol. Scand. 25:186-191.

Porter, K. R. 1964. Chromatographic comparisons of the parotoid gland secretions of six species in the *Bufo valliceps* group, p. 451-456. *In* C. A. Leone, [ed.], Taxonomic biochemistry and serology. Ronald Press, New York.

_______. 1968. Evolutionary status of a relict population of *Bufo hemiophrys.* Cope. Evolution 22:583-594.

Porter, K. R., and W. F. Porter. 1967. Venom comparisons and relationships of twenty species of New World toads (genus *Bufo*). Copeia 2:298-307.

Poulik, M. D. 1960. Immunoelectrophoresis, p. 60-69. *In* L. A. Lewis, [ed.], Electrophoresis in physiology. C. C. Thomas, Springfield, Ill.

Poulik, M. D., and O. Smithies. 1958. Comparison and combination of the starch-gel and filter paper electrophoretic methods applied to human sera: two-dimensional electrophoresis. Biochem. J. 68:636-643.

Reichlin, M., E. Bucci, C. Fronticelli, J. Wyman, E. Antonini, C. Ioppolo, and A. Rossi-Fanelli. 1966. The properties and interactions of the isolated alpha and beta chains of human hemoglobins. IV. Immunological studies involving antibodies against the isolated chains. J. Molec. Biol. 17:18-28.

Reichlin, M., M. Hay, and L. Levine. 1964. Antibodies to human A_1 hemoglobin and their reaction with A_2, S, C, and H hemoglobins. Immunochemistry 1:21-30.

Richmond, J. E. 1968. Changes in serum proteins during the development of bullfrogs *(Rana catesbeiana)* from tadpoles. Comp. Biochem. Physiol. 24:991-996.

Rossi, A. 1960. Osservazioni preliminari sul frazionamento delle proteine del siero nei due sessi di *Bufo bufo* (L.) con il metodo dell' elettroforesi sul gel d'amido. Ricera Sci. 30:141-144.

Salthe, S. N. 1969. Geographic variation of the lactate dehydrogenases of *Rana pipiens* and *Rana palustris.* Biochem. Genet. 2:271-303.

Salthe, S. N., and N. O. Kaplan. 1966. Immunology and rates of enzyme evolution in the amphibia in relation to the origins of certain taxa. Evolution 20:603-616.

Salthe, S. N., and E. Nevo. 1969. Geographic variation of lactate dehydrogenase in the cricket frog, *Acris crepitans.* Biochem. Genet. 3:335-341.

Sarich, V. M., and A. C. Wilson. 1966. Quantitative immunochemistry and the evolution of primate albumins: microcomplement fixation. Science 154:1563-1566.

Shaw, C. R., and A. L. Koen. 1968. Starch gel zone electrophoresis of enzymes, p. 325-364. *In* I. Smith, [ed.], Chromatographic and electrophoretic techniques, Vol. II. Wiley, New York.

Sibley, C. G. 1964. The characteristics of specific peptides from single proteins as data for classification, p. 435-439. *In* C. A. Leone, [ed.], Taxonomic biochemistry and serology. Ronald Press, New York.

Smithies, O. 1959. Zone electrophoresis in starch gels and its application to studies of serum proteins. Adv. Protein Chem. 14:65-113.

Uzzell, T. M., Jr., and S. M. Goldblatt. 1967. Serum proteins of salamanders of the *Ambystoma jeffersonianum* complex, and the origin of the triploid species of this group. Evolution 21:345-354.

Wasserman, E., and L. Levine. 1961. Quantitative microcomplement fixation and its use in the study of antigenic structure by specific antigen-antibody inhibition. J. Immunol. 87:290-295.

WILLIAMS, R. J. 1956. Biochemical individuality. Wiley and Sons, New York. 214 p.

WITTLIFF, J. L. 1962. Parotoid gland secretions in two species groups of toads (genus *Bufo*). Evolution 16:143-153.

_______. 1964. Venom constituents of *Bufo fowleri, Bufo valliceps* and their natural hybrids analyzed by electrophoresis and chromatography, p. 457-464. *In* C. A. Leone, [ed.], Taxonomic biochemistry and serology. Ronald Press, New York.

WRIGHT, D. A., AND F. H. MOYER. 1966. Parental influences on lactate dehydrogenase in the early development of hybrid frogs in the genus *Rana*. J. Exptl. Zool. 163:215-230.

_______. 1968. Inheritance of frog lactate dehydrogenase patterns and the persistence of maternal isozymes during development. J. Exptl. Zool. 167:197-206.

PART I DISCUSSION

Tihen

It is our intent to use this discussion period primarily for questions and comments from the floor, but first, I would like to ask Dr. Osvaldo Reig if he wishes to make a few brief comments at this time.

Reig

One of the first things I want to point out is that the contributions in this session show an overwhelming amount of new information about the subject of phylogeny and the evolutionary history of frogs. Needless to say, we have now much more information on this topic than Noble had at his time. Nevertheless, I believe that there is a lack of proportion between the theoretical extensions of recent discoveries and what these discoveries offer for the elucidation of previous ideas.

We are fortunate that improved techniques in the study of the fossil material have been developed, as Dick Estes told us. These new techniques have enabled a better interpretation of such early frogs as *Vieraella* and *Notobatrachus*. We are also fortunate to have some new material about early frogs. Most important, new and precise methodologies have been recently applied to the study of the phenetic relationships among living families of anurans, as in the contribution by Kluge and Farris. Moreover, data from the fields of chemosystematics and comparative cytology have contributed lately their own share to a better understanding of anuran relationships. However, information gathering and elucidation of evidence is not enough for scientific explanation, and the sum of old and new available data does not answer the two main questions I have about frog origins and early history, which are, What is the ancestry of frogs? and What is the main pattern of the early diversification of frogs?

As to the first question, I want to insist, in agreement with Estes, that there is no fossil record indicating the very early history of frogs. The earliest record of true anurans—in the Liassic—belongs to a step in salientian history much in advance of the presumed time of appearance of the first members of the order. The presence of a probable proanuran in the Eotriassic, *Triadobatrachus*, indicates that anuran origins must be placed somewhere in the long span from the lower to the upper Triassic, but no fossil frog has yet been discovered from those times. We therefore have to rely on indirect evidence to approach the question of their origins, a procedure that demands extreme caution about what we might conclude.

Of primary relevance to the question of the origin of the Anura is the validity of the lissamphibian hypothesis, which claims that the three orders of modern amphibians are more closely related to each other than any of them are to the Paleozoic groups of amphibians and, therefore, that the former shared a common ancestor among the latter. Parsons and Williams have been good advocates of this hypothesis, which was discussed recently by Bolt in connection with the discovery of a supposed lissamphibian ancestor in the Permian, *Doleserpeton*. In the paper Estes and I submitted to this symposium, we have analyzed the validity of the lissamphibian hypothesis and the probable meaning of *Doleserpeton.* I want only to add here that, appealing as it is, the lissamphibian hypothesis is far from being a corroborated hypothesis (*sensu* Bunge). The test results are contradictory, and probably the hypothesis is to be kept as a plausible (but not corroborated) one and limited to a description of the relationships between Anura and Urodela. Recent results by Beçak and collaborators on the DNA content of members of the three orders of living amphibians can also be applied under this restriction.

Taking into account the present state of the lissamphibian hypothesis, and the known record of the history of the Amphibia, what, then, can we conclude about the origin of the frogs? The most plausible, yet limited, conclusion at this time is that frogs, probably along with urodeles, originated from a dissorophid-like labyrinthodont during Permian times and existed in the form of proanurans in the early Triassic. Probably the true anurans originated from something like a proanuran during middle or upper Triassic times.

The answer to the other main questions regarding what we might conclude from the fossil record about the pattern of the early diversification of anurans is, Not very much. At the time of the earliest record of true anurans, in the early Jurassic, one family of frogs that is living today was present—the Ascaphidae. This is also the most primitive family of frogs according to morphological evidence. The Mesozoic record of frogs is very scanty, but it is within the limits of statistical probability that other families were already present by those times. The Jurassic frogs of Patagonia, important as they are, have not necessarily the ancestral significance claimed by me and Casamiquela some time ago. In the contribution that Estes and I present here, we attempt to formulate the few inferences we are authorized to draw from the early fossil record of frogs regarding the pattern of the first differentiation of anuran families. These are mostly guesses—or rather, isolated conjectures—supported by some kind of empirical data but lacking theoretical validation. By awaiting more information to validate hypotheses the scientist can contribute to a broader area by trying to discover what other problems, different from phylogeny construction, are posed by the known evidence.

Anurans, we can conclude, are a very slowly evolving group of vertebrates. Several living families are known to go back to the Jurassic or Cretaceous; even some living species are old rooted in the geological times, as Dick Estes told us when he referred to the discovery of the living species *Bufo spinulosus* in a Ricochican (upper Paleocene) deposit of Brazil.

The realization of slow rates of taxonomic evolution in the Anura poses intriguing problems, the analyses of which have received little attention by evolutionists—Why are frogs and toads so slowly evolving? and What sort of mechanisms enforce a slow rate of evolution? It is my belief that trying to answer questions of this kind may be a more profitable current endeavor than discussing weak inferences from a poor record.

In connection with the above questions, I want to point out the correlation of slow taxonomic rates of evolution with conservatism in the chromosome system and with ethological mechanisms of reproductive isolation. Work by Morescalchi, Bogart, Beçak, Barrio, and other scholars has shown that the anurans have a widespread intrataxon uniformity among karyotypes. Most of the species of the polytypic genera share similar chromosome complements. Too, most of the genera within a given family repeat chromosome numbers and structure. There are, however, interesting exceptions to this rule. Bogart noted, for instance, that in some species of the neotropical genus *Eleutherodactylus* and in some members of the Hylidae of the same biogeographical region, chromosome multiformity has been demonstrated. The pattern is also complicated by the surprising discovery by Saez, confirmed recently by Beçak, Bogart, Bianchi, and Barrio, of polyploidy in at least one genus of leptodactylids and one genus of ceratophrynids. However, these exceptions do not jeopardize the statement of a generalized occurrence of chromosome uniformity in anuran taxa. Actually, they rather help to clarify the meaning of that uniformity.

In rapidly diversified taxa of mammals, such as cricetids and *Ctenomys,* I have found a great deal of diversity in the chromosome complement, whereas in slowly evolving taxa, such as didelphids and dasypodids, conservatism in karyotypes is the rule. Gross chromosome changes, when fixed in a monomorphic state within different populations, may act as effective cytological mechanisms of reproductive isolation and thus promote speciation. This is probably the reason for the high chromosome multiformity found in such mammalian genera as *Ctenomys, Proechimys,* and *Akodon.* The fact that taxa showing such gross chromosome diversity are rapidly evolving ones also suggests the rapid appearance of isolating mechanisms through chromosome changes. Turning back to the Anura, the rarity of intrataxon chromosome rearrangements might indicate that this mechanism, as an isolating device, has not been an important operating factor in frog speciation. In most anurans, the main isolating mechanism is ethological and belongs to the allaesthetic characters of the mating call. It is suggestive that in the one species group of *Eleutherodactylus* that seems to be deprived of a mating call, the chromosome complement is no more conservative and shows a wide range of interspecies repatterning. One might assume that the mating call as an isolating mechanism is associated with slow evolution in the Anura and that the rates are rapidly increased when that mechanism is suppressed and replaced by "rapid-effect" cytological isolating mechanisms. I offer this assumption as a working hypothesis. But even if the above correlations are corroborated by more extensive data, many problems will still remain unanswered. Among the more important is the implication that some kind of genetic mechanism—one that simultaneously controls chromosome stability and enhances changes in the allaesthetic characters of the mating call—could be fixed under the impact of natural selection. Stability of chromosomes that interfere with the role of cytological isolating mechanisms might be partially responsible for the slow evolutionary rates of the Anura, but at the same time, the widespread occurrence of relatively low chromosome numbers in the group may also be identified with slow evolutionary rates. Reduced numbers of linkage groups would favor a high rate of neither genetic recombination nor genetic variability. Both are requisite for evolutionary changes.

When Dr. Blair mentioned in his address the problem of the geographic center of origin of frogs, he made a statement with which I was, and still am, very sympathetic. Dr. Blair pointed out that frogs are probably of Gondwanian origin, a hypothesis that I advanced in 1957 and developed in 1959. Blair arrived at this conclusion after thorough studies of the genus *Bufo.* Previous support for this view was afforded by the presence of early frogs in the Jurassic of Patagonia and by the presence of *Triadobatrachus* in the Triassic of Madagascar. These were the main facts supporting my belief in the Gondwanian origin of frogs.

It now seems to me that these facts are not better proof of the southern origin than the reverse argument by Matthew's followers, who claimed that because of the presence of the earliest frogs in the Jurassic of North America and Spain (before the discovery of *Notobatrachus*) and the absence of such old fossils in the Southern Hemisphere, the center of origin of frogs ought to be placed in the Northern Hemisphere. In both cases, too much was concluded from too little evidence. It can be argued, however, that there are different lines of evidence giving a stronger support now to the Gondwanian hypothesis of frog origins. I am ready to agree with this argument, and I support that hypothesis, but I see it as an empirical hypothesis not yet adequately corroborated by the facts. We are far from a thorough understanding of the geographic history of the anurans.

The problem of the major classification of frogs has been analyzed here by Dr. Lynch, and I see with a great sympathy many of the ideas he has proposed. As you know, I am the author of a major classification of anurans in which I grouped the different families in several suborders. Well, I guess that this proposal had some heuristic value in the sense that it emphasized the fact that some families are more closely related to each other than with other ones. One of these groupings I named the Neobatrachia, and further work by Hecht, Griffiths, Estes, and other authors recognized the naturalness of such a taxon. However, I now believe that we have not an adequate base upon which to construct suborders among the more primitive families and that, therefore, it would be better to discard the suborders in frog classification. I agree with Lynch that it is probably more convenient to use only superfamilies in such a system. One must keep in mind, however, that superfamilies are also taxa and that taxa must represent clades, not grades. On this basis, I do not think it justifiable to make a new superfamily for the Myobatrachidae and the Pelobatidae if it is not proved that they are connected phyletically. Probably most of the evolutionary taxonomists agree in the monophyletic requirement of a taxon and concur that taxa are not to be created for evolutionary grades, even when definition of a grade can be very useful in order to understand the evolution of the group.

Tihen

We're open now for questions and comments from the floor. If you have a question directed to a particular panel member, please so state.

Salthe

Dr. Lynch, I was wondering about your correlation of the inguinal clasp and pond sites of egg deposition. I don't think it really holds up in the sense that a good proportion of the more primitive frogs don't deposit eggs in ponds—that is, *Sooglossus, Barbourula, Ascaphus,* and as you pointed out, *Leiopelma*—while many of what you called the cycloranids—that is, the Australian forms—deposit eggs in streams. In fact, I think that maybe a third of them are nonpond breeders, contrary to your statement of the situation. That modern groups deposit eggs in many different places is perhaps true only because there are so many more of such frogs. Therefore, while this suggested correlation is not a very important part of your entire discussion, I think that it's worth pointing out that it is not quite all that it might be.

Lynch

Part of the problem is that there are relatively few species of what I've termed in my grouping "nonadvanced" frogs, the archaic and transitional groups. I will agree that there are a number associated with streams, in fact a number that deposit their eggs in streams, but I don't know if I would go as far as calling it one-third. A number have stream-adapted tadpoles, but as I recall, they deposit their eggs in torrential situations, which again, are not ponds; however, I'm not entirely remiss in calling them principally pond-type frogs. I still think that advanced frogs have invaded a broader range of habitats, in spite of the fact that there are, indeed, many more advanced frogs. I think the percentage would bear out the statement that there is a greater percentage of advanced frogs than of nonadvanced frogs that utilize streams, either reasonably slow streams or torrents. The nonadvanced frogs have not been able to enter the torrent habitats, except when they deposit their eggs in terrestrial situations along the stream (e.g., *Heleophryne*).

Edward H. Taylor, University of Kansas

Would Dr. Estes comment further on Piveteau's Madagascar amphibian as an ancestral anuran?

Estes

I could comment on it quite a bit. I might first say something that I've already said to Romer. His sketch restoration has the head two times as big as it actually is, and I mention this because a lot of people photograph these restorations and show them to classes.

I think that Hecht's discussion of *Protobatrachus* is very useful and, as Osvaldo Reig has already said about his suborders, has heuristic value. I think it forced a lot of us to reconsider, perhaps, the nice pleasant picture that having an ancestral frog in the Triassic gave us. I've not studied the actual specimen, but I have studied some excellent latex casts of the animal, and while I don't like this kind of cut-and-paste morphology, if you lowered the presacral vertebral number of fourteen for *Protobatrachus* to nine or ten, you'd have a lot of trouble keeping it out of the frogs. It has a body form, as indicated by pigment marks, that closely approaches the body form of frogs. This may or may not mean anything, but those Paleozoic amphibians with which I'm familiar and that are represented by body outlines don't have that body shape. It certainly has the elongated ilium that is otherwise unique to frogs. One point that Hecht made was that the sacral diapophyses are separate from the sacral vertebra, which would cause problems with an animal that was basically a jumping animal. He is going on the premise, already suggested by Eaton, that the primitive adaptation is for jumping in frogs. Taking that fact and putting it together with the suggestion of Griffiths that *Protobatrachus* is a metamorphosing tadpole, I would just say that I don't know if it's possible to assume that an animal so ancient and unlike frogs in some respects as *Protobatrachus* had a frog-like tadpole. We have no idea whether there was anything more than the sort of metamorphosis that now occurs in salamanders. There is no indication that this is a metamorphosing tadpole, but I would say that I think it's very possible that the failure of fusion of sacral diapophyses and a few other characters indicate that it is probably a young animal. Very possibly as the animal matured, its sacrum might enlarge and become somewhat more frog-like. This is hopeful, of course, on my part.

One further thing: If you look at the larvae of frogs as they ossify, one of the last things to ossify is the sacral diapophyses. In even very late larvae that have their limbs pretty well developed in many forms, there are no sacral diapophyses, and the sacral vertebrae are very like that of others. That is, the sacral diapophyses are rather late in forming. This would seem to me to be advantageous in an animal that is swimming, and basically a larva is a swimming animal. What it does when it grows up, if it hops off a bank back into the water again, is all very fine and good, but I see the basic adaptation of frogs as an aquatic one.

Other than that, I think that one can only say that *Triadobatrachus* (as we have to call it now, since the name *Protobatrachus* is preoccupied) doesn't indicate in any way whatsoever closer relationship to the dissorophids. I've been fond of the dissorophid hypothesis for some years, and I think it's perhaps significant that every piece of new evidence that has so far come up supports a dissorophid-like ancestry for Lissamphibia, but the interesting thing is that I don't see any special relationship of *Triadobatrachus* to dissorophids, which one might possibly expect, since they're not all that far apart in time.

Inger

I would like to ask Shelly Guttman to comment on a problem. It seems to me that there are two sources of difficulty in the kinds of data that you collect, and I know that you are aware of them. One is the effect of the physiological state of the individual on the pictures you get in your gels, and the other is the effect of isozymes in adding to the confusion when you try to determine the genotypic relationship.

Guttman

Well, to respond to the part about physiological variation, we had an unfortunate situation in Texas: the toads were dying of redleg. When I bled them, I found quantitative differences, not qualitative. So far I have not found anything in the patterns that I can consider a qualitative difference caused by the physiological variations in the organisms. With regard to the isozyme problem, physiological variations will complicate the situation. With regard to esterases and to transferrins and albumins, for example, we don't have an isozyme problem, but I am aware of the complications caused by isozymes.

Duellman

I'd like to direct a question to Frank Blair. You commented, Frank, about the utilization of various sets of criteria of karyology as being one set, osteology another, and you seem to give considerable weight to hybridization. I would like you to answer a question about whether you consider hybridization to be of equal value to any of these other sets of criteria.

Blair

It's rather invidious to try to single out any system. In fact, one of the points I was trying to make is that we must look at all systems and then make a judgment. If I had to select certain systems that I might give extra weight to, I think the two would be karyotypes and hybridization. Speaking to the latter, I think that any time you have positive results from hybridization in such a group as *Bufo* or any time you've got positive results in any group from which you are actually producing a true hybrid and not simply stimulating development with all the various kinds of things that can happen in the way of doubling the chromatin material and so on, any time you get positive results, you can be sure of one thing, and that is that you're dealing with genotypes that are so similar that they can produce a viable offspring or produce a hybrid that will survive to some given level.

The negative results are not necessarily so good, and I think that this should be kept in mind: There are many pitfalls in looking at negative results. Another point I made rather quickly in the interest of time is that you cannot extrapolate from one taxon to another. The kind of thing you can do in *Bufo* you cannot do in *Rana*. *Rana* is, for some reason of evolutionary strategy, a group in which genetic incompatibility evolves very early, and beyond what I would call the species-group, it's useless to even think about using hybridization techniques.

Duellman

You talked about convergence in a number of different disciplines. Do you have any indication of convergence insofar as genetic compatibility is concerned? Is there any way to determine this?

Blair

Well, I think that perhaps when we get the DNA sequences—when we get sufficient biochemical information so that we actually know down to the last molecule what the chemical composition is—we might answer this question. Right now I think it would probably be one of the most improbable kinds of convergence, since you're dealing with such an enormous number of combinations of genes or characters or whatever you want to call them. But I don't think you can rule out any event as a possibility, but the probability would be such that I'd be prone to put my faith more in hybridization than in almost any other set of characters you could come up with.

Duellman

I would like to direct another question to both you and John Lynch. You commented, Frank, on possible relationships between pelobatids, you used *Scaphiopus* as an example, and *Lepidobatrachus* in South America. In Lynch's scheme, these are rather widely separated, and I would like to hear a little discussion between the two of you about this and see if there is any meeting of the minds.

Blair

Largely I'm quoting a work by someone in the audience who can probably speak to this much better than I. I would suggest that you call on Jim Bogart for this side of the picture.

Bogart

I don't really deserve the credit for it because Morescalchi showed it before I did. Chromosome similarities between pelobatids and *Ceratophrys calcarata* from Colombia are very close. I followed this up because I had the other so-called ceratophrid frogs from Argentina and we had *Scaphiopus* in Texas. The chromosomes are, indeed, very similar. Following this up, I got some eggs of *Lepidobatrachus* and crossed them with *Scaphiopus,* and it did produce a little, wiggly tadpole, which was diploid but died very shortly. I didn't have any control group, so I can't conclude a whole lot. But I think these are in the group of leptodactylids I would put as most primitive. I don't know about you, John. Would *Ceratophrys* and *Lepidobatrachus,* the ceratophryines, be more similar to the Australian leptodactylids and pelobatids?

Lynch

I regard the Old World tropical Megophryinae as the more primitive representatives of the family, with the Holarctic pelobatids as a derived group. The distance that Dr. Duellman mentioned between pelobatids and *Ceratophrys,* which I include in the leptodactylids is, I think, in part, an illusion. We're forced to use family-group names, and I think we tend to magnify in our minds how far apart these units are. I contend that on the basis of my studies of the leptodactylids, the Australian leptodactylids are derived from pelobatids, most likely megophryine-type pelobatids, and that one of the subfamilies of Australian leptodactylids, namely, the cycloranines, gave rise to the New World leptodactylids. As Dr. Bogart mentioned, *Ceratophrys* and *Lepidobatrachus* are among the more primitive members of the New World leptodactylids and thus the distance is really not that great. However you want to measure phyletic distances, I don't know, but I contend they are perhaps more closely related than the constraints imposed by the use of families, subfamilies, and the like would imply.

PART II

REPRODUCTIVE, DEVELOPMENTAL AND BEHAVIORAL CONSIDERATIONS

5

EVOLUTIONARY ASPECTS OF THE REPRODUCTIVE BEHAVIOR OF FROGS

George B. Rabb

The evolutionary and ecological intermediacy of amphibians among the vertebrates has long attracted students to their biology, and particularly to their reproductive activities. In recent years various aspects of the natural history of reproduction in frogs have been intensively studied. Reviews of the works pertinent to the following discussion include those on taxonomic patterns in breeding behavior (Jameson, 1955, 1957; Goin, 1960), breeding-site ecology (Goin and Goin, 1962), vocalizations (Bogert, 1960; Blair, 1963, 1968; Schneider, 1966*b*), and territoriality (Heusser, 1969*b*). The present eclectic survey is not a comprehensive review, but it is intended chiefly as a stimulant to further studies. I propose to describe briefly the general patterns of reproductive behavior in relation to recent material on the underlying hormonal and neurological bases and to discuss facets of the variations in reproductive behavior among taxa. Reproductive behavior is here considered to include all those behaviors that operate in the propagation of the species.

Like many behavioral patterns, reproductive behavior depends on both a set of internal motivating factors and a set of external inducers or stimuli. In anurans the relationships among these factors are often close, and particular temporal sequences of interactions are part of normal successful reproduction.

INTERNAL FACTORS IN REPRODUCTIVE BEHAVIOR

The main internal factors have to do with the priming of receptor and effector systems by gonadotropic, gonadal, and probably other hormones. Sex-related growth factors have not been experimentally investigated, but their influence is reflected in sexual dimorphism in body size, hearing organs, vocal sacs, integumental glands, and various anatomical features of the limbs (Noble, 1931). The development of some male secondary characteristics, including color, brachial musculature, and cornifications of the integument of limbs and chest, has been directly correlated with the presence of male gonadal hormones (e.g., Greenberg, 1942). The maturation of the gametes themselves is well known to be under direct hormonal control (see reviews by Smith, 1955; Forbes, 1961).

The relationship of hormonal levels to specific behavioral acts is not known precisely or directly. However, in inducement of egg-laying, clasping assault, and mating calling responses with foreign gonadotropic hormones, investigators have observed progressive effects with increasing dosages (e.g., Russell, 1954, 1960). Stages of arousal, presumably reflecting hormonal levels, have been recognized in *Hymenochirus* (Rabb and Rabb, 1963*a*) and probably are involved in behavioral intensity differences in other frogs (e.g., Brown and Pierce, 1967). On the other hand, although the gonadal hormones normally act as neural priming agents (see below), they are not absolutely essential to performance of particular behavioral acts, such as calling and clasping assault (Ber and Zieleniewski, 1959).

The central neural machinery responsible for reproductive behaviors is only grossly known, although considerable knowledge now exists regarding the input functions of visual and aural receptors (e.g., Frischkopf et al., 1968; Ewert, 1967*a;* Lettvin et al., 1959, 1961). The ablation experiments of Aronson and Noble (1945) in localizing behavioral functions in the brain of *Rana pipiens* are still the most extensive in scope. The more recent studies of Hutchison (1964, 1967) on *Xenopus laevis, R. angolensis,* and *Bufo regularis,* and of Schmidt (1966, 1968*a,* 1969, 1971) on *R. pipiens, B. terrestris,* and various North American *Hyla* have considerably refined our knowledge in certain respects, particularly in regard to the two outstanding features of frog courtship and mating behavior—calling and clasping.

The initial nonvisual mating orientation and clasping assault by the male is under control of hormone-sensitive areas in the preoptic region, in particular the dorsal magnocellular nuclei. A self-regulatory clasping-maintenance center is located in the anterior medulla. This center acts independently of hormonal level. Hutchison views the medullary clasp-maintenance system as a specialization developed in anurans in connection with extended mating periods. Schmidt (1972*b*) located control of normal unclasping (releasing) in the inferior collicular area.

Schmidt's detailed work on the calling mechanisms by ablations and by electrical stimulation indicates that the ventral magnocellular nucleus in the preoptic area controls the initiation of mating calling. Mating calling itself appears to stem actually from a more posterior center in the medulla, about the main sensory nucleus of the trigeminal nerve, where release calling and warning cries can be evoked directly. The vocal mechanisms are closely related to, and apparently derived from, respiratory mechanisms in the same region. The actual effector center for these mechanisms is in the vagal-hypoglossal area of the medulla.

The preoptic activating center for mating calling is sensitive to androgen levels. Schmidt suggests that when activated, this center provides a tonic stimulus to the posterior vocal center and that mating calling evolved as connections were established between the two. The general model for the neural basis of vocalizations in male frogs given by Schmidt (1971) is shown in Figure 5–1. This kind of hierarchical arrangement is in accord with current ideas on the functional organization of innate behavior (Tinbergen, 1952; Holst and St. Paul, 1963); that is, there appears to be an anterior mood center that influences a posterior coordination center that responds to filtered sensory inputs ("innate releasing mechanism") and activates a posterior motoric unit ("fixed action pattern"). Such a model may be generally useful in analyzing reproductive behaviors. As suggested above, what is known of the clasping behaviors roughly fits this scheme. Schmidt (1969) has also shown that the female's orientation to the mating call is controlled by the ventral magnocellular nucleus and that this response occurs only when she is ovulating, presumably a period of high hormonal release.

Another aspect of the neural control systems involves relationships of reproductive behavior with other general patterns of behavior. This has been roughly approached in naturalistic observations of the suppression of other activities by mating behavior (Birkenmeier, 1954; Heusser, 1958; Rabb and Rabb, 1963*a*). It seems that both feeding and flight patterns of behavior are often involved. Stimuli that would normally induce flight are not very effective at peak sex hormonal levels (Heusser, 1969*b*). Feeding, an "attack" system, appears to be expressed or involved in at least some mating behaviors. In *Xenopus laevis,* the initial feeding action involves a swooping motion of the forearms. The control of this action is

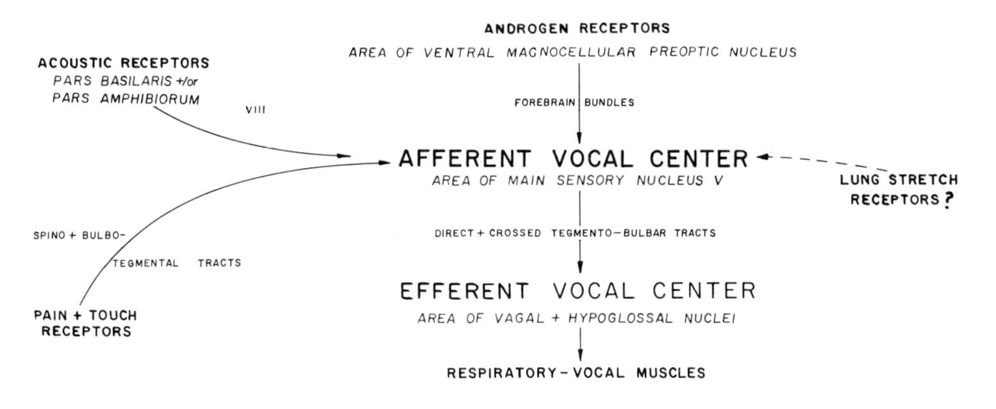

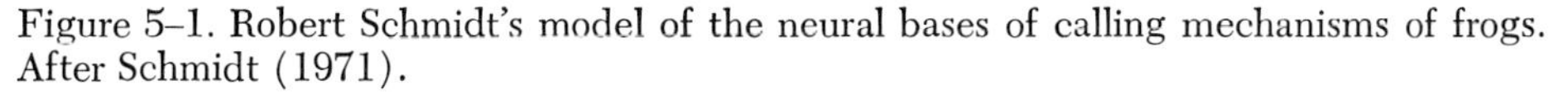

Figure 5–1. Robert Schmidt's model of the neural bases of calling mechanisms of frogs. After Schmidt (1971).

in the medulla, and close connections appear with clasping and other mating actions, such as pumping and stroking (Hutchison, 1964). Stroking the female with the hindlimbs is practically identical to the action involved in "high-kicking" a food item (Rabb and Rabb, 1963*a*). In other frogs, feeding does not involve the limbs initially. Even so, the clasping, calling, and feeding reactions brought on in a group situation either through feeding or acoustic stimulation (Schmidt, 1968*b*) suggest close relationship of control centers in the brain.

In regard to such commonality of behaviors, the minor behavior of toe-twitching deserves notice. Reported by Firschein (1951) as a normal behavior of the clasping male of *Bufo nayaritensis*, it is well known in the German literature as an accompaniment of the fixation stage of feeding behavior in *Bufo bufo* (Ewert, 1967*b*) and occurs in other *Bufo*, *Rana*, and *Alytes* (Heusser, 1960*a*). Brattstrom and Yarnell (1968) recorded it in the preliminary stage of aggressive display in *Leptodactylus melanonotus*. Toe-twitching in the latter instance can be regarded as simply an intention movement, but it may be generally classifiable as a displacement activity generated by the conflicting tendencies to flee or attack that may occur in feeding, mating, and maintaining territory. Similar motivational argument has been given for the occurrence of dancing, which may be a homologue of toe-twitching, by courting *Hymenochirus* and *Discoglossus* (Rabb and Rabb, 1963*a*).

EXTERNAL FACTORS IN REPRODUCTIVE BEHAVIOR

The external stimuli to reproductive behavior can be put in three categories in the fashion of Beach (1951): (1) those affecting physiological readiness, (2) critical features of the immediate surroundings, and (3) actions of the sexual partner.

The remote stimuli of the first category include many environmental variables. Light and temperature are prominent influences on the annual reproductive cycles in higher vertebrates and undoubtedly are important to many frogs from temperate locales. However, rainfall is perhaps the single most important feature throughout the anuran group. In many temperate-zone frogs, individual females are reproductively ready only for a short period once a year. However, the European discoglossids appear to be an exception (Hellmich, 1962; Knoepffler, 1962). In more tropic areas, females, as well as males, apparently have an extended, if not year-round, reproductive capability (Inger and Bacon, 1968). Pyburn (1966) has recorded an individual tropical hylid frog *(Hyla phaeota cyanosticta)* laying eggs at intervals throughout the year, and we have noted bimonthly cycles in individual *Pipa parva* in captivity.

What role gonadal ripening plays in motivating migration to the breeding site is an open question. Various external stimuli have been implicated (e.g., algal smells for *Rana temporaria;* Savage, 1961), but none seems universal. Actual long-range orientation ability, however, seems to have a common basis throughout the Anura (Ferguson, 1963; Landreth and Ferguson, 1967; but see also Grubb, 1970). Heusser (1960*b*) considered migration to be an independent behavior, not just the appetitive stage of reproductive behavior, but the sequential relations of the two indicate some common causal factors.

The effect of conspecifics as remote stimuli has not been well studied, although laboratory observations on *Pseudacris triseriata* and *Xenopus laevis* indicate that the calling of males can induce ovulation and oviposition in ripe but inactive females (Rabb and Rabb, unpublished data). Territorial assertion is now known for several frogs, and indubitably motivational states are affected by calling interactions and aggressive contacts among males (and females in dendrobatids; Senfft, 1936; Test, 1954; Sexton, 1960). The behavioral structure of choruses varies from those based simply on distance relations (Foster, 1967) to others controlled by strong dominance hierarchies (Brattstrom, 1962). The latter situation amounts to a lek formation, and the dominant males apparently are selectively sought out by females (Brattstrom, 1962; see also Emlen, 1968).

Physical characteristics of the breeding site are evidently important to most frogs (Heusser, 1961), and ecologic segregation of species at common sites for calling and laying has been noted by several authors (e.g., Bogert, 1960; Duellman, 1967). However, the precise parameters involved have rarely been experimentally investigated (*Engystomops pustulosus;* Sexton and Ortleb, 1966). At least in the case of xeric breeders (Bragg, 1960), climatic precipitants, including precipitation (Balinsky, 1969), and atmospheric pressure (Beck, *fide* Oliver, 1955; see also Knepton, 1952) should also be included in this proximate stimulus category. The commonly observed daily cycle in reproductive activity points to a set of regular, if not precise, interactions between external physical stimuli, such as light, and internal states. Again, studies are few (one exception is Flindt and Hemmer, 1967).

The interactions of sexual partners can be, and have been, set out in simple stimulus-response diagrams (e.g., Figure 5–2). Generally, in the usual sequence of events, a ripe female orients to and approaches a courting (calling) male, is clasped by him, then in a number of separate bouts lays her eggs; each bout is heralded by postural adjustments apparently coordinating fertilization (Figure 5–3); and finally by rejection signals, the female effects release from the male's grasp. These interactions of partners are largely species-specific, with relatively little variation. There is, however, an enormous range of variation among species and higher taxa—so much so, indeed, that as the following section exemplifies, there is no simple answer to the question of whether we can discern progressive evolutionary changes in the reproductive behavior of frogs.

BEHAVIORAL VARIATIONS AS EVOLUTIONARY EVENTS

As has often been recited, trends to terrestrialism or protection of eggs and larvae occur within many groups of frogs (Lutz, 1947; Jameson, 1957; Goin, 1960). The differences within family groups reflect ecological shifts, such as Heyer (1969) has sketched in the trend to terrestrial development within *Leptodactylus,* where the basic behavioral preadaptation is the formation of a tough foam nest for eggs by churning actions of the male. But gross mating behavior, as well as ecology, may change. In the Pipidae, there appears to have been a shift

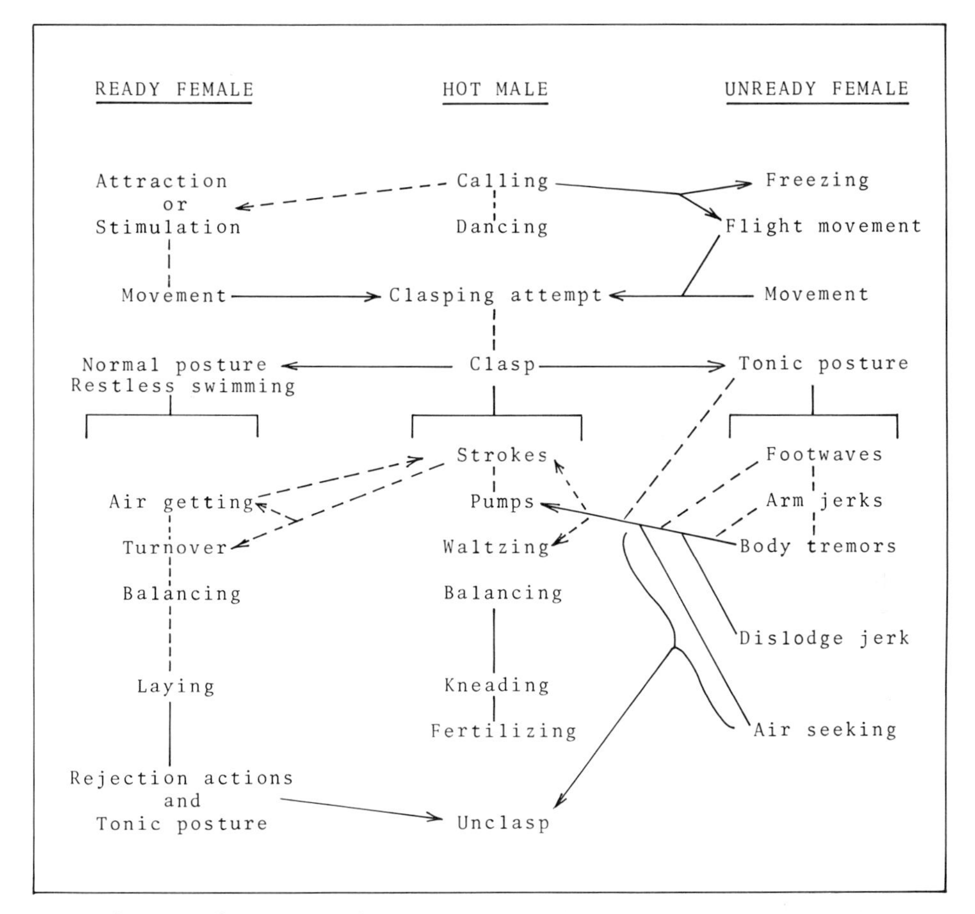

Figure 5–2. Stimulus-response diagram for mating behaviors in *Hymenochirus boettgeri*. Reprinted, by permission of Paul Parey Press, from G. B. Rabb and M. S. Rabb, On the behavior and breeding biology of the African pipid frog *Hymenochirus boettgeri*, Zeitschrift für Tierpsychologie, 20 (1963).

from an ordinary midwater oviposition maneuver attaching eggs to plants *(Xenopus)* to upside-down egg-laying on plants *(Xenopus)* to egg-laying turnovers at the surface film *(Hymenochirus, Pseudhymenochirus)* to egg-laying midwater turnovers resulting in attachment of eggs to the female's back *(Pipa)* (see Rabb and Rabb, 1961, 1693*a*, 1963*b*; Österdahl and Olsson, 1963; for a variant interpretation, see Swisher, 1969). These stages are illustrated in Figures 5–4–5–6. Similar pictures could probably be produced for many families in which there are both aquatic egg-layers and those with terrestrial development, but in general, detailed documentation of the behavior of enough taxa is lacking.

When we consider particular components of behavior among different groups, we find that frequently there are similarities. For instance, Licht (1969) reported unusual sequences of release-seeking behavior in female *Rana aurora* and an amplectic call by males of this species as situations unparalleled in other American *Rana*. However, the basic extension of the hind legs by rejecting females is known in *R. nigromaculata* (Liu, 1931), and an amplectic call does occur in other *Rana;* further, the tonic posture is known in several primitive groups (Rabb and Rabb, 1963*a*). In *Xenopus laevis, R. aurora* behavior is prac-

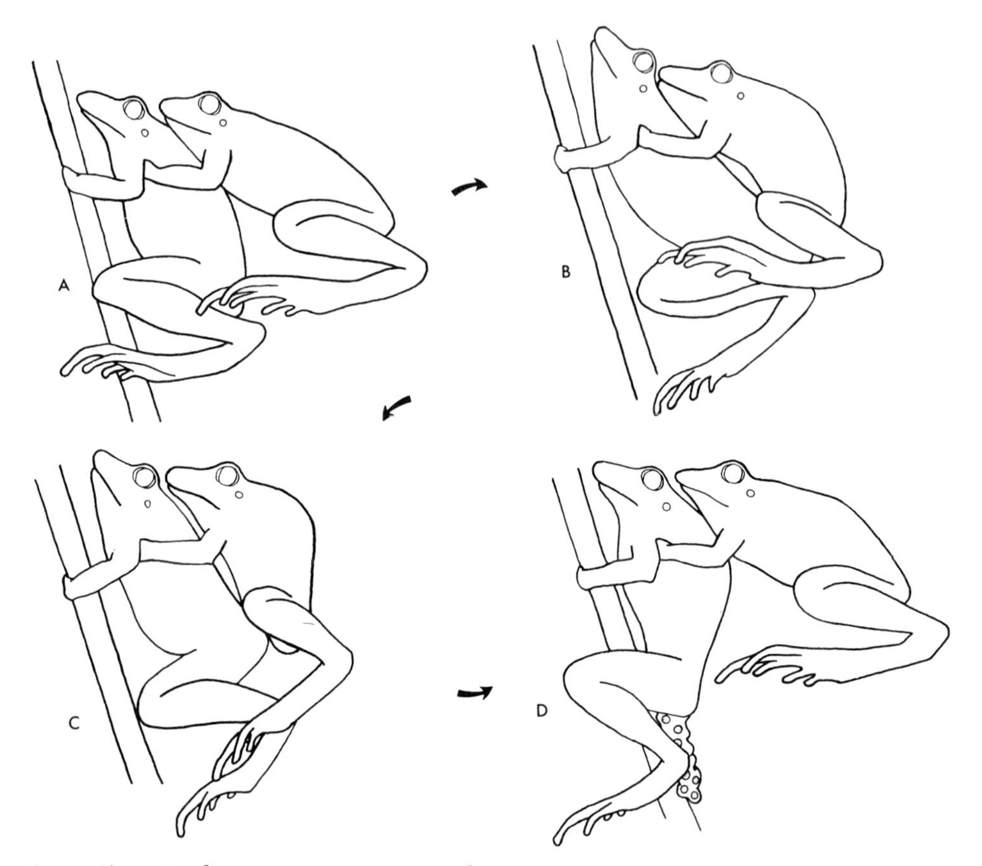

Figure 5–3. Egg-laying sequence in *Pseudacris triseriata:* A, shielding posture of male as female rests between egg-laying bouts; B, concave arching of back by female, a signal posture common to many advanced frogs; C, the male downthrust as eggs start to leave the vent; D, the adpressing of eggs to the stem. After Gosner and Rossman (1959).

tically duplicated, including male amplectic calls, as well as rolling and tonic immobility by females seeking release.

Another example concerns combat behavior, the most aggressive part of territoriality. Emlen (1968) has described the chest-to-chest struggles of *Rana catesbeiana* and associated postures of contending animals. In *Hymenochirus boettgeri* and *Pipa parva,* similar fierce struggles occur (Österdahl and Olsson, 1963; Rabb and Rabb, 1963*a*; Rabb, 1969). Territorial fighting by males is known also in hylids (Lutz, 1960), dendrobatids (Duellman, 1966), pelobatids (Whitford, 1967), and leptodactylids (Brattstrom and Yarnell, 1968). Occurrence of pronounced territoriality in such a range of ecological types weakens the notion that territoriality is a corollary of long-term occupation of sites (Rabb and Rabb, 1963*b*; Duellman, 1966). Defensive fighting has long been known to occur in most frogs. The clasping male of many advanced taxa adopts a characteristic shielding position to kick off attackers (Figure 5–3; Eibl-Eibesfeldt, 1953, 1956). Similar kicking (and biting) is present in primitive forms that lack the postural response (Rabb and Rabb, 1963*a*, 1963*b*). This type of competition (and the amplectic feedback system) seems to have been modified to allow a communal fertilization pattern by trios of males in *Chiromantis rufescens* (Coe, 1967).

Sexually diphasic behavior is not genetically bound. In other words, males

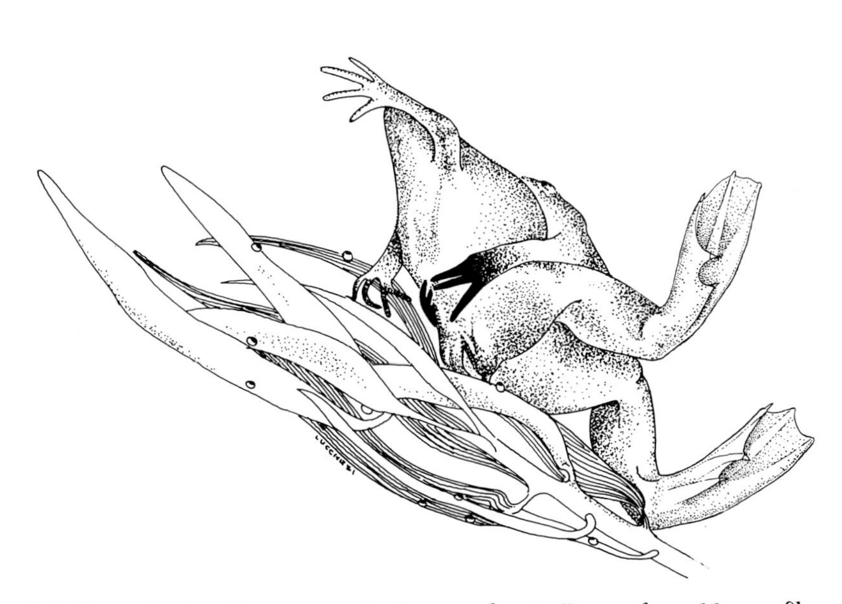

Figure 5–4. Normal ovipositing posture in *Xenopus laevis*. Drawn from 16-mm. film.

Figure 5–5. Sequence of positions in egg-laying of *Hymenochirus boettgeri*. A, leaving the substrate; B, swimming to the surface; C, inspiration by female (and sometimes male); D, dropping back beneath surface; E, turnover egg-laying position with vents above water; F, return to bottom. Drawn from 16-mm. film.

Figure 5–6. The egg-laying turnover in *Pipa pipa:* A, sideways push-off from bottom; B, mid-water upside-down position; C, return to bottom, male in fertilizing thrust posture. Drawn from 16-mm. film.

and females can show behavior normally characteristic of the opposite sex, although most demonstrations concern females. The absence of neural differentiation by sex is true of all taxa thoroughly investigated (examples are cited in Hutchison, 1967; Schmidt, 1966; Noble and Aronson, 1942; Heusser, 1968; and others). Most sexually inappropriate behavior is rare, but a lack of differentiation, if not partial reversal of courtship roles, is apparently the norm in the dendrobatids and *Tomodactylus angustidigitorum.* In the latter species, females call from fixed positions, and males orient and move to them (Dixon, 1957). If Schmidt's neurological schemata apply, there must have been an evolutionary shift in the hormone-sensitive preoptic control of these behaviors.

Again, the coordination of behavior with physiological events seems no more perfect in advanced forms than in others. For example, ovulation and extrusion of eggs may occur without ovipositing, and ovipositing with or without eggs may occur without amplexus in rhacophorids, bufonids, ranids, and hylids, as well as pipids (examples in Pope, 1931; Noble and Aronson, 1942; Eibl-Eibesfeldt, 1956; Heusser, 1968; personal observations on *Pseudacris triseriata, Xenopus laevis, Pipa parva*).

But surely there has been some over-all advance and change? Perhaps there has, in respect to vocalization. It seems likely that the discoglossids are primitive in regard to accessory vocal apparatus (Liu, 1935; but see Blair, 1963) and structure of the calls (Schneider, 1966*a*, 1966*b*). Neurophysiological investigation of call mechanisms in discoglossids would be valuable in discerning progress in other groups. Knoepffler (1962) reported that the warning cry in *Discoglossus* is used as a protest and release call by the females; this appears to confirm Schmidt's idea about the origin and interrelations of release calling and warning crying. Further, there seems to be a major divergence in the primary structure that produces calling pulses. *Scaphiopus* and some *Bufo* use the vocal cords, according to McAlister (1959, 1961), whereas *Hyla*(?), *Rana pipiens,* and some *Bufo* use glottal movements (Schmidt, 1966, 1972*a*). A wider sampling is obviously in order. Schmidt (1972*a*) believes that four or more calling types may exist when all effector structures are considered (body cavity, vocal cords, glottis, nares, throat).

On the receptor side, the degenerate nature of the distal auditory apparatus of *Ascaphus, Leiopelma,* and *Bombina* is paralleled in other frogs (Slabbert and Maree, 1945) and is probably a secondary evolutionary phenomenon. As Schmidt (1970) has shown, the mute *Ascaphus* has retained the basic auditory receptors. *Bombina* is quite vocal (Schneider, 1966*a*) and *Leiopelma* has at least a warning cry (Sharell, 1966). Nevertheless, the disappearance of functional aural communication must have consequences for the social system and reproductive behavior of a species. Heusser (1969*a*) has remarked on such consequences in the practically silent *Bufo bufo* of Switzerland. North American *Bufo* seem to provide stages in reduction of accessory vocal apparatus and correlated voicelessness (Blair and Pettus, 1954; Inger, 1958; McAlister, 1961), and close analysis of their behavioral biology might be fruitful. An interesting situation possibly representing a compromise between selective forces has already been reported by Axtell (1959) and Brown and Pierce (1967): some males of *B. speciosus* and *B. cognatus* exercise fairly powerful voices, while other males sit around them as sexual parasites, not calling but ready and able to mate with females that approach the calling males. Unfortunately, there is no precise information on the motivational states of such males; administration of hormones is one obvious experimental approach to finding out what part of the information system is changing.

Aside from vocal mechanisms, perhaps there has also been general progress in respect to amplexus and to clasp position of the male. It has been proposed that inguinal or pelvic amplexus (known to be common to ascaphids, pipids, pelobatids, discoglossids, and some leptodactylids) is relatively less efficient for fertilization than the axillary-region clasp of more advanced frogs (Rabb and Rabb, 1963*a*). An example in regard to this idea is the midwife toad *Alytes,* which shifts from a posterior clasp to a neck-grip just before fertilization (e.g., Joly, 1958; Marquet and Salverda, 1964; Figure 5–7). Another possible factor in clasping-position variation lies in the use of the hind legs in stimulating the female. Stroking of the head or vent is a constant feature of most primitive frogs but is rarely noted in advanced frogs. Many of the latter, however, use the feet in other ways, such as churning foam nests for the eggs.

Evolutionary evaluation of such differences suffers a want of data. Here, for instance, it would be very helpful to have fecundity data, along with comprehensive descriptions of the mating acts of Australian leptodactylids, some of which have an anterior clasp and some posterior (Fletcher, 1889; Moore, 1961; Littlejohn, 1963). Unfortunately, little is known about *Heleophryne,* the only South African leptodactylid genus; it apparently lays eggs on land, but the mating act has not been seen. A most interesting case is *Chrysobatrachus,* which Laurent (1964) has reported to differ from its hyperoliid relatives in having a lumbar amplexus.

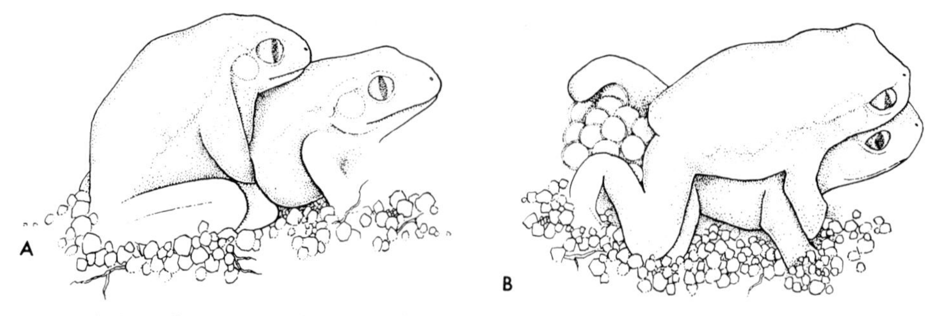

Figure 5–7. Mid-mating shift of amplectic position in the midwife toad, *Alytes obstetricans.* After photographs by Jakob Forster and Walter Goetz (1960).

Elucidation of the behaviors associated with the principal nonclasping mating mechanism would also be of considerable interest. I refer to the gluing phenomenon in microhylids, which has been reported in *Gastrophryne* (Fitch, 1956; Conaway and Metter, 1967), *Kaloula* (Inger, 1964), and *Breviceps* (Wager, 1965; Figure 5–8).

Another radical variation in mating habits is found in the dendrobatids, two of which have been reported to omit actual physical union of male and female (*Dendrobates auratus;* Senfft, 1935; *D. granuliferus;* Crump, 1972). Senfft reported the male fertilizing the eggs after they were laid; Crump observed frogs opposing vents, apparently laying and fertilizing simultaneously. Mudrack (1969) has recorded amplexus during a fertile mating in another dendrobatid, a species of *Phyllobates* from Ecuador. However, in his captive frogs, the clasp was extremely brief—10 to 20 seconds, with different males participating. An interesting, possibly precursory, stage has been noted in *Rhinoderma darwini,* which is said to have a very brief, gentle embrace (Pflaumer, 1935; Cei, 1962).

Adaptational explanations of such gross differences in behavior can only be speculative at this stage. However, a few ideas seem plausible and applicable. Amplectic movements in water are generally extremely well coordinated. This may reflect the need for prompt fertilization of the eggs, whose properties quickly change in water (Salthe, 1963; Savage, 1961; Pyburn, 1963). The absence or brevity of the clasp in dendrobatids and in *Rhinoderma* may be related to less-stringent fertilization requirements of eggs deposited on land. Perhaps directly or indirectly related is the shift of attention and energy from a lengthy amplexus

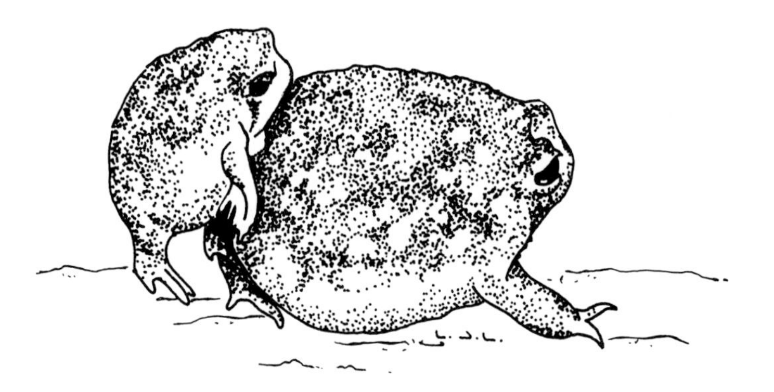

Figure 5–8. Amplexus by gluing in *Breviceps.* After Wager (1965).

to care of terrestrial eggs and larvae. It is difficult to imagine the nature of the mental reward that sustains interest in the inanimate eggs, but obviously a shift in information processing has to take place. Perhaps the prey-watching feeding pattern has been taken over and tuned to an egg sign-stimulus (see Lettvin et al., 1961). In the case of the gluing microhylids, the mechanical difficulties of sustaining even a brief clasp may be the paramount factor (Conaway and Metter, 1967). A similar argument was given by Laurent for the adoption of a pelvic clasp by *Chrysobatrachus.*

PROSPECTS

Parsimony is the general rule in evolutionary change, and we should expect it to be reflected among the frogs in their various behavioral adjustments to semiterrestrial life. Although the development of qualitatively new behaviors may involve only small quantitative shifts in a neural or neurohormonal system (Du Brul, 1958, 1967), it is still economic to have each new behavior serve in more than one function—thus the apparent takeover or modification of respiratory mechanisms in the sound-production patterns and the possible incorporation of feeding mechanisms in the mating pattern. From the research standpoint, this is a pleasing proposition, since materials on one system may be directly applicable to another.

At this time it is still easy to despair of using reproductive behavior as independent data in phylogenetic analyses. The panorama of existing discrete modes of reproduction, from aquatic spawning to terrestrial viviparity, is simply overwhelming in variety. Yet, by taking a holistic approach, grounding finer behavioral observations in more extensive knowledge of functional neurohormonal systems, and sorting out environmental factors with more precision, we may at least come to a confident understanding of the courses of evolution of the vast diversity in the anuran world.

Acknowledgments

Thanks for bibliographic help are extended to Philip Silverstone; Serge Daan; H. Heusser; Robert Schmidt; Karel Liem; Field Museum of Natural History Library; and the Committee on Evolutionary Biology, University of Chicago. The drawings are the work of Lido Lucchesi and Marge Moran. Studies on the pipid frogs mentioned here were supported by National Science Foundation Grant GB477.

REFERENCES

Aronson, L. R., and G. Noble. 1945. The sexual behavior of Anura. 2. Neural mechanisms controlling mating in the leopard frog, *Rana pipiens.* Bull. Am. Mus. Nat. Hist. 86(3):89-139.

Axtell, R. W. 1959. Female reaction to the male call in two anurans (Amphibia). Southwest. Naturalist 3:70-76.

Balinsky, B. I. 1969. The reproductive ecology of amphibians of the Transvaal highveld. Zool. Africana 4:37-93.

Beach, F. A. 1951. Instinctive behavior: reproductive activities, p. 387-434. *In* S. S. Stevens, [ed.], Handbook of experimental psychology. John Wiley & Sons, Inc., New York.

Ber, A., and J. Zieleniewski. 1959. Wplyw czynnikow nerwowych i psychicznych na jajeczkowanie, spermaiogeneze i kopulacje u zab *Xenopus laevis* Daudin. Endokrynol. Polska 10:83-90.

Birkenmeier, E. 1954. Beobachtungen zur Nahrungsaufnahme und Paarungsbiologie der gattung *Bombina*. Verh. Zool.-Botan. Ges. Wien 94:70-81.

Blair, W. F. 1963. Acoustic behaviour of Amphibia, p. 694-708. *In* R. G. Busnel, [ed.], Acoustic behaviour of animals. Elsevier Publ. Co., Amsterdam.

———. 1968. Amphibians and reptiles, p. 289-310. *In* T. A. Sebeok, [ed.], Animal communication. Indiana Univ. Press, Bloomington.

Blair, W. F., and David Pettus. 1954. The mating call and its significance in the Colorado River toad (*Bufo alvarius* Girard). Texas J. Sci. 6:72-77.

Bogert, C. M. 1960. The influence of sound on the behavior of amphibians and reptiles. *In* W. E. Lanyon and W. N. Tavolga, [ed.], Animal sounds and communication. Am. Inst. Biol. Sci. Publ. 7:137-320.

Bragg, A. N. 1960. Ovulation and breeding patterns in salientia. Herpetologica 16:124.

Brattstrom, B. H. 1962. Call order and social behavior in the foam-building frog, *Engystomops pustulosus*. Am. Zool. 2:394.

Brattstrom, B. H., and R. M. Yarnell. 1968. Aggressive behavior in two species of leptodactylid frogs. Herpetologica 24(3):222-228.

Brown, L. E., and J. R. Pierce. 1967. Male-male interactions and chorusing intensities of the great plains toad, *Bufo cognatus*. Copeia 1:149-154.

Cei, J. M. 1962. Batracios de Chile. Ed. Univ. Chile, Santiago. 128+cviii p.

Coe, M. J. 1967. Co-operation of three males in nest construction by *Chiromantis rufescens* Gunther (Amphibia: Rhacophoridae). Nature 214(5083):112-113.

Conaway, C. H., and D. E. Metter. 1967. Skin glands associated with breeding in *Microhyla carolinensis*. Copeia 3:672-673.

Crump, M. L. 1972. Territoriality and mating behavior in *Dendrobates granuliferus* (Anura: Dendrobatidae). Herpetologica 28:195-198.

Dixon, J. R. 1957. Geographic variation and distribution of the genus *Tomodactylus* in Mexico. Texas J. Sci. 9:379-409.

Du Brul, E. L. 1958. Evolution of the speech apparatus. C. C. Thomas, Springfield, Ill. 103 p.

———. 1967. Pattern of genetic control of structure in the evolution of behavior. Perspect. Biol. Med. 10:524-539.

Duellman, W. E. 1966. Aggressive behavior in dendrobatid frogs. Herpetologica 22: 217-221.

———. 1967. Courtship isolating mechanisms in Costa Rican hylid frogs. Herpetologica 23(3):169-183.

Eibl-Eibesfeldt, I. 1953. Vergleichende Verhaltensstudien an Anuren; 1. Zur Paarungsbiologie des Laubfrosches, *Hyla arborea* L. Z. Tierpsychol. 9:383-395.

———. 1956. Vergleichende Verhaltensstudien an Anuren; 2. Zur Paarungsbiologie der Gattungen *Bufo*, *Hyla*, *Rana* und *Pelobates*. Zool. Anz. (Suppl.) 19:315-323.

Emlen, S. T. 1968. Territoriality in the bullfrog, *Rana catesbeiana*. Copeia 2:240-243.

Ewert, J. P. 1967*a*. Aktivierung der Verhaltensfolge beim Beutefang der Erdkröte (*Bufo bufo* L.) durch Elektrische Mittelhirn-Reizung. Z. Vergl. Physiol. 54:455-481.

———. 1967*b*. Untersuchungen uber die Anteile zentralnervoser Aktionen an der taxisspezifischen Ermudung beim Beutefang der Erdkröte (*Bufo bufo* L.). Z. Vergl. Physiol. 57:263-298.

Ferguson, D. E. 1963. Orientation in three species of anuran amphibians. Ergebn. Biol. 26:128-134.

Firschein, I. 1951. A peculiar behavior pattern in a Mexican toad, *Bufo nayaritensis*, during amplexus. Copeia 1:73-74.

Fitch, H. S. 1956. A field study of the Kansas ant-eating frog, *Gastrophryne olivacea*. Univ. Kansas Publ. Mus. Nat. Hist. 8:275-306.

Fletcher, J. J. 1889. Observations on the oviposition and habits of certain Australian batrachians. Proc. Linn. Soc., New South Wales, Ser. 2, 4:357-387.

Flindt, R., and H. Hemmer. 1967. Die Parameter für das Einsetzen der Paarungsrufe bei *Bufo calamita* Laur. und *Bufo viridis* Laur. Salamandra 3:98-100.

FORBES, T. R. 1961. Endocrinology of reproduction in cold-blooded vertebrates, p. 1035-1087. *In* W. C. Young, [ed.], Sex and internal secretions, Vol. II. Williams & Wilkins Co., Baltimore.

FORSTER, J., AND W. GOETZ. 1960. Frosche, Kröten, Unken. Du 235:52-57.

FOSTER, W. A. 1967. Chorus structure and vocal response in the Pacific tree frog, *Hyla regilla*. Herpetologica 23:100-104.

FRISHKOPF, L. S., R. R. CAPRANICA, AND M. H. GOLDSTEIN, JR. 1968. Neural coding in the bullfrog's auditory system—a teleological approach. Proc. Inst. Elec. Electron. Engr. 56:969-980.

GOIN, C. J. 1960. Amphibians, pioneers of terrestrial breeding habits. Smithsonian Rept. 1959:427-445.

GOIN, O. B., AND C. J. GOIN. 1962. Amphibian eggs and the montane environment. Evolution 16:364-371.

GOSNER, K. L., AND D. A. ROSSMAN. 1959. Observations on the reproductive cycle of the swamp frog, *Pseudacris nigrita*. Copeia 3:263-266.

GREENBERG, B. 1942. Some effects of testosterone on the sexual pigmentation and other sex characters of the cricket frog *(Acris gryllus)*. J. Exptl. Zool. 91:435-451.

GRUBB, J. C. 1970. Orientation in postreproductive Mexican toads, *Bufo valliceps*. Copeia 4:674-680.

HELLMICH, W. 1962. Reptiles and amphibians of Europe. Blandford Press, London. 160 p.

HEUSSER, H. 1958. Zum Häutungsverhalten von Amphibien. Rev. Suisse Zool. 65(38): 793-823.

———. 1960*a*. Instinkterscheinungen an Kröten, unter besonderer Berücksichtigung des Fortpflanzungsinstinktes der Erdkröte (*Bufo bufo* L.). Z. Tierpsychol. 17:67-81.

———. 1960*b*. Über die Beziehungen der Erdkröte (*Bufo bufo* L.) zu ihrem Laichplatz II. Behaviour 26.93-109.

———. 1961. Die Bedeutung der äusseren Situation im Verhalten einiger Amphibienarten. Rev. Suisse Zool. 68(1):1-39.

———. 1968. Die Lebensweise der Erdkröte, *Bufo bufo* (L.); Laichzeit: Umstimmung, Ovulation, Verhalten. Vierteljahres. Naturforsch. Ges. Zurich 113:257-289.

———. 1969*a*. Der rudimentäre Ruf der männlichen Erdkröte *(Bufo bufo)*. Salamandra 5:46-56.

———. 1969*b*. Ethologische Bedingungen für das Vorkommen von Territorialität bei Anuren. Salamandra 5:81-160.

HEYER, W. R. 1969. The adaptive ecology of the species groups of the genus *Leptodactylus* (Amphibia, Leptodactylidae). Evolution 23:421-428.

HOLST, E. VON, AND U. VON ST. PAUL. 1963. On the functional organisation of drives. Animal Behaviour 11:1-20.

HUTCHISON, J. B. 1964. Investigations on the neural control of clasping and feeding in *Xenopus laevis* (Daudin). Behaviour 24:47-66.

———. 1967. A study of the neural control of sexual clasping behaviour in *Rana angolensis* Bocage and *Bufo regularis* Reuss with a consideration of self-regulatory hindbrain systems in the anura. Behaviour 28:1-57.

INGER, R. F. 1954. Systematics and zoogeography of Philippine amphibia. Fieldiana: Zool. 33: 181-531.

———. 1958. The vocal sac of the Colorado River toad (*Bufo alvarius* Girard). Texas J. Sci. 10:319-324.

INGER, R. F., AND J. P. BACON, JR. 1968. Annual reproduction and clutch size in rain forest frogs from Sarawak. Copeia 3:602-606.

JAMESON, D. L. 1955. Evolutionary trends in the courtship and mating behavior of salientia. Systematic Zool. 4:105-119.

———. 1957. Life history and phylogeny in the salientians. Systematic Zool. 6(2):75-78.

JOLY, J. 1958. Les moeurs nuptiales du crapaud accoucheur. Sci. et Nat. 29:33-36.

KNEPTON, J. C. 1952. The response of male salientia to human chorionic gonadotropic hormone. Quart. J. Florida Acad. Sci. 14 (4):255-265.

KNOEPFFLER, L. P. 1962. Contribution a l'étude du genre *Discoglossus* (Amphibiens, Anoures). Vie et Milieu Paris 13:1-94.
LANDRETH, H. F., AND D. E. FERGUSON. 1967. Movements and orientation of the tailed frog, *Ascaphus truei.* Herpetologica 23:81-93.
LAURENT, R. F. 1964. Adaptive modifications in frogs of an isolated highland fauna in Central Africa. Evolution 18:458-467.
LETTVIN, J. Y., H. R. MATURANA, W. H. PITTS, AND W. S. MCCULLOCH. 1959. What the frog's eye tells the frog's brain. Proc. Inst. Radio Engr. 47:1940-1951.
———. 1961. Two remarks on the visual system of the frog, p. 757-776. *In* W. A. Rosenblith, [ed.], Sensory communication. M.I.T. Press and John Wiley & Sons, Inc., New York.
LICHT, L. E. 1969. Unusual aspects of anuran sexual behavior as seen in the red-legged frog, *Rana aurora aurora.* Canadian J. Zool. 47:505-509.
LITTLEJOHN, M. J. 1963. The breeding biology of the baw baw frog *Philoria frosti* Spencer. Proc. Linn. Soc., New South Wales 88:273-276.
LIU, C. C. 1931. Sexual behavior in the Siberian toad, *Bufo raddei,* and the pond frog, *Rana nigromaculata.* Peking Nat. Hist. Bull. 6:43-60.
———. 1935. Types of vocal sac in the salientia. Proc. Boston Soc. Nat. Hist. 41: 19-40.
LUTZ, B. 1947. Trends towards non-aquatic and direct development in frogs. Copeia 4:242-252.
———. 1960. Fighting and an incipient notion of territory in male tree frogs. Copeia 1:61-63.
MARQUET, P. L., AND I. Z. SALVERDA. 1964. De Paring en Eiverzorging van de Vroedmeesterpad *(Alytes obstetricans).* Overdr. Naturh. Maandblad 53:119-123.
MCALISTER, W. H. 1959. The vocal structures and method of call production in the genus *Scaphiopus* Holbrook. Texas J. Sci. 11:60-77.
———. 1961. The mechanics of sound production in North American *Bufo.* Copeia 1:86-95.
MOORE, J. A. 1961. The frogs of eastern New South Wales. Bull. Am. Mus. Nat. Hist. 121(3):149-386.
MUDRACK, W. 1969. Pflege und Zucht eines Blattsteigerfrosches der Gattung *Phyllobates* aus Ecuador. Salamandra 5:81-84.
NOBLE, G. K. 1931. The biology of the amphibia. McGraw Hill Book Co., New York. 577 p.
NOBLE, G. K., AND R. ARONSON. 1942. The sexual behavior of Anura. 1. The normal mating pattern of *Rana pipiens.* Bull. Am. Mus. Nat. Hist. 80:127-142.
OLIVER, J. A. 1955. The natural history of North American amphibians and reptiles. Van Nostrand, Princeton. 359 p.
ÖSTERDAHL, L., AND R. OLSSON. 1963. The sexual behaviour of *Hymenochirus boettgeri.* Oikos 14:35-43.
PFLAUMER, C. 1935. Observaciones biológicas acerca de la *Rhinoderma darwinii* D. & B. Rev. Chil. Hist. Nat. 39:28-30.
POPE, C. H. 1931. Notes on amphibians from Fukien, Hainan, and other parts of China. Bull. Am. Mus. Nat. Hist. 61:397-611.
PYBURN, W. F. 1963. Observations on the life history of the treefrog, *Phyllomedusa callidryas* (Cope). Texas J. Sci. 15:155-170.
———. 1966. Breeding activity, larvae and relationship of the treefrog *Hyla phaeota cyanosticta.* Southwest. Naturalist 11:1-18.
RABB, G. B. 1969. Fighting frogs. Brookfield Bandarlog. Chicago Zool. Soc. 37:4-5.
RABB, G. B., AND M. S. RABB. 1961. On the mating and egg-laying behavior of the Surinam toad, *Pipa pipa.* Copeia 4:271-276.
———. 1963*a*. On the behavior and breeding biology of the African pipid frog *Hymenochirus boettgeri.* Z. Tierpsychol. 20:215-241.
———. 1963*b*. Additional observations on breeding behavior of the Surinam toad, *Pipa pipa.* Copeia 4:636-642.
———. 1965. Effects of isolation on reproductive behavior in the pipid frog *Xenopus laevis.* Am. Zool. 5:685.

RUSSELL, W. M. S. 1954. Experimental studies of the reproductive behaviour of *Xenopus laevis.* I. The control mechanisms for clasping and unclasping, and the specificity of hormone action. Behaviour 7:113-188.

———. 1960. Experimental studies of the reproductive behaviour of *Xenopus laevis.* II. The clasp positions and the mechanisms of orientation. Behaviour 15:253-283.

SALTHE, S. N. 1963. The egg capsules in the amphibia. J. Morphol. 113:161-171.

SAVAGE, R. M. 1961. The ecology and life history of the common frog *(Rana temporaria temporaria).* Pitman & Sons, Ltd., London. 221 p.

SCHMIDT, R. S. 1966. Central mechanisms of frog calling. Behaviour 26:251-285.

———. 1968*a*. Preoptic activation of frog mating behavior. Behaviour 30:239-257.

———. 1968*b*. Chuckle calls of the leopard frog *(Rana pipiens).* Copeia 3:561-569.

———. 1969. Preoptic activation of mating call orientation in female anurans. Behaviour 35:114-127.

———. 1970. Auditory receptors of two mating call-less anurans. Copeia 1:169-170.

———. 1971. A model of the central mechanisms of male anuran acoustic behavior. Behaviour 39:288-317.

———. 1972*a*. Release calling and inflating movements in anurans. Copeia 2:240-245.

———. 1972*b*. Mechanisms of clasping and releasing (unclasping) in *Bufo americanus.* Behaviour 43:85-96.

SCHNEIDER, H. 1966*a*. Die Paarungsrufe einheimischer Froschlurche (Discoglossidae, Pelobatidae, Bufonidae, Hylidae). Z. Morphol. Ökol. Tiere 57:119-136.

———. 1966*b*. Bio-Akustik der Froschlurche. Ein Bericht über den gegenwärtigen Stand der Forschung. Stuttgarter Beiträge Naturkunde. 152:1-16.

SENFFT, W. 1936. Das Brutgeschäft des Baumsteigerfrosches (*Dendrobates auratus* Girard) in Gefangenschaft. D. Zool. Garten 8:122-131.

SEXTON, O. J. 1960. Some aspects of the behavior and of the territory of a dendrobatid frog, *Prostherapis trinitatis.* Ecology 41:107-115.

SEXTON, O. J., AND E. P. ORTLEB. 1966. Some cues used by the leptodactylid frog, *Engystomops pustulosus,* in selection of the oviposition site. Copeia 2:225-230.

SHARELL, R. 1966. The tuatara, lizards, and frogs of New Zealand. Collins, London. 94 p.

SLABBERT, G. K., AND W. A. MAREE. 1945. The cranial morphology of the Discoglossidae and its bearing upon the phylogeny of the primitive anura. Ann. Univ. Stellenbosch 23A(2-6):91-97.

SMITH, C. L. 1955. Reproduction in female amphibia. Mem. Soc. Endocrinol. 4:39-56.

SWISHER, J. E. 1969. Spawning turnovers in *Xenopus tropicalis.* Am. Zool. 9:573.

TEST, F. H. 1954. Social aggressiveness in an amphibian. Science 120:140-141.

TINBERGEN, N. 1952. The study of instinct. Oxford Univ. Press, London. 221 p.

WAGER, V. A. 1965. The frogs of South Africa. Purnell & Sons, Capetown Johannesburg. 242 p.

WHITFORD, W. G. 1967. Observations on territoriality and aggressive behavior in the western spadefoot toad, *Scaphiopus hammondii.* Herpetologica 23:318.

6

QUANTITATIVE CONSTRAINTS ASSOCIATED WITH REPRODUCTIVE MODE IN ANURANS

Stanley N. Salthe and
William E. Duellman

INTRODUCTION

There are a variety of factors affecting reproduction in anurans. Some of these factors can be quantified and are thus parametric in nature. Parameters included in the study were female snout-vent length, ovum diameter, clutch size, and site of development. The purpose of this paper is to present the results of our investigations of the interrelationships of these parameters and to discuss their adaptive values and limitations. Previous works of interest are those of Terentyev (1960) on temperate Eurasian members of the genus *Rana* and Salthe (1968, 1969) on salamanders. These studies concluded that positive correlations exist between female body size and both ovum size and clutch size, whereas negative correlations prevail between ovum size and clutch size. Salthe noted that these parameters are more closely correlated if the comparison is restricted to species sharing a single mode of reproduction. Mode of reproduction, as used herein, is a concept combining the site of development with the mode of development. Mode of development is a complex of features, including such factors as rate of development, stage at hatching, and others. More detailed descriptions of anuran reproductive modes are forthcoming (Salthe and Mecham, 1973). Due to the diversity of reproductive modes in the large numbers of species of frogs, in contrast to the genus *Rana* and all salamanders, many of the relationships reported herein are both incomplete and, in the aggregate, less conclusive than those in the other studies cited. Nevertheless, it can be shown that the principles operating in frogs of the genus *Rana* and in salamanders are applicable to the Anura as a whole.

MATERIALS AND METHODS

Much of the data on which this study is based—female snout-vent length, ovum diameter, clutch size, and site of development—were obtained from a variety of literature sources, none of which has been specifically cited when only measurements or descriptions of the above items were used in this study. Additional data were collected by Duellman. The data on ovum diameter represent measurements of preserved material almost exclusively; a few measurements were made on living material. Although we are aware that some shrinkage may take place upon preservation, we have dismissed this consideration because of the small number of measurements made on living material and because we found no significant deviations between these measurements.

In most cases clutch size is based on measurements of clutches found in the field. Only in exceptional cases displaying some unique combination of body size and reproductive mode or ovum size, as in *Barbourula* (Figure 6–4), was an ovarian egg count used for clutch size.

In our classification of sites of development, distinctions are made between completely aquatic, completely terrestrial, and partially aquatic sites. Partially aquatic sites are those used by species with modes of development taking place sequentially in more than one site. In frogs, this sequence invariably involves a terrestrial nesting site (or its functional equivalent, from this point of view, of parental pouches or other cavities in which eggs are carried) followed by an aquatic tadpole stage. Distinctions also are made between sites in lotic water and lentic water. These distinctions, even though more or less traditional, are reflected in our data on ovum size and clutch size, whereas other distinctions—for example, arboreal versus fossorial nests—arc not, unless accompanied by differences in mode of development. Distinction is made between hatching of embryonic tadpoles (earlier than stage 21 of Gosner [1960]) and advanced tadpoles.

For the purposes of this paper we recognize three major modes of development. The first is relatively rapid intraoval development, with hatching occurring at some relatively early embryonic stage (with external gills and suckers); this is typically associated with totally aquatic sites of development and with some partially aquatic ones, notably those involving arboreal nesting sites. The next is relatively slow intraoval development leading to hatching at fairly advanced tadpole stages, for example, with closed or partially closed operculum; this mode is usually associated with partially aquatic sites of development involving either fossorial nests (sometimes with parental care) or brood pouches. The third is direct development, which is associated with entirely terrestrial nesting sites or, in some cases, brood pouches. Our definition of this last mode is basically an ecological one: this is the mode of development wherein all energy needed to complete development is provided by the female. We include here species in such genera as *Anhydrophryne, Eleutherodactylus, Hemiphractus,* and *Hemisus.* The mode of development found in *Nectophrynoides,* which is associated with viviparity, is not treated in this paper.

Although our definitions of various modes of development are heuristic and there are some uncontrolled variables in the data on ovum size and clutch size, a coherent picture of reproductive adaptations in anurans can be shown.

It would have been desirable to estimate body volume of female frogs, as was done for a salamanders by Salthe (1969); however, the body shapes of frogs are far less uniform than are those of salamanders. There is no mathematically simple shape that can be used to generalize the body shape of all frogs, nor in most cases is there any means of assigning to a given species any one of a series of mathematically simple, idealized body shapes. Therefore, volumetric data on adult frogs has not been utilized.

RESULTS

All of the parameters of reproductive adaptations treated in this paper will be shown to be interrelated and subject to an interpretation of reproductive strategies.

On the basis of the ovum size–body size relationship in salamanders, Salthe (1969) predicted that in salamanders ovum size would be found to vary intraspecifically, or intrapopulationally, most noticeably in the smallest species. If intrapopulational variability is the ultimate source of raw material for adaptive evolution, an intraspecific correlation between body size and ovum size must precede an interspecific correlation of the same kind. McAlister (1962) and Hönig

(1966) showed that in at least some temperate species of *Rana,* ovum size is a function of female body size within a given population (Table 6–1). This scant evidence on medium-sized frogs suggests that differences in ovum size are detectable in frogs of different sizes.

Table 6–1. Ovum Size and Female Body Size in Two Species of *Rana*

Rana pipiens—Texas		*Rana temporaria*—Germany	
Snout-vent length (*mm*)	Ovum diameter (*mm*)	Weight of females (*gr*)	Weight of 1000 eggs (*gr*)
>90	2.0–2.2	90–92	14
		81–88	12
60–70	1.6–1.8	70–75	11
		60–67	13
		50–59	10
		40–48	9
		31–34	5

Sources: McAlister (1962); Hönig (1966).

We have found little evidence of a positive correlation between ovum size and female snout-vent length when data are plotted for frogs from a variety of families having diverse modes of reproduction (Figure 6–1). However, when mode of reproduction is emphasized, instead of taxonomic position, a more positive correlation is apparent (Figure 6–2). Re-examination of Figure 6–1 from the point of view of general reproductive mode reveals some meaningful relationships. The data indicate that typically aquatic pond-breeders (some members of the genera *Hyla, Bufo,* and *Rana*—the largest *Rana* eggs are deposited by stream-breeders) differ from species having other reproductive modes. On the other hand, such reproductive specialists as species that have direct development (*Eleutherodactylus* and *Platymantis*), those that have terrestrial nests and produce advanced tadpoles (*Helioporus* and *Pseudophryne*), and those that are stream-breeders (*Ansonia, Ascaphus,* and *Barbourula*) differ from the pond-breeders. Dividing the species into these groups allows us to perceive two broadly different ovum size–body size relationships among these frogs. Three such relationships, much more clearly divided, were noted in salamanders by Salthe (1969). The data for *Agalychnis* are suggestive of a possible third relationship in frogs. These anurans deposit clutches in trees; the tadpoles hatch at a relatively early stage and are not as well developed as the tadpoles of most *Pseudophryne* and *Helioporus.* The data presented in Figure 6–2 also fall into two categories: (1) lentic breeders and species that have terrestrial nests in the vicinity of lentic water in which the tadpoles subsequently develop and (2) species having direct development and those with terrestrial nests in the vicinity of lotic water in which the tadpoles develop. There is broad overlapping at the smaller body sizes, and some overlapping occurs even at larger sizes in the genus *Helioporus* (the six triangles farthest to the right in Figure 6–2). Indeed, the most overlapping species, *Helioporus barycragus,* utilizes lotic water (Main, 1968); thus, it is an outstanding exception to the relationship thus far discovered.

Both *Bufo* and *Rana* display the same fundamental mode of reproduction, but it is evident that the ova of *Bufo* are consistently smaller at all body sizes

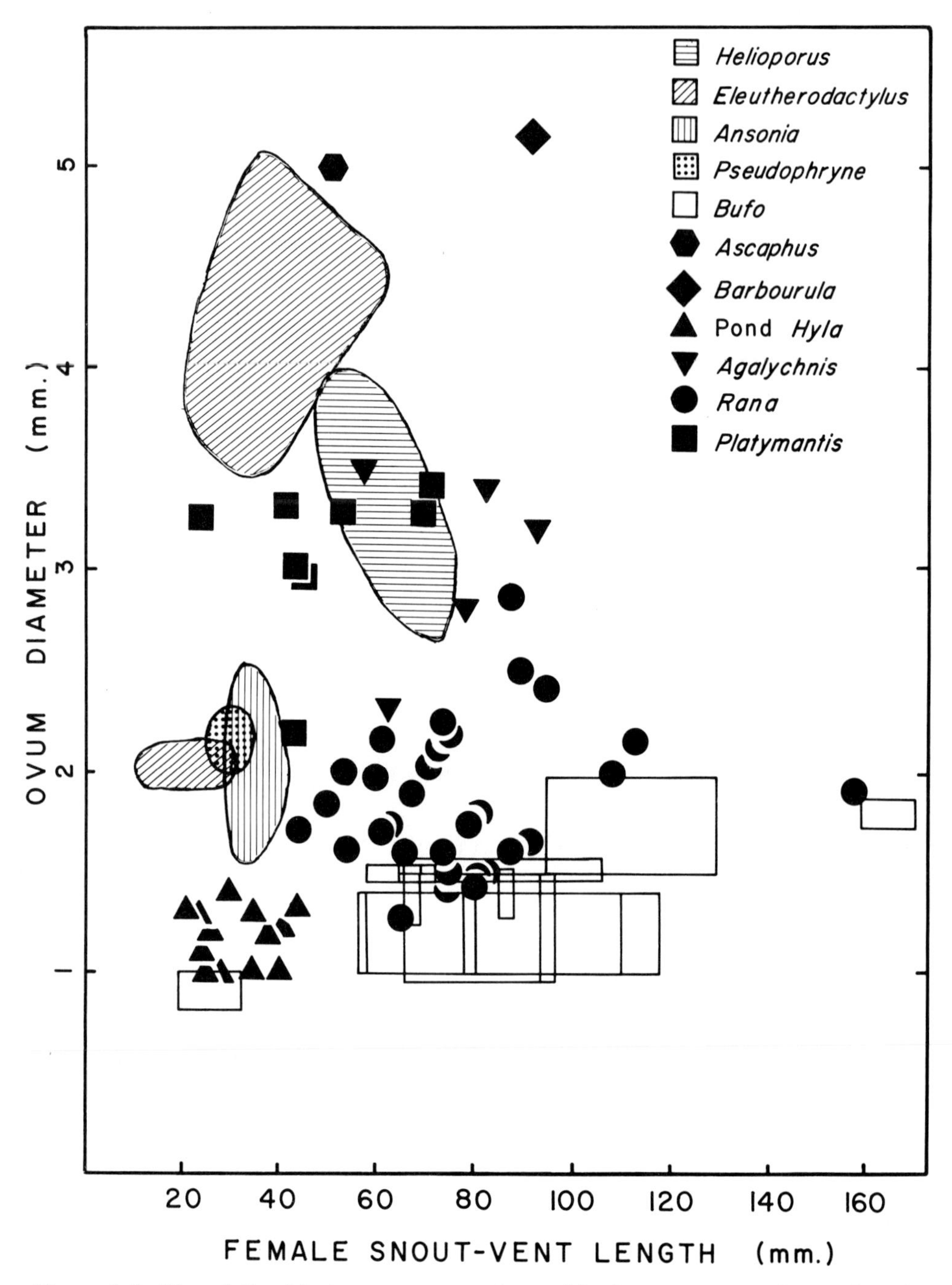

Figure 6–1. The relationship between ovum size and body size in some frogs, plotted as taxa. The data for various species of *Bufo* are plotted as unshaded areas for individual species.

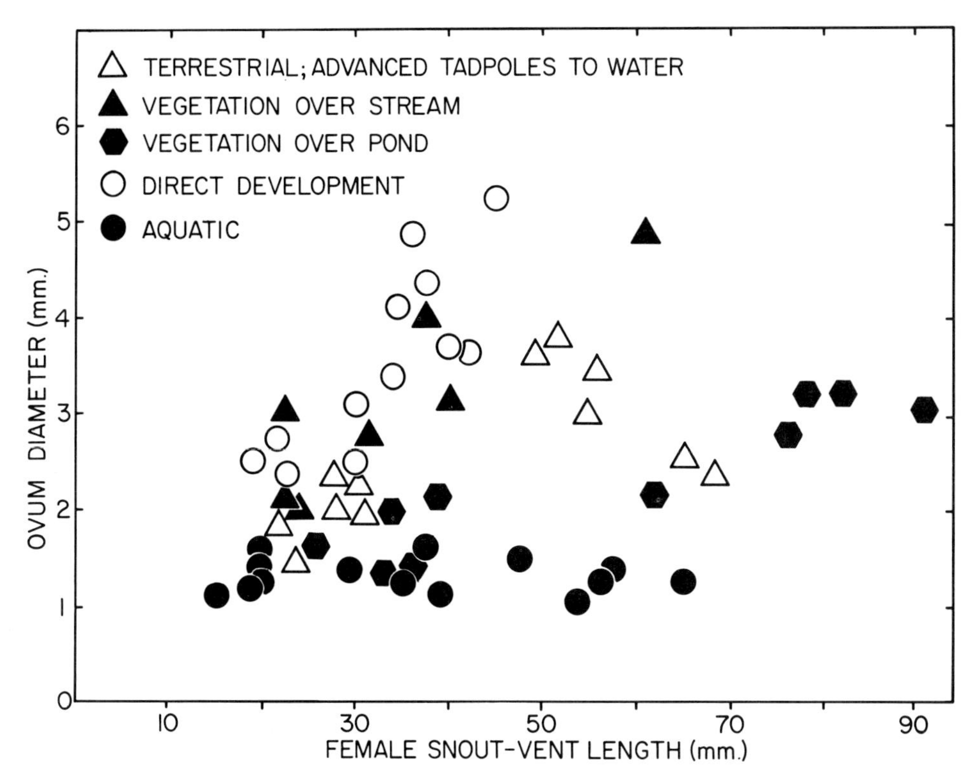

Figure 6–2. The relationship between ovum size and body size in some frogs, plotted as mode of reproduction; each point represents one species. This figure is based on Figure 5 in Main (1968), to which new data has been added.

than those of *Rana* (Figure 6–1). This difference, resulting in a broad ovum size–body size relationship for this mode of reproduction, is explicable in terms of clutch size. Thus, as shown in Figure 6–3, *Bufo* clutches are consistently larger at all body sizes than those of *Rana* (also see Smith, 1947). Calculations of the total clutch volume at the smallest, an intermediate, and the largest body sizes show *Rana* with 870, 7500, and 71,000 mm^3 and *Bufo* with 1000, 4300, and 90,000 mm^3, respectively, thereby suggesting that plotting total body volume against clutch volume would have shown a much closer relationship among species in these two genera. It is for this reason that we have cause to regret our inability to estimate body volume for more than an insignificant number of frogs at this time. It is of some interest to note that *Bufo* has adopted a reproductive strategy involving many more, much smaller eggs than has *Rana*. Whether such differences will ultimately justify subdivision of the general mode of reproduction shared by these genera is at present not clear. So few frogs have been examined, relative to the total number of species, that there is as yet little justification for such refinement.

The data just discussed imply that within a given range of body size, species with relatively large ova produce relatively fewer eggs at each reproduction. This kind of relationship occurs in salamanders (Salthe, 1969), and the same kind of relationship is apparent among the few species of frogs for which data are available (Figure 6–4). The data on *Agalychnis annae* (Figure 6–4) suggest that the relationship does not hold for this species. However, *Agalychnis* is peculiar in that females deposit their ovarian complement in more than one clutch (Pyburn, 1970; Duellman, 1970); thus, the sizes of the individual clutches (see data pre-

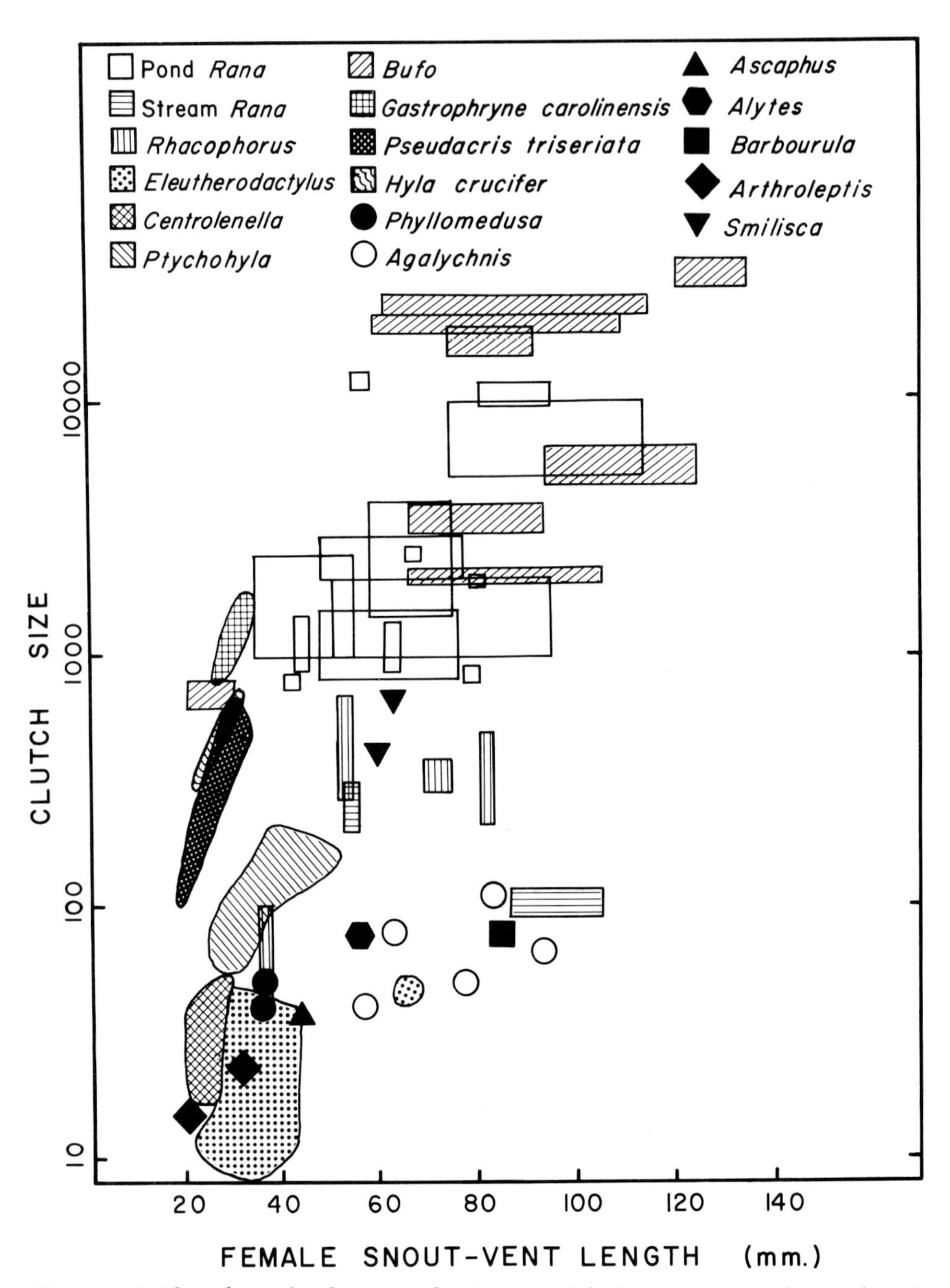

Figure 6–3. The relationship between clutch size and body size in some frogs, plotted as taxa. The data for *Rana* and *Bufo* show the ranges for each parameter. Data for *Gastrophryne carolinensis, Hyla crucifer,* and *Pseudacris triseriata* show the relationship within one population of each species. The data for *Gastrophryne carolinensis* came from Henderson (1961); for *Hyla crucifer,* from Oplinger (1966); and for *Pseudacris triseriata,* from Pettus and Angleton (1967).

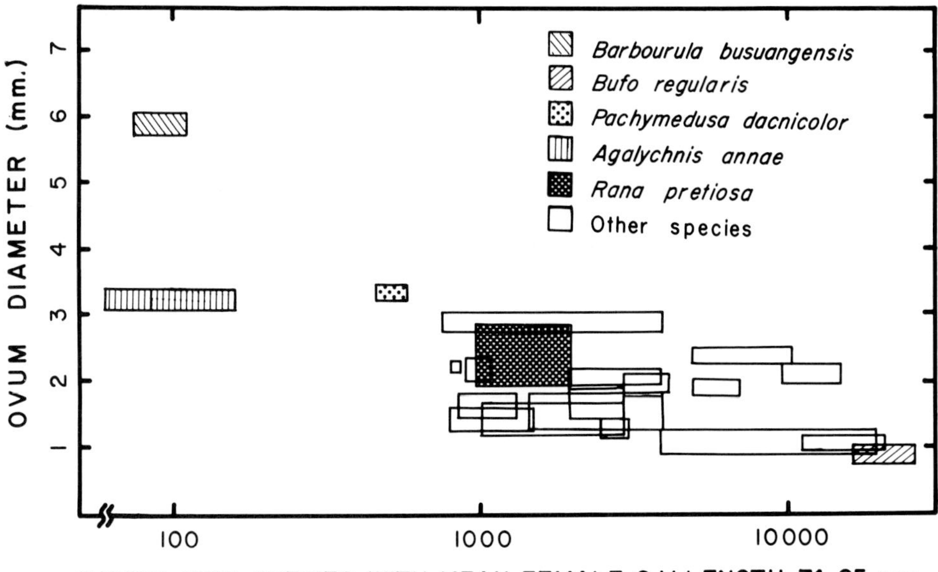

Figure 6–4. The relationship between ovum size and clutch size within a limited adult size range. The range of each parameter is shown for each species, only some of which are identified by name.

sented in Figure 6–4) are smaller than the total ovarian complement. With the exception of *Agalychnis annae,* the data presented in Figure 6–4 show the relationship that would be expected if total clutch volume were positively correlated with body volume. Although we cannot provide unquestionable data on body volume at present, our data suggest that this relationship prevails in *Rana* and *Bufo* and probably among most frogs in general, just as it does in salamanders.

Our data also suggest that there are relationships between ovum size, clutch size, body size, and mode of reproduction of the same kind found in salamanders by Salthe (1969). The relationships are not clear unless all four parameters are considered simultaneously, a feat that we cannot accomplish here in the absence of data on body volume. Even without these volumetric data, trends suggestive of these realtionships can be, and have been, detected. Thus, a consideration only of site of development and ovum size would lead to the tentative conclusion that sites of development involving terrestrial nests or streams tend to be associated with relatively large ova (Tables 6–2, 6–3, 6–4). These kinds of data have led several authors to make general statements associating large ovum size with increasingly terrestrial modes of reproduction (Lutz, 1948; Main et al., 1959; Alcala, 1962; Goin and Goin, 1962; Poynton, 1964; Heyer, 1969).

Tyler's data (Table 6–4) include division of ova by presence or absence of pigmentation. When found, pigmentation consists of a thin layer of melanin in the cortex of the animal hemisphere of the ovum, giving a color ranging from black to pale tan. An association between the presence of this pigment and egg deposition sites exposed to sunlight has been noted many times in frogs, most recently by Tyler (1968) and Heyer (1969). It seems that ova in eggs deposited in situations exposed to sunlight invariably are pigmented to some degree. However, not all ova in eggs deposited in situations hidden from sunlight are unpigmented—for example, *Arthroleptella* (de Villiers, 1929), *Notaden nicholsii* (Slater and Main, 1963), *Colostethus subpunctatus* (Stebbins and Hendrickson, 1959);

Table 6–2. The Relationship of Ovum Size (in *mm*) and Development Site

	Lentic water	Terrestrial eggs, tadpoles to lentic water	Lotic water	Terrestrial development
Philippine frogs	1.0–1.3	1.7–3.0	2.0–2.7	3.1–3.5
Leptodactylus	1.0–1.5	1.6–2.5		2.1–3.0
Australian frogs	1.1–2.1	1.9–3.8		2.3–5.4

Sources: Main (1958); Alcala (1962); Heyer (1969).

Table 6–3. Ovum Diameter (in *mm*) and the Lentic-Lotic Water Dichotomy in Three Species of the Genus *Smilisca*

Species	Streams	Pools in streams	Temporary rain pools
S. sila	2.4		
S. cyanosticta		2.3	
S. baudinii			1.3

Source: Duellman and Trueb (1966).

Rana spinosa (Pope, 1931). The functions of cortical color in sunlight may range from protection from ultraviolet radiation (Beudt, 1930; Sergeev and Smirnov, 1939) to absorption of infrared radiation in cold climates (Bernhard and Bratuschek, 1891; Douglas, 1949; Guyetant, 1966).

Within a population, increasing size of females is usually associated with increased clutch size (Henderson, 1961; Oplinger, 1966; Pettus and Angleton, 1967). Within a single reproductive mode, this relationship also exists interspecifically in frogs (Figures 6–3, 6–5). The data shown in Figure 6–3 suggest the following dichotomy: (1) species having typical aquatic development in lentic water and (2) species breeding in, or nesting above, lotic water, or having direct development. Species nesting above lentic water tend to be intermediate between these groups, which is the same position in the relationship between ovum size and body size (Figure 6–1). Data based entirely on neotropical hylids emphatically demonstrate that there is no body size–clutch size relationship outside single modes of reproduction (Figure 6–5), but within each mode there is an increase in

Table 6–4. The Relationship of Ovum Size and Pigmentation to Vertical Distribution in Papuan Hylids

Relative size and pigmentation	Altitude <3500	(*feet*) >3500
Small, pigmented ova	13	6[a]
Large, unpigmented ova	0	8

[a]Includes four species in lentic pools.
Source: Tyler (1968).

clutch size with increased body size. It should be noted that species depositing eggs in bromeliads tend to have clutches as small as those of species that deposit eggs on vegetation over ponds. The small clutch size of bromeliad breeders may be due to spatial restrictions imposed by the oviposition site. The clutch size–body size relationships in neotropical hylids segregate into four groups: (1) typically aquatic pond-breeders, (2) stream-breeders, (3) species using some sort of arboreal site of development, and (4) species having direct development.

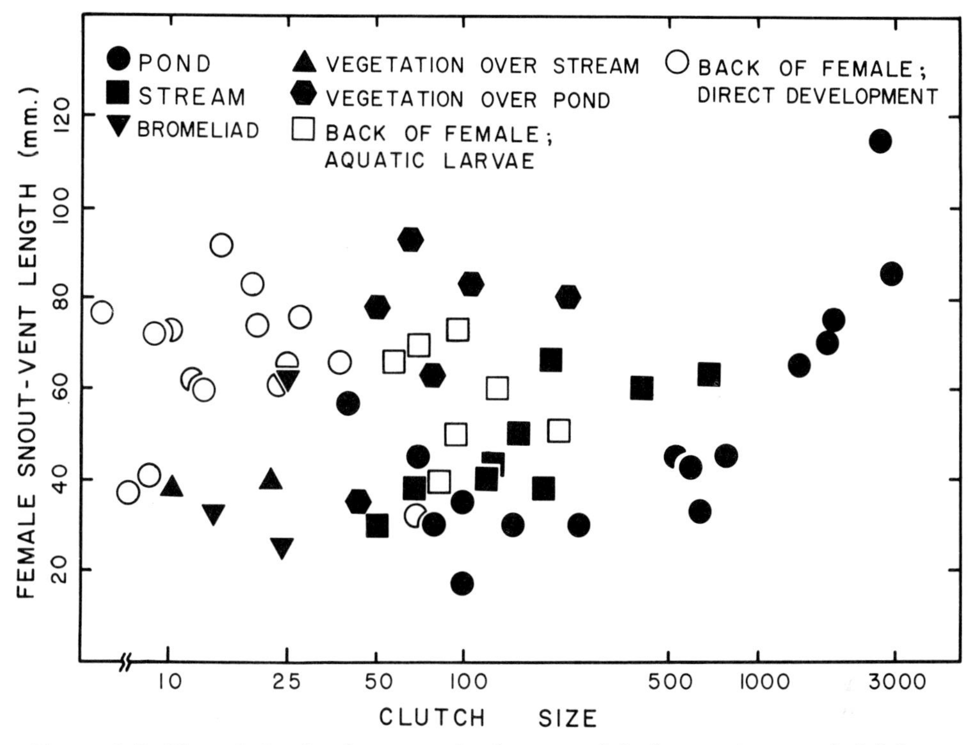

Figure 6–5. The relationship between clutch size and body size in some hylid frogs, plotted as modes of reproduction.

There is a direct relationship between ovum size and the size of the hatchlings regardless of reproductive mode (Figure 6–6). This includes species having relatively embryonic hatchlings *(Bufo, Discoglossus)*, species that hatch at relatively advanced stages *(Ascaphus, Alytes)*, species having direct development *(Eleutherodactylus, Syrrhophus)*, and species characterized by various intermediate reproductive modes. Evidently, the yolk and metabolic traits involved in development are roughly comparable from one reproductive mode to another (see modification below). Therefore, the only way to increase the size of the hatchling is to increase the size of the ovum. This increase can happen only if there is either an increase in body size or a modification in reproductive mode, specifically in the site of development. For example, adaptation to stream life apparently includes the production of larger hatchlings, which are more capable of locomotion in lotic water. This would explain why stream-associated reproductive modes involve generally larger ova than do pond-associated modes. This relationship also seems to explain why ova deposited in arboreal nests tend to be larger than those deposited in water. Thus, at hatching, a larger (older) tadpole would be more successful in dropping into the water below simply because the pull of gravity would be more effective. Moreover, the shape of these larva (as in *Agalychnis, Chiromantis, Phyllomedusa,* and some *Rhacophorus*) with a large, round, yolk-filled gut region also probably is adaptive to plummeting into the water below the nest. This general shape is characteristic of embryos and early larvae developing from very large ova; the yolk-filled gut at later stages is probably a consequence of the need for increased energy for maintenance in larger organisms. However, this explanation fails in the cases of the arboreal nesting frogs just discussed; in these groups the ova are not appreciably larger than those of frogs that have embryos and larvae that never accumulate enough yolk to form a bulging belly.

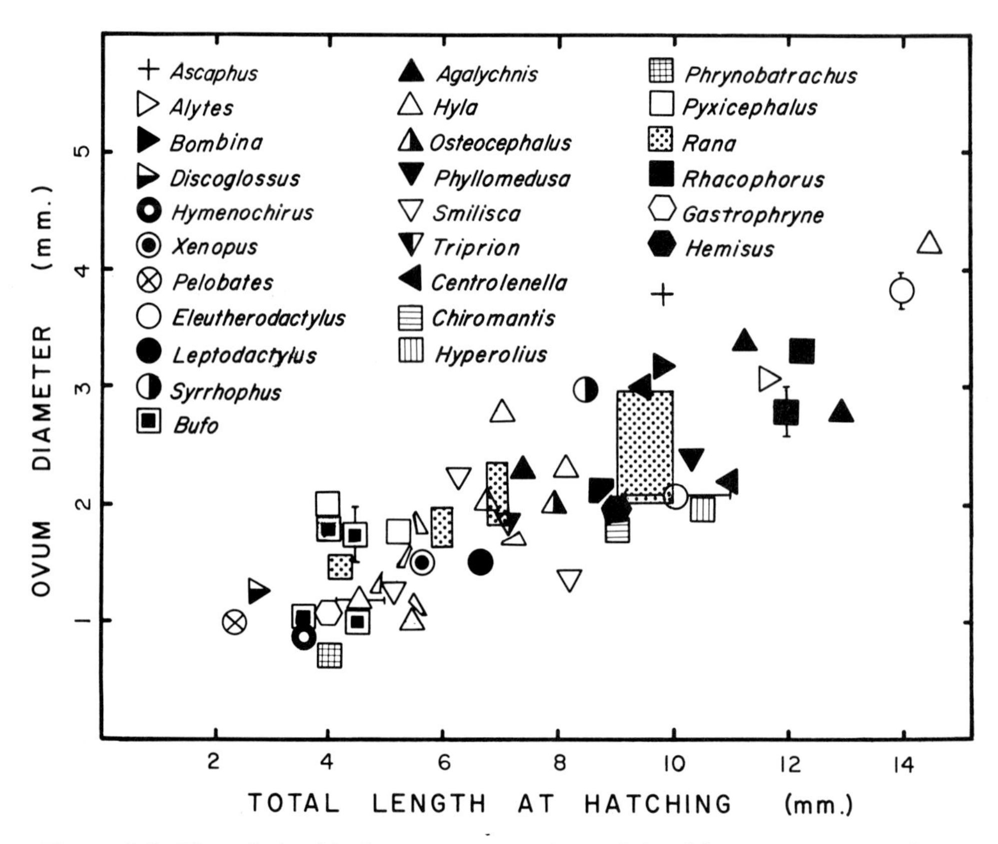

Figure 6–6. The relationship between ovum size and hatchling size in some frogs, plotted as taxa. Measurements of hatchlings undergoing direct development are from the tip of the snout to the tip of the outstretched hind limbs.

The eggs of *Centrolenella* are deposited on vegetation above torrential streams. At hatching the larvae do not have huge, yolk-filled gut regions like the larvae of *Agalychnis* and *Phyllomedusa,* which drop into ponds. Instead, the larvae of *Centrolenella* are relatively advanced and have rather small bodies with functional mouths and long muscular tails. Upon hatching the larvae may drop onto boulders in the streams, in which case they flip about vigorously until they reach the water. Obviously, the relatively embryonic and yolk-filled larvae of *Agalychnis* and others would meet disaster in such lotic conditions.

Certain developmental consequences of ovum size must be considered. Ovum size apparently cannot be increased (intraspecifically or interspecifically) without slowing the rate of early development (Figure 6–7). Stages later than neurula cannot be compared among the different modes of development, so we present no data on later stages. It is not evident why this relationship exists, because respiratory rate increases with ovum size intraspecifically (Table 6–5). Stefanelli's (1938) data (Table 6–5) can be interpreted to indicate simply that an increased amount of living material per ovum results in an increased respiratory rate, but Barth and Barth's (1954) data (Table 6–5) suggest a true increase in metabolic rate in larger ova. In view of the fact that larger ova develop more slowly (Figure 6–7), the increased respiratory rate apparently represents the need for greater levels of energy for maintenance in larger ova, thereby indicating a price paid for increased ovum size. Another contributing factor may be the increased resistance to cleavage and morphogenetic movements entailed by the yolk mass of increasingly larger ova. Thus, if a species is adapting to stream life for its tadpoles, increased ovum size may be selected for, but only at the expense of slower development, which also is a consequence of lower ambient temperatures in lotic environments. Assuming that slower development increases the probability of predation on the eggs or random environmental changes capable of damaging the clutch, selection will favor sites of development less exposed to the usual hazards faced by anuran eggs. This may explain the production of foam nests and various other arboreal and fossorial nesting sites commonly found among frogs adapted for stream reproduction and, indeed, among frogs with relatively large ova that still produce tadpoles.

McLaren (1965) investigated the physical meanings of the various constants in Belehádrek's (1957) formula for metabolic function and noted that ovum size in *Rana pipiens* was proportional to over-all developmental rate until hatching, when the assumption was made that the function describing the relationship departs from linearity to the same degree in different populations. When this assumption was not made, he found no statistically significant relationships between ovum size and developmental rate. Our data, derived in part from Bachmann (1969) and using many more species of amphibians, clearly shows that ovum size is inversely proportional to developmental rate to the neurula stage. The contradiction between our conclusions and those of McLaren (1965) perhaps can be

Table 6–5. Ovum Size and Respiratory Rate: Intraspecific Comparisons at Various Segmentation Stages

Species	Ovum diameter (*mm*)	O_2 *mm*^{3}/4 min./embryo
Bufo viridis	1.44	0.008
	1.00	0.006
B. bufo	2.50	0.013
	2.20	0.008
	2.00	0.002
	Ovum dry weight (*mg*)	O_2 ul/*mg* dry weight hour
Rana pipiens	1.86	0.184
	1.47	0.150

Sources: Stefanelli (1938); Barth and Barth (1954).

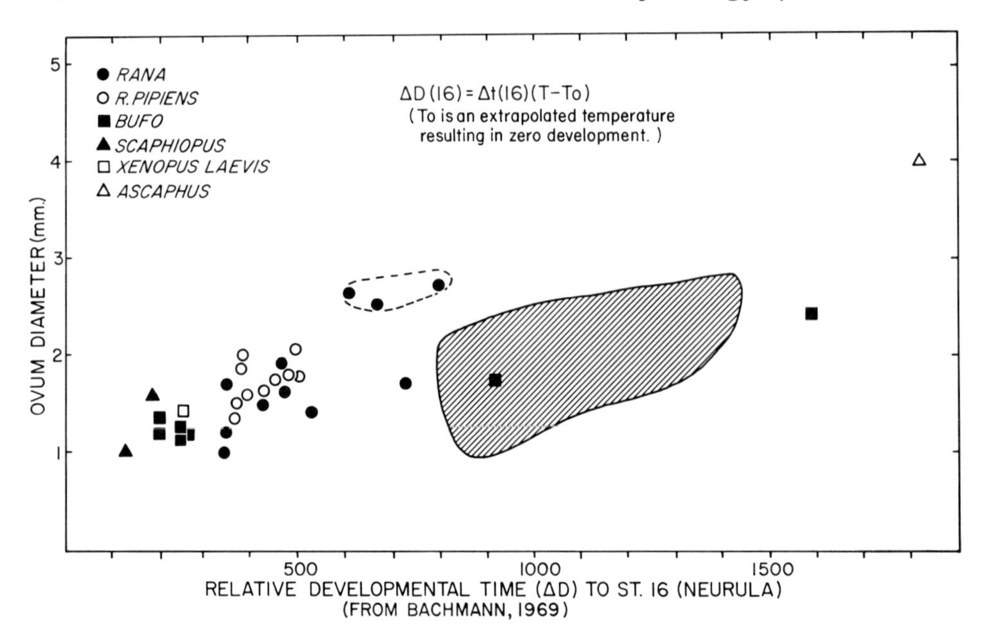

Figure 6–7. The relationship between ovum size and developmental time to neurula stage, regardless of temperature, for some frogs and six species of salamanders *(shaded area). Dashed line* encloses symbols for *Rana temporaria.*

explained by the different stages of development taken as the end point of study. Within a given species, embryos developing from slightly larger ova possibly can develop relatively faster after basic organogenesis has taken place than can individuals from slightly smaller ova, after they have developed somewhat more slowly up to that time. However, on the basis of subjectively scanning many tables for amphibian normal growth, it seems unlikely that the postulated postorganogenesis acceleration of development, if it occurs in all species, would show increasing rates in species with larger ova. Increased size alone would almost certainly impose on increased energetic "drag" throughout development. Moore (1949*b*) showed that in *Rana pipiens* a reciprocal relationship exists between ovum size and developmental rate to stage 20 at 28° C and that quite possibly the same relationship was evident at 12° C. This conclusion renders McLaren's assumption somewhat suspect. Moore (1949*a*) noted a direct relation between developmental rate to stage 20 at 20° C and ovum size in five species of *Rana.*

Different races of *Rana pipiens* have slightly different average ovum sizes, and ovum size is roughly correlated positively with latitude and altitude or with the temperature to which the populations are adapted (Moore, 1949*b*). Limited data on geographically adjacent populations living at different altitudes indicate that ova from frogs living at cooler climates (higher elevations) are somewhat larger than those from warmer climates (lower elevations) (Table 6–6). Because developmental rate has been correlated with latitude and altitude (Moore, 1949*a*, 1949*b*) and with ovum size, it can be expected that all three parameters are interdependent.

There are several ecological, as well as developmental, consequences of ovum size. For example, in the absence of an increase in body size, any increase in ovum size probably must entail a decrease in clutch size or in fecundity. Parental care has been cited as an adaptation to this problem, and indeed, this behavior is always associated with relatively smaller clutches of relatively larger ova. Such

Table 6–6. The Relationship of Ovum Size to Altitude of Deposition Site in Two Species of Frogs

Species	Ovum diameter (*mm*)	Altitude (*feet*)
Rana pipiens[a]	1.35	7150
	1.40	9000+
Pseudacris triseriatus[b]	0.80	5000
	1.15	9300

[a]In Wyoming, prior to deposition.
[b]In Colorado.
Sources: Baxter (1952); Pettus and Angleton (1967).

care may involve solely one parent remaining at the nesting site, as in *Hemisus* and *Centrolenella,* or the brooding of the clutch by one parent, as in *Sphenophryne* and some *Eleutherodactylus.* Myers (1969) demonstrated that the habit of female *Eleutherodactylus caryophyllaceus* sitting on eggs adherent to leaves of bushes and low trees is important protection against the desiccating effects of winds. *Sphenophryne* sits on small clutches of eggs in covered depressions in the ground; this behavior possibly acts to prevent desiccation but probably is a deterrent to potential predators. The same explanation can be inferred for *Hemisus* and *Centrolenella,* which remain by the eggs but do not sit upon them.

The habit of carrying the eggs is present in various groups of anurans. Males of *Alytes* entwine the eggs on their legs and carry them about on land until the eggs are about ready to hatch, at which time the males move to water. By means of a complex series of aquatic acrobatics, the completely aquatic *Pipa* manages to deposit fertilized eggs on the back of the female (Rabb and Rabb, 1963). The egg-carrying habits of both of these frogs seem to be associated only with protection of the developing eggs; those in *Pipa pipa* undergo direct development, whereas the eggs of other species of *Pipa* hatch into aquatic larvae.

Males of *Rhinoderma, Sooglossus,* and dendrobatids carry tadpoles. *Rhinoderma* picks up eggs as they are hatching, and the tadpoles develop in the vocal pouch. In *Sooglossus* and the dendrobatid genera *Dendrobates, Colostethus,* and *Phyllobates,* tadpoles adhere to the backs of males. Eggs undergo direct development on the backs of females of the hylid genera *Hemiphractus, Cryptobatrachus,* and *Stefania* or in dorsal pouches in *Amphignathodon* and some *Gastrotheca.* In *Flectonotus* and some species of *Gastrotheca,* the eggs hatch into tadpoles in the brood pouch and complete their development in bromeliads or ponds.

It seems unlikely that parental care evolved directly as a mechanism to compensate for reduced fecundity, because this would entail invoking a period of relative nonadaptiveness. It is more likely that parental care evolved independently in various phyletic lines in response to a variety of environmental stresses, including both physical and biotic factors. Such adaptations allow development to take place in unconventional environments (Salthe and Mecham, 1973); a similar hypothesis has been suggested for lizards by Tinkle (1969). In this formulation, increased ovum size would be an indirect result of selection working toward reduced clutch size commensurate with parental care and also possibly a direct result of selection acting to produce the smallest possible surface relative to volume as an adaptation for desiccating environmental conditions. The increased size of individual hatchlings in the smaller clutches presumably provides a sufficient competitive advantage to offset their relative paucity of hatchlings in the clutch. Also it is possible that due to predation on smaller tadpoles the numbers of tadpoles reaching a given size from frogs producing large clutches may be no greater than in those with smaller clutches in which the tadpoles hatch at the larger size.

In the tropics and subtropics, where small clutches are quite common, another reproductive strategy may compensate for small clutch sizes. The number of times a given female breeds each year is potentially modifiable in these regions. Thus, Chibon (1962) reported that in the laboratory, female *Eleutherodactylus martinicensis* can breed every two to three months throughout the year. Data on two species of Australian *Pseudophryne* demonstrate different fecundity but possibly similar yearly reproductive efforts (Table 6–7). Both species are about the same size, and the total clutch volume is about 20 per cent of the volume of the body cavity in both. In *Pseudophryne australis* a given volume of egg material is divided into fewer ova than in *P. bibroni;* thus, the fecundity of *P. australis* is about one-fifth that of *P. bibroni* for each reproduction. However, it is known that populations of *P. australis* breed throughout the year, whereas those of *P. bibroni* breed only once a year. It would be important to know if a given female of *P. australis* can reproduce more than once each year. Duellman (1970) noted that one population of *Hyla pseudopuma* in Costa Rica had three breeding seasons in a given year, but data on the frequency of reproduction of a given female are lacking. Indirect evidence for multiple breeding in the Central American *Agalychnis callidryas* was presented by Duellman (1970), who noted that many females of that species contained ovarian eggs of two distinct size classes. Furthermore, relatively large ovarian eggs were found in females after they had deposited clutches of eggs. Large ovarian eggs were noted in females of *Flectonotus pygmaeus* that were carrying a complement of eggs in the dorsal pouch (Abraham Goldgewicht, personal communication).

In the case of the *Pseudophryne,* if a given female of *P. australis* bred every two or three months, as does *Eleutherodactylus martinicensis* in the laboratory, there would be no difference in over-all fecundity between *P. australis* and *P. bibroni.* We are not proposing a direct causal relationship between the occurrence of multiple breeding and reduced clutch sizes, although multiple breeding may in fact compensate for reduced clutch size. Perhaps ovum size could not increase without increasing body size unless the lineage in question had the potential for breeding more than once each year, thereby allowing a decrease in clutch size only because over-all fecundity was not affected. We are considering the importance of fecundity to reside in the amount of genetic variability generated per unit time. If this amount is less than some minimal level, adaptive evolution could not occur in a changing environment.

As noted earlier, smaller clutch sizes are common in the tropics and subtropics (see also Turner, 1962). The disparity in average clutch sizes between

Table 6–7. Size Relationships to Developmental Period and Breeding Frequency in Two Species of *Pseudophryne*

	Female snout-vent length (*mm*)	Ovum diameter (*mm*)	Clutch size	Weeks to metamorphosis	Breeding frequency
P. bibroni	28	2.0–2.1	100	24	Once/year
P. australis	25	2.6–2.7	20	4	All year, after rain

Sources: Harrison (1922); Moore (1961).

tropical and temperate species having the same mode of reproduction may be due to greater frequency of breeding in tropical species. Thus, the annual fecundity rate in a tropical species having a relatively small clutch size may be equal to, or even greater than, that in a temperate species having a much larger clutch size but breeding only once each year. However, quantitative data are not available to prove this supposition. We have noted that clutch size decreases with specialization of reproductive mode. In general, reproductive specialists among anurans occur in the tropics. Three exceptions can be noted: (1) the egg-carrying habits of the European discoglossid genus *Alytes,* (2) the tadpole-brooding behavior of the Chilean *Rhinoderma,* and (3) direct development in the New Zealand ascaphids of the genus *Leiopelma.* Comparisons of modes of reproduction in temperate and tropical anuran faunas (Table 6–8) reveals that the greatest diversity in reproductive modes occurs in the tropics. Much of the discrepancy in relative numbers of species having various reproductive modes in the tropical regions is due to differences in topography and, to a lesser extent, climate. Relatively high numbers of species of stream-breeding frogs in Guatemala and frogs having direct development in Guatemala and on Barro Colorado Island are due to relief and presence of humid forests and near absence of ponds on Barro Colorado Island.

Table 6–8. Diversity in Anuran Reproductive Modes

(Expressed as percentage of total)

Reproductive mode	United States	Guatemala	Barro Colorado Island, Panamá	Belém, Brasil	Santa Cecilia, Ecuador
Eggs and tadpoles in ponds	90%	44%	20%	53%	37%
Eggs over ponds; tadpoles in ponds		5	11	10	10
Eggs in foam nests; tadpoles in ponds	1	5	11	23	8
Eggs and tadpoles in streams	3	18	11		4
Eggs over streams; tadpoles in streams		2	11		4
Eggs and tadpoles in bromeliads		2			
Eggs terrestrial; tadpoles carried to water			6	6	7
Eggs terrestrial; direct development	6	24	30	6	21
Eggs carried by female; direct development				2	3
Unknown					6
Total number of species in fauna	70	59	29	38	78

The diversity in mode of reproduction may be highly important in the coexistence of large numbers of species in lowland tropical rain-forests. Seasonal differences in breeding times, coupled with different sites of development, are important features in the partitioning of available resources. Different modes of reproduction permit species to utilize various habitat resources in different ways and thereby reduce interspecific competition, an important factor in the concepts of niche breadth and overlapping. Hence, it is significant to note that in regions having high species diversity, we find the greatest diversity of reproductive modes. The evolution of diverse reproductive modes in the tropics may have been in response to competition for available resources; such diversity was possible because of equable climatic conditions. In contrast, in temperate regions climatic conditions and associated limited numbers of niches preclude a diversity of reproductive modes, thereby placing most coexisting species in competition for available resources.

DISCUSSION

The production of an adequate amount of genetic variability per unit time may be viewed as the goal of all reproductive strategies. Some strategies achieve this end with the production of remarkably few eggs, as in certain lizards and in birds and mammals. Most analyses of selective pressures leading to reduced clutch sizes suggest adaptations of eggs, nests, or larvae to microclimates that are rigorous for the eggs or larvae (for examples, see Marshall, 1953; Salthe, 1969; Tinkle, 1969). In response, the ovum increases in size, adapting to extreme cold or drought, or becoming a means of placing a larger hatchling in the environment. This in turn allows, or forces, a smaller clutch size, either because the total amount of energy available for reproduction in a season is limited or because body size is limited.

In amphibians, clutch size in these circumstances might remain unchanged, or it might increase if female body size increased simultaneously. In this case the evidence (Figures 6–1—6–5; Salthe, 1969) suggests that there is little or no change in reproductive mode beyond producing a larger hatchling. If body size does not change, or if it decreases simultaneously with increased clutch size, we can predict that there will be an evolutionary change in reproductive mode. In cases where environmental change is such that it allows the continuation of the original reproductive mode, increased body size can compensate for the change. In cases of more drastic environmental change, increased body size cannot adequately compensate; instead it is necessary for a change in the mode of reproduction, which may involve the site of development, mode of development, or both.

Graphs, such as those presented in this paper, raise interesting possibilities for their use as tools in predictive analyses. Of course, they do not indicate which are primal or causal factors; for example, they do not tell us which is the dependent variable and which is the independent one. For this distinction we must rely upon our knowledge of biological phenomena and common sense. For example, we could use the graphs to propose that in a given phyletic line, body size in frogs might decrease as a response to arid conditions. Standing water then becomes ephemeral with the advancement of desert conditions, and under such conditions, the most successful frogs are those that metamorphose in the shortest possible time. Thus, individuals producing the smallest ova in a population tend to have an advantage because these develop fastest through hatching (Figure 6–7). In salamanders the time from hatching to metamorphosis is directly correlated with adult body size (Salthe, 1969). Both sets of data demonstrate that, other things being equal, the smallest individual females in the population would produce the most metamorphosed young in the shortest possible time. Our graphs also suggest that it would not be possible to achieve the kind of desert adaptation common to many Australian frogs (Main et al., 1959) without a decrease in body size. If the environment is not quite so arid but water still is ephemeral, another option is available by moving to a new mode of reproduction; thus, we can derive another reproductive mode known in Australian anurans. The time between hatching and metamorphosis can be decreased by placing a more advanced hatchling in the water (Pettus and Angleton, 1967). This necessitates a larger ovum (Figure 6–6), but not more ova, because as such they could not be deposited in the water. Therefore, this increase in ovum size does not favor an increased adult body size. In order to achieve a larger ovum without a change in body size, selection must favor a change in site of intraoval development. This intraoval development can be achieved by depositing the eggs out of water but in the vicinity of an incipient pond, so that hatching of advanced tadpoles occurs with the filling of the pond.

It can be argued that, contrary to the situation characterizing salamanders (Salthe, 1969), we do not have adequate data from a sufficiently large array of frogs in order to justify such analyses. However, the fact that the data we do have agrees in general with the more nearly complete data on salamanders suggests that we have before us a fairly true picture of quantitative aspects of anuran reproductive modes.

If, as we believe, the data presented in this paper are representative of anuran reproductive modes, it should be possible to utilize these data in describing the limits of some aspects of anuran reproductive strategies, but first we need to know the shapes of the relationships involved. The data presented in Figures 6–1 and 6–2 suggest that the pattern of the ovum size–body size relationship is more or less the same as it is in salamanders; that is, as the body size increases interspecifically, ovum size increases at an ever decreasing rate. However, inclusion of existing data on various egg-brooding hylids disrupts the pattern (Figure 6–5); consequently, the determination of a general relationship of this nature in anurans is not possible. Moreover, there may be more than a single pattern to this relationship, since there is a diversity of reproductive modes.

The shape of the clutch size–body size relationship (Figures 6–3, 6–5) most probably should be interpreted to be the same as it is in salamanders, with clutch size increasing at a greater rate with increased body size in lineages having absolute larger body sizes. Indeed, it would be difficult to interpret the data in any other way. If this is so, and as we can demonstrate a reciprocal relationship between clutch size and ovum size (Figure 6–4), then the general relationship between body size and ovum size must also approximate that found in salamanders.

In the absence of data on certain egg-brooding hylid frogs, and solely on the basis of data on other anurans and salamanders, we would have concluded that no amphibian ovum can be larger than about 7 mm in diameter. However, the ova of two species greatly exceed the expected maximum. One *Hemiphractus scutatus* having a snout-vent length of 74 mm carried on its back 10 eggs; the diameter of each ovum was about 10.5 mm. A female *Gastrotheca ceratophrys* having a snout-vent length of 72 mm contained in its brood pouch nine eggs; the average diameter of three ova was 12.0 mm. In both of these species, development is direct. Other marsupial frogs having direct development for which measurements of ova are available are *Gastrotheca griswoldi* (6.2 mm–7.0 mm) and *G. ovifera* (6.6 mm). In salamanders the largest ova recorded are for *Aneides lugubris* (Stebbins, 1951), *Phaeognathus hubrichti* (Brandon, 1965), and *Andrias japonicus* (Kerbert, 1904); all are between 6 and 7 mm. Apparently, the egg-brooding hylids with enormous ova represent a mode of reproduction different from any other found in amphibians; consequently, data on other frogs did not allow us to predict the size of ova in these hylids. This raises the problem of generality in systematic biology: Will we ever have sufficient data to derive empirically a relationship from which prediction is possible, or will a newly examined taxon frequently show something fundamentally different in its adaptive zone?

Our data on body size–clutch size and on body size–ovum size relationships in anurans provide the basis for certain hypotheses regarding the evolutionary consequences of reproductive modes. Beginning with the increase in phylogenetic size, individuals of species having small absolute body sizes apparently gain less reproductive advantage by adding a few small eggs to their complement than they do by slightly increasing the size of each ovum, thereby gaining a somewhat larger hatchling. Individuals of species with large body sizes seem to gain more reproductive advantage by increasing the clutch size than they do by producing larger hatchlings. The increment in egg numbers with increased body size in these already large animals evidently is more significant than it would be in small animals. This is indicated by our data with or without consideration of reproductive modes.

A different situation exists in the case of phylogenetically decreasing body size. In small species it seems to be more advantageous to maintain a certain minimal number of eggs in the clutch than it is to maintain the size of the hatchling. Presumably, clutch sizes are already at, or near, a lower limit allowable for the production of an adequate amount of genetic variability per unit time. In large species, clutch size diminishes rapidly with decreased body size—much more rapidly than does ovum size. Evidently, selection is operating to stabilize hatching size or to minimize reduction of hatching size with decreased adult body size. Thus, in general, selection in small species seems to function by maintaining minimal clutch sizes and maximizing the sizes of relatively tiny hatchlings. In species with large body sizes, selection tends to maximize clutch sizes while minimizing reduction in hatchling size.

These statements probably would have to be modified when applied to a specific reproductive mode; also, they are inapplicable to the egg-brooding hylids, which seem to be an enigma to the anuran reproductive scheme. Furthermore, in a transition from one mode to another, some or all of these strictures may be obviated. Thus, if a relatively small-sized lineage moves from a completely terrestrial mode of development to a completely aquatic mode, then there will have to be an increase in the number of ova and a decrease in their size. This example was chosen deliberately in order to trigger the following question: In view of the relationships just discussed, can selection ever proceed in a contrary fashion, depending upon specific ecological conditions? If it is true that small-sized species tend to change ovum size faster than clutch size, then since they are already restricted to more or less small clutch sizes, does this not preclude their venturing into new adaptive zones that would necessitate increasing clutch size? Is not a transition from an aquatic to a terrestrial mode of reproduction the only possibility open to small frogs, and is not this possibility closed to large frogs? The data that we have (again, excluding the egg-brooding hylids) does indicate that only small to middle-sized frogs occupy adaptive zones involving fully terrestrial reproductive modes.

At this point it is important to recognize that the data suggest these patterns within each separate reproductive mode throughout the range of body sizes within each mode. In summary, some small frogs seem to tend toward terrestrial modes of reproduction because clutch sizes are already comparatively small and great reduction in clutch size is not favored in them. It is easier for small frogs to develop modes of reproduction involving selective advantage for larger hatchlings; this most frequently occurs by entering new habitats where typical aquatic modes of reproduction are difficult to maintain. Thus, small frogs seem to be more prone to experiment with new reproductive modes (assuming that the typically aquatic mode is primitive), and intraspecific competition tends to favor individuals that are able to produce the largest tadpoles without drastically reducing the number of tadpoles. In large frogs, selection tends to favor individuals producing the largest numbers of tadpoles without drastically reducing their size, which keeps the large frogs within the primitive aquatic reproductive mode. Thus, we tentatively suggest that small body size in frogs is a preadaptation for reproductive experimentation. In an altered environment, where such experimentation is important, the small frogs have a better chance for survival.

CONCLUSIONS

Although our data are far from being complete, certain quantitative relationships and qualitative phenomena associated with anuran reproduction seem to be evident. These are enumerated below.

1. Within a given reproductive mode there is a positive correlation between ovum size and female snout-vent length.
2. Within a given reproductive mode there is a positive correlation between clutch size and female snout-vent length.
3. Regardless of reproductive mode, there is a negative correlation between clutch size and ovum size.
4. Regardless of reproductive mode, there is a positive correlation between ovum size and size of hatchlings.
5. There is a negative correlation between ovum size and rate of development.
6. Increased ovum size is associated with lotic and terrestrial sites of development.
7. Egg-brooding hylids display quantitative relationships at variance with those of most other frogs.
8. The greatest diversity of reproductive modes occurs in tropical regions.
9. It is hypothesized that selection favors increased clutch size in larger frogs and increased ovum size in smaller frogs. Furthermore, it is suggested that only small to middle-sized frogs are able to enter adaptive zones involving fully terrestrial reproductive zones.
10. Moreover, it is tentatively suggested that small body size in frogs is a preadaptation for reproductive experimentation.

Acknowledgments

Salthe's research was done under the auspices of grants from the National Science Foundation (GB7749) and The City University of New York (1108). Much of the new data incorporated herein was obtained by Duellman in the course of field work supported by the National Science Foundation (GB1441) and by Watkins Grants, administered by the Museum of Natural History, University of Kansas. We are indebted to Abraham Goldgewicht for permission to include some of his original data on *Flectonotus* and to Martha L. Crump for data on reproductive modes of frogs at Belém, Brasil. The graphs were drawn by Leland D. Bowen. Finally, we thank Linda Trueb for critically reading the manuscript.

REFERENCES

ALCALA, A. C. 1962. Breeding behavior and early development of frogs of Negros, Philippine Islands. Copeia 4:679-726.

BACHMANN, K. 1969. Temperature adaptations of amphibian embryos. Am. Naturalist 103:115-130.

BARTH, L. D., AND L. J. BARTH. 1954. The energetics of development. Columbia Univ. Press, New York. 117 p.

BAXTER, G. T. 1952. Notes on the growth and the reproductive cycle of the leopard frog, *Rana pipiens* Schreber, in southern Wyoming. J. Colorado-Wyoming Acad. Sci. 4:91.

Belehádrek, J. 1957. Physiological aspects of heat and cold. Ann. Rev. Physiol. 19:59-76.

Bernhard, H., and K. Bratuschek. 1891. Der nutzen der Schleimhullen die Froscheier. Biol. Zentr. 11:691-694.

Beudt, E. L. 1930. Der einfluss der Lichtes der Quarz-Quecksilber Lampe auf der Furschungs und Larvenstadien verschieden Amphibien. Zool. Jahrb. (Zool.) 47:623-684.

Brandon, R. A. 1965. Morphological variation and ecology of the salamander *Phaeognathus hubrichti.* Copeia 1:67-71.

Chibon, P. 1962. Différentiation sexuelle de *Eleutherodactylus martinicensis* Tschudi, batracien anoure à dévelopement direct. Bull. Soc. Zool. France 87:509-515.

de Villiers, C. G. S. 1929. The development of a species of *Arthroleptella* from Jonkersttoek, Stellenbosch. South African J. Sci. 26:481-510.

Douglas, R. 1949. Temperature and rate of development of eggs of British anura. J. Animal Ecol. 17:189-192.

Duellman, W. E. 1970. The hylid frogs of Middle America. Monog. Univ. Kansas Mus. Nat. Hist. Monograph 1. 753 p.

Duellman, W. E., and L. Trueb. 1966. Neotropical hylid frogs of the genus *Smilisca.* Univ. Kansas Publ. Mus. Nat. Hist. 17(7):281-375.

Goin, O. B., and C. J. Goin. 1962. Amphibian eggs and the montane environment. Evolution 16:364-371.

Gosner, K. L. 1960. A simple table for staging anuran embryos and larvae with notes on identification. Herpteologica 16:183-190.

Guyetant, R. 1966. Observations écologiques sur les pontes de *Rana temporaria L.* dans la région de Besançon. Ann. Sci. Univ. Besançon (Zool., Physiol., Biol. Anim.) 1966:12-18.

Harrison, L. 1922. On the breeding habits of some Australian frogs. Australian Zool. 3:17-34.

Henderson, C. G., Jr. 1961. Reproductive potential of *Microhyla olivacea.* Texas J. Sci. 13:355-356.

Heyer, W. R. 1969. The adaptive ecology of the species groups of the genus *Leptodactylus* (Amphibia, Leptodactylidae). Evolution 23:421-428.

Hönig, J. 1966. Über eizahlen von *Rana temporaria.* Salamandra 2:70-72.

Kerbert, C. 1904. Zur fortpflanzung von *Megalobatrachus maximus* Schlegel. Zool. Anz. 27:305-310.

Lutz, B. 1948. Ontogenetic evolution in frogs. Evolution 2:29-35.

Main, A. R. 1968. Ecology, systematics and evolution of Australian frogs. Adv. Ecol. Res. 5:37-86.

Main, A. R., M. J. Littlejohn, and A. K. Lee. 1959. Ecology of Australian frogs. Monographs Biol. 8:396-411.

Marshall, J. A. 1953. Egg size in Arctic, Antarctic, and deep sea fishes. Evolution 7:328-341.

McAlister, W. H. 1962. Variation in *Rana pipiens* Schreber in Texas. Am. Midland Naturalist 67:334-363.

McLaren, I. A. 1965. Temperature and frog eggs: a reconsideration of metabolic control. J. Gen. Physiol. 48:1071-1079.

Moore, J. A. 1949*a*. Patterns of evolution in the genus *Rana,* p. 315-338. *In* G. L. Jepsen, G. G. Simpson, and E. Mayr, [ed.], Genetics, paleontology and evolution. Princeton Univ. Press, Princeton.

_______. 1949*b*. Geographic variation of adaptive characters in *Rana pipiens* Schreber. Evolution 3:1-24.

_______. 1961. The frogs of eastern New South Wales. Bull. Am. Mus. Nat. Hist. 121(3):149-386.

Myers, C. W. 1969. The ecological geography of cloud forest in Panama. Am. Mus. Novitates 2396:1-52.

Oplinger, C. S. 1966. Sex ratio, reproductive cycles, and time of ovulation in *Hyla crucifer* Wied. Herpetologica 22:276-283.

Pettus, D., and G. M. Angleton. 1967. Comparative biology of montane and piedmont chorus frogs. Evolution 21:500-507.

Pope, C. H. 1931. Notes on amphibians from Fukien, Hainan, and other parts of China. Bull. Am. Mus. Nat. Hist. 61:397-612.

Poynton, J. C. 1964. Relationships between habitat and terrestrial breeding in amphibians. Evolution 18:131-132.

Pyburn, W. F. 1970. Breeding behavior of the leaf-frogs *Phyllomedusa callidryas* and *Phyllomedusa dacnicolor* in Mexico. Copeia 2:209-218.

Rabb, G. B., and M. S. Rabb. 1963. Additional observations on breeding behavior of the Surinam toad, *Pipa pipa*. Copeia 4:636-642.

Salthe, S. N. 1968. The reproductive biology of salamanders. (Abstr.) J. Herp. 1:111.

———. 1969. Reproductive modes and the numbers and sizes of ova in the urodeles. Am. Midland Naturalist 81:467-490.

Salthe, S. N., and J. S. Mecham. 1973. Reproductive biology of the amphibia. *In* B. Lofts, [ed.], Physiology of the amphibia, vol. 2. Academic Press, New York. In press.

Sergeev, A. M., and K. S. Smirnov. 1939. The color of eggs of the Amphibia. Voprosy Ekologii i Biotzenologii 5:319-322.

Slater, P., and A. R. Main. 1963. Notes on the biology of *Notaden nichollsi* Parker (Anura, Leptodactylidae). West. Australian Naturalist 8:163-166.

Smith, H. M. 1947. On the number of eggs laid by certain species of *Bufo*. Science 105:619.

Stebbins, R. C. 1951. Amphibians of western North America. Univ. California Press, Berkeley and Los Angeles. 539 p.

Stebbins, R. C., and J. R. Hendrickson. 1959. Field studies of amphibians in Colombia, South America. Univ. California Publ. Zool. 56:497-532.

Stefanelli, A. 1938. Il metabolismo dell'uovo e dell'embrione studiato negli anfibi anuri: l'assonzione di ossigeno. Arch. Sci. Biol. 24:411-441.

Terentyev, P. V. 1960. Some quantitative peculiarities of frog eggs and tadpoles. Zool. Zhur. 39:779-781.

Tinkle, D. W. 1969. The concept of reproductive effort and its relation to the evolution of life histories of lizards. Am. Naturalist 103:501-516.

Turner, F. B. 1962. The demography of frogs and toads. Quart. Rev. Biol. 37:303-314.

Tyler, M. J. 1968. Papuan hylid frogs of the genus *Hyla*. Zool. Verh., Leiden 96:1-203.

7

EVOLUTIONARY PATTERNS IN LARVAL MORPHOLOGY

Priscilla H. Starrett

INTRODUCTION

Although life history and larval morphology appealed to early naturalists in that both provided significant information to use for forming anuran classifications (Heron-Royer, 1884, 1887; Lataste, 1876, 1877), these features invariably were considered of lesser importance than osteological characters and have not had wide acceptance. Noble (1927) recognized the importance of using larval characters but confined them primarily to intrageneric relationships. Orton's (1953, 1957) use of larval characters as meaningful elements for defining major groups of anurans has been attacked by Griffiths and Carvalho (1965) and by Kluge and Farris (1969), but I believe this controversy has resulted from interpreting external characters without adequate knowledge of the internal morphology. My own investigations into the development and structure of the branchial chambers, chondrocranium, and musculature, particularly that of the jaw (Starrett, 1968), have been recently expanded to include a total of 68 species of tadpoles, representing 13 families. Additional observations were made of 23 species of living tadpoles.

Orton (1953) used roman numerals to designate the types of tadpoles she described. I am proposing names for these groups of frogs. Type I, including the Pipidae and Rhinophrynidae, I identify as the Xenoanura, a term derived from the Greek *xenos* ("stranger"), which refers to the bizarre habitus of these animals. These two families also differ from other frogs by being tongueless or by having a posteriorly attached tongue. I call Orton's Type II the Scoptanura; this group is represented by only the Microhylidae. The Greek word *skoptikos* ("mocking, scoffing, or jesting") seems to be an appropriate source from which to derive the name for Type II, for the microhylids superficially resemble other groups of frogs and their true relationships have confounded many students of anuran biology. Type III—the Ascaphidae and Discoglossidae—I classify as the Lemmanura, from the Greek *lemma* ("an assumption or something taken for granted"), because it has long been assumed that this group of frogs is the most primitive. Type IV contains the remaining frog families, which I collectively treat as the Acosmanura, a name derived from the Greek *a* ("absence") and *kosmos* ("order") referring to the large numbers of frogs in this category, as well as their baffling familial relationships.

BRANCHIAL CHAMBERS

Structure and Development

The position of the spiracular openings in anuran tadpoles has increased importance when it is related to the structure and function of the branchial chambers and the position of the forelegs. The first type of branchial chamber is characteristic of the Xenoanura (Figure 7–1A). There are two separate branchial chambers with two separate external openings. The forelimbs develop posterior to the chambers and lateral to the spiracles. In this situation the eruption of the forelimbs does not interfere with water intake and outflow.

The second type of branchial chamber occurs in the Scoptanura. Two separate branchial chambers that each have a medial internal opening are connected to the exterior by an opercular tube that continues posteriorly to a ventral spiracular tube (Figure 7–1B). The forelimbs are posterior to the branchial chambers in this group also.

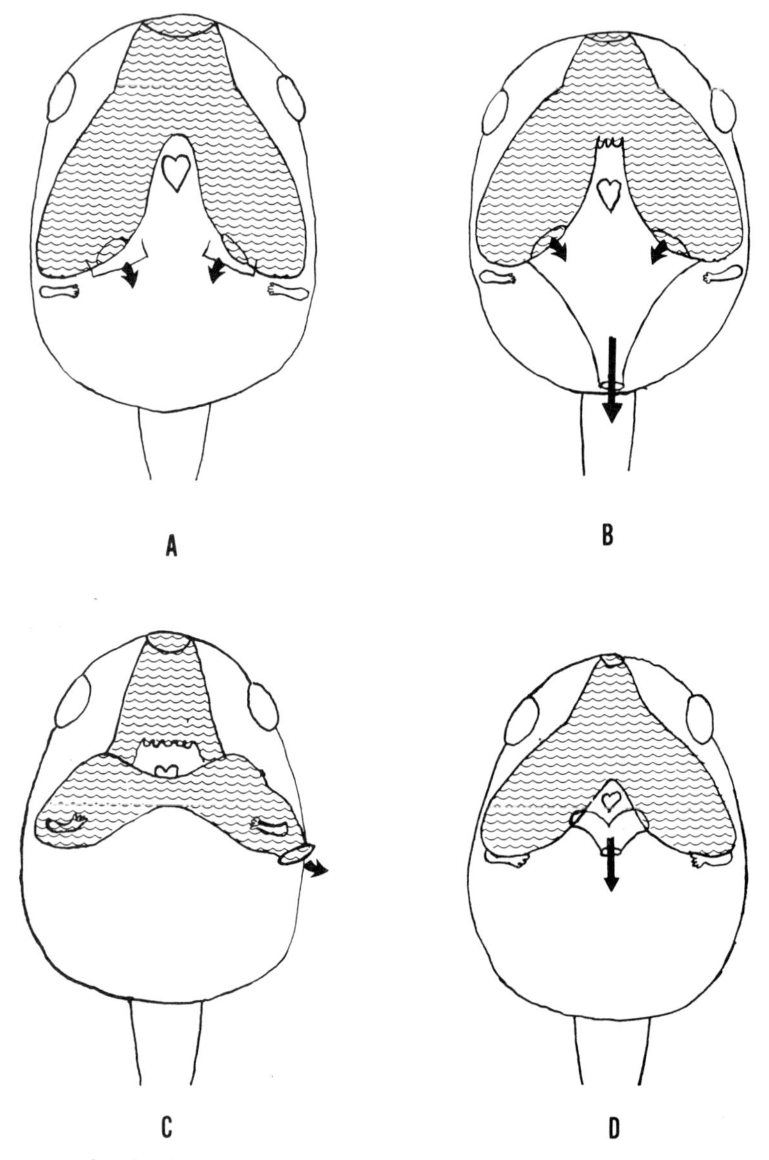

Figure. 7–1. Stylized relationship of the branchial chambers, spiracles, heart, and forelimbs: A, Xenoanura; B, Scoptanura; C, Acosmanura; D, Lemmanura. *Hatched areas,* branchial chambers; *arrows,* direction of water.

A third type of branchial chamber occurs in the Lemmanura. Two separate chambers are joined at the midline by two very short opercular tubes that immediately open as a single spiracle at the posterior level of the branchial chambers (Figure 7–1D). The forelimbs develop very close to the peribranchial ectoderm, and it is possible that they break through this tissue during metamorphosis. Stephenson's (1951) photographs of *Ascaphus* show the front legs emerging from the spiracle.

The fourth type of branchial chamber occurs in the Acosmanura. The branchial chambers are joined internally by a communicating chamber ventral to the heart. Usually there is a single sinistral external opening (Figure 7–1C). The forelimbs actually develop within the branchial chambers. This is the condition even in tadpoles with a more ventral spiracle, such as *Agalychnis* and *Phyllomedusa*, and also in *Lepidobatrachus*, which has bilateral openings.

All frog larvae have two openings at some stage in the development of the operculum. In the Xenoanura two separate opercular folds grow posteriorly over the external gills without fusing at the midline. The branchial chambers remain separate (Figure 7–2A). Kluge and Farris (1969) quote Nieuwkoop and Faber (1967), stating that the opercular fold is a composite tissue originating primarily from the third visceral arch and having only the basal portion hyoidean. Bles (1905) states that the operculum grows out from the hyoid arch in *Xenopus*. My own observations on early larvae of *Xenopus* support Bles's contentions. One of the difficulties posed by Nieuwkoop's and Faber's (1967) interpretation is that at stage 50 (Gosner, 1960), when they believe the hyoid outgrowth occurs, the tadpole is at a very advanced stage in which it is already feeding and exhibiting anterior limb buds. Determining primordial origins at this stage in development would seem risky; also, the origin of the operculum in most anuran larvae has not been studied, and there seems to be very little support for considering the opercula of pipids nonhomologous with that of other anuran larvae.

The formation, development, and growth of the operculum in scoptanuran frogs was studied in a complete series of *Gastrophryne carolinensis* by observing living embryos, as well as preserved material gathered every two hours during early development. The operculum grows posteriorly over the external gills and at the same time grows mediad very slowly so that the opercular folds of each side meet at the midline at the time of complete coverage of the external gills. The opercular growth continues posteriorly to form a tubular opening near the vent. The branchial chambers remain separate and have two internal apertures emptying into the opercular tube (Figure 7–2B).

In lemmanuran tadpoles (Figure 7–2C) the operculum apparently grows in a similar manner but stops at about the posterior extent of the gills (Gallien and Houillon, 1951; Heron-Royer, 1887).

The operculum in acosmanuran tadpoles grows posteriorly over the external gills and simultaneously grows medially very rapidly so that a single confluent tissue, which extends posteriorly, is formed (Figure 7–2D). It fuses with the belly ectoderm medially just prior to fusion on the right side, so for a very brief period there are two openings (Sedra and Michael, 1961). An anomalous *Pelobates* (Heron-Royer, 1884) retained two tubular spiracles. In animals exhibiting direct development the operculum merges with the yolk midventrally, and the elongated gills emerge on each side to surround the yolk, thus suggesting the original bilateral spiracles.

Respiratory and Feeding Functions

The stream of water enters the oral cavity primarily through the mouth, flows along the filter organs and gills, and leaves the branchial chamber via the spira-

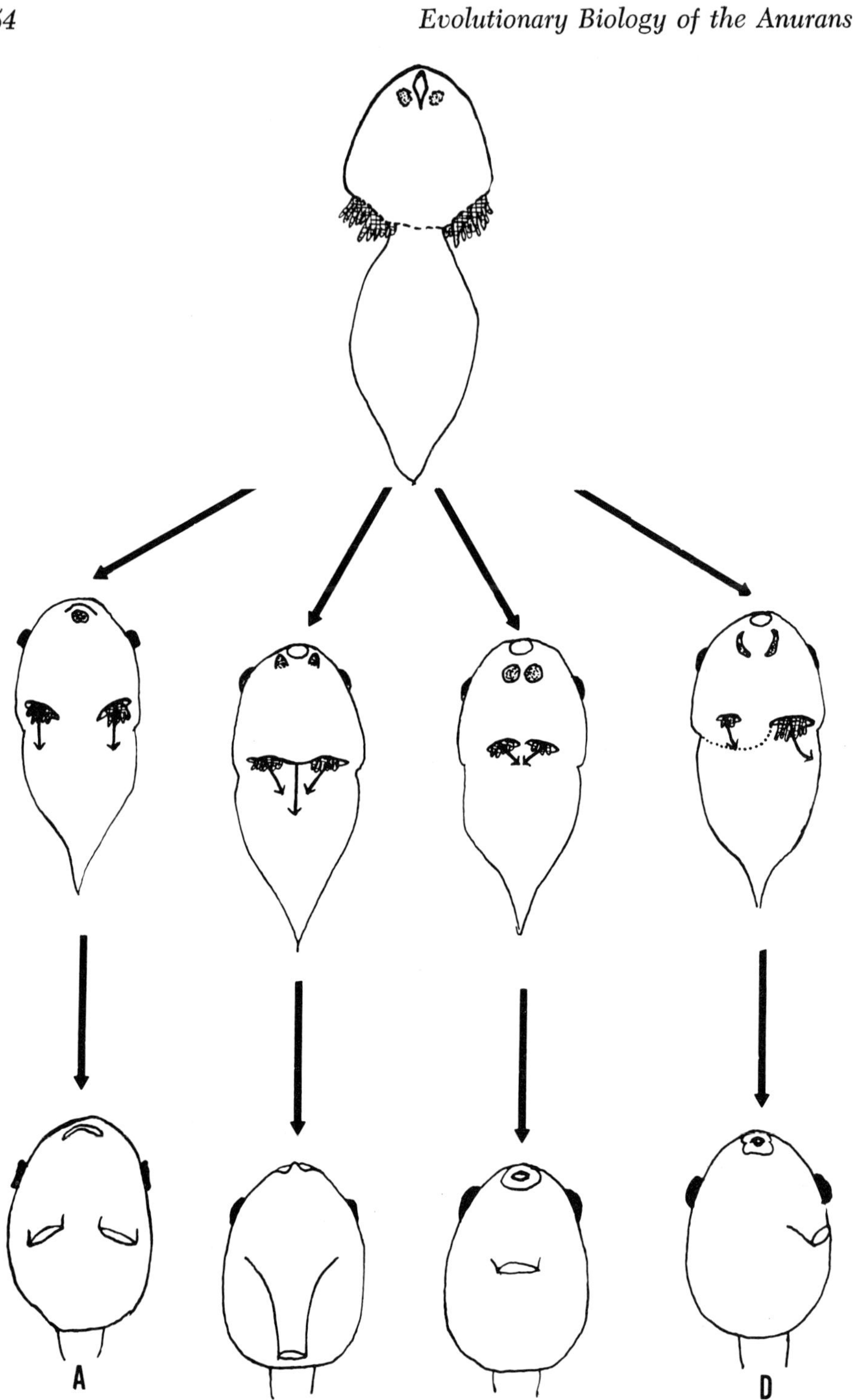

Figure 7–2. Comparisons of embryological development of operculum and spiracles: A, Xenoanura; B, Scoptanura; C, Lemmanura; D, Acosmanura. *Upper figure,* stage with external gills and beginning of opercular growth.

cles. Basically the water current functions in both feeding and respiration. Food particles are removed by the filter organs, and respiration takes place in the gill cavities. In some tadpoles, the water enters solely through the mouth, while in others, it comes in through the nostrils as well. Although the external nares open into the pharynx in *Xenopus* and *Rhinophrynus,* water does not normally enter by this means, according to my observations, which were made with suspended carmine particles. The external nares do not break through until late in larval life in the Microhylidae. Water enters the nostrils in *Ascaphus,* according to Noble (1927). He believed this was also a method of obtaining food; however, I have observed *Ascaphus* tadpoles scrape paths through boiled lettuce while feeding in captivity.

Observations with carmine particles showed water entering the nostrils freely in *Bufo boreas* and *B. terrestris, Hyla regilla, H. ebraccata, H. rufitela, Agalychnis moreleti,* and *Leptodactylus melanonotus.* Small amounts of water may enter the nostrils in *Rana pipiens* and *R. catesbeiana,* but the greatest volume entered through the mouth. Possibly there is a trend to separate feeding and respiratory streams. Observations on tadpoles of *Hyla boulengeri* demonstrate the extreme condition of this separation. The jaws and mouth are tightly closed and show no visible movement while a continual stream of water enters the nostrils and leaves from the spiracle. Small particles of carmine dropped into the nostrils immediately exited from the spiracle, whereas large particles and pieces of debris entering the nostril were spat out from the mouth.

Many tadpoles have lungs and may use them for supplementary respiration. Air bubbles are swallowed rather than taken in, as they are in adult respiration. Brock (1929) mentions that the gills are still used at metamorphosis. Actually the respiratory function of the gills is very significant at this stage of development. Observations on *Rana catesbeiana* and *R. boylii* showed that after the forelegs have erupted, the froglet remains under water, continuously gulping in water and passing it out through the two openings at the base of the arms. During this period—about two days—the frog does not rise to the surface to take in a bubble of air. At the end of this interval the frog floats at the surface and breathes in through the nostrils as in adulthood. A metamorphosing *Bufo boreas* also stayed under water, gulping continuously. At this stage of life, water was not entering the pharynx via the nostrils as it did during tadpole life. Water remained in the narial openings when the head was out of water, so apparently air was not entering through this passageway either. In *Bufo* tadpoles functional lungs do not develop until metamorphosis. Cutaneous respiration undoubtedly continues during this period, but it seems that the tremendous reconstruction of the pharynx and jaws involved in changing to aerial respiration requires a dependence on aqueous respiration during this critical time of life. *Rana* tadpoles have no obvious advantage over the lungless *Bufo* tadpoles at this time. The hydrostatic function of lungs may be the more significant one. *Gastrophryne* tadpoles take a bubble of air from the surface of the water and are so buoyant that they float to the surface unless they are actively swimming or anchored beneath some object. *Xenopus* tadpoles have in the lung a dorsal diverticulum that helps to maintain a stable position in the water (Van Bergijk, 1959).

Hydrostatic Functions

A strong current of water flows from the spiracle during the excurrent phase of respiration. The direction of this current can affect the position of the tadpole in water. This is especially important in tadpoles that are surface feeders or midwater feeders. The excurrent flow of water can also affect the amount of drag on a swimming tadpole. Tadpoles that feed in midwater have two ventrolateral spiracles *(Xenopus)*, one posterior ventral spiracle *(Calluella)*, or a midventral, slightly sinistral spiracle (*Phyllomedusa* and *Agalychnis*). A stream of water leaving a single sinistral spiracle would tend to rotate the animal and would also increase the drag (Figure 7–3B). *Xenopus* is a filter feeder that utilizes food in suspension. Feeding *Xenopus* are suspended head down at about a 45-degree angle. This position is maintained by the rapid vibration of an extended tail tip. The tail shape may contribute to the stability of this position, and the deeper lower tail fin that occurs in *Xenopus* has the effect of an upward lifting tail, as it does with the heterocercal tail of a shark (Alexander, 1968). The enlarged, posteriorly projecting lung of *Xenopus* produces a center of buoyancy posterior to the center of gravity, which therefore supports the head-down, tail-up position. Minute maneuvers are executed by movement of the barbels and later of the webbed hind feet.

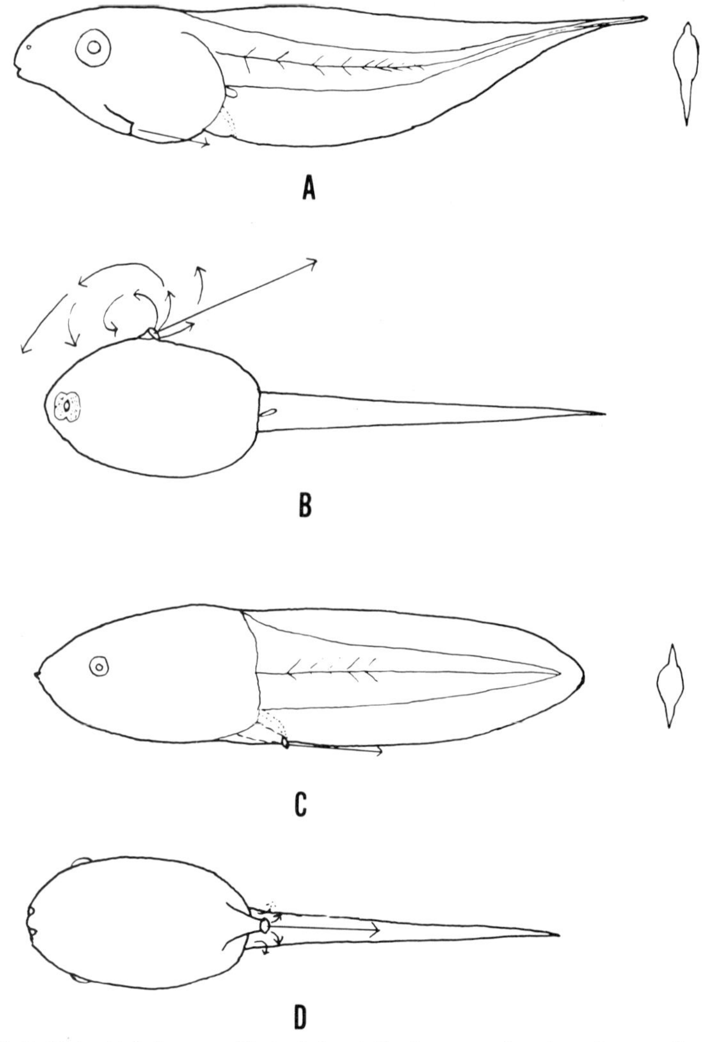

Figure 7–3. Potential forces affected by tail shape and spiracular outflow: A, lateral view of *Agalychnis* showing deep lower tail fin and ventral spiracle; B, ventral view of acosmanuran tadpole showing spiracular outflow; C, lateral view of *Hypopachus* showing equal height of upper and lower tail fins and posterior ventral spiracle; D, ventral view of *Hypopachus*, as in C.

Many microhylids seek out the source of the infusoria and move toward it. One of the problems facing these animals, which feed on free-swimming infusoria, is efficient acquisition of food. It is important in this type of feeding not to stir up and disperse the food organisms. The tail movements of *Gastrophryne* are not as vigorous as those of a *Bufo*. The flow from the spiracle may even aid in moving the tadpole forward and in maintaining its suspended position (Lagler, Bardach, and Miller, 1962). *Gastrophryne* tadpoles create much less water disturbance than do tadpoles with a sinistral spiracle (Figure 7–3D). *Gastrophryne* tadpoles can remain near the food source, moving the lower jaw in and out without creating great disturbances in the water that would scatter the food organisms. They do this with a minimum of body movements; they can float at the surface of the water apparently motionless, except for the minimal mouth action. The animals have been observed releasing bubbles at the water's surface; after so doing, they are capable of remaining on the bottom to feed, which indicates the hydrostatic function of the lungs. Normally the tadpoles are so buoyant that they need to be partially anchored beneath stones or other objects to keep from floating to the surface. These tadpoles spend much of their time in a horizontal position, and the upper and lower tail fins are about equal in size (Figure 7–3C).

The sinistral spiracle of the acosmanuran tadpoles in which the excurrent siphon is not obstructed is an adaptation for bottom feeding. Such tadpoles as *Bufo* and *Rana* swim by undulating the tail laterally. These vigorous movements propel the animal forward and at the same time stir up particles that have settled on the bottom. *Rana* tadpoles gulp the food particles made available in this manner.

Agalychnis and *Phyllomedusa* tadpoles are adapted for feeding in several positions. The extended tail filament, as well as the shape of the tail, is similar to that of *Xenopus* (Figure 7–3A). The spiracle is in a sinistro-ventral position. These animals continuously vibrate the tip of the tail and can feed in a head-down position (Kenny, 1969). Changing the position of the tail tip enables the tadpoles to surface feed in a head-up position. The locations of the center of gravity and center of buoyancy appear to be close and contribute to the ability of the tadpole to feed in several positions. The *Hylambates maculatus* tadpole feeds on water plants at the surface and has a tail with a large dorsal fin. The feeding position is about a 45-degree angle but in a tail-down, head-up position (Wager, 1965). The lift in this animal would be expected to be opposite from that in *Xenopus*. The vibrating tail tip aids in maintaining this position also.

CHONDROCRANIUM

Tadpoles of Xenoanura and Scoptanura lack horny beaks and denticles, and their chondrocrania are correspondingly simple. In the Xenoanura (Figure 7–4A) an elongate Meckel's cartilage forms the base of the lower jaw. The infrarostrals are small and indistinctly separate, forming the anterior symphysis of the lower jaw. The ceratohyals are large and project anteriorly, parallel to Meckel's cartilage. The basihyal lies between the anterior ceratohyal projections and therefore occupies a rather forward position. Both Meckel's cartilage and the ceratohyal articulate with the palatoquadrate cartilage. The upper jaw is formed by the trabecular horns and the ethmoid flange of the palatoquadrate cartilage. The mouth opening is wide.

The lower jaw of scoptanuran tadpoles consists of Meckel's cartilages that are thickened but still elongate. Anteriorly, the small but separate infrarostrals form the characteristic protrusible lower lip. The wide ceratohyals lie obliquely to the longitudinal axis of the head. The prominent basihyal lies between the anterior ceratohyal projections. An elongate arm of the palatoquadrate articulates

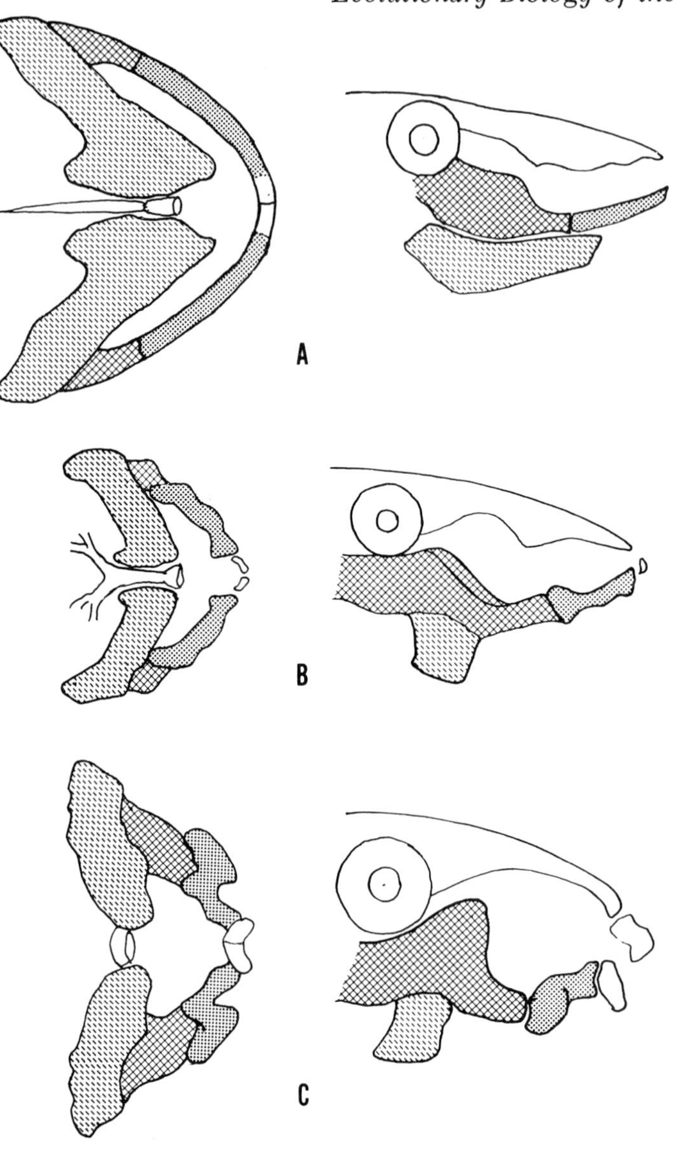

Figure 7–4. Lateral and ventral views of three types of chondrocrania: A, Xenoanura; B, Scoptanura; C, Acosmanura.

with the ceratohyal. The upper jaw is formed by the trabecular horns (Figure 7–4B).

Lemmanuran and acosmanuran tadpoles have horny beaks. The chondrocranium supporting these jaws is characteristic. The lower jaw is supported by well-developed infrarostral cartilages. These articulate with irregularly shaped Meckel's cartilages, which in turn articulate with the anterior projections of the orbital process of the palatoquadrate. The ceratohyals are broad but lie in a plane almost perpendicular to the longitudinal axis of the head. The basihyal, which lies between the ceratohyals, therefore occupies a more posterior position than it does in the two previous types. When the mouth is closed, the upper jaw lies on, and is actually anterior and ventral to, the lower jaw. The upper beak is supported by a separate element, a suprarostral cartilage, which may be single or divided into three separate cartilages as in *Ascaphus,* a lemmanuran frog. The suprarostral cartilages articulate with the trabecular horns (Figure 7–4C).

MUSCULATURE

The larval jaw and throat musculature parallels the development of the jaw cartilages. In order to simplify the following discussion, some muscles are grouped. Table 7–1 lists the specific muscles in each group, their origins and insertions, and the general reference letters. In the Xenoanura the mandibular-arch musculature of the jaw consists of three primary groups. The first (MJ) are muscles originating primarily on the posterior palatoquadrate and inserting on Meckel's cartilage. These muscles close the mouth by adducting the lower jaw. The second group (MR) is represented by a single muscle that inserts on the base of the tentacular cartilage in *Xenopus,* on connective tissue at the angle of the jaw in *Rhinophrynus,* and on the supralabial pad in *Hymenochirus.* This muscle aids in closing the mouth. The third group is represented by the intermandibularis posterior (MM). (The intermandibularis anterior develops later in larval life so it is excluded from my treatment.) In *Xenopus* the MM muscle lies transversely across the chin, joining the Meckel's cartilages from one side to the other. In *Hymenochirus* the muscle lies in a more oblique position and apparently inserts more posteriorly on the basihyal (Sokol, 1962). In *Rhinophrynus* the muscle also has a firm attachment on the basihyal cartilage but the course of the fibers is generally more transverse (Figure 7–5A).

The hyoid muscles that open the mouth by lowering the lower jaw are the JH series. The interhyoideus (HH) runs from the ceratohyals to a median raphe. The floor of the mouth is lowered by the action of the orbitohyoideus and, if present, the suspensoriohyoideus (CH). The geniohyoideus, a hypobranchial spinal muscle, runs from the branchial cartilage to the anterior part of Meckel's cartilage, which is fused with the infralabial cartilages.

Feeding and respiration involve essentially the same actions. The mouth is opened by the JH and GH muscles, and the buccal cavity is expanded by contraction of the CH muscles. Water flows into the pharynx. The mouth is closed by the MJ and MR muscles. The floor of the buccal cavity is raised by the MM and HH, and the water is forced back over the gills and out the spiracle. There are also branchial muscles involved, but these are not included in the present discussion.

The scoptanuran tadpole represented by Figure 7–5B is more complicated in that there is some separation of feeding from respiration. With the exception of MM, the muscles involved in respiration follow the pattern described above. The MM is not connected by a median raphe but has a definite insertion on the prominent basihyal. Slips from the interhyoideus also attach here. The feeding specialization is the peculiar U-shaped, spoon-like lower jaw that moves in and out when the mouth is opened, catching microorganisms. The muscles that extend and move the lower jaw forward are the GH and MM. The lower lip is retracted and the mouth closed by a slip of the HH and MI muscles. The MI is a short muscle joining the infralabial cartilage to Meckel's cartilage. It may possibly be homologous to the mandibulolabialis muscle in acosmanuran frogs. At this point it is considered separately because there is evidence that this muscle is retained in the adult as an accessory slip running parallel to the mandible. Sedra (1950) believes the mandibulolabialis fuses with the intermandibularis posterior in *Bufo,* and De Jongh (1968) found that it disappeared during metamorphosis in *Rana temporaria.* Until the fate of this muscle is determined in other acos-

Table 7–1. Muscle Groups and Arrangements Associated with Anuran Larval Types

Reference letters (in text)	Individual muscles	Associated arch	Origin	Insertion
Xenoanura				
MJ	Adductor mandibulae anterior	Mandibular	Palatoquadrate	Meckel's cartilage
	Adductor mandibulae posterior profundus	Mandibular	Palatoquadrate	Meckel's cartilage
	Adductor mandibulae posterior articularis	Mandibular	Orbital quadrate	Meckel's cartilage
MR	Adductor mandibulae externus	Mandibular	Orbital quadrate	Tentacular cartilage, supralabial pad, connective tissue at angle of jaw
MM	Intermandibularis posterior	Mandibular	Meckel's cartilage	Median aponeurosis, basihyal
JH	Quadrato-hyoangularis	Hyoid	Orbital quadrate and ceratohyal	Meckel's cartilage
HH	Interhyoideus	Hyoid	Ceratohyal	Median aponeurosis
CH	Orbitohyoideus	Hyoid	Orbital quadrate	Ceratohyal
	Suspensoriohyoideus	Hyoid	Orbital quadrate	Ceratohyal
GH	Geniohyoideus	Branchial	Hypobranchial plate	Infrarostral
Scoptanura				
MJ	Adductor mandibulae anterior	Mandibular	Palatoquadrate	Meckel's cartilage
	Adductor mandibulae posterior profundus	Mandibular	Palatoquadrate	Meckel's cartilage
	Adductor mandibulae posterior articularis	Mandibular	Orbital quadrate	Meckel's cartilage
	Adductor mandibulae externus	Mandibular	Orbital quadrate	Meckel's cartilage
MR	None			
MM	Intermandibularis posterior	Mandibular	Meckel's cartilage	Basihyal
MI	Infralabial retractor	Mandibular	Meckel's cartilage	Infrarostral
HH	Interhyoideus anterior	Hyoid	Ceratohyal	Basihyal
	Interhyoideus posterior	Hyoid	Ceratohyal	Median aponeurosis
JH	Hyoangularis	Hyoid	Ceratohyal	Meckel's cartilage
	Quadratoangularis	Hyoid	Orbital quadrate	Meckel's cartilage
CH	Orbitohyoideus	Hyoid	Orbital quadrate	Ceratohyal
GH	Geniohyoideus	Branchial	Hypobranchial plate	Infrarostrale

Reference letters (in text)	Individual muscles	Associated arch	Origin	Insertion
Lemmanura				
MJ	Adductor mandibulae anterior	Mandibular	Palatoquadrate	Meckel's cartilage
	Adductor mandibulae posterior profundus	Mandibular	Palatoquadrate	Meckel's cartilage
	Adductor mandibulae posterior articularis	Mandibular	Orbital quadrate	Meckel's cartilage
MR	Adductor mandibulae posterior subexternus	Mandibular	Orbital quadrate	Suprarostral
	Adductor mandibulae externus	Mandibular	Orbital quadrate	Suprarostral
MM	Intermandibularis posterior	Mandibular	Meckel's cartilage	Median aponeurosis
ML	Mandibulolabialis	Mandibular	Infrarostral	Skin of lip
HH	Interhyoideus	Hyoid	Ceratohyal	Median aponeurosis
JH	Hyoangularis	Hyoid	Ceratohyal	Meckel's cartilage
	Suspensorioangularis	Hyoid	Orbital quadrate	Meckel's cartilage
	Quadratoangularis	Hyoid	Orbital quadrate	Meckel's cartilage
CH	Orbitohyoideus	Hyoid	Orbital quadrate	Ceratohyal
	Suspensoriohyoideus	Hyoid	Orbital quadrate	Ceratohyal
GH	Geniohyoideus	Branchial	Hypobranchial plate	Infrarostral
Acosmanura				
MJ	Adductor mandibulae anterior	Mandibular	Palatoquadrate	Meckel's cartilage
	Adductor mandibulae posterior superficialis	Mandibular	Palatoquadrate	Meckel's cartilage
	Adductor mandibulae posterior articularis	Mandibular	Orbital quadrate	Meckel's cartilage
MR	Adductor mandibulae posterior profundus	Mandibular	Palatoquadrate	Suprarostral
	Adductor mandibulae posterior subexternus	Mandibular	Orbital quadrate	Suprarostral
	Adductor mandibulae externus	Mandibular	Orbital quadrate	Suprarostral
MM	Intermandibularis posterior	Mandibular	Meckel's cartilage	Median aponeurosis
ML	Mandibulolabialis	Mandibular	Infrarostral	Skin of lip
HH	Interhyoideus	Hyoid	Ceratohyal	Median aponeurosis
JH	Hyoangularis	Hyoid	Ceratohyal	Meckel's cartilage
	Quadratoangularis	Hyoid	Orbital quadrate	Meckel's cartilage
	Suspensorioangularis	Hyoid	Orbital quadrate	Meckel's cartilage
CH	Orbitohyoideus	Hyoid	Orbital quadrate	Ceratohyal
	Suspensoriohyoideus	Hyoid	Orbital quadrate	Ceratohyal
GH	Geniohyoideus	Branchial	Hypobranchial plate	Infrarostral

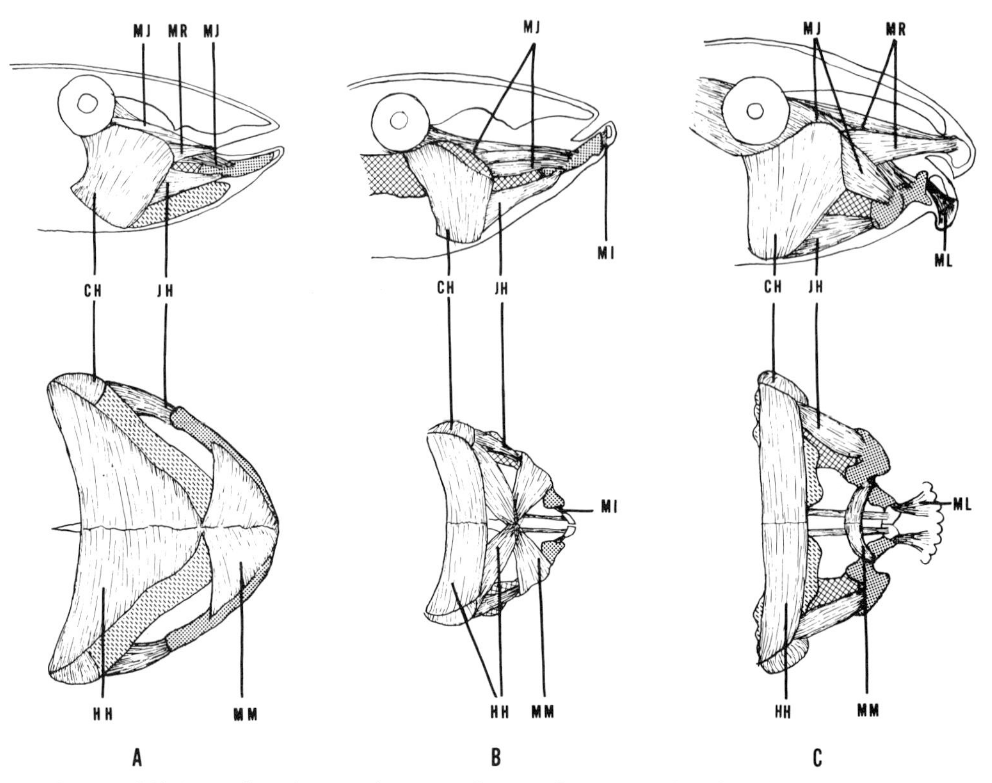

Figure. 7–5. Lateral and ventral views of musculature on the chondrocrania of Figure 7–4: A, Xenoanura; B, Scoptanura; C, Acosmanura. For explanation of reference letters, see text and tables.

manuran frogs, I prefer to call the scoptanuran muscle the infralabial retractor and to retain mandibulolabialis for the acosmanuran larval muscle.

There are many feeding modifications in acosmanuran tadpoles but the basic muscle relationship remains constant (Figure 7–5C). The muscles used for respiration are the same as for xenoanurans and scoptanurans. Shallow respiration can occur with only the lower jaw opening (Gradwell, 1968). Deeper ventilation occurs when the mouth is opened wider by the abduction of both upper and lower jaws. The feeding specializations enter in with this action. Both jaws can be closed in a rasping action by the MR, as well as the MJ muscles. Scraping with papillae and denticles is aided by the opposing action of the GH and MJ, as well as the ML. The lower beak is spread to make a better shearing surface by the MM. The great many individual specializations are made possible by this arrangement of cartilage and muscles. As previously mentioned, *Hyla boulengeri* is an example of a tadpole in which feeding and respiration are separate. Water enters through the nostrils, passes over the gills, and exits through the spiracle. The HH and CH muscles are involved in this function. The buccal cavity is expanded by the contraction of the CH, which lowers the floor of the mouth. Water enters the nostrils because of the difference in pressure. The contraction of the HH forces water over the gills as the pressure of the water closes the narial valves, preventing backflow through the nostrils. The mouth remains tightly closed during this process. It is significant that the tadpoles of acosmanuran frogs have the MM and HH widely separated. (The typical condition is two narrow parallel muscles.)

The MM has shifted from its original function of aiding the HH in raising the floor of the mouth in respiration to the function of aiding in feeding actions only.

TYPES OF VARIATION

Aquatic Adaptations

External mouth structures reflect the internal morphology to a certain degree, but there has been a tendency to see superficial structures without exploring the internal anatomy. Table 7–2 summarizes the variation in the various frog families.

Xenoanuran and scoptanuran larvae never have beaks or denticles and have a limited range of adaptations. They are primarily dwellers of quiet waters, feeding on infusoria and other suspended organisms. *Rhinophrynus* may be predaceous, although it is primarily a filter feeder (Starrett, 1960; Stuart, 1961); however, the same kind of random gulping is used for both types of feeding. *Hymenochirus* is specialized for taking small aquatic arthropods. The microhylid larvae are confined to infusoria-rich pools primarily in the tropics, and their present distribution would be much more restricted were it not for their successes in nonacquatic development. Most of the Malagasy, New Guinean and Australian microhylids apparently do not have free-living and feeding aquatic larvae. Of the African microhylids, *Phrynomerus* is the only genus with typical tadpoles; *Hoplophryne* has a larval stage in which it lives inside bamboo or between leaves. It is a feeding tadpole but has very reduced gills and in many ways has nonaquatic characters. One *Microhyla sp.* tadpole has an expanded lower lip that enables it to become a better surface feeder. Only the lower lip forms the funnel, in contrast with acosmanuran tadpoles in which both upper and lower lips are modified to form the funnel.

The development of upper-jaw cartilages, which are present in lemmanuran and acosmanuran tadpoles, has permitted many specializations in jaw function. All the free-living tadpoles in these groups have beaks, although there is variation in the number of tooth rows (see Table 7–2).

Ascaphus (family Ascaphidae-Lemmanura) tadpoles live by means of a mouth with 2–3/7–12 tooth rows and movable upper and lower beaks. The tadpoles inch over rocks, scraping off algae for food. The nostrils probably function as an inlet for the respiratory stream of water.

The tadpoles in the family Discoglossidae have 2/3 tooth rows and live in ponds.

Pelobatids have three types of tadpoles. The first, represented by mountain-stream forms, has expanded oral discs, no teeth, and weakly developed beaks. Both other types of pelobatid larvae have increased numbers of tooth rows from 3/4 to 6/7. One group lives in streams and is carnivorous. Members of this group are not equipped with holdfast organs but hide under rocks or occur in side pools, thus avoiding strong currents. Many of the others are pond-living tadpoles that are also carnivorus and have increased tooth rows.

Bufonid tadpoles occur in still water, streams, and bromeliads. In spite of this range of adaptation all larvae that are free living and feeding have 2/3 tooth rows. *Ansonia* is a stream-adapted tadpole with an adhesive mouth. *Atelopus* is also adapted for fast water but has an abdominal disc posterior to the mouth.

The Leptodactylidae is a large and diverse family. Many species have complete terrestrial (direct) development in which miniature frogs hatch from the eggs. Others are intermediate in nonaquatic development, hatching with the tail present and sometimes retaining tooth rows. Many species have 2/3 tooth rows and live in ponds; often their eggs are protected in a foam nest.

Lepidobatrachus has a very unusual tadpole that has a spiracular opening

Table 7–2. Relationship of Mouth Structure and Habitat in Families of Anura

Family	Beaks	Tooth rows				Aquatic habitat			Nonaquatic development
		0/0	< 2/3	2/3	> 2/3	Still water	Stream	Bromeliads	
Pipidae	O	X				F, C			
	—	—							B
Rhinophrynidae	O	X				F, C			
Microhylidae	—	—							S, W, D
	O	X				F	X	X	
Ascaphidae	—	—							W
	X				X		AM		
Discoglossidae	X			X		X			
Pelobatidae	X	X				X	NM		
	X				X	X, C	NM		
Bufonidae	—	—							B
	X		X						S
	X			X		X	AM, AO	X	
Leptodactylidae	—	—							B, D
	X	X				C			S, W
	X		X			X			W
	X			X		X	NM		
	X				X	X, C	AM		
Dendrobatidae	X	X					AM		
	X		X					X	
	X			X		X	X	X	
Rhinodermatidae	X		X						B
Centrolenidae	X			X			NM		
Hylidae	—	—							B
	X	X				X	AM	X	S
	X		X			X	AM	X	
	X			X		X	X, AM	X	
	X				X	X	AM		
Ranidae	—	—							W, D
	X	X				X	NM		
	X		X			X	X		
	X			X		X			
	X				X	X	X, NM, AM, AO		
Hyperoliidae	X	X				X			
	X		X			X			
	X				X	X			

O Absence in free-living tadpole.
X Presence or occurrence.
— Absence in nonaquatic tadpole.
F Filter feeder.
C Carnivorous.
AM Adhesive mouth.
NM Nonadhesive mouth.
AO Abdominal adhesive organ.
B Brood pouch.
S Swimming, nonfeeding.
W Nonswimming, nonfeeding, hatching with a tail.
D Hatching without a tail.

on each side of its body. However, as in all other acosmanuran tadpoles, the branchial chamber is connected by a ventral communicating chamber. The forelegs also develop within the branchial chambers. The mouth is very wide, and no tooth rows are present. The beaks are weakly developed and serrated (Parker, 1931; Cei, 1968). Close examination and serial sections show that at stage 36 (Gosner, 1960) maxillary and premaxillary teeth are already beginning to form. It seems that this animal has telescoped its development so that some primitive larval characteristics are combined with adult features.

Most leptodactylid larvae are found in quiet waters. The most common tooth row formula is 2/3. A number have increased numbers of tooth rows and in some cases are carnivorous, while a few are adapted for stream life. *Heleophryne* in Africa and *Mixophyes* in Australia have adhesive mouths.

Young dendrobatid tadpoles are carried by the parents and are usually deposited in small bodies of adventitious water, such as bromeliads or depressions in logs. At least one species occurs in shallow streams and has a toothless, expanded oral disc that is used both for surface feeding and for adhesion.

Centrolenids live in torrential streams and have an expanded mouth with 2/3 tooth rows, but they apparently do not use the mouth for attachment.

Hylidae is another large family with many types of tadpoles. Some without tooth rows occur in quiet water, streams, or bromeliads or have transitional nonaquatic development. The *Hyla microcephala* group has tadpoles with small mouths that lack tooth rows but have beaks. The mouth forms a protrusible tube, and the tadpoles apparently feed on the bottom. Tadpoles of the *Hyla leucophyllata* group are also small mouthed, but they have one lower tooth row and feed by plucking material from submerged leaves or other debris. There are stream-adapted tadpoles with suctorial mouths; some have the basic 2/3 tooth-row pattern, e.g., *Hyla angularis* (Tyler, 1963), but others have a large number of tooth rows, e.g., *Hyla claresignata.*

There are many larval adaptations in the family Ranidae. Some tadpoles—those lacking tooth rows but having well-developed beaks, e.g., *Ooeidozyga*—are carnivorous pond dwellers. Other lentic tadpoles have an increased number of tooth rows, e.g., *Nyctibatrachus* and *Rana palmipes.* Stream-adapted tadpoles, such as *Rana magna* (Alcala, 1962), may have the basic tooth-row pattern; some, such as the genera *Conraua, Petropedetes,* and *Staurois,* have increased tooth rows; and others, such as *Amolops* (Inger, 1966), have both an abdominal disc and increased tooth rows.

Tadpoles of the family Hyperoliidae have tooth rows numbering either less than or more than 2/3. Apparently they are primarily pond dwellers. Some have heavy beaks and few tooth rows and feed on aquatic plants, e.g., *Kassina* and *Hylambates. Trichobatrachus* lives in swift torrents and attaches to rocks with its large mouth.

Nonaquatic Development

As mentioned before with special reference to the leptodactylids, the free-swimming larval stage may be completely eliminated from the life cycle; however, transitional stages in nonaquatic development appear in several groups of frogs. (See Table 7–2 for characters and legend.) These stages may be grouped as a swimming, nonfeeding tadpole (S); a nonswimming, nonfeeding larva that hatches with a tail (W); and finally, a miniature frog hatching without a tail from the egg (D). In the latter type, larval characters are suppressed to the extent that no upper-jaw or lower-jaw cartilages are developed, and the opercular fold is rudimentary (Lynn, 1942). Unfortunately, information on the morphology of many intermediate types is very meager.

Extensive parental care is associated with development in a brood pouch, which may be a dorsal, inguinal, or gular structure in the adult. Sometimes the young are retained in the pouch during early development and at later stages leave it to continue further development as tadpoles. In other cases, the young complete their development within the pouch and emerge as minute frogs.

In the family Pipidae, *Pipa pipa* young undergo development in depressions on the back of the female and emerge as froglets. Paired spiracles are present during the pouch life of these young.

Kalophrynus, a microhylid, has a swimming, nonfeeding larva (Inger, 1966) with external gills covered by a complete operculum that has a ventral spiracle. Several microhylid genera (*Asterophrys, Platypelis, Plethodontohyla, Cophixalus, Oreophryne* and *Sphenophryne*) have "wriggler" larvae. These hatch with a tail but are nonswimming and nonfeeding (Tyler, 1963). No spiracle is recorded for any of these. *Breviceps* larvae hatch without a tail. They have internal gills only. There is a complete operculum but no spiracle.

Leiopelma, an ascaphid, has "wriggler" larvae that have no gills but gill clefts. The operculum is represented by a gular fold.

The family Bufonidae has two types of nonaquatic development. *Pelophryne* has a swimming, nonfeeding larva with a complete operculum but no spiracle. Ridges, which represent tooth rows, are present (Inger, 1960). *Nectophrynoides* develops within the enlarged oviducts of the female parent. External gills develop, as does a tiny spiracle. The jaw cartilages and muscles conform to the bufonid larval pattern (Orton, 1949).

Several patterns of nonaquatic development occur in the family Leptodactylidae. *Philoria* can be considered a swimming, nonfeeding larva. A spiracle is present, as well as mouth structures consisting of tooth ridges, beaks and papillae (Littlejohn, 1963). *Kyrranus* larvae are nonswimming and nonfeeding, but they also have beaks, oral papillae, and a sinistral spiracle (Moore, 1961). Another Australian leptodactylid with unusual development is *Crinia darlingtoni;* the young develop inside inguinal pouches of the male parent and emerge as miniature frogs (Straughan and Main, 1966). The South American *Zachaneus* has terrestrial and nonfeeding "wriggler" larvae. However, the tooth rows are well developed and probably used as an aid in locomotion. The operculum is complete, but no spiracle is present (Lutz, 1944). *Eleutherodactylus* is a large genus in which all the known life cycles exhibit direct development. The gill condition varies from external gills to the absence of even gill clefts. Also, the operculum is rudimentary, occurring only as a gular fold. Larval jaw cartilages are absent as well.

Rhinoderma (Rhinodermatidae) undergoes development in the vocal sac of the male. The larvae develop tooth ridges, papillae, and a complete operculum with spiracle.

Some species of the family Hylidae have developed parental care. In *Hemiphractus, Cryptobatrachus,* and *Cerathyla,* the eggs develop on the back of the female, but in *Gastrotheca,* the dorsal skin is modified into a pouch that completely covers the eggs. The young either hatch as froglets or leave the pouch as tadpoles. In the larval forms that complete their development on the parent, gills are well developed and extend over the yolk, but the operculum does not complete its development until late and may not close entirely. The jaw cartilages suggest the larval form in *Cryptobatrachus* but are more simplified in *Hemiphractus* (Orton, 1949).

Sooglossus (Sooglossidae) of the Seychelle Islands family has nonaquatic development. The young are carried on the back of the male and complete their development there.

Arthroleptella and *Androphryne* are ranids that hatch with the tail present. *Arthroleptella* has jaw cartilages and a complete operculum. *Arthroleptis* and *Cornufer* hatch without a tail. In *Cornufer*, no gills develop, but the operculum is complete and covers the forelimbs (Alcala, 1962).

EVOLUTION OF THE TADPOLE TYPES

The primitive filter-feeding mode of procuring food suggests that tadpoles of early frogs may have been similar to the recent xenoanuran type in which two separate branchial chambers occupy about one half the body length. In this recent type, water entering the mouth passes through the branchial chambers, where food particles are filtered and passed to the esophagus. During the process of gulping and water intake, a steady stream of water passes out through the two spiracles. The jaw cartilages are simple, with no separate infralabial cartilages. Both feeding and respiration involve essentially the same actions, so the same mandibular and hyoid muscles are utilized.

In the second primitive group, Scoptanura, the two branchial chambers, each of which has an internal aperture, are connected to the exterior by a single, ventral opercular tube. The forelimbs develop posterior to the gill chambers. Respiration and feeding are similar, but a small, separate infralabial cartilage, along with a unique mandibular muscle, enables these tadpoles to feed on infusoria by means of a protrusible lower jaw. These two groups feed on microorganisms. Basically, however, this group is also adapted to filter feeding.

The other two groups of tadpoles are characterized by the development of a long coiled gut, an adaptation related to the special digestive action associated with algae feeding. Also, there is the development of an upper-jaw cartilage and modification of the lower-jaw cartilages, which produce an extra joint and a more flexible functional mouth. Modification of the mandibular and hyoid musculature accompany this feeding specialization. The branchial area is reduced and appears to have been "pushed" anteriorly by the much larger gut. The presence of either a single median or a sinistral spiracle increases the efficiency of the respiratory and filtering surface of the gills (Figure 7–6).

The development of accessory mouth structures—shearing beaks, denticles, and papillae—has enabled the tadpoles of acosmanuran frogs to occupy a myriad of niches. The adaptations are usually related to specific habits, and any generalizations should be made with caution. For example, expanded lips can produce an adhesive mouth structure that enables tadpoles to live in fast-moving water; however, rheophilous tadpoles occur with abdominal adhesive discs or without any modifications for clinging. Large lips are also used for surface feeding and for sweeping in food. Increased numbers of denticle rows occur in some forms with adhesive mouths but also may be used for grasping prey. Other tadpoles have greatly reduced lips. The many specializations allow many aquatic habitats to be utilized.

In contrast, the more primitive tadpole of the scoptanuran frogs is successful primarily in the tropics and has developed very few major variations in structure. Various types of nonaquatic development occur in many families of frogs. In the microhylids this has been an especially important factor in overcoming the tadpole's limitations.

The internal morphology of anuran larvae continues to be one of the more constant basic patterns, although there is considerable variation in the nature of more specific adaptations. From the evidence at hand, I have concluded that three divergent lines of evolution from the primitive, filter-feeding tadpole type gave rise to the Xenoanura and Scoptanura as early, separate lineages, while the third line of evolution, the characteristics of which were initially shared by the

SCOPTANURA

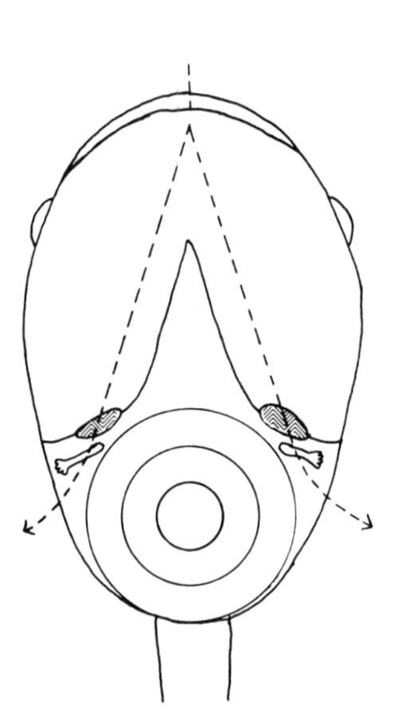

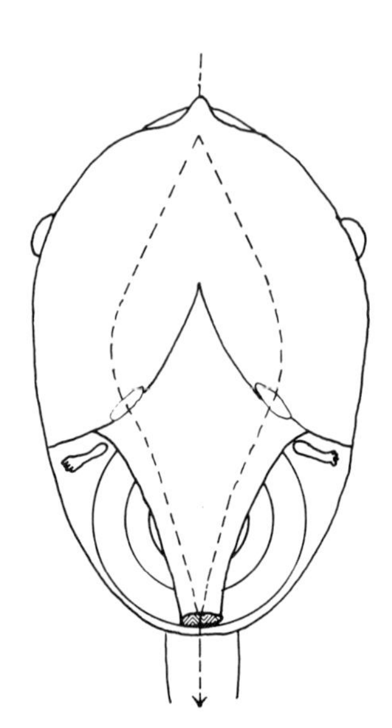

LEMMANURA

ACOSMANURA

Figure 7–6. Four basic tadpole types showing relationship of mouth, branchial chambers, spiracle, forelimbs, and gut. Ventral views.

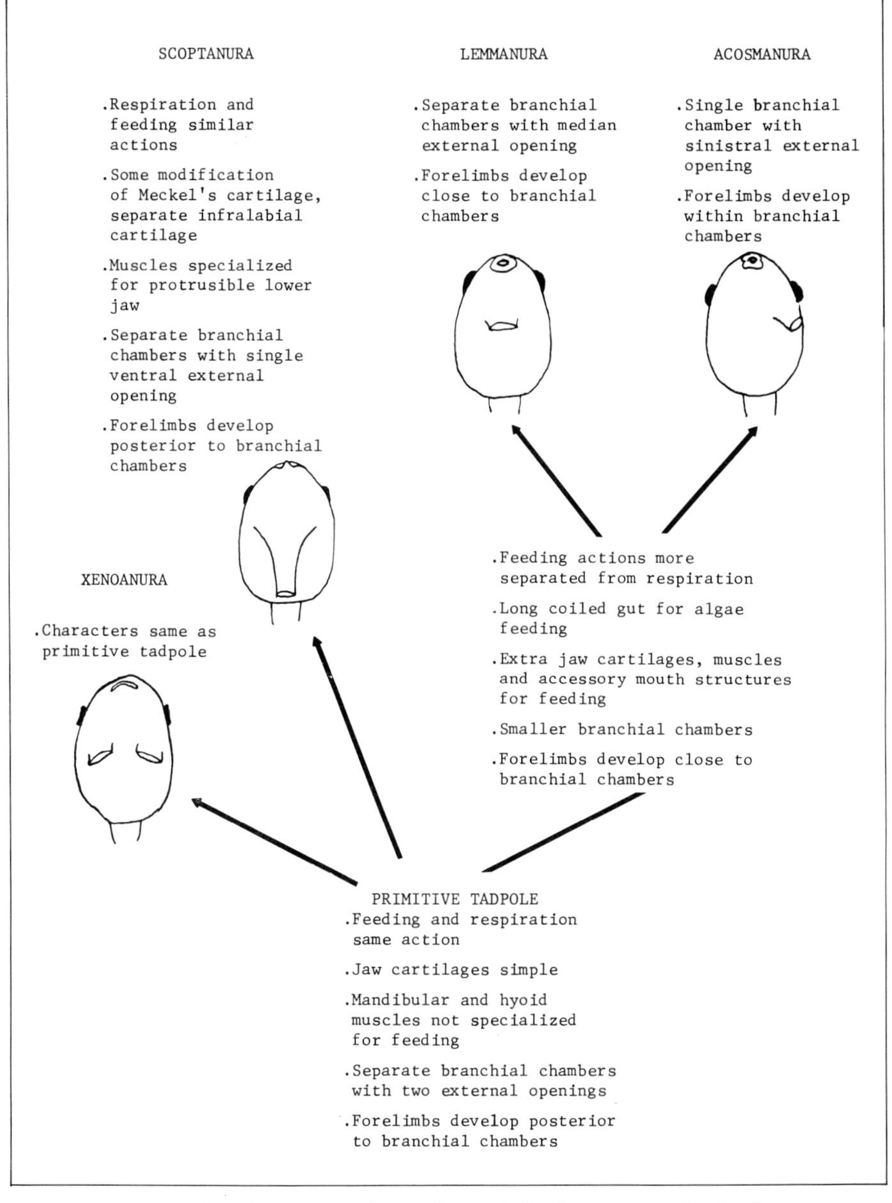

Figure 7–7. Proposed evolutionary relationships of the four types of tadpoles.

remaining types, subsequently emerged as the Lemmanura and Acosmanura. Figure 7–7 outlines these directions and summarizes the characters upon which they are defined.

REFERENCES

ALCALA, A. C. 1962. Breeding behavior and early development of frogs of Negros, Philippine Islands. Copeia 4:679-726.

ALEXANDER, R. MCNEILL. 1968. Animal mechanics. Univ. Washington Press. 346 p.

BLES, EDWARD J. 1905. The life history of *Xenopus laevis,* Daud. Trans. Roy. Soc. Edinburgh 41:789-822.

Brock, Gwendolen. 1929. The formation and fate of the operculum and gill-chambers in the tadpole of *Rana temporaria*. Quart. J. Microscop. Sci. 73:335-343.

Cei, J. M. 1968. Notes on the tadpoles and breeding ecology of *Lepidobatrachus* (Amphibia, Ceratophryidae). Herpetologica 24:141-146.

De Jongh, H. J. 1968. Functional morphology of the jaw apparatus of larval and metamorphosing *Rana temporaria* L. Netherlands J. Zool. 18:1-103.

Gallien, L., and C. Houillon. 1951. Table chronologique du developpement chez *Discoglossus pictus*. Bull. Biol. France et Belg. 85:373-375.

Gosner, K. L. 1960. A simplified table for staging anuran embryos and larvae with notes on identification. Herpetologica 16:183-190.

Gradwell, Norman. 1968. The jaw and hyoidean mechanism of the bullfrog tadpole during aqueous ventilation. Canadian J. Zool. 46:1041-1052.

Griffiths, I., and A. L. de Carvalho. 1965. On the validity of·employing larval characters as major phyletic indices in amphibia, salientia. Rev. Brasileira Biol. 25:113-121.

Heron-Royer, L. F. 1884. Cas tératologiques observés chez quelques têtards de batraciens anoures et de la possibilité de prolonger méthodiquement l'état larvaire chez les batraciens. Bull. Soc. Zool. France 9:1-7.

———. 1887. Observations sur le développement externe et l'état adulte des batraciens du genre *Bombinator*. Bull. Soc. Zool. France 12:640-655.

Inger, R. F. 1960. Notes on the toads of the genus *Pelophryne*. Fieldiana: Zool. 39:415-418.

———. 1966. The systematics and zoogeography of the amphibia of Borneo. Fieldiana: Zool. 52:1-402.

Kenny, J. S. 1969. Feeding mechanisms in anuran larvae. J. Zool. London 157:225-246.

Kluge, A. G., and J. S. Farris. 1969. Quantitative phyletics and the evolution of anurans. Systematic Zool. 18(1):1-32.

Lagler, Karl, John Bardach, and Robert Miller. 1962. Ichthyology. John Wiley and Sons, New York. 545 p.

Lataste, M. Fernand. 1876. Sur la position de la fente branchiale chez le têtard du *Bombinator igneus*. Act. Soc. Linn. Bordeaux 31:1-3.

———. 1877. Quelques observations sur les têtards des batraciens anoures. Bull. Soc. Zool. France 1-6.

Littlejohn, M. J. 1963. The breeding biology of the baw baw frog *Philoria frosti* Spencer. Proc. Linn. Soc., New South Wales 88:273-276.

Lutz, B. 1944. Biologia e taxonomia de *Zachaenus parvulus*. Bol. Mus. Nac. Rio de Janeiro 17:1-66.

Lynn, W. Gardner. 1942. The embryology of *Eleutherodactylus nubicola*, an anuran which has no tadpole stage. Carnegie Inst. Washington Publ. 541:27-62.

Moore, J. A. 1961. The frogs of eastern New South Wales. Bull. Am. Mus. Nat. Hist. 121(3):149-386.

Nieuwkoop, P. D., and J. Faber, ed. 1967. Normal table of *Xenopus laevis* (Daudin). North-Holland Publ. Co., Amsterdam. 252 p.

Noble, G. K. 1927. The value of life history data in the study of the evolution of the amphibia. Ann. New York Acad. Sci. 30:31-128.

Orton, G. L. 1949. Larval development of *Nectophrynoides tornieri* (Roux), with comments on direct development in frogs. Ann. Carnegie Mus. 31:257-276.

———. 1953. The systematics of vertebrate larvae. Systematic Zool. 2(2):63-75.

———. 1957. The bearing of larval evolution on some problems in frog classification. Systematic Zool. 6(2):79-86.

Parker, H. W. 1931. Report of an expedition to Brazil and Paraguay in 1926-27 . . . amphibia and reptilia. J. Linn. Soc. London 37:285-289.

Sedra, S. N. 1950. The metamorphosis of the jaws and their muscles in the toad, *Bufo regularis* Reuss, correlated with the changes in the animal's feeding habits. Proc. Zool. Soc. London 120:405-449.

Sedra, S. N., and Milad Michael. 1961. Normal table of the Egyptian Toad, *Bufo regularis* Reuss, with an addendum on the standardization of the stages considered in previous publications. Ceskoslovenska Morfol. 9:333-351.

Sokol, Otto. 1962. The tadpole of *Hymenochirus boettgeri*. Copeia 2:272-284.

STARRETT, P. 1960. Descriptions of tadpoles of middle american frogs. Misc. Publ. Mus. Zool. Univ. Michigan 110:1-37.

———. 1968. The phylogenetic significance of the jaw musculature in anuran amphibians. Ph.D. Dissertation, Univ. Michigan.

STEPHENSON, N. G. 1951. Observations on the development of the amphicoelous frogs *Leiopelma* and *Ascaphus*. J. Linn. Soc. London 42(283):18-28.

STRAUGHAN, I. R., AND A. R. MAIN. 1966. Speciation and polymorphism in the genus *Crinia* Tschudi (Anura, Leptodactylidae) in Queensland. Proc. Roy. Soc. Queensland 78:11-28.

STUART, L. C. 1961. Some observations on the natural history of tadpoles of *Rhinophrynus dorsalis* Dumeril and Bibron. Herpetologica 17:73-79.

TYLER, M. J. 1963. A taxonomic study of amphibians and reptiles of the Central Highlands of New Guinea, with notes on their ecology and biology. 1. Anura: Microhylidae. Trans. Roy. Soc. South Australia 36:11-34.

VAN BERGIJK, WILLEM A. 1959. Hydrostatic balancing mechanism of *Xenopus* larvae. J. Acoust. Soc. America 31:1340-1347.

WAGER, V. A. 1965. The frogs of South Africa. Purnell and Sons, Capetown Johannesburg. 242 p.

8

ASPECTS OF SOCIAL BEHAVIOR IN ANURAN LARVAE

Richard J. Wassersug

Tadpoles can be treated as independent, free-living vertebrates and studied directly. Although they do not breed, tadpoles move, feed, and grow, and they face all the environmental pressures confronting any aquatic animal. The consideration of tadpoles as aquatic vertebrates is maintained in this study of social behavior, a broad, comparative approach is adopted, and hypotheses relating to adaptive patterns and convergence are developed.

Emphasis is placed on the usefulness of several techniques for studying social behavior, and many topics of importance, such as metamorphosis and direct development, are not considered. Adult breeding behavior is introduced where it directly relates to the adaptive patterns of the larvae. If there is one theme to this paper, it is that methodologies applicable to the study of the comparative behavior and ecology of aquatic vertebrates are available and should be utilized in the study of anuran larval biology.

TERMINOLOGY

Many accounts of group behavior by tadpoles are hidden in the herpetological literature. Most are brief field observations buried in taxonomic treatises, and terms descriptive of group behavior are rarely defined. Such expressions as "school," "shoal," and "aggregate" are often used interchangeably. In assuming common usage for these terms, for which there are no standard definitions, the various writers have made it difficult to compare their observations. Furthermore, definitions of fish behaviorists are ignored.

Arthur Bragg (1948, 1954, 1960, 1965, 1968), the one herpetologist who attempted precise definitions, recognized tadpole groups as either metamorphic or feeding aggregations. He further made the distinction between *asocial* and *social* aggregates. A metamorphic aggregate is a clustering of individuals as they approach metamorphosis. An asocial aggregation results from the attraction of individuals to favorable environmental conditions. Bragg suggests that the attraction is usually to food and sometimes to a more suitable temperature. Social aggregations result from the attraction of individuals solely to the presence of others. The distinction made by Bragg between social and asocial aggregates was not a new one in the behavioral literature. The two terms are broadly equivalent to Wheeler's "associations" and "societies" as discussed by Allee (1931). Bragg used the terms "school" and "aggregate" interchangeably (e.g., 1954, p. 102), but earlier in the same paper (p. 100) he said schooling exists when animals "move much from place to place." This statement could be interpreted to mean that schooling is a special subset of aggregation, one that requires coordinated movement. There is a multitude of obvious problems with these definitions; they often imply some clairvoyant knowledge of tadpole motivation. At what stage in ontogeny tadpoles stop feeding and a feeding aggregate becomes a metamorphic aggregate is not clear. To complicate matters, tadpoles are often nondiscriminating, microphagous suspension-feeders (Dickman, 1968; Jennsen, 1967). If one cannot readily determine if a tadpole is feeding, it is not possible to tell if a feeding aggregate is operative.

Shaw (1970) reviewed the major ichthyological papers in which "school" and related terms have been defined. In Table 8–1, I have attempted to condense this information into a chart for comparison with Bragg's definitions. As Shaw notes, there is agreement among all the authors that a school is a social grouping "based on biosocial mutual attraction." The term "biosocial" needs further explanation. Shaw took it from the recent work of Tobach and Schneirla (1968), who used it to mean a biotaxis in which the external stimulus is another individual. Whether a school need be polarized, as claimed by Breder (1959), or not, as implied by Morrow (1948), Williams (1964), Keenleyside (1955), and others, is a point of contention. Shaw (1970) suggests that "schooling should be considered as a two order system, one order, mutual attraction-approach, and the other order, parallel orientation."

Table 8–1. Categories Used by Bragg to Describe Social Behavior of Tadpoles and Approximate Equivalents Used to Describe Social Behavior of Fish

Bragg (1948, 1954, 1960, 1965, 1968)	Barr (1927)	Spooner (1931)	Breder & Halpern (1946)	Morrow (1948), Keenleyside (1955)	Williams (1964)	Shaw (1970)
Asocial aggregations (reacting to stimuli other than conspecifics)	Aggregations other than schools (temporary)	Aggregations ("chance")	Nonsocial aggregates	Aggregations	Aggregations	Aggregations (attracting other than conspecifics)
Social aggregations ("reacting to each other's presence")	Schools (stable)	Schools	Social aggregates	Schools	Schools	Schools = biosocial aggregates (mutual attraction): A. nonpolarized B. polarized
Schools	("habitual spatial relationship")		Schools (parallel orientation; uniform spacing and velocity)			

Other terms that have been found in the literature but are not included in Table 8–1 are "shoal," "pod," "facultative schools," and "obligate schools." A shoal may be considered equivalent to a school (Breder, 1967). A pod is the limiting case of an aggregate when the mean spatial distance between fish reaches zero (see Breder, 1954, for his mathematical model of schooling). Pods can be either polarized or nonpolarized. Because there are reports of vast numbers of tadpoles in contact with each other, the term "pod" seems worth introduction into the herpetological literature. The other terms do not seem applicable to tadpoles; for example, Breder distinguishes an obligate schooler from a facultative schooler as a fish that is always found in a school and that shows characteristic signs of stress when separated from the school. I know of no example of such a phenomenon for anuran larvae.

In this paper, two categories of active aggregations in tadpoles are recognized. The first category is simple aggregates, based on biotaxis other than biosocial mutual attraction (Bragg's asocial aggregates). Simple aggregates may be either polarized or nonpolarized. The second is schools of biosocial aggregates (Bragg's social aggregates). These also may be polarized or nonpolarized. Certain tadpoles are known to form distinct groups, with all the individuals oriented in the same direction (polarized) but lying still on the bottom or suspended stably in midwater; hence, movement is not part of the definitions.

TAXIC BEHAVIOR–SIMPLE AGGREGATES

Thermotaxy. Thermotaxy is the most commonly offered mechanism to account for natural tadpole aggregates. Experiments have shown that tadpoles respond to thermal gradients (Beiswenger and Test, 1966; deVlaming and Bury, 1970) and may form aggregations as individual responses to a thermal preferendum. Species of the following genera form aggregates in warm parts of natural thermal gradients: *Bufo* (Mullally, 1953; Karlstrom, 1962; Brattstrom, 1962; Tevis, 1966), *Rana* (Carpenter, 1953; Brattstrom, 1962; Bragg, 1968; Ashby, 1969), *Hyla* (Brattstrom and Warren, 1955; Brattstrom, 1962; Ashby, 1969), *Pseudacris* (Bragg, 1968), and *Scaphiopus* (Bragg, 1968). Carpenter, Karlstrom, and Brattstrom and Warren argued that metabolic rate is increased and that the time needed for larval development and metamorphosis is decreased by this behavior. Not all tadpoles that show thermotaxic behavior aggregate at the highest available temperature. Given a choice, *Ascaphus* tadpoles tend to cluster near or below 10° C and avoid higher temperatures (deVlaming and Bury, 1970), while *Bufo americanus* prefers temperatures near 30° C in a similar experimental apparatus (Beiswenger and Test, 1966). Beiswenger and Test found that natural populations of *B. americanus* in Michigan avoided temperatures above 37° C. *Bufo marinus,* from the lowlands of Costa Rica, aggregates in water as warm as 37° C but moves to cooler water when the temperature reaches 37.4° C (M. Mares, personal communication 1970). The similarity in maximum temperature tolerance of *Bufo* species suggests that evolutionary history, as well as individual acclimation, may contribute to the thermal preferendum and tolerance of anuran larvae.

Rheotropism. Perhaps the most visible environmental factor that can affect aggregating is the force of the current. Many tadpoles are adapted specifically to stream life (Noble, 1927; Orton, 1953), and such tadpoles as the highly specialized *Ascaphus truei* can be found in fast-flowing currents (Myers, 1931). Fast currents may isolate aggregates in quiet refugial pools on the edges of streams, as Brattstrom's (1962) account of *Rana boylii* aggregating in a cove of a swift river illustrates. Inger (1966) observed aggregates of from 25 to more than 100 stream-adapted *Amolops jerboa* larvae at night on shallow rocks bathed by less than 1 cm of water. This may have resulted from a preferred current and depth or perhaps from a response by individual tadpoles to latent warmth in the protruding rocks or to algal food on those surfaces.

Tadpoles are rheotropic whether or not they normally live in streams. By heading upstream in an otherwise overpowering current, a tadpole will be deflected to the side, out of the swifter water. Thus, rheotropism may be a mechanism by which tadpoles avoid being swept downstream. The physiological mechanism for rheotropism is discussed briefly in the section below on parallel orientation. Both tactile and optical stimuli may be involved, and the same mechanism need not be operative in all tadpoles.

Phototaxy. Noble (1931) stated that *Hyla versicolor* larvae were positively phototropic and suggested that the basking of *Rana pipiens* and *Rana sylvatica* in shallow water may be a photic response. He admitted, however, that such behavior could just as well be thermotaxic and took to task previous investigators for not controlling this variable. There is good evidence for phototaxy in certain tadpoles. Bragg (1964) found positive phototaxy in *Scaphiopus bombifrons;* Starrett (1960) noted that *Rhinophrynus dorsalis* were attracted to the beam of her flashlight at night; and Ashby (1969) observed *Rana temporaria* concentrating in the areas where flecks of sunlight came through the foliage and hit the pool bottom. Also, Brattstrom and Warren (1955) described *Hyla regilla* oriented in a way that exposed "the maximum amount of dorsal surface area to the sun's rays, providing the greatest amount of heat absorption from the sun." While these accounts suggest that tadpoles are generally positively phototropic and may aggregate in light, Noble (1931) noted that "species differ enormously in the phototropic responses of their larvae." DeVlaming and Bury (1970) claim that *Ascaphus truei* is negatively phototropic, avoiding light. *Ascaphus* (deVlaming and Bury, 1970) and *Rana clamitans* (Cole and Dean, 1917) show varying photic responses, which depend upon the age of the larvae.

Olfaction. The olfactory ability of tadpoles was recognized early (Risser, 1914) and is considered important in food location. Bragg emphasized that food, more than all other factors, could account for tadpoles aggregating. Hungry tadpoles will quickly swarm about available food. For large food items there are advantages to individuals participating in collective feeding. By applying opposing forces, scavengers, such as most tadpoles, can disproportionately increase their mechanical efficiency in shredding food down to ingestible size. Richmond (1947), in his excellent paper on the group behavior of *Scaphiopus holbrooki* larvae, suggested that the stimulus for an individual to follow the "school" was the stream of microscopic food stirred up by the front of the procession. Food-laden currents may be a factor in group adhesion in *Bufo marinus* aggregation (M. Mares, personal communication 1970).

Oxygen Concentration. Dissolved oxygen concentration is another ecological variable to which tadpoles are sensitive and to which they react (Costa, 1967). The common response of individual tadpoles in oxygen deficient water is to come to the surface and gulp air. Deanesly, Emmett and Parkes (1945) showed that for *Xenopus laevis* the rate of surfacing correlated with the amount of oxygen in the water. *Rhinophrynus* tadpoles in poorly oxygenated standing pools characteristically show this surfacing behavior. Individuals collected from such pools will continue this surfacing behavior until their aquarium water is aerated, then, within moments, the surfacing movement will cease. *Ascaphus truei* tadpoles that rarely swim freely and live in highly oxygenated water will expose their nostrils, and often much of their bodies, above water if kept in poorly aerated or warm water. Oxygen deficiency serves as the stimulus for negative geotaxis in newly hatched *R. temporaria* tadpoles, and the dense, dark, carpetlike aggregates formed by these tadpoles just beneath the surface could be a result of limited oxygen supply (Savage, 1961). This explanation could be applied to aggregates of other species that develop in shallow water or immediately below the surface.

Miscellaneous. Other factors could lead to the formation of simple aggregates. Tadpoles are attracted to the anode in an electric field (Scheminsky, 1924), but this artificial situation tells us little about the natural behavior of tadpoles. Wiens (1970) has shown that there is species-specific selection of substrate patterns by *Rana aurora* and *R. cascadae.* If this is a common phenomenon it may account for aggregating in certain species but would be particularly difficult to discern. In one sense, recognition of a conspecific and approach to it may be considered a special case of Wiens' pattern selection and the basis for any biosocial aggregates that may be found among anuran larvae.

THE FORMATION OF SCHOOLS—BIOSOCIAL AGGREGATES

The question remains as to whether tadpoles will form biosocial aggregates or true schools (*sensu* Shaw, 1970). Almost anyone who has watched *Xenopus*, *Scaphiopus*, or *Bufo* tadpoles would subjectively claim that these larvae exhibit biosocial mutal attraction. Beiswenger and Test (1966) found dense aggregates of *Bufo* tadpoles at all temperatures below 37° C in an otherwise uniform environment and emphasized that temperature could not be the controlling factor. Black (1969) offered support for a biosocial foundation of group behavior based on the development of *Bufo americanus* aggregates. He observed that "aggregational behavior develops initially from the interaction of two . . . tiny tadpoles a few days after hatching" and briefly related the behavior of two such tadpoles on approaching each other.

An experimental procedure, using quarter-sectioned clear plastic trays, has been developed to test for the biosocial mutual attraction responses of tadpoles (Wassersug and Hessler, 1971). In that study it was found that tadpoles of *Xenopus laevis* (from an early free-swimming stage to near metamorphosis) exhibit visually mediated mutual attraction. *Xenopus* larvae are known to form natural aggregates so that this result is not too surprising. Seven species from five families, not counting the pipid *X. laevis*, have now been tested in the apparatus to determine how typical biosocial mutual attraction is and whether it occurs in tadpoles that do not regularly aggregate. In all cases, the tadpoles were free swimming or "mature" (stages 26–36; Gosner, 1960) and matched to size and stage for each experiment. The experimental design was the same as that used with *Xenopus*, and only the parts necessary to understanding the results are here reiterated.

In each experimental run one tadpole of a given species was placed in each of the four sections of a tray. The tray was then photographed every two or three minutes, depending on the species, until at least 20 exposures were made. Controls were run by arbitrarily removing all but one of the four tadpoles and repeating the procedure.

In analyzing the photographs each tray section was divided into lettered quadrants, which were used to denote the positions of all the tadpoles (see Figure 8–1). A tadpole was considered to lie in the quadrant in which its eyes were located. All quadrants with the same letter were considered a quadrant group. For each exposure the number of tadpoles located in each quadrant group was recorded. The total score was then calculated for each quadrant group by summing up all the exposures in each experiment.

If there were no visually mediated social interactions, no one position would be preferred within any tray section. The tadpoles would presumably position themselves randomly in relation to each other. On the other hand, if social interaction—specifically, biosocial mutual attraction—were operative, then we would expect a disproportional clustering of the tadpoles in the common corner, quadrant group "d." Likewise, we would expect the individual tadpoles to spend the least time in quadrant "a," where they would be most isolated from each other.

The total scores for each quadrant group were compared by the chi-square test for each experiment. This test can be valid only if there is independence between samples, in this case, photographs. The tadpoles must move between

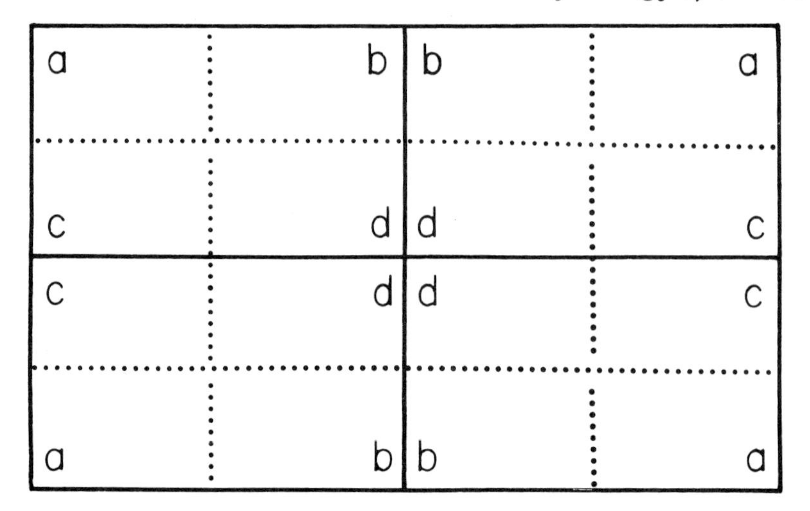

Figure 8–1. Schematic view of quarter-sectioned tray used to measure aggregating tendencies. *Solid lines,* tray walls; *dotted lines,* quadrant boundaries in each tray section.

photographs and not just freeze in one position for most of the experiment. Where there was any question of sample independence, this assumption was tested separately by comparing the number of duplications on consecutive photographs against those predicted by a binomial expansion.

Sample results for five of these seven species—the ranids *Rana pipiens* and *R. boylii,* the hylid *Hyla regilla,* the pelobatid *Scaphiopus hammondi,* and the bufonid *Bufo boreas*—are presented in Figure 8–2a.

Of these five species only *Bufo boreas,* a tadpole collected in dense aggregates in Livermore, California, deviated greatly from random. The common corner, "d," was the most often occupied by *B. boreas,* but because the corner scoring next highest was the uncommon corner, "a," the results are slightly ambiguous. The controls, however, were not significantly different from random. Although there can be little doubt that *Bufo boreas* actually schools, the study deserves repeating to resolve this irregularity. It is possible that the mutual attraction of this and other schooling species is not totally visually mediated. Olfactory or mechanical stimuli may be equally important (it may be appropriate to note that *Bufo* larvae generally have larger nasal capsules than do other tadpoles), but the height of the tray walls above the water surface in each tray section prevented the transmission of chemical or pressure clues in these experiments.

Rana catesbeiana (for which a histogram is not presented), *R. pipiens, R. boylii, Hyla regilla,* and *Scaphiopus hammondi* demonstrate little or no visually mediated biosocial mutual attraction. Apparently the occasional reports of aggregates in these species are largely the result of biotic attractants other than conspecifics, as suggested by Brattstrom (1962). In some of the experimental runs *R. pipiens* larvae showed a slight bias toward quadrant "d" ($0.95>P>0.90$), so biosocial mutual attraction may be operative to a degree.

A slight behavioral difference between *R. boylii* and *R. pipiens,* which were of equal size and stage in these tests, is reasonable when one considers the differences in habitat of these two species. In the quieter waters, where *R. pipiens* is likely to be found, it would be relatively easy to aggregate. In the flowing waters, where *R. boylii* occurs, it would require a great deal of energy to maintain group integrity. In this regard Brattstrom (1962) found a "thermal" aggregate of *R. boylii* in a calm refugial pool out of the current.

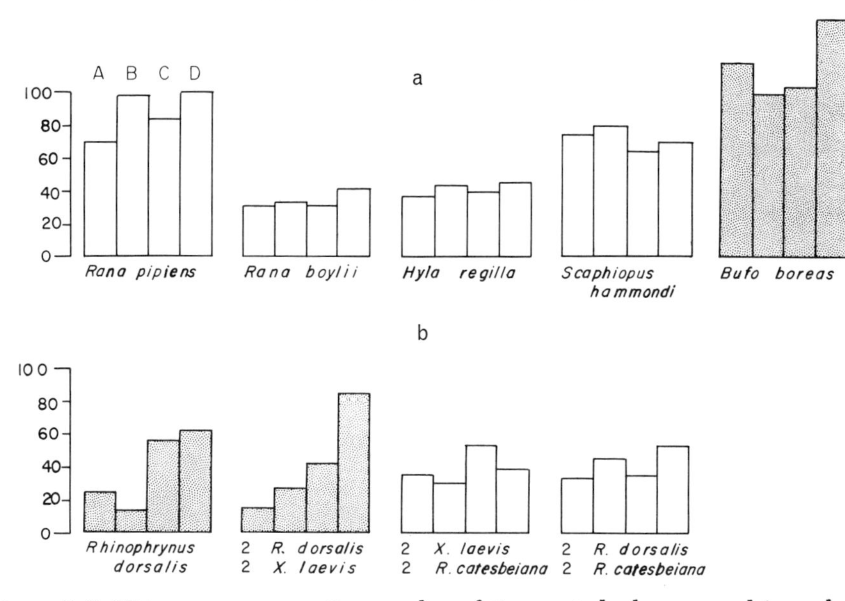

Figure 8–2. Histograms representing number of times a tadpole appeared in each quadrant group. *Stippling,* P>.95. A, B, C, and D are quadrants.

Scaphiopus hammondi did not exhibit mutual attraction in repeated runs of the experiment. This is contrary to expectation when one considers the vast amount of literature, in particular by Bragg, on natural aggregates of *Scaphiopus* larvae. There are two possible explanations of this discrepancy. The first is that visual clues do not mediate the biosocial behavior in this genus; the second hypothesis is that there are significant species or population differences in aggregating tendencies of *Scaphiopus* larvae. While these theories are not mutually exclusive, there is some circumstantial evidence in support of the latter. None of the field accounts of aggregating in *Scaphiopus* actually deal directly with S. *hammondi,* and in the streamlets around Livermore, California, from which I collected the tadpoles studied here, the animals were not found aggregating.

Ascaphus truei tadpoles did not exhibit enough movement in the experimental apparatus to make the assumption of sample independence between photographs valid; consequently, no histograms are presented for the tests on this species. When placed in the trays *Ascaphus* tadpoles would attach to the walls or bottoms of the sections and not move for ten to twenty minutes at a time. Experiments were run at room temperature (23° C) and 8° C. Even at the lower temperature, activity did not increase measurably. Some of the photographs show two *Ascaphus* tadpoles in a shared corner. Often these tadpoles would be attached by their oral discs to the opposite sides of the section wall. In this position, however, the tadpoles could not see each other, so there is no evidence here for biosocial aggregating.

Aggregating is well documented in *Rhinophrynus dorsalis* (Taylor, 1942; Starrett, 1960; Stuart, 1961). Tadpoles of this species, collected in Tehuantepec, Mexico, showed a strong bias toward quadrant group "d" (Figure 8–2b) when tested in the laboratory.

The remarkable resemblance of *Rhinophrynus* and *Xenopus* tadpoles was the basis for the Orton (1953, 1957) Type I tadpoles (Xenoanura of Starrett, this volume). This raises the question of whether tadpoles of these two genera, which are so similar in gross, external morphology, can distinguish between each other visually. Although both are gregarious, there are noticeable differences in their schooling patterns. *Xenopus* tadpoles can maintain a stable position suspended in midwater by using their rapidly beating tail filaments. *Rhinophrynus* tadpoles lack the *Xenopus*-type tail filament and are constantly moving, using a full tail stroke. Also, schooling *Rhinophrynus* larvae are in contact with each other much more often; the mean spatial distance is considerably less between individuals in a *Rhinophrynus* school than it is in a *Xenopus* school. Subjectively, it appeared that *Rhinophrynus* tadpoles, when placed in a mixed-species tank with *Xenopus*, approached them more closely than *Xenopus* tadpoles would approach their own kind. As the *Rhinophrynus* approached closer and closer, the *Xenopus* tadpole dashed away to take up a new position.

When two *Xenopus* tadpoles were tested with two *Rhinophrynus* tadpoles in the partitioned trays (arranged with the conspecifics diagonally opposite each other), the strongest approach response of any of the experiments was recorded (Figure 8–2b). Apparently, different spacing of these tadpoles in their respective schools does not relate to differences in the strength of their visual approach response. Schooling for any species requires not only a specific approach response of all individuals but also a balanced withdrawal response that need not be solely visual. Differences in schooling between *Xenopus* and *Rhinophrynus* may in part derive from their different thresholds for withdrawal from mechanical or chemical stimuli.

A final set of experiments was run using combinations of schooling and nonschooling tadpoles. The data for two of these experiments are presented in Figure 8–2b. In one experiment two small *Rana catesbeiana* tadpoles were placed in the trays with two *Xenopus* tadpoles so that, again, the two tadpoles of the same species were in diagonally opposite sections. In the other experiment *Rhinophrynus* was used instead of *Xenopus*. In both experiments there was no significant tendency to aggregate by the participating tadpoles. These results emphasize the fact that the approach response of certain schooling tadpoles is induced by rather specific forms, not just any visible object, tadpole or otherwise, that is in the same size range.

The aggregating response of the two *Xenopus* and *Rhinophrynus* tadpoles involved in each of these mixed experiments was never as great as when they were with four schooling tadpoles. The intensity of the mutual attraction response of schooling tadpoles may be, in part, density dependent—the more tadpoles present, the more intense the response.

POLARIZATION AND THE OPTOMOTOR RESPONSE IN TADPOLE BEHAVIOR

As mentioned at the beginning of this paper, Shaw (1970) makes the distinction between the biosocial aspects of schools and the polarization of schools. By means of biosocial mutual attraction, an aggregate may form. In addition, these biosocial aggregates may exhibit parallel orientation in which the individuals are polarized (see Breder, 1959, for a detailed discussion of polarization in fish). In a polarized school, individuals may shift their positions in relation to one another, but the total orientation is not affected. Polarized aggregates have been reported for the larvae of diverse frogs from various families.

The optomotor response, in which an individual moves in order to stabilize the image of its environment on its retina, is a well-known phenomenon in both schooling and nonschooling fish. The results of studies by Lyon (1904), Clausen (1931), Harden Jones (1963), and Harder (1963) suggest that the rheotropic responses of certain fish may be the result of an effort on the part of the fish to maintain a stationary visual image, rather than a mechanical response to water pressure. Shaw and her associates (Shaw and Tucker, 1965; Shaw and Sachs, 1967) have investigated optomotor behavior as the mechanism by which fish shift their position in a school, change velocity, and yet maintain their parallel orientation. A survey of optomotor behavior was undertaken in tadpoles to see whether a similar mechanism operates in anurans.

From early investigations it was assumed that the optomotor response was ubiquitous in anurans. In their work on vision, Birukow (1950) and Burgers (1952) demonstrated the presence of the optomotor response in adult *Bufo, Rana, Bombina, Alytes, Hyla,* and *Xenopus*. The only tadpoles, however, that have been tested are those of *Rana temporaria* (Birukow, 1949) and *Xenopus laevis* (Cronly-Dillan and Muntz, 1965); in those tadpoles the response was present. In the experiments reported here, tadpoles from eight species and six families were tested for the presence of optomotor behavior.

An optomotor apparatus was constructed similar to Shaw's (1965; Shaw and Sachs, 1967). It consisted of a small circular tank inside a larger tank (see Figure 8–3). Both tanks had a layer of clean gravel at the bottom. The inner tank, which was 9 cm in diameter, was filled with water from the tadpole's stock tank. The outside of the outer tank was covered with heavy white paper and this tank was filled with water to the same level as the inner tank. Between the two tanks was a rotating stimulus cylinder 11 cm in diameter and 15 cm high; it was made from a sheet of frosted polyethylene film. Vertical black stripes one cm wide and two

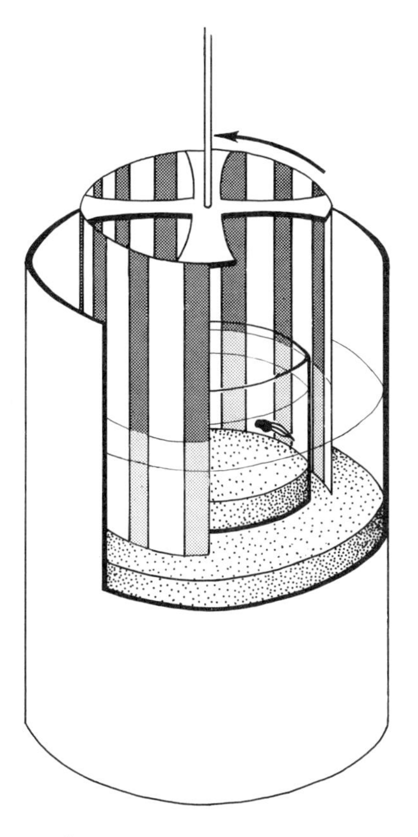

Figure 8–3. Optomotor apparatus.

cm apart were painted on the cylinder. This cylinder was attached by a cardboard spindle to a variable-speed motor that permitted revolutions of from 4 to 20 per minute. The greater thickness of the gravel in the inner tank permitted the stimulus drum to extend down 2 cm below the actual gravel surface in the inner tank. A single tadpole was placed in the inner tank. Because of the height of the stimulus drum, not only above but also below the gravel bottom of the inner tank, the tadpole was fairly well surrounded by the stimulus pattern. One can calculate that a tadpole postioned 1 cm from the wall of the inner tank could have as much as 120 degrees of its lateral visual field in the vertical axis filled by the pattern. Lighting for the apparatus was supplied by two 100-watt light bulbs suspended a meter or so above the bottom of the inner tank. The light did not shine directly on the tadpoles. All tests with the apparatus were made at room temperature (23° C), except where noted.

It would have been preferable to use tadpoles of the same size and stage, but this was rarely possible. The design for the stimulus drum—one cm black stripes two cm apart—was chosen because the width of the black stripes was of the same order of magnitude as the length of the majority of tadpoles tested (see Table 8–2).

Table 8–2. Sizes and Stages of Tadpole Species Tested in Optomotor Apparatus

Species	Snout-vent length (in *cm*)	Developmental stage[a]
Rana pipiens	2.1	30
R. catesbeiana	2.1	30
R. boylii	1.4	31
Hyla regilla	1.2	33
Bufo boreas	1.3	37
Xenopus laevis	1.2	34
Rhinophrynus dorsalis	1.8	35
Ascaphus truei	1.8	34

[a]Numbers refer to stages of development as defined by Gosner (1960).

Testing sessions consisted of one five-minute run with the cylinder rotating at eight revolutions per minute, followed by two different five-minute control runs. One individual was tested at a time. During the first control, the motor was off but the drum was in place. This kind of run was done to determine if the tadpoles had any stimulus-unrelated constancy in orientation, that is, a left- or right-handedness. For the second control, the motor was on, but the drum was disconnected from the spindle, a precaution against the possibility that the tadpoles were responding to the shadows cast by the light coming through the spindle arms or the noise of the motor. Four of these 15-minute testing sessions were run in succession for each tadpole, and for the final analysis, the total was taken from the data collected in all four 15-minute sets. The tadpoles were initially allowed 20 minutes to acclimate before the experiment was begun, and between each 15-minute test, the water in the inner tank was aerated through a bubble stone for at least 2 minutes.

In the apparatus, a tadpole was always considered to be facing either in the same direction or the opposite direction of drum rotation. If a tadpole remained perpendicular to the glass wall so that it could not be readily determined which way it was oriented, then the time spent in this position was divided equally between the two choices. Because in natural schools of fish or tadpoles polarization may be present even when the total group is stationary, orientation alone, not movement, was considered a response.

One would expect tadpoles that do not show a preferential orientation to spend an equal amount of time facing toward and away from the direction of drum rotation. Figure 8–4 gives the percentage of the time that each species spent oriented in the direction of drum rotation. The controls for all the species, except *Ascaphus truei* (discussed below), were within 5 per cent of spending equal time heading with and against the drum. This percentage indicates that the tadpoles do not have an innate left- or right-handedness when undisturbed. They did not respond to the shadows cast by the rotating spindle arms or the noise of the motor, but during the experimental runs, all of these tadpoles showed a significant tendency to orient in the direction of drum rotation.

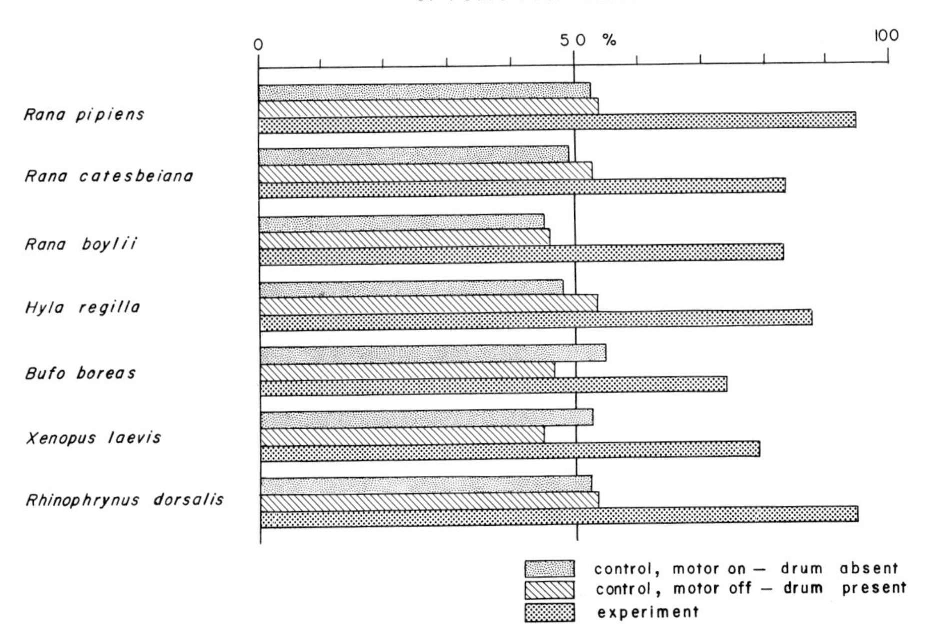

Figure 8–4. Optomotor responses given as percentages of time during experiments that the tadpoles oriented in the same direction as rotation of stimulus drum.

There may be differences in the strength of the response among the different species, but these can not be evaluated considering the seven species tested and differences in size and stage among the various tadpoles. The important point is that tadpoles of the Ranidae, Hylidae, Bufonidae, Pipidae, and Rhinophrynidae and probably most free-swimming anuran larvae have the ability to orient in a visual field. As with teleosts, the ability to orient in a parallel manner seems to be more general than does biosocial mutual attraction (Shaw, 1965). Shaw speculated that the mutual attraction and polarization in schooling fish had separate neurological mechanisms. This may also be true for tadpoles.

Beyond the gross optomotor response there were notable variations in several of the species in the apparatus. Harder (1963) recognized ten categories for the behavior of different fish species in his optomotor apparatus. I am considering only the three broad patterns outlined below. Harden Jones (1963) believed that the differences in optomotor behavior had ecological meaning; this is doubtless true for tadpoles. *Ascaphus truei,* as one extreme, did not respond in the apparatus at all. As soon as an *Ascaphus* tadpole was placed in the inner tank, it anchored itself to the glass wall with its suctorial mouth and rarely moved. When it moved, it moved by inching along with its mouth and did not swim freely. This behavior was the same during both the controls and the experimental runs and did not change when the experiment was rerun at 8° C.

Ascaphus tadpoles have relatively reduced dorso-laterally directed, nonprotruding eyes in conjunction with their torrent-adapted body form (Hora, 1930), and they can be found in areas of extremely heavy shade (Stebbins, 1951). *Ascaphus* tadpoles may be unable to discern the pattern on the stimulus drum when they are attached to the glass of the inner tank. The absence of the optomotor response in *Ascaphus,* however, could be related to factors other than visual perception. Harden Jones (1963) found the response absent in six species of fish that live in contact with the bottom. He suggested that optomotor behavior, which is otherwise used to maintain a visual fix and, consequently, stability in currents, may be absent or inhibited in forms that have strong contact with the substratum. This may be true for *Ascaphus,* whose tadpoles depend largely on tactile, rather than visual, clues for maintaining their position in a current. It is no surprise that tadpoles of this genus have a conspicuously well developed lateral-line system. The absence in *Ascaphus* of an optomotor response, when it is present in all other tested tadpoles, underlines the specialized nature of this anuran larva. There is little in the behavior of *Ascaphus* tadpoles to support the contention that they represent the primitive adaptive type of anuran larvae, as Schmalhausen (1964) suggested.

The remaining tadpoles tested, with the exception of *Xenopus,* exhibited a second behavioral pattern. They spent most of their time during test runs swimming near the wall of the inner tank in the direction of drum rotation. Individual variation occasionally included resting on the bottom, reversing direction, and crossing toward the middle of the tank. The modal behavior, nevertheless, was a steady swimming around the perimeter of the tank in close synchrony with the speed of the stimulus drum. Tadpoles, which otherwise appear to be poorly coordinated swimmers, could go dozens of times around the inner tank always in line with the same few stripes on the stimulus drum. When in this path, a tadpole would seemingly strive to maintain a lateral visual fix between two black stripes. If it started to advance on this position or fall behind, it would either slow down or accelerate in order to maintain this intermediate station. Such behavior is identical to what has been observed for many fish (Harder, 1963; Shaw, 1965). The advantage of such a response is readily evident for the tadpoles that truly school, such as *Rhinophrynus dorsalis* and *Bufo boreas.* For them, other individuals in the school may be considered equivalent to the dark stripes of the stimulus pattern.

For a stream tadpole, such as the nonschooling *Rana boylii,* this precision behavior would help it maintain a visual fix in a current. The rotation of the drum not only initiated a strong swimming response in *Rana boylii,* but also induced a rolling motion after every few tail beats. At that time, the tadpole brought its mouth in contact with the glass surface. I interpret this as an attempt by the tadpole to secure itself in its visual field. In streams and aquaria *Rana boylii* tadpoles often hang on to the surfaces by their oral discs, which are apparently modified for increased adhesive ability (Zweifel, 1955). In the optomotor apparatus, no action of the oral disc could ever succeed in stabilizing the visual field; consequently, after a fraction of a second, the tadpole would dash to catch up with the last advancing stripe on the stimulus pattern. The optomotor responses discussed so far are consistent with the broad spectrum of specialization to stream life found among tadpoles (Noble, 1927; Orton, 1953). *Rana boylii* tadpoles live in more open and gentle streams that *Ascaphus* and are far less specialized in their morphology and behavior. The optomotor responses of the tadpoles that do not exhibit biosocial mutual attraction or do not normally occur in streams, such as *Rana pipiens, Rana catesbeiana,* and *Hyla regilla,* may still be rheotactic adaptations that prevent these species from being caught in areas of strong current.

Rarely did *Xenopus* show the typical optomotor response of the other tadpoles. While the drum was starting to rotate, a *Xenopus* tadpole might swim near the edge of the inner tank for short spurts. Soon, however, the tadpole would take up a position in the middle of the tank. It would then pivot about this point in a characteristic head-down posture. A *Xenopus* tadpole in this position would rotate at a speed considerably less than the drum. If it migrated from this central position, it would seem to swim randomly, but it would soon return as if either attracted to this central point or repelled by the stimulus drum.

The differences of the behavior between *Xenopus* and the other tadpoles in the optomotor apparatus underscores the uniqueness of this tadpole. *Xenopus* larvae are both behaviorally and morphologically specialized suspension-feeders (Kenny, 1969*b*; Wassersug, 1972). As in all pipids, they lack the keratinized mouth parts that would otherwise allow them to masticate large food items. They subsist on pelagic microscopic particles that they extract from the water. The position of the eyes in *Xenopus* is fully lateral. The combination of a symmetrical spiracle pattern and a rapidly beating tail filament allows these tadpoles to suspend themselves stably in midwater. They have greatly reduced pigmentation, except for a band of melanophores distally on the tail. These melanophores are independent of the remaining pigmentation and unique in that they are directly photosensitive (Bagnara, 1957, 1960).

MODES OF SCHOOLING

Schools formed by *Xenopus* are clusters of strongly polarized individuals suspended in midwater, heads tipped downward. Lateral eyes may be seen as an adaptation to midwater schooling, giving the tadpole a maximized, three-dimensional field of vision. *Xenopus* avoid contact with each other and maintain a mean spatial distance of at least a full body's width, while most other schooling tadpole's form much tighter clusters and have high levels of contact between individuals. Consistent with the fact that these tadpoles have a relatively high mean spatial distance is their retreat toward the center of the optomotor apparatus. This behavior may also be viewed in the light of the approach-withdrawal theory of Schneirla (1965), which suggests that the stimulus of the drum is of such high intensity that it causes withdrawal in *Xenopus.*

Frog larvae display more than one distinctive mode of schooling. The biosocial aggregates formed by *Bufo* tadpoles differ radically from the schools formed by *Xenopus* larvae. The former appear as dense, black mats in shallow water or near the bottom, with hundreds to thousands of individuals participating. The tadpoles are in physical contact with each other much of the time and may be polarized. Polarized schools have been described as moving in an amoeboid fashion, sending out "pseudopodial" projections of individuals. Rarely do such schools move with perfect symmetry. As with fish schools, when the school is disrupted individuals may briefly scatter, only to regroup shortly thereafter.

Schools of this nature do not seem restricted to any particular zoogeographic or climatic region. I have observed *Bufo* schools in the wet lowlands of the Osa Peninsula, Costa Rica, at less than 3 meters elevation *(B. marinus)* and in a 2800-meter pool above Quetzaltenango, Guatemala *(B. bocourti)*, as well as throughout temperate North America (*B. boreas* and *B. americanus*). Moreover, one can find accounts of *Bufo* aggregates in areas as distant and diverse as South Africa (*B. carens;* Wager, 1965) and China (*B. bufo;* Liu, 1950). Apparently, this behavior is also independent of subgeneric relationships, transcending species groups.

The differences in the modes of schooling of *Bufo* and *Xenopus* reflect the extreme morphological differences of these genera. *Bufo* larvae tend to be subspheroidal in body form and have dorso-laterally directed eyes, subequal dorsal-ventral tail fins, normal 2/3 denticle pattern, and dense melanophores on the body. Considering the conservative morphology of *Bufo* (Zweifel, 1970) and the great number of species that are known to aggregate, one may conclude that schooling is a generalized behavior pattern for this genus.

Many aspects of the morphology and ecology of *Bufo* larvae are consistent with their mode of schooling. These tadpoles, so often in contact with the bottom and each other, have their eyes dorso-laterally directed and are not particularly efficient at suspension feeding, compared to *Xenopus* in terms of both volume per unit time and particle size (unpublished data). Instead, *Bufo* tadpoles are scavengers that depend on bottom detritus and occasional large food items, rather than uniformly suspended food, for most of their nutrition.

The fright reaction, in which organisms flee from the odor of a dead or injured conspecific, is strong in *Bufo* (*B. bufo* and *B. calamita*) but absent in *Xenopus laevis, Alytes obstetricans, Bombina variegata, Hyla arborea, Rana temporaria,* and *R. esculenta* (Pfeiffer, 1966). Its presence in *Bufo* may be useful in dispersing the individuals at the time when aggregate behavior would be most disadvantageous, specifically, once they have been attacked by a predator. The fright reaction also suggests good olfaction, which may otherwise aid in group integrity and location of food items.

The dark color of *Bufo* tadpoles and their aggregates is usually assumed to be an adaptation for absorbing the maximum available solar radiation. Indeed, that there are metabolic advantages to dark color and group behavior seems most plausible. However, in speculating on the adaptive value of melanism and social behavior to tadpoles, one must not overlook the obvious disadvantages. For one thing, the dark color of *Bufo* tadpoles and their tendency to aggregate make them highly conspicuous and easy prey for larger predators. The fact that any *Bufo* tadpoles reach metamorphosis suggests that they have some protection from such potential attack. One suggestion is that they are unpalatable. Fisher (1929) noted that conspicuous insect larvae that aggregate are often unpalatable. In controlled experiments, the bluegill, *Lepomis macrochirus,* preferred the tadpoles of *Pseudacris triseriata* (a hylid) over those of *Bufo americanus* and rejected the latter (Voris and Bacon, 1966). Comparative palatability tests on live tadpoles of eight species from four families (Hylidae, Leptodactylidae, Dendrobatidae, and Bufonidae) have been made (Wassersug, 1971). *Colostethus nubicola* (Dendrobatidae) and *Smilisca sordida* (Hylidae), the two species found most palatable, are unlikely to be attacked by large predators. They are cryptically colored and are isolated, secreted between rocks or vegetation in flowing water, where it would be most difficult for a predator to find or capture them. On the other hand, the tadpoles found least palatable belonged to the one bufonid species tested, *Bufo marinus.* These tadpoles aggregate in shallow, open areas and are remarkably conspicuous, hence vulnerable to attack. It was therefore suggested that palatability in tadpoles may correlate inversely with vulnerability. More precisely, it now appears that the conspicuous aggregating tadpoles of *Bufo* are also unpalatable. Heusser (1971*b*) has substantiated this by showing that four species of European newts (*Triturus alpestris, T. helveticus, T. vulgaris,* and *T. cristatus*) eat the larvae of *Rana esculenta, Rana temporaria,* and *Bombina variegata* (Discoglossidae), but only *T. cristatus* also eats *Bufo bufo* and *B. calamita* larvae.

Because there are many reports of *Bufo* tadpoles being attacked by natural predators, such as *T. cristatus,* to say that *Bufo* larvae are unpalatable is a generalization for which there are undoubtedly exceptions, depending on the toxicity of the various *Bufo* species and the tolerance of the potential predators. Invertebrate predators, specifically those that suck the body fluids from their victims rather than masticate them, and vertebrate predators that swallow indiscriminately, are unlikely to be deterred by a superficial unpalatability.

A definite mechanism for this unpalatability is not known, although the subjective opinions of the human tasters suggest that it lies in a skin secretion. Pfeiffer (1966) has described the skin of *Bufo* larvae. Large oval or pear-shaped glandular cells are present, which may be considered forerunners of the adult poison glands. Pfeiffer suggests that these glands are the source of the alarm substance in the fright reaction for he could not find them in the other tadpoles that he studied. The glands could equally well serve as the source for the transmitter of tadpole unpalatability.

In contrast to *Bufo, Xenopus* larvae are palatable, at least to *Xenopus* adults (Savage, 1963) and the author. Although aggregating *Xenopus* are more likely to be seen by predators than solitary individuals, their relatively translucent nature and immotile stance probably afford them some protection that *Bufo* does not have.

Hierarchial behavior is a common component of social behavior among vertebrates, and in many "societies," a single individual is the leader or superior to the other individuals. On theoretical grounds, however, we would not expect to find hierarchial behavior among schooling organisms. Aggression should be low, and there should be no appreciable "peck order" in a school (Breder, 1954, 1959). If individuals are to work as a school, they should be relatively equipotent. This has not been tested. Savage (1961), Heusser (1970), Bragg (1965), and others have reported intraspecific aggressive behavior leading to cannibalism in species that are known to aggregate. Such behavior, however, appears in food-limited situations. Heusser (1971*a*) suggests that interspecific and intraspecific predation among anuran larvae is a common density-regulating mechanism. Dimorphic cannibalistic larvae of spadefoot toads have received some attention (e.g., Orton, 1954) but are still not well understood, either ecologically or taxonomically.

The *Xenopus* Mode

Certain microhylids and hylids, especially in the subfamily Phyllomedusinae, are strikingly similar to *Xenopus.* While all microhylids lack the keratinized mouth parts and have a single medial (symmetrical) spiracle, such species as *Glyphoglossus molossus* and *Calluella guttalata* resemble *Xenopus* in pigment pattern and body and tail morphology as well. Their posture while swimming is the same as *Xenopus'*. Other microhylids, such as certain species of *Microhyla* and *Phrynomerus,* share some of the distinctive characteristics of *Xenopus* tadpoles.

The Phyllomedusine tadpole, although Orton's Type IV (Acosmanura of Starrett, this volume), has a medial or nearly medial spiracle, a dorso-ventrally flattened head, lateral eyes, a filamentous tail, and reduced pigmentation. It can maintain a stationary midwater position by rapidly vibrating its tail filament. Unlike *Xenopus,* it swims with its tail tipped downward and its head tipped upward about 45 degrees above the horizontal. Otherwise, the resemblance to *Xenopus* is extensive. At least one species, *Agalychnis dachnicolor,* has the independently photosensitive melanophore patch on the tail (Bagnara, 1972). Certain species of *Hyla* (e.g., *H. misera* and *H. minuta,* Kenny, 1969*c*) have the tail filament, but all members of this genus retain the strongly sinistral spiracle position.

Microhyla has enlarged gill filters (personal observation), and *Phyllomedusa trinitatis* can effectively extract particles as small as 2 μ from the water (Kenny, 1969*a*), so it appears that these tadpoles are not only similar in gross morphology but may be convergent with *Xenopus* in their suspension-feeding ability. A final and significant point is that tadpoles of the above-mentioned genera, which morphologically approximate *Xenopus,* also form dense polarized aggregates suspended in midwater. In all these forms the lungs develop early and serve a hydrostatic function. These tadpoles tend to live in calm habitats; in lotic conditions, it would be difficult to maintain stability and group adhesion. One can view the midwater suspension-feeding, gregarious tadpoles, exemplified by *Xenopus,* as representatives of a distinctive adaptive mode for anuran larvae.

The *Bufo* Mode

There are species in other genera that form *Bufo*-like aggregates. *Rana chalconota* in Asia forms mobile aggregates that are considered quite conspicuous to large predators (Liem, personal communication 1970). Tadpoles of this species have poisonous skin glands (Liem, 1959) and are never eaten by the carnivorous *Fluta alba* and *Tilapia mossambica,* even though these fish would readily eat other local ranids (Liem, 1961). Annandale (quoted in Boulenger, 1920) commented that *Rana alticola* larvae were "gregarious" and "very conspicuous in lateral view." *R. alticola* tadpoles are unusual in having both parotoid glands on their bodies and distinct ocelli on their tails, and "from above they appear entirely dark, of a deep olivaceous shade. . . . The parotoid glands pour out a profuse milky secretion when the tadpole is irritated." *Aubria subsigillata* is an African ranid with characteristically velvet black tadpoles. Schiøtz (1963) described a ball of several hundred to several thousand *A. subsigillata* tadpoles rolling slowly over the bottom of a shallow pool. I have collected the conspicuously melanistic larvae of *Leptodactylus melanonotus,* which form dense aggregates in roadside pools in Tehuantepec, Mexico. Dowling's (1960) initial reaction on sighting tadpoles of the hylid *Hyla geographica* (not *H. maxima,* as originally published) in Trinidad was that they belonged to *Bufo marinus. The tadpoles* were "jet-black" and "acted like a school of fishes." Kenny (1969*c*) described the larvae of *Hyla geographica* from Trinidad as being uniformly black and forming dense schools that "can be seen as a compact black mass floating at the surface." Tadpoles of *Philautus gherrapunjiae,* an Indian rhacophorid, have been described by Roonwal and Kripalani (1961) as "blackish all over dorsum" and "swimming in compact, irregular-shaped schools . . . seen as dark patches on the water." The dark bodies of *Rana heckscheri* give them a *Bufo*-like appearance. Wright (1924) remarked, "They travel in big schools as no other big tadpoles do. They remind me of a school of mature *Bufo* tadpoles."

Although it is not known whether any of these last few genera are unpalatable, they nevertheless represent examples of convergences to the *Bufo* behavior and morphology. The possible reasons for such convergences are many; among the most intriguing, however, is the possibility that various mimicry complexes exist. Such complexes need not be limited to anuran larvae. Smallmouthed-bass fry in the Middle West of the United States are very black until they grow larger than

24 mm. During this period they closely associate with the *Bufo* tadpoles that they resemble (R. Weldon Larimore, personal communication 1970). Speculations of this kind await comparative palatability tests using natural predators in the field.

Many aspects of the *Bufo* pattern—conspicuousness, unpalability, and gregariousness—suggest that kin selection may be operative. The degree of genetic similarity among proximal individuals could be determined in *Bufo* because the eggs adhere in strands. Clutches could be individually stained with vital dyes to determine the amount of sibling relationship in schools that form once the eggs hatch.

Various authors (Akin, 1966; Licht, 1967; Pourbagher, 1967; Richards, 1958, 1962; Rose, 1960; Rose and Rose, 1961, 1965; and earlier workers cited therein) have shown that crowding inhibits the growth of tadpoles—a phenomenon that is difficult to understand, considering the density of certain schools. Most authors believe in some species specificity for this inhibition. Licht (1967) has argued that growth inhibition is nonspecific and that the inhibitor, which has been shown to be water-carried and, in part, to involve an "alga-like" unicellular intermediate, could be transmitted from one species to another. Simple explanations to account for inhibition, such as competition for food (Savage, 1961), can now be discredited in light of laboratory experiments by the above authors.

Bufo larvae form compact schools with high density yet somehow manage to grow. A parsimonious explanation, supported by the work of Akin (1966) and Pourbagher (1967) is that different species vary in their sensitivity to, and production of, the inhibitory factor. Indeed, Pourbagher states that *Bufo bufo* larvae show less growth retardation when crowded than *R. esculenta, R. dalmatina,* or *R. temporaria.* Pourbagher, however, did not deal with the question of specificity or interspecific transmission of inhibition. To get at this question Akin grew tadpoles in water previously conditioned by crowding other tadpoles in it for at least 24 hours. The weight gained by an assay group of tadpoles was compared with a control group that was allowed to grow in unconditioned water. In one experiment in which *Rana pipiens* tadpoles were grown in water conditioned by *R. pipiens* tadpoles, the assay group showed 65 per cent less growth than the control group. When the experiment was repeated, with *Bufo valliceps* as the species used to condition the water, the *R. pipiens* assay tadpoles showed only a 10 per cent decrease in growth. In a battery of similar experiments, Licht (1967, Table 2) studied the inhibiting effect of water conditioned by *R. pipiens* on the growth of twelve species and one hybrid. Only the weight gained was considered in these experiments. Licht emphasized the commonness of the inhabitory phenomenon, and although he did not compare the intensity of the inhibition between species, his data are applicable for this purpose. For comparing inhibition I have used an "index of inhibition" equal to the final weight of the control minus the final weight of the assay, divided by the final weight of the control minus the initial weight of the control (see Figure 8–5). (The control and the assay groups always started with the same weight and number of tadpoles.) The salient properties of such an index (here called I) are as follows: $I = 0$ if there is no inhibition and both groups grow equally well; $I = 1$ if there is total inhibition and the assay group does not grow at all; $I > 1$ if the assay group loses weight; $I < 0$ if the assay group actually gains more weight than the control. A reanalysis of the data in Licht's Table 2 is presented graphically in Figure 8–5. The one assay species that showed weight loss ($I > 1$) was *R. pipiens* in the *R. pipiens*–conditioned water, suggesting some species specificity of the inhibitory agent, which is counter to Licht's own conclusion from the same data. Most surprising is that of the six species that showed the least inhibition (for convenience, the *B. woodhousei–B. valliceps* hybrid is here considered a species), five belonged to the

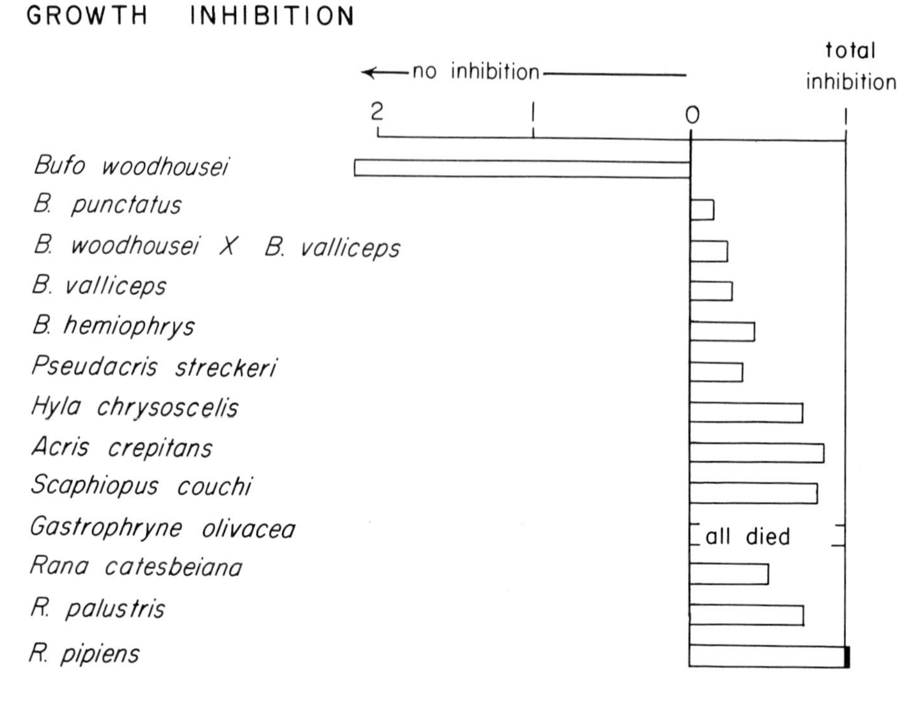

$$\text{Index of Inhibition} = \frac{\text{control}_{\text{finish}} - \text{assay}_{\text{finish}}}{\text{control}_{\text{finish}} - \text{control}_{\text{start}}}$$

Figure 8–5. Index of Growth Inhibition. Data originally from Licht (1967).

genus *Bufo*. In fact, *Bufo woodhousei*, an enigma to Licht, has a negative index value and actually gained more weight in the assay than in the control group.

Licht could not relate phylogenetic distance to level of inhibition. However, it appears that larvae likely to form biosocial aggregates show the least susceptibility to inhibition, in this case, the *Bufo* tadpoles. Inasmuch as larval aggregation—and, therefore, resistance to growth inhibition—is characteristic of *Bufo*, the intensity of inhibition may indeed relate to phylogeny.

Some novel views have recently appeared on the problem of specificity in growth inhibition. Berger (1968) claimed that the sensitivity to inhibition was related to genetic similarity. He found that pure lines of *Rana lessonae*, when grown with *R. lessonae–R. ridibunda* hybrids, were more inhibited in their growth than were the hybrids. Although Berger's work is suspect, especially because of his reliance on unnatural hybrids, Shvarts and Pyastolova (1970*a*, 1970*b*) have extended his ideas. They go so far as to claim that high specificity in growth inhibition is, itself, an adaptive mechanism "ensuring . . . the increase of genetic diversity of the population." Their ideas are based on numerous experiments using *Rana arvalis* and are centered around their discovery that siblings are more likely to exhibit inhibition than are individuals mixed from different broods. They demonstrate that crowding of late-stage tadpoles stimulates some to metamorphose, but the tadpoles do so at a smaller size and with less energy reserves than normal. One regulating system is thought to control both the inhibiting and accelerating processes. Shvarts and Pyastolova (1970*a*, 1970*b*) produce the following model: for an individual brood under conditions of increasing density, such as the drying of the pool, intragroup competition is limited by growth inhibition. Thus, certain individuals stop growing, which speeds up the development of other, less closely related individuals. As the accelerated individuals emerge from the pool (it is argued, although hardly proved, that there is no increased mortality on individuals that metamorphose at exceptionally small size), the density drops, which permits remaining larvae to continue growth. While not all may metamorphose, those that do will have emerged over a broad period of time and should be the individuals genetically most dissimilar. The specificity in growth inhibition is thus seen as the mechanism sustaining genetic diversity.

Even if such a system could in any way maintain genetic diversity, it is not clear how it could increase it. Furthermore, the great propensity toward cannibalism among anuran larvae (Heusser, 1971*a*) of unequal size was ignored by Shvarts and Pyastolova. A larva sacrificed as food can scarcely contribute to the genetic variability of a population.

Shvarts and Pyastolova readily admit that their scheme implies intergroup selection. The whole problem, however, can be circumvented if one does not assume that specificity in growth inhibition is necessarily a positive adaptation. Growth inhibition may be an event, like starvation, from which organisms incur no benefits but evolve mechanisms to avoid. Two such possible mechanisms are a resistance to inhibition, as suggested above for *Bufo,* and an ability to metamorphose in a reduced period of time and survive as a frog of diminutive size.

Shvarts' and Pyastolova's work does raise some interesting questions about the relationship of density to total emergence time and genetic variability in natural tadpole populations. They, however, join other ecologists who have failed to appreciate the possible range of behavioral patterns among anuran larvae. For example, the significance of Brockelman's (1969) study on density effects and predation in *Bufo americanus* must be reassessed with the realization that tadpoles of this species are exceptionally tolerant of high density and unpalatable to certain potential predators.

OTHER MODES OF SCHOOLING

Arguments have been presented for recognition of two distinctive patterns for schooling tadpoles, the midwater schoolers (the *Xenopus* mode) and the bottom aggregators (the *Bufo* mode). Many tadpoles that school do not fit either of these classes, although some share characteristics with both the *Xenopus* and *Bufo* modes.

Tadpoles of the genus *Scaphiopus* form aggregates that resemble those described for *Bufo* (Richmond, 1947; Bragg, 1968). Characteristically, the individual tadpoles are closely packed, and the level of activity within an aggregate may be very high. *Scaphiopus* larvae are voracious and readily cannibalistic (Bragg, 1965). At the same time, Richmond (1947) has found that S. *holbrooki* are as efficient at suspension-feeding as the microhylid *Gastrophryne carolinensis* and much more efficient than *Rana sphenocephala (R. pipiens).* There is no evidence that *Scaphiopus* tadpoles are unpalatable; however, one observation by Richmond suggests that *Scaphiopus holbrooki* may exhibit a short-term fright reaction. Black (1970) suggests that when S. *bombifrons* school, they are somehow protected from cannibalistic tadpoles and beetle larvae. Morphologically *Scaphiopus* tadpoles have strong muscular tails, heavy beaks, and squarish bodies; otherwise, they are typical Orton's (1953, 1957) Type IV tadpoles (Acosmanura of Starrett, this volume).

Larvae of *Rhinophrynus* are morphologically similar to *Xenopus,* but their schooling behavior is radically different (see Stuart, 1961, and earlier discussion in this report). Like *Scaphiopus,* they form active swarming pods. I have watched individuals in such aggregates being actively attacked and devoured by grackles, egrets, and chickens. Subjectively, I found them no less palatable than *Rana pipiens.* Like *Xenopus, Rhinophrynus* tadpoles are highly efficient suspension-feeders (Wassersug, 1972). Unlike *Xenopus,* they may also prey on active, large food items, including conspecifics (Starrett, 1960).

Several of the characteristics that distinguish *Rhinophrynus* and *Scaphiopus* larvae from either the *Xenopus* or *Bufo* pattern are shared by the two genera. Although *Scaphiopus* and *Rhinophrynus* are dissimilar in external morphology, both school and are highly motile, strong swimmers. They are tolerant of extensive physical contact, conspicuous, and may be palatable to some potential predators. Both are exceptionally omnivorous in terms of food-particle size and quality. I believe that these features correspond to similarities in the ecology of these genera. Adult *Rhinophrynus* and *Scaphiopus* are fossorial. They are adventitious breeders that mate in temporary rain pools (James, 1966; Bragg, 1964). These pools can dry quickly, and it is important that the larvae exploit a rapidly depleting environment. Similar aggregations occur in the larvae of other fossorial frogs that inhabit regions with low or unpredictable rainfall, such as *Pyxicephalus adspersus* in southern Africa (Wager, 1965), *Pelobates fuscus* in southeastern Europe (Bragg, 1961) and many Australian anurans (Ian Straughan, personal communication 1970). *Scaphiopus* tadpoles may metamorphose within two weeks of hatching (Bragg, 1966). I have found metamorphosing *Rhinophrynus* larvae within eighteen days of the first rain of the season, and before that their pools had been dry. These are remarkably fast developmental rates. The sociality of these tadpoles may, in part, reflect past selection for high density tolerance. Bragg (1961, 1965) has detailed these arguments for *Scaphiopus,* and his line of reasoning may apply equally well to *Rhinophrynus* and other xeric-adapted anurans with schooling larvae. Apparently the specializations for ephemeral larval existence have been converged upon by lineages as separate as the pelobatids, the primitive pipoids, and possibly more advanced groups.

Although we could now designate a mode for larval behavior modeled on *Scaphiopus,* the best-known fossorial anuran, I believe such action premature. We do not know why these tadpoles school. Surprisingly little is understood about larval ecology in this genus and less in *Rhinophrynus* or the other genera that could share this pattern. There is a definite need for comparative studies of the social behavior in *Scaphiopus,* and at present, the high inhibitory index for *S. couchi* (see Figure 8–5) is an enigma, Shvarts and Pyastolova (1970*a*, 1970*b*) notwithstanding. The one trait that makes these tadpoles so intriguing—their ephemeral existence—also makes them particularly inconvenient for field or laboratory investigation.

CONCLUSION

Social behavior is used in this paper as a means of recognizing discrete ways of life, or adaptive patterns, of anuran larvae. Significant ecological and behavioral differences distinguish tadpoles. Awareness of these differences can provide clues to interpretation of the few morphological characters used in larval taxonomy. Often these are treated in isolation, devoid of ecological or behavioral qualifications. Interpretations of spiracle position illustrate this point. A medial position of the spiracle in the microhylid tadpoles is readily comprehensible as one expression of an adaptation to a stable midwater, filter-feeding mode of life, where symmetry is essential for balance. Without knowledge of the ecological and behavioral correlates, it would be impossible to interpret the medial spiracle position in phyllomedusine hylids as a likely convergence with that of the beakless tadpoles of other families. In fact, one might be led to question the evolutionary and taxonomic significance of spiracle structure, as Griffiths and Carvalho (1965) have done.

A major advantage of an adaptive-pattern approach is the uniting of seemingly unrelated aspects of larval and adult biology. This can be especially helpful in understanding the evolution of anurans, organisms in which pronounced caenogenesis obscures correlations between larval and adult life histories and morphology. Frequently, unusual tadpoles are found in species with generalized adults. Similarly, the most typical tadpoles may become the most bizarre frogs. It is at this point that a search for adaptive patterns becomes most appropriate. Biosocial aggregation, the specific phenomenon emphasized here, may be a key to the elucidation of such a unifying pattern.

Tadpole behavior is constrained ultimately by the reproductive patterns of adult anurans. Within this broad framework, there are several considerations relating to modes of selection. For example, homing ability of adult toads serves to reduce gene flow and could maximize genetic similarity at any breeding site. Displaced *Bufo marinus* males will return to a specific pool over distances of a kilometer (unpublished data), even if other breeding sites are closer. *B. bufo* adults show lifelong fidelity to the same pond and can return to that pond if displaced up to 5,000 m (Heusser, 1969*a*, 1969*b*). Such adult behavior is consistent with the model of kin selection that I have proposed for the evolution of the aggregating, conspicuous, and unpalatable tadpoles.

Selection for increased clutch size might also be implicated. With increased clutch size there would be an increase in hatchling density and concomitant crowding of the larvae—conditions conducive to selection for biosocial aggregation. Coincidentally, large clutch sizes, as well as the deposition of the clutch as a cohesive unit, increase the chance that individuals in an aggregate will be siblings, and this is consistent with a kin selection model.

Acknowledgements

I am most grateful to Dr. David Wake, under whose guidance I first began to study tadpoles. He has been a constant source of encouragement throughout the present study.

The following researchers have graciously read parts or all of manuscript drafts, and I thank them for their critical discussions: David Wake, Robert Inger, George Rabb, Ronald Heyer, Marvalee Wake, R. Eric Lombard, and C. H. F. Rowell. It was not possible to satisfy all these critics, and they should not be held responsible for any flaws in my arguments. Many other people, too many to list, have made valuable contributions to the development of my ideas on tadpole ecology and behavior. My thanks to all.

Dr. Richard Zweifel and Dr. Herndon Dowling graciously re-examined Margaret Dowling's Trinidad *Hyla* collection at the American Museum for me. I am grateful to Dr. O. P. Pearson, Director of the Museum of Vertebrate Zoology, for making available all of the museum's facilities. Financial support for field and laboratory work was provided by the Hinds Fund, University of Chicago. Assistance in illustration was kindly given by Gene M. Christman and R. Eric Lombard.

REFERENCES

AKIN, G. W. 1966. Self inhibition of growth in *Rana pipiens* tadpoles. Physiol. Zool. 39:341-356.

ALLEE, W. C. 1931. Animal aggregations. Univ. Chicago Press, Chicago. 431 p.

ASHBY, K. R. 1969. The population ecology of a self-maintaining colony of the common frog *(Rana temporaria)*. J. Zool. London 158:453-474.

BAGNARA, J. T. 1957. Hypophysectomy and the tail darkening reaction in *Xenopus*. Proc. Soc. Exptl. Biol. Med. 94:572-575.

_______. 1960. Tail melanophores of *Xenopus* in normal development and regeneration. Biol. Bull. 118:1-8.

_______. 1972. Color change. *In* B. Lofts, [ed.], Physiology of the Amphibia, vol. 2. Academic Press, New York. In press.

BEISWENGER, R. E., AND F. H. TEST. 1966. Effects of environmental temperature on movements of tadpoles of the American toad, *Bufo terrestris americanus*. Papers Michigan Acad. Sci. 51:127-141.

BERGER, L. 1968. The effect of inhibitory agents in the development of green-frog tadpoles. Zool. Poloniae 18(3):381-390.

BIRUKOW, G. 1949. Die Entwicklung des Tages– und Dammerungssehens im Ange des Grasfrosches (*Rana temporaria* L.). Z. Vergl. Physiol. 31:322-347.

_______. 1950. Vergleichende untersuchungen über das Helligkeits– und Farbensehen bei Amphibien. Z. Vergl. Physiol. 32:348-382.

BLACK, J. H. 1969. Ethoecology of tadpoles of *Bufo americanus charlesmithi* in a temporary pool in Oklahoma. J. Herp. 3:198.

_______. 1970. A possible stimulus for the formation of some aggregations in tadpoles of *Scaphiopus bombifrons*. Proc. Oklahoma Acad. Sci. 49:13-14.

BOICE, R., AND D. W. WITTER. 1969. Hierarchical feeding behavior in the Leopard frog *(Rana pipiens)*. Animal Behaviour 17:474-479.

BOULENGER, G. A. 1920. A monograph of the South Asia, Papuan, Melanesian and Australian frogs of the genus *Rana*. Records Indian Mus. 20:1-266.

BRAGG, A. N. 1948. Additional instances of social aggregation in tadpoles. The Wasmann Collector 7:65-79.

_______. 1954. Aggregational behavior and feeding reactions in tadpoles of the Savannah spadefoot. Herpetologica 10:97-102.

_______. 1960. Aggregational and associated behavior in tadpoles of the Plains Spadefoot. Wasmann J. Biol. 18:273-289.

_______. 1961. A theory of the origin of spade-footed toads deduced principally by a study of their habits. Animal Behaviour 9:178-186.

_______. 1964. Further study of predation and cannibalism in spadefoot tadpoles. Herpetologica 20:12-24.

_______. 1965. Gnomes of the night: The spadefoot toads. Univ. Pennsylvania Press, Philadelphia. 127 p.

_______. 1966. Longevity of the tadpole stage in the plains spadefoot (Amphibia: Salientia). Wasmann J. Biol. 24:71-73.

_______. 1968. The formation of feeding schools in tadpoles of spadefoots. Wasmann J. Biol. 26:11-16.

BRATTSTROM, B. H. 1962. Thermal control of aggregation behavior in tadpoles. Herpetologica 18:38-46.

BRATTSTROM, B. H., AND J. W. WARREN. 1955. Observations on the ecology and behavior of the pacific treefrog, *Hyla regilla*. Copeia 3:181-191.

BREDER, C. M., JR. 1954. Equations descriptive of fish schools and other animal aggregations. Ecology 35:361-370.

_______. 1959. Studies on social groupings in fishes. Bull. Am. Mus. Nat. Hist. 117(6): 397-481.

_______. 1967. On the survival value of fish schools. Zoologica 52:25-40.

BREDER, C. M., JR., AND F. HALPERN. 1946. Innate and acquired behavior affecting aggregation of fishes. Physiol. Zool. 19:154-190.

Brockelman, W. Y. 1969. An analysis of density effects and predation in *Bufo americanus* tadpoles. Ecology 50:632-644.

Burgers, A. C. J. 1952. Optomotor reactions of *Xenopus laevis.* I. Physiol. Comparata et Oecol. 2(4):263-281.

Carpenter, C. C. 1953. Aggregation behavior of tadpoles of *Rana p. pretiosa.* Herpetologica 9:77-78.

Clausen, R. G. 1931. Orientation in fresh water fishes. Ecology 12:541-546.

Cole, W. H., and C. F. Dean. 1917. The photokinetic reactions of frog tadpoles. J. Exptl. Zool. 23:361-370.

Costa, H. H. 1967. Avoidance of anoxic water by tadpoles of *Rana temporaria.* Hydrobiologia 30:374-384.

Cronly-Dillan, J. R., and W. R. A. Muntz. 1965. The spectral sensitivity of the gold fish and the clawed toad tadpole under photopic conditions. J. Exptl. Biol. 42:481-493.

Deanesly, R., J. Emmett, and A. S. Parkes. 1945. The preparation and biological effects of iodinated proteins. 8. Use of *Xenopus* tadpoles for the assay of thyroidal activity. J. Endocrinol. 4:324-355.

deVlaming, V. L., and R. B. Bury. 1970. Thermal selection in tadpoles of the tailed-frog, *Ascaphus truei.* J. Herp. 4(3-4):179-189.

Dickman, N. 1968. The effect of grazing by tadpoles on the structure of a periphyton community. Ecology 49:1188-1190.

Dowling, M. 1960. Interlude at Simla. Animal Kingdom 63:137-139.

Fisher, R. A. 1929. The genetical theory of natural selection. Clarendon Press, Oxford. Reprinted, 1958. Dover, New York. 291 p.

Gosner, K. L. 1960. A simplified table for staging anuran embryos and larvae with notes on identification. Herpetologica 16:183-190.

Griffiths, I., and A. L. de Carvalho. 1965. On the validity of employing larval characters as major phyletic indices in amphibia, Salientia. Rev. Brasileira Biol. 25:113-121.

Harden Jones, F. R. 1963. The reactions of fish to moving backgrounds. J. Exptl. Biol. 40:437-446.

Harder, W. von. 1963. Ueber das Verhalten verschiedener fischarten in Streifeuzylinder und gegenueber rotierenden Figuren. Ichthyologica 11(1-2):37-43.

Heusser, H. 1969*a*. The ecology and life history of the European common toad *Bufo bufo* (L.) (Abstr. of a five-year study.) Druckerei der Zentralstelle der Studentenschaft, Zurich 1-8.

———. 1969*b*. Die Lebensweise der Erdkrote, *Bufo bufo* (L.); Das Orientierungsproblem. Rev. Suisse Zool. 76:443-518.

———. 1970. Laich-Fressen durch Kaulquappen als mögliche Ursache spezifíscher Biotoppraferenzen und kurzer Laichzeiten bei europäischen Froschlurchen (Amphibia, Anura). Oecologia (Berlin) 4:83-88.

———. 1971*a*. Laich-Räubern und -Kannibalismus bei sympatrischen Anuran-Kaulquappen. Experientia 27(4):474-475.

———. 1971*b*. Differenzierendes Kaulquappen-Fressen durch Molche. Experientia 27(4):475-476.

Hora, S. L. 1930. Ecology, bionomics and evolution of the torrential fauna, with special reference to the organs of attachment. Phil. Trans. Roy. Soc. London, B, 218:171-288.

Inger, R. F. 1966. The systematics and zoogeography of the amphibia of Borneo. Fieldiana: Zool. 52:1-402.

James, P. 1966. The Mexican burrowing toad, *Rhinophrynus dorsalis,* an addition to the vertebrate fauna of the United States. Texas J. Sci. 18:272-276.

Jennsen, T. A. 1967. Food habits of the green frog, *Rana clamitans,* before and during metamorphosis. Copeia 1:214-218.

Karlstrom, E. L. 1962. The toad genus *Bufo* in the Sierra Nevada of California: ecological and systematic relationships. Univ. California Publ. Zool. 62:1-104.

Keenleyside, M. H. A. 1955. Some aspects of the schooling behavior of fish. Behaviour 8:183-248.

Kenny, J. S. 1969*a*. Feeding mechanisms in anuran larvae. J. Zool. London 157:225-246.

———. 1969*b*. Pharyngeal mucous secreting epithelia of anuran larvae. Acta Zool. 50:143-153.

———. 1969*c*. Amphibia of Trinidad. *In* P. Wagenaar Hummelinck, [ed.], Studies on the fauna of Curaçao and other Caribbean islands, vol. 29. Publication 54 of the Foundation for Scientific Research in Surinam and the Netherlands Antilles. Martinus Nijnoff, The Hague. 78 p.

Licht, L. E. 1967. Growth inhibition in crowded tadpoles: intraspecific and interspecific effects. Ecology 48:737-745.

Liem, K. F. 1959. The breeding habits and development of *Rana chalconota* (Schleg.) (Amphibia). Treubia 25:89-111.

———. 1961. On the taxonomic status and the granular patches of the Javanese frog *Rana chalconota* (Schlegel). Herpetologica 17:69-71.

Liu, C. C. 1950. Amphibians of western China. Fieldiana, Zool. Mem. 2:1-400.

Lyon, E. P. 1904. On rheotropism. I.—Rheotropism in fishes. Am. J. Physiol. 12(2):149-161.

Morrow, J. E., Jr. 1948. Schooling behavior in fishes. Quart. Rev. Biol. 23(1):27-38.

Mullally, D. P. 1953. Observations on the ecology of the toad *Bufo canorus*. Copeia 3:182-183.

Myers, G. S. 1931. *Ascaphus truei* in Humboldt County, Calif., with a note on the habits of the tadpole. Copeia 2:56-57.

Noble, G. K. 1927. The value of life history data in the study of the evolution of the Amphibia. Ann. New York Acad. Sci. 30:31-128.

———. 1931. The biology of the Amphibia. McGraw-Hill Book Co., New York. 577 p.

Orton, G. L. 1953. The systematics of vertebrate larvae. Systematic Zool. 2(2):63-75.

———. 1954. Dimorphism in larval mouthparts in spadefoot toads of the *Scaphiopus hammondi* group. Copeia 2:97-100.

———. 1957. The bearing of larval evolution on some problems in frog classification. Systematic Zool. 6(2):79-86.

Parr, A. E. 1927. A contribution to the theoretical analysis of the schooling behaviour of fish. Occ. Pap., Bingham Oceanogr. Coll. 1:1-32.

Pfeiffer, W. 1966. Die Verbreitung der schreckreaktion bei kaulquappen und die herkunft des schrechstoffes. Z. Vergl. Physiol. 52:79-98.

Pourbagher, N. 1967. Sur l'effet de groupe chez les têtards de divers Amphibiens. Compt. Rend. Acad. Sci. Paris, D, 265:1244-1247.

Richards, C. M. 1958. The inhibition of growth in crowded *Rana pipiens* tadpoles. Physiol. Zool. 31:138-151.

———. 1962. The control of tadpole growth by alga-like cells. Physiol. Zool. 35:285-296.

Richmond, N. D. 1947. Life history of *Scaphiopus holbrookii holbrookii* (Harlan). Part I: Larval development and behavior. Ecology 28:53-67.

Risser, J. 1914. Olfactory reactions in amphibians. J. Exptl. Zool. 16:617-652.

Roonwal, M. L., and M. B. Kripalani. 1961. New frog *Philautus gherrapunjiae* from Assam, India, with field observation on its behaviour and metamorphosis. Records Indian Mus. 59:325-333.

Rose, S. M. 1960. A feedback mechanism of growth control in tadpoles. Ecology 41:188-199.

Rose, S. M., and F. C. Rose. 1961. Growth-controlling exudates of tadpoles. *In* F. L. Milthorpe, [ed.], Mechanisms in biological competition. Soc. Exptl. Biol. Symposium 15:207-218.

———. 1965. The control of growth and reproduction in freshwater organisms by specific products. Mitl. Internat. Verein. Limnol. 13:21-35.

Savage, R. M. 1961. The ecology and life history of the common frog *(Rana temporaria temporaria)*. Pitman & Sons, Ltd., London. 221 p.

———. 1963. A speculation on the pallid tadpoles of *Xenopus laevis*. Brit. J. Herp. 3:74-76.

Scheminsky, F. 1924. Versuche über Elektrotaxis und Elektronarkose. Arch. ges. Physiol. 202:200-216.

SCHIØTZ, A. 1963. The amphibians of Nigeria. Videnok, Medd. fra Dansk naturls. Foren. 125:1-92.
SCHMALHAUSEN, I. I. 1964. Proishoženie nazemarjh Pozvonočynh. Izdatel'stvo Nauka, Moscow. Trans. by Leon Kelso, 1968, as The origin of terrestrial vertebrates. Academic Press, New York. 314 p.
SCHNEIRLA, T. C. 1965. Aspects of stimulation and organization in approach/withdrawal processes underlying vertebrate behavioral development, p. 1-71. *In* D. S. Lehrman, R. A. Hinde, and E. Shaw, [eds.], Advances in the study of behavior. Academic Press, New York.
SHAW, E. 1965. The optomotor response and the schooling of fish. Intern. Comm. N. W. Atlantic Fisheries Special Publication 6.
———. 1970. Schooling in fishes: critique and review. *In* L. R. Aronson, Ethel Tobach, D. S. Lehrman, and J. S. Rosenblatt, [eds.], Development and evolution of behavior, essays in memory of T. C. Schneirla. W. H. Freeman & Co., San Francisco. 656 p.
SHAW, E., AND B. D. SACHS. 1967. Development of the optomotor response in the schooling fish, *Menidia menidia*. J. Comp. Physiol. Psychol. 63(3):385-388.
SHAW, E., AND A. TUCKER. 1965. The optomotor reaction of schooling carangid fishes. Animal Behaviour 13(2-3):330-336.
SHVARTS, S. S., AND O. A. PYASTOLOVA. 1970*a*. Regulators of growth and development of amphibian larvae. I. Specificity of effects. [translated from Russian] Ecology 1:58-62.
———. 1970*b*. Regulators of growth and development of amphibian larvae. II. Diversity of effects. [translated from Russian] Ecology 2:122-134.
SMITH, M. A. 1917. On tadpoles from Siam. J. Nat. Hist. Soc. Siam 2(4):262-274.
SPENCER, G. M. 1931. Some observations on schooling in fish. J. Marine Biol. Assoc. United Kingdom 17:421-448.
STARRETT, P. 1960. Descriptions of tadpoles of middle american frogs. Misc. Publ. Mus. Zool. Univ. Michigan 110:1-37.
STEBBINS, R. C. 1951. Amphibians of western North America. Univ. California Press, Berkeley and Los Angeles. 539 p.
STUART, L. C. 1961. Some observations on the natural history of tadpoles of *Rhinophrynus dorsalis* Dumeril and Bibron. Herpetologica 17:73-79.
TAYLOR, E. H. 1942. Tadpoles of Mexican anura. Univ. Kansas Sci. Bull. 28:37-55.
TEVIS, L., JR. 1966. Unsuccessful breeding by desert toads *(Bufo punctatus)* at the limit of their ecological tolerance. Ecology 47:766-775.
TOBACH, E., AND T. C. SCHNEIRLA. 1968. The biopsychology of social behavior in animals, p. 68-82. *In* R. E. Cooke, [ed.], The biologic basis of pediatric practice. McGraw-Hill, New York.
VORIS, H. K., AND J. P. BACON. 1966. Differential predation on tadpoles. Copeia 3:594-598.
WAGER, V. A. 1965. The frogs of South Africa. Purnell & Sons, Ltd., Capetown Johannesburg. 242 p.
WASSERSUG, R. 1971. On the comparative palatability of some dry season tadpoles from Costa Rica. Am. Midland Naturalist 86:101-109.
———. 1972. The mechanism of ultraplanktonic entrapment in anuran larvae. J. Morphol. 137:279-288.
WASSERSUG, R., AND C. M. HESSLER. 1971. Tadpole behavior: aggregation in larval *Xenopus laevis*. Animal Behaviour 19:386-389.
WIENS, J. A. 1970. Effects of early experience of substrate pattern selection to *Rana aurora* tadpoles. Copeia 3:543-548.
WILLIAMS, G. C. 1964. Measurement of consociation among fishes and comments on the evolution of schooling. Publ. Mus. Michigan State Univ. 2:351-383.
WRIGHT, A. H. 1924. A new bullfrog *(Rana heckscheri)* from Georgia and Florida. Proc. Biol. Soc. Washington 37:141-152.
ZWEIFEL, R. G. 1955. Ecology, distribution, and systematics of frogs of the *Rana boylei* group. Univ. California Publ. Zool. 54(4):207-292.
———. 1970. Descriptive notes on larvae of toads of the *debilis* group, genus *Bufo*. Am. Mus. Novitates 2407:1-13.

PART II DISCUSSION

Inger

It's customary, on occasions like this for a group to congratulate itself on how well it's been doing and how bright its future is. I think what we can say with reasonable accuracy and objectivity is that we are now beginning to see what we can do. It's clear from the people who have spoken this morning that we are witnessing a multidirectional approach to the evolution of frogs. Now, to speak out against a multidisciplinary attack on the problem is like a politician speaking out against God or some such thing, and I'm certainly not going to do that. But I do think it's important to keep in mind that having a number of individual investigators, each working on his own single approach to the problem, does not constitute a multidisciplinary approach to problem solving.

I'm not a pessimist by nature, so I suspect that we will see more and more people individually attacking a piece of this large problem from various directions, and I want to state clearly that the people who spoke this morning are neither blind nor showing tunnel vision. The restrictions of this kind of a meeting, I think, necessarily impose on them this outward aspect.

Before opening the discussion I'd like to make some comments about the paper presented by Dr. Salthe and Dr. Duellman on various aspects of clutch size–ovum size. A picture has developed here that attempts to unify a lot of data, but it may appear to do so because it presents a lot of bits of information that in effect cancel one another out. I think it can be very dangerous to put together data from tropical and temperate zone communities. The energetic relations are radically different in these two situations, and here we have to be rather careful. There is some work, for example, that indicates that nesting birds in the tropical rain forest feed their young much less frequently than those in the temperate zone. There is also some evidence to indicate that tropical frogs have smaller clutches than their similar-sized relatives in the temperate zone. These are elements that have to be sorted out and put aside from differences in modes of development.

Another element that has to be continuously in mind is the impact of the historical or phylogenetic factor. It apparently makes very little difference where the species of *Bufo* come from in the world. Their mode of reproduction, egg size, and clutch size all seem to fit together regardless of the place of origin of the species. On the other hand, if we look at the genus *Rana* we find a great deal of diversity between major species groups, and this is not just in mode of reproductive behavior, but in such things as egg size, egg coloration, and clutch size.

Let's consider just pond tadpoles. If, for example, we are collecting in Thailand, we're likely to get mostly microhylid tadpoles, and all of them will fit into a particular size range. They will all have pretty much the same developmental relations. On the other hand, if you collect in a pond in another place I'm familiar with—Borneo—you're likely to get a lot of *Polypedates* tadpoles, and they are quite different. They are very large, and their eggs are different in size. In this case there is a phylogenetic influence on the patterns we see regardless of the type of habitat within a general climatic zone. Well, what I should do at this point is to ask Stan Salthe or Bill Duellman if they want to respond to my comments.

Duellman

I would like to add a couple of comments to what Dr. Inger just said. While he was commenting about ponds in Thailand and in Boreno, I was trying to recall some information about some ponds in Ecuador. My impression of this situation, where we would find tadpoles of as many as fifteen or sixteen species in one pond at one time, is that we have a great variety not only in kinds of tadpoles, but also in size and their habitat utilization within the pond; some are pelagic, some are littoral, some are bottom feeders, some are surface feeders. One of the most interesting aspects of tadpoles to me is that we do have these aquatic communities. You are all familiar with the various kinds of ecological studies that have been done on frogs or lizards or snakes, but no one has yet gone to the ponds and studied the ecological community that exists there in the way of anuran larvae.

I think here we could come up with some new understandings and interpretations of the kind of information that Dr. Starrett has presented insofar as the morphological adaptations, and we certainly could gain considerable insight into some of the behavioral adaptations that Richard Wassersug has presented.

Inger

Well, I was making a comment about the validity of some of the generalizations in your paper. Do you want to say anything, Stan [Salthe]?

Salthe

I want to say one thing. Tropical frogs may breed more than once a year, and what we have not systematically taken into account is the number of times a year a female breeds, because these data are not available.

I suspect that when we start plotting this material, it will look pretty good. Maybe we didn't run into forms that are pulling this kind of trick, but I think that we would have seen serious disturbances in our relationships. If we had, my first gambit would be to question the number of times that the female might breed. I do tend to generalize quite a bit, but I have in mind the possibility that there's a lower limit to the amount of genetic variability that can be generated per unit time by a species. I think so far we have not run into a serious disturbance of the correlations to the degree that I am now worried about it. But it is certainly a real possibility.

Rabb

Stan [Salthe] just brought up something on the point of fecundity. There really aren't many data in the litereature, but I'd like to call attention to at least two examples of multiple clutches. The tropical *Pipa parva* carries eggs on its back. At least in captivity, it will breed up to five times a year, including about a thirty-day incubation period of the eggs each time. The same is true of the temperature zone *Discoglossus* that Knoepffler studied. It lays eggs six times a year, with a total output of maybe 6,000 eggs in the course of a year, as opposed to from one hundred to fifteen hundred during each laying.

George Drewry, Puerto Rico Nuclear Center

I have a comment inspired by Dr. Salthe's talk about the *Pseudophryne* species. If we look at this from a system analyst's viewpoint, we might have a practical approach to the problem. This is the question of nutritional input to the species. Those species with the large eggs and short incubation period might actually be deriving much less of the total nutritional requirements from the aquatic system, whereas *Pseudophryne bibronii*, with small eggs and a long incubation period, is utilizing the aquatic habitat for much more of its nutritional input.

Salthe

I think this is an extremely interesting point. In other words, you have two possibilities: In one case, the female uses some of her food supply to produce a new little one; in the second, some of the female's food supply is divided up a little more cheaply and produces more individuals, and then *they* get their own energy from a different environment. These alternatives summarize the entire meaning of amphibian larvae in my view. Distribution becomes very important in larval marine organisms, but I think in amphibians the larval stage is primarily a feeding stage and a clear specialization for that purpose.

Drewry

Do these two cases occur in sympatric forms?

Salthe

I believe they're found in the same general region, but I don't believe they're found breeding at the same time and place.

Drewry

Well, of course, there's always a host of other things that are potential compatagens with both adults and larvae, but here you could have avoidance of two much niche overlap.

Reig

I was very impressed by the paper of Dr. Priscilla Starrett, especially because it went beyond the previous knowledge and typification of the morphology of anuran larvae. One question I have is whether the names she proposed are really an implication of taxonomic names for definite taxa.

Starrett

They could be. I suppose I was hedging, seeing that you have such good reasons for not naming suborders. It is hard to say that these could be considered suborders, but it is certainly better than using just numbers, and I feel they are indicative of evolutionary lines. I do intend to propose them as formal taxonomic categories.

Bogart

After listening to Dr. Salthe's talk, I recalled many personal observations that he may not be aware of. Some of these may tend to muddy the water around his theories, but within the bufonid toads are several instances that require explanations. For instance, *Melanophryniscus* is a wee beast that has tiny, highly pigmented eggs, whereas *Bufo holdridgei* in Costa Rica has very large, yolky eggs, and *Pedostibes* in Borneo has very large, nonpigmented eggs. Also, regarding the part of his discussion about pigmentation, I can think of some species of *Phyllobates* that have pigmented eggs that are hidden from solar radiation. Some species in the *Physalaemus-Edalorhina* complex have unpigmented eggs, and even though they're in foam nests the eggs are exposed to the ultraviolet light of the sun.

As far as his comments on bromeliad species, it is true that many of the bromeliad frogs lay small clutches, but *Corythomantes* in Brazil, which by all accounts is a bromileaceus species, has up to 2,000–4,000 eggs. Obviously, the female must hop from one bromeliad to the other if many of these are going to survive. I wonder if you have any explanation for some of these problems.

Salthe

I may not have stated my comments about pigmentation exactly the way I should have. You will apparently find pigmented eggs, or slightly pigmented eggs, that are hidden; this is quite common in *Hyperolius* and a number of other forms. However, I know of no examples where you find unpigmented eggs entirely exposed to sunlight. My feeling about this is that the pigment, for some reason, is lost when it is no longer needed. By the way, the pigment may serve a number of functions. There is some evidence that in *Rana temporaria* and some very early spring-breeding, northern forms, pigment will actually warm up the clutch by absorbing heat. It's a general proposition that those developing in pond situations have pigmented eggs of the same type (even primitive fishes, such as *Polypterus,* show this). I would like to refine what I said in that it seems pigment will begin to be lost when the site of development is no longer entirely exposed to the sun. Egg clutches deposited on leaves and the like are always in a sort of understory situation; they aren't really "out" in the way a *Rana temporaria* clutch is.

The frog that has a very large clutch and lays small, pigmented eggs but spreads the eggs out among several bromeliads is quite believable. Some of these plants have very large tanks, and a small tadpole could live there for a very long time. This would be a clever way of increasing the clutch dimensions in space and time. I would have to make this a different mode of reproduction than the ones we were talking about before—an entirely new type that we haven't dealt with.

Duellman

I'd like to add something here too, but first, let me ask Dr. Bogart if his number of 4,000 eggs [in *Corythomantes*] is the number of ovarian eggs or the number of eggs in the clutch.

Bogart

Ovarian.

Duellman

Well, of course, these data are not comparable to what we are calling clutch size. There are many instances we know of in which the animal contains a large number of ovarian eggs. A number of these will be ovulated, perhaps not all at once, and then this number in itself may be further subdivided into various clutches. Perhaps the best-documented case of this is in some of the phyllomedusine frogs. I know of at least four [phyllomedusine] species in which, at the time of ovulation, the female retains another clutch of equal size. This is a kind of "clutch." Another, approximately equal, number of ovarian eggs may be ovulated a week or two later and, with the first batch that were ovulated, can be counted as two, three, or more clutches. The number of ovarian eggs is not comparable to what we are calling clutch size. These can be two entirely different things, although they may, in some species, be one and the same.

Roy W. McDiarmid, University of South Florida

Stan [Salthe], I know you're looking for some support, and I think I can offer something.

Within the centrolenids, for example, eggs that are deposited underneath leaves are, in fact, white or nonpigmented, and those that are deposited at the tips of leaves in the sun or at least potentially in the sun are dark and pigmented in all instances that I know of. Heyer has reported this in *Leptodactylus* recently, I think. So this does lend support to what you are suggesting as far as pigmentation is concerned.

I would like to initiate some discussion about another topic. It's apparent from listening to some of the talks here that larval characteristics, whether we call them 1, 2, 3, 4, or by those names that Holly [Starrett] proposed, are in contradiction to the literature. Bob [Inger], you've been involved in the utilization of larval characteristics for phylogenetic considerations at various levels of the over-all evolutionary patterns among anurans. Some of the things that John Lynch has considered are quite different than what Holly [Starrett] was talking about. There must be other people in the audience who can contribute to these ideas, and I would like to hear some discussion in terms of what these mean from the phylogenetic viewpoint.

Blair

I would like to point out something that I mentioned in our initial talk; it is that we have to look at all systems. I think that we have to recognize the fact that in the larval stage, whatever taxon you're talking about, individuals are subjected to entirely different evolutionary pressures than they are in the adult stage. To achieve any kind of phylogeny, we're going to have to look at all the evidence in the larval stage and in the adult stage and try to reconcile any differences. I think the old concept in the literature that the larval stage is the more indicative of ancestry has pretty well been shown to be nonsensical. We have to look at the whole organism throughout its life history and look at it in as many ways as possible.

Rabb

That reinforces Bob Inger's championing the ecologic approach in general, and I'd like to urge that Stan [Salthe] and Bill [Duellman] consider the selective forces that are operating to produce large eggs and terrestrial-type nests. Predation factors have been mentioned in regard to tropical frogs that go the terrestrial route, and dessication factors probably operate in marginal savannah-type habitats. The latter situation was explored in a way by the Goins when they were looking at amphibian eggs in relation to reptilian evolution, in particular, when they were looking at amphibian eggs in montane environments where this aquatic resource is not really avilable. Tadpoles are, by and large, detritus, plankton, and plant feeders, so perhaps the montane environment forces them to go to complete terrestrialism.

Duellman

May I reply to this? One of the most obvious things is that the greater the amount of parental care, the fewer in number are the eggs, and the larger the eggs, the fewer they are in number. For example, in the species of *Gastrotheca* and *Hemiphractus*, in which the females carry the eggs in a pouch in the case of *Gastrotheca* or on the back in the case of *Hemiphractus*, there are relatively few eggs, clutches down to eight in number, and very large eggs; the largest amphibian egg that we have found is that of *Gastrotheca ceratophrys*.

Obviously, these animals must have a very high percentage of ova that survive to become froglets. Now, how can we compare this percentage with that of the normal pond-breeding aquatic egg of *Rana* or *Bufo*? What percentage of a clutch of 2,000 *Rana catesbiana* eggs survive to metamorphose I can't answer, but I'm quite sure that it is far less than it is in these in which we have parental care. On the other hand, in regard to terrestriality, I think you hit the nail on the head, George [Rabb], when you noted that for the most part, these are in montane situations—habitats we can roughly group together as cloud forests, where we have highly equitable habitat conditions for frogs, where there's continual moisture throughout most, if not all, of the year. The absence or scarcity of ponds precludes a large variety of pond-breeding frogs, so they are forced into one of three developmental situations: in streams; in bromeliads; or terrestrial.

Estes

I'd like to fire a double-barreled blast at the two "larval" people up there and the view of trying to correlate the adult and the larva and the utility of various characters. In view of the relatively generalized pattern that the adults of both discoglossids and ascaphids show and the overlap of character states that occur in these two groups as adults, I wonder if the larval morphology of all discoglossids and all ascaphids allows a familial separation. Do you see the same kind of overlap in these stages, a sort of mirror image? Do the Acosmanura indicate that parallel developments may have occurred in the formation of the character-state groupings that you used to form the Acosmanura? Do they show this kind of parallel development? You can fight it out there between you about who is going to answer what.

Starrett

Well, I'm not sure of your question, exactly. Are you asking if the Acosmanura and the characters they have, and the Lemmanura could be derived in parallel? I assume this was a very ancient split in the trend toward feeding specializations of those groups. As far as the similarities in the lemmanurans, there is *Ascaphus,* which is quite specialized, and *Leiopelma,* without any tads, and then the discoglossid tads, which are really quite similar to those of the Acosmanura.

Wassersug

There has been very little work that has been done on the histology of any of these discoglossid or ascaphid tadpoles. *Ascaphus* itself is pretty well known, but I myself am looking for *Leiopelma* material to see whether we can find, for instance, remnants of structures related to filter feeding. This would help to answer your question. Before we can decide how close, how significant, the ascaphids are to the discoglossids, for example, we have to find out how good the ascaphids are as a group.

Drewry

Dr. Duellman made a statement that the greater the amount of parental care, the larger and fewer are the eggs. I would suggest that because the eggs are large and few, the greater is the opportunity for parental care to contribute noticeably to the survival. However, the ranids with large clutch size and so forth are, it seems to me, exhibiting a form of parental care that's subtle. Perhaps it's only at the intraspecific level, but the recently demonstrated acoustic territoriality in ranid frogs is a form of parental care in that it prevents the deposition of eggs by other members of the same species within the area that is defended by the male. In this line, I'd like to report some recently finished research in which I have definitely documented territoriality in Puerto Rican *Eleutherodactylus.* I would suggest that we look for territoriality in any situation where the adults live in the same habitat that the young are going to occupy. This is the kind of situation that would allow territoriality to make a real contribution to the survival of the species.

Inger

It seems to me you just mentioned a general kind of thing similar, analogous really, to what Dick Wassersug was talking about. Perhaps we have paid entirely too little attention to the aspects of social behavior of adult frogs.

Drewry

I would agree with that 100 per cent.

Reig

I want to put another question to Dr. Priscilla Starrett, since we are speaking about larvae so much. From a lot of morphological data, I have concluded that ceratophrynids

—a group comprising the genera *Ceratophrys, Lepidobatrachus,* and *Chacophrys*—is to be separated from the leptodactylids. This is substantiated in the paper I have now in press on Dr. Blair's book on *Bufo.* From what I recall, there is some evidence in the literature that the larvae of *Lepidobatrachus* differ strongly from other larvae of leptodactylids. I want to know if Dr. Starrett has some information on larvae to help elucidate the relationships of ceratophrynids. Are they really a different group, or are they more closely related to bufonids or to leptodactylids?

Starrett

The *Lepidobatrachus* tadpole is the one that's so strange that, unfortunately, *Ceratophrys* doesn't seem as unusual. I mentioned the very early ossification in *Lepidobatrachus,* and I think that John Lynch mentioned it also occurring in *Ceratophrys,* although not nearly as early as in *Lepidobatrachus.* On the basis of the tadpoles I can separate them off as a very distinct group. It was interesting to note that when the relations of the pelobatids were mentioned yesterday, I saw an illustration of a *Pelobates* with two spiracles, which is the condition in *Lepidobatrachus.* This was obviously an anomaly, but it was interesting to know that it happened.

Kraig Adler, Cornell University

I'd like to direct a question to Mr. Wassersug. Because of the equipment that you used, you restricted your animals to visual receptors for the aggregating response. I would like to know if you have any experimental evidence for other kinds of sensory input. I'm thinking of olfactory input, which might be most important among tadpoles living in slow-moving or still bodies of water, as in the case of *Bufo.* Might this not be a confounding factor in your arrangement of aggregation types?

Wassersug

One of the things I was considering was whether vision was a basis of mutual attraction in aggregations—and it is. I admit that this does not rule out lateral lines or olfaction, which could be equally important. In a paper about twenty years back, Richmond discussed schooling in *Scaphiopus.* It's about the best paper that's been done on *Scaphiopus* tadpoles, and he felt very strongly that olfaction, feeding characters, and food disturbed and raised up by the tadpoles at the front of a moving aggregate were a stimulus for the tadpoles behind to follow. I think that's very important. I don't think it negates anything that I said. A different aspect of this, which I really didn't get into my talk, is that there are different distances that animals have between them, so if you can find aggregates in very muddy ponds, like you can find with *Scaphiopus,* it is possible to suggest that they just can't see each other and that olfaction or lateral line responses might be very important. I didn't mention *Rhinophrynus* tadpoles in my talk, and I'm surprised no one questioned me on where I put *Rhinophrynus* in this scheme because it is so much like *Xenopus* in that it doesn't school, it lacks the filament on the tail, and it can't hang in midwater. I think it's sort of a generalized tadpole that perhaps could be basic or leading to the type of specializations that you find in *Xenopus.* The big difference between *Xenopus* and *Rhinophrynus* is that the mean spatial difference between *Rhinophrynus* tadpoles is much, much less. I think that they live in cloudier water than *Xenopus,* which could account for the shorter distance. It would be nice if experiments were done to see to what extent olfaction might be important here. It would be difficult, but worth testing.

Blair

I'd like to come back to the matter of egg size and the relation of egg size to body size and to ecological factors. The only group I have any real competence in speaking about is *Bufo,* and I'd like to make three points about *Bufo* that I think are inconsistent with some of the generalizations that were made by Dr. Salthe. First, in at least *Bufo valliceps,* there is a great deal of intrapopulational variation in egg size. I don't know the basis of it—whether it's physiological, with some kind of stimulus inducing ovulation at an earlier stage of development—but the eggs are quite equally viable and are a great deal different in size, which is quite unrelated to body size of the female.

Second, there are additional cases to those that Jim Bogart mentioned in which toads of approximately equal size have enormously different sized eggs.

Third, the one generalization that I've seen in working with the eggs of perhaps seventy-five to eighty species of *Bufo* is that there is a general kind of correlation relating to the early Tertiary dichotomy, in what I call thin-skulled and thick-skulled, or narrow-skulled and broad-skulled, toads. The narrow-skulled toads—for example, *boreas,* the *americanus* group, *Bufo bufo,* and others—in general have larger eggs relative to body size than do the broad-skulled toads, represented by the *valliceps* group. Also, there is, in general, heavier pigmentation of the eggs of the narrow-skulled toads, which are in cooler climates, as well as often in higher elevations. So, I think there's much that is yet unknown about what these generalizations mean, and I would urge that while it is useful to make the kind of survey we have seen here, the next step is to pick out specific examples and go then to a very detailed ecological observation of these individual cases to see what is really back of the kind of character states we see.

Rabb

Obviously, this should be done on the holistic basis that Stan [Salthe] was urging a while ago.

I'd like to back up to a comment that Bill [Duellman] made about the relative success of the large-egg forms versus such aquatic breeders as *Rana* and *Bufo.* If we look at things from a total-energy viewpoint, taking into account such factors as the social organization, which may function to spread utilization of resources, as Drewry pointed out, are there any differences in what goes into producing a breeding aggregation in different species, for example, an aquatic breeder versus a terrestrial one, in the kind of energies that go into yolk and into calling and into everything else? Is there no difference, or are the very successful taxa like *Bufo* and *Rana,* which are successful on the basis of number of species, distribution, and so on, actually more efficient in terms of energy management?

Inger

I'm going to use my prerogatives and let another speaker in here.

Max A. Nickerson, Arkansas State University

I would like to refer to Dr. Blair's comments on the high pigmentation of the eggs of the narrow-cranium toads that he mentioned as being found in high mountain areas. Might this not be an ecological situation relating to clarity of the water? Do we have a difference with regard to the amount of pigmentation of anuran eggs in clear versus murky water?

Blair

I'm afraid that this is a question to which I can't give you an answer based on any data. It's a very interesting possibility. Of course, you do have other variables. You have more incoming cosmic radiation in higher elevations. The clarity of the water is one thing, and another is the temperature difference. In the sub-Andes, I've seen *Bufo spinulosus* in very large ponds in which all the tadpoles were actually crowded so close into shallow water that their backs were out of the water, and if you put your finger in the water and tested the temperature differential, you would see a very good reason for their crowding.

Salthe

I'd like to talk to the point about whether there really are any differences in these reproductive strategies. Do they all wind up being the same thing? Recently I was reading a book on ecology by Margalef, and he talks about mature ecosystems as opposed to immature ones—the immature ones have more net flow of energy through them. It may well be that there is a tropical-temperate difference on this basis. There is some indication in our preliminary calculations that ranid-type frogs that breed very early in the spring in the northern systems have a much larger total clutch volume than tropical frogs do; the northern frogs are way off the scale. This is something I didn't mention before because we're just getting at it.

This kind of ecosystem difference may have to be ultimately taken into account. I agree with Dr. Blair's comment that all these things have to be examined in much more detail, but there are no others at this point.

Konrad Bachmann, University of South Florida

I think there are two basic approaches to explaining a generalization like the one about egg size–clutch size. There is an ecological approach, in which one studies the selective factors that have acted on all the eggs together, but I think there's another way to attack the same problem, and that way has not been mentioned in detail here. There may be inherent restrictions to the genetic system and the physiological system that force such correlations. It may be that not every frog can do whatever he wants. A frog cannot very easily make a big egg unpigmented or a small egg pigmented or any combination. One of the factors that we have latched onto lately that may link up quite a few of those correlations is the specific DNA amount, the diploid DNA amount that enters into a species' reproductive strategy at several levels and ties these together. For instance, species with more specific DNA tend to have larger eggs. *Bufo bufo* has a large egg that may be correlated with the extremely high DNA amount, but we don't have enough *Bufo* species measured to see if that correlation is closely linked with the skull correlation.

Another level that the DNA affects is the speed of development. There are pretty good indications that species with large DNA amounts develop more slowly, and this correlation is not directly dependent on egg size. It seems that this is one physiological factor that may tie up some of those things at the genetic or molecular level.

Salthe

I would like to comment on the last point, that the speed of development is correlated [with egg size]. There is, perhaps, a more pedestrian way of considering the slower development of larger eggs. Remember that our correlation only goes up to neurula. We can't make correlations beyond that. Consider the fact that you have cleavage, gastrulation, and the morphogenetic movements concerning neurulation. All of these could be impeded by sheer weight and amount of inert yolk. I think that is a simpler explanation than the one just proposed [by Bachmann].

Bachmann

May I answer that right away? Undoubtedly it is true that a bigger egg takes more time to divide and that the egg size is dependent on the DNA amount, very indirectly. But there seem to be indications that aside from the egg size, even the later developmental speed depends on the DNA amount and cell size. For instance, I'm at the moment doing some initial measurements of growth rates of maximally fed, newly metamorphosed frogs—hylids that have quite a scale of variability in their DNA amounts. And there is a possibility (I don't want to cite this officially yet) that those hylids with low DNA amounts, for instance, *Hyla septentrionalis*, grow very, very fast, compared to other species that have relatively higher DNA amounts. The DNA amount may enter twice or several times into this same problem; once by determining egg size, and then, via egg size, by determining developmental speed, and still another time by determining metabolic level of somatic cells, once these are established. It's a very complex system, with this one factor entering in at several levels.

Salthe

The evidence, which is very poor concerning the metabolic level, indicates that intraspecifically, as an ovum gets larger it metabolizes more intensely and yet interspecifically, as ova get larger, development slows down. It's possible that intensified metabolism does not necessarily mean more rapid development. You may simply have more maintenance energy needed in the larger system.

Bachmann

This is very possible and that has to be sorted out in the specific instance, which has not been done yet.

PART III

ACOUSTIC, GENETIC, AND GEOGRAPHIC CONSIDERATIONS

9

EVOLUTION OF ANURAN MATING CALLS ECOLOGICAL ASPECTS

Arne Schiøtz

FUNCTION OF MATING CALL

The mating call is the voice the male emits at the breeding locality in the breeding season. Much has been written and many experiments have been done in order to elucidate the significance of this voice. Even so, in 1960 Bogert concluded that although there seemed to be a positive response from the females to the voice of their own males, when visual and olfactory impressions were eliminated (for instance, by transmitting the voice through a loudspeaker) the experiments were not very convincing.

Since Bogert's paper, however, a number of experiments and observations seem to show that at a certain time during the reproductive cycle the female responds positively to the voice of a conspecific male and is indifferent to a male belonging to another species.

One reason for the many negative experimental results is undoubtedly that the period in which the relevant response is given by the female is very brief. During recent years Heusser has worked with European *Bufo*, particularly *Bufo bufo*, and has shown (Heusser, 1968) that there are two distinct phases prior to copulation and spawning. First, there is a breeding migration during which both sexes move towards the breeding locality. This breeding migration seems to be initiated by climatic and hormonal factors in close interaction, supplemented by the animals' familiarity with the environment and possibly other factors. The voice plays apparently no part at this stage. Later comes what Heusser terms a prespawning period; in *Bufo bufo* its duration is from six to fourteen days, and in other European anura it is apparently briefer. During this period the behavior of the animals is altered; the threshold for mating behavior is lowered and the threshold for flight becomes greater; they stop eating and are active both day and night. The migrating impulse fades and egg ripening in the females takes place. The toads are very sensitive in this period so that unnatural conditions, the laboratory for instance, may alter their behavior.

A similar prespawning period has been mentioned by Savage (1961) for *Rana temporaria*. In Africa I have observed that the breeding migration is not immediately followed by spawning behavior in *Hyperolius occidentalis* but that a prespawning period of at least a couple of days exists after the animals have arrived at the breeding locality when no vocal activity is heard.

The prespawning period is followed by a spawning period, when copulation and egg laying occurs. In many species it is during this time that the female is positively attracted to the voice of a conspecific male.

An indication of the mechanism that may be involved in the specific response of *Rana catesbeiana* is given by Frischkopf and Goldstein (1963). The ear of this species is so constructed that it has maximum sensitivity in the same frequency ranges as the call.

Females will only react positively to the voice of a male for a very brief period (possibly only for a few hours) during which they are attracted to a male, copulate, lay the eggs, and then leave the breeding locality. The very brief span in which this response is given explains why so many experiments have produced such confusing and unconvincing results: females kept under laboratory conditions for some days, or even a few hours, cannot be expected to respond positively to a voice even if they were in the right physiological condition at the moment of capture.

Positive responses by females to the call of other species, even closely related ones, have only been observed a few times and are explicable either by similar voices in allopatric forms or by the voice playing no role as isolating mechanism in the species-group in question (Straughan, personal communication 1970).

I think it is therefore safe to conclude that the voice in anurans plays an important part as isolating mechanism, at least for a large number of species. There are observations (Aubrey, unpublished data) that seem to show, however, that in some Ranidae the voice is of no importance in this respect.

Some of my observations indicate that the mating call also plays a territorial rôle. That males, even in dense breeding aggregations, have territories is obvious, and one can often observe fights, sometimes only vocal ones, between males that have approached each other too closely. Capranica (1965) has shown that male *Rana catesbeiana* countercall in response to the mating calls of other males of their own species but not to those of males from other species. The chorus structure that Duellman (1967) found in many of the Central American frogs and that is also found among African frogs is probably territorial in origin.

Observations by Brattstrom (1962) of *Engystomops pustulosus* seem to show that the dominant male is preferred by the female. If this is a general tendency it is highly interesting, for while the function of the voice as an isolating mechanism will tend to keep it as stable as possible within a taxon, an active selection by the female of a single male might induce the development of the voice within the taxon and might mean competition among males within a population in this respect.

A very tempting assumption, namely that a breeding chorus attracts members of the species, male or female, to the breeding locality has never been confirmed, although scattered reports may support this assumption. In fact, as mentioned before, factors *other* than voice have been dominant in all species studied during the migrating period. It is thus strange that the voices of many species are so loud that they can be heard from half a mile or more. When observing a small tree-frog calling hour after hour, night after night, one is struck by the waste of energy it must be, especially if it is of no biological significance, to have a voice that can be heard more than a few feet away—that is, the distance from the female that has been attracted to the breeding site for other reasons. It may be explained as a result of a selection by females of males with the loudest voice, the voice that is best heard in the chorus.

As the voice is generally an isolating mechanism, it follows that it would be an advantage if voices of sympatric species, taxonomically related or unrelated, were always different. This seems to be the case. There are no reports of two sympatric species having identical voices. We might find identical voices among allopatric species, which is of no biological significance, and this also seems to be the case. For instance, in the genus *Hyperolius* most species have a simple voice consisting of clicks. In all cases in this genus the voices of sympatric forms are different, but in allopatric forms some are identical or very similar.

There are a few records of reinforcement, namely by Littlejohn (1965, 1968) in the *Hyla ewingi* group and (1959) in *Pseudacris* and by Blair (1962) in the *Bufo americanus* group, (1955) in *Microhyla,* and (1958) in *Acris.* In all these cases the difference in voice between two closely related forms is greater in zones of overlap than between allopatric populations.

MATING CALL AND TAXONOMY

If we look at the more general rules governing the voices, it will be seen that the voice follows the taxonomy to a certain extent, which is quite natural as the voice is a product of the morphology of the species. This is not the case on the family level. In most instances it is impossible to point out characters common to genera in one family that cannot be found in genera in other families. On the genus level, however, there is normally a conspicuous similarity between species within the genus—a rule with a number of exceptions (Schiøtz, 1967; see also Figure 9–1). At the species level (species in a superspecies and subspecies in a species) the problem is different. Such categories will, of course, always be allopatric, and there is no apparent reason for different voices. Generally speaking, the voices within infraspecific categories seem to be identical or at least very similar.

We have, roughly, three sets of characters that are used in the taxonomy of frogs on the species-subspecies level: habitat preference, morphology—including size and coloration—and voice. The emphasis given the three characters will vary according to different groups and probably to the points of view of different workers. It is my impression from my work on the African frogs that habitat preference (whether the animals are associated with dense forest, open forest, savanna, running water, stagnant water, or other environmental factors) is normally constant within a species or superspecies. The voice is generally less constant, and the morphology even less so. It has thus been a working rule to regard allopatric populations differing in morphology as members of the same species if not only the habitat preference but also the voice is identical. If, however, morphology and voice are different, I have been reluctant to regard them as conspecific, and I have not done so in a single case where the habitat preference is also different. The usage of habitat preference, voice, and morphology as taxonomic characters is not based on any deep theoretical consideration, although it can be said that if the voice between two populations differs significantly, it is probably proof of reproductive isolation. In practice this usage seems to provide a reasonably sound taxonomy.

In some groups the voice apparently differs more than the morphology. Littlejohn's (1959) work on the genus *Crinia* in Australia is well known. In that genus, a number of morphologically very similar forms, sympatric and allopatric, differ mainly in voice. The difference in voice of sympatric, similar forms is easy to explain as a necessity for maintaining reproductive isolation. Littlejohn (1959) asserts that the differences between allopatric, related forms are the result of a different sound environment or are incidental by-products of other adaptive processes.

Nevo's (1969) work on the cricket frogs, *Acris,* has also shown that there is a considerable variation in voice, especially of *A. crepitans,* throughout their range. This variation can be correlated with the difference in size, which can in turn be correlated with greater or lesser humidity of habitats of the different

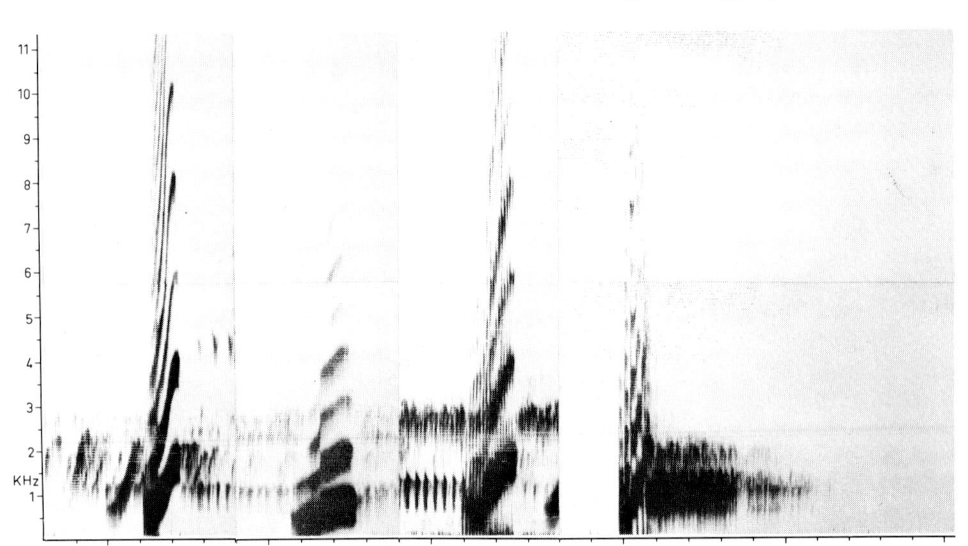

Figure 9–1. Example of vocal similarity between members of the same genus. From left to right: *Kassina senegalensis, K. cassinoides, K. fusca,* all from Walewale, Ghana; *Kassina cochranae,* Adawso, Ghana.

populations. In this species there is a cline possibly of such magnitude that the ends are reproductively isolated from each other because of difference in voice.

MATING CALL AND SOUND ENVIRONMENT

It is a question of the degree to which the voice of a species is influenced by the sound environment. One could postulate that, because it is important for the voice of each species to be easily recognized in the general chorus, there will be a tendency for the voices of the different species to become as different as possible so that the full acoustical spectrum is used. As shown earlier (Schiøtz, 1967), in each habitat there is a tendency among genera to have a full "spectrum" present, in the sense that most genera will have species represented in all habitats.

One could also suggest that selective forces in certain habitats or certain vocal communities would tend to produce voices with some common characters that are useful in the particular sound environment.

There seems to be some indication, at least among the African frogs, that this may be the case. It is my impression that the species connected with stagnant water in dense forest generally have much more quiet voices than species from more open situations, such as farmbush or savanna (Figure 9–2). No attempt was made to measure the intensity of the voices, but the difference between a "quiet" and a "noisy" form is often very great, the former being audible only from a few feet away.

Table 9–1 shows the comparison between two well-explored localities in Ghana. The Bobiri Forest Reserve has a well-developed fauna of species associated with stagnant water in dense forest, together with some species from farmbush; the farmbush species are probably recent introductions, for they were very rare on this locality. Three species utilizing flowing water in forest—*Leptopelis macrotis, Hyperolius viridigulosus,* and *H. laurenti*—have been omitted because they were not taken in stagnant pools. These forms are compared with the fauna in Walewale, Ghana, a typical rich savanna breeding locality. It can be seen

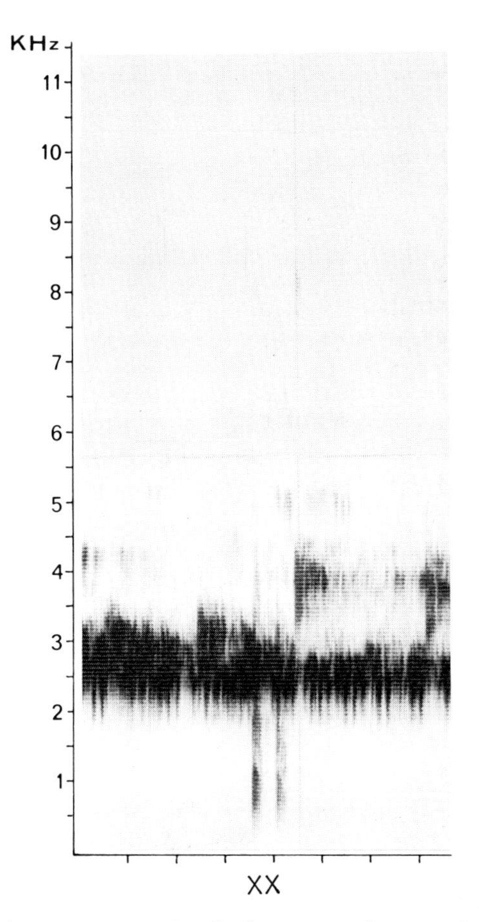

Figure 9–2. Example of a quiet voice. *Ptychadena aequiplicata,* Bobiri, Ghana.

that the proportion of quiet species to noisy species is entirely different in the two localities.

A closer examination will reveal that in some cases, for example, *Leptopelis hyloides* versus *L. viridis* or *L. bufonides,* or *Afrixalus congicus* versus *Afrixalus laevis,* the forest form is quieter because of a lower intensity of voice, which is otherwise very similar. In other cases, for example, *Hyperolius bobirensis* versus other *Hyperolius,* the voice of the high-forest form has a different structure (see sonagrams in Schiøtz, 1967).

Although it is difficult to give an explanation of this, it may be that the forest is so much quieter than the savanna that it isn't necessary to call so loudly. It is not my impression that this is the case, however. If insect voices are included, I find the forest just as noisy as the savanna by night. A very convincing explanation would be that the breeding localities in the savanna are very far apart, while in the humid rain forest they are often very close together. Therefore it would be necessary for the savanna frogs to call loudly to attract other males or females to the breeding locality, but the quiet forest voices are sufficient for attraction over a much shorter distance. This explanation has the one serious drawback; no one has been able to show definitely that there is an attraction to the breeding locality caused by the voices of calling males. A more probable explanation is that the breeding locality in the forest is generally small; in the savanna, often

Table 9–1. "Quiet" and "Noisy" Species on a High Forest and a Savanna Locality

Species	Quiet	Noisy
A. High forest locality Bobiri Forest Reserve, Ghana		
High forest		
Chiromantis rufescens	x	
Leptopelis hyloides	x	
L. occidentalis	x	
Afrixalus c. nigeriensis		x
A. laevis	x	
Hyperolius bobirensis	x	
Ptychadena aequiplicata	x	
Transition high forest-farmbush		
Kassina cochranae		x
H. sylvaticus		x
Farmbush (rare)		
H. f. burtoni		x
H. concolor		x
H. picturatus		x
Breeding habitat not known		
P. longirostris		x
B. Savanna locality Walewale, Ghana		
L. viridis		x
L. bufonides		x
K. senegalensis		x
K. cassinoides		x
K. fusca		x
A. weidholzi	x	
H. nitidulus		x
Bufo regularis		x
P. maccarthyensis		x
Phrynobatrachus natalensis		x
P. francisci		x
Phrynomerus microps		x

very large. In order to be heard throughout the single breeding locality, the savanna frogs must therefore call more loudly.

Another ecological community of animals that has certain common characters in the voices are those forest species that do not form breeding aggregations. These are mainly species that either develop independently of water, such as the genus *Arthroleptis*, or species that, as far as we know, place their eggs in extremely small accumulations of water in the forest, for example, *Cardioglossa.* Typically, such species have a very high pitched voice (Schiøtz, 1964) with a distinct insect-like quality (Figure 9–3). A field collector receives the strong impression that these voices are much more difficult to localize than other frog voices. It is therefore tempting to assume that predators have difficulty in localizing frogs with this type of voice—a great advantage for the single males that are scattered over the forest floor and that may be much easier prey than those in large pond-breeding aggregations.

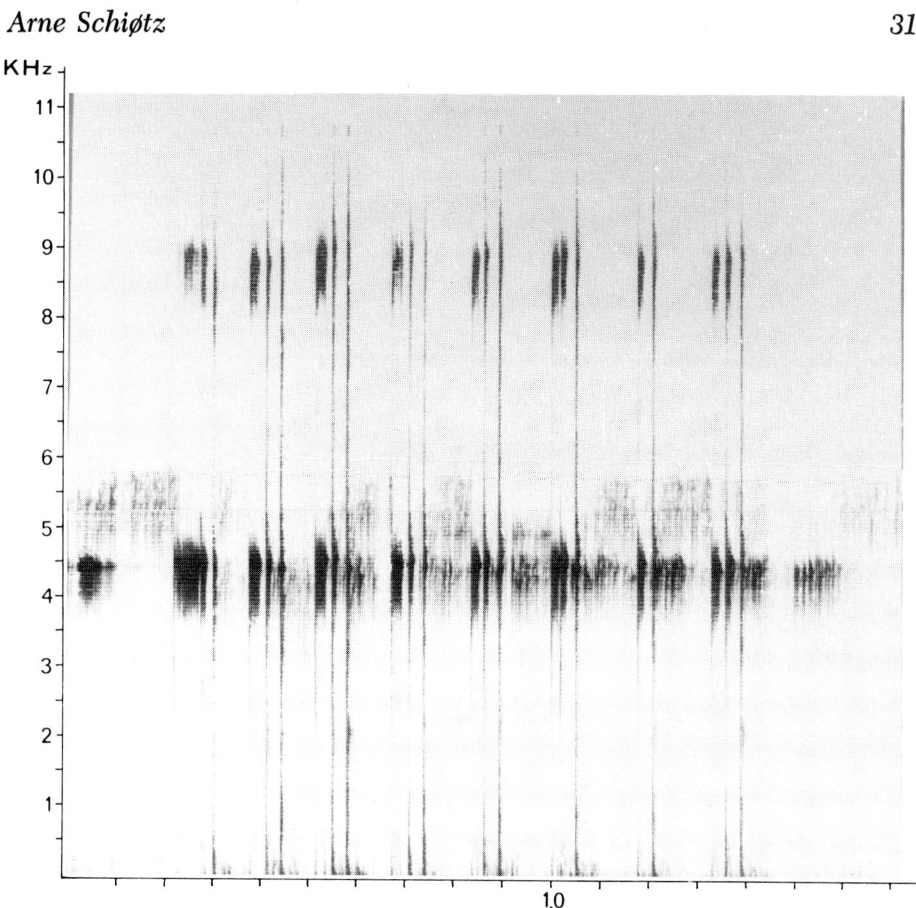

Figure 9–3. Example of an insect-like voice, *Cardioglossa aureoli,* Freetown, Sierra Leone.

In contradiction to this is Konishi's (1970) statement that higher frequencies are better suited for exact localization by binaural animals. It may be that in small animals having a distance between the ears of the same magnitude as the wave length of the call, it is easy to localize such a voice; for larger animals it may be difficult. The acoustical reason for the distinct insect-like quality has not been analyzed.

In western Africa there is apparently only one frog that does not form breeding aggregations or have an insect-like voice, namely the forest-dwelling *Phrynobatrachus alticola* (Schiøtz, 1964). *Hyperolius laurenti* has an insect-like voice, but it calls high up in trees. It is one of the few *Hyperolius* that do not form breeding aggregations.

The mute frogs are peculiar because the lack of voice seems to occur without any systematic relevance in species within genera that have a normal voice and possibly, in some cases, in single populations within species that have normal voices, as shown by Inger (1954) in the Philippines. Blair (1956) has mentioned that the American *Bufo alvarius* is almost mute and explained that because there are no similar and closely related species in the same areas, voice is not necessary as an isolating mechanism. The same explanation may apply to the few African species without voice. *Bufo rosei* (a montane form from South Africa), *Bufo superciliaris* (a large forest form), *Acanthixalus spinosus* (a tree frog), and *Rana goliath* (the giant frog from Cameroon) are all species that do not occur together with similar, closely related forms.

EVOLUTION OF MATING CALLS

Fossil records that could give information about the evolution of mating calls are lacking, and the living families of anurans offer very little help. The most primitive family, Leiopelmidae, lacks vocal sacs and a mating call; the latter may be a truly primitive character. The Pipidae, another primitive family, also lack vocal sacs but have what is probably a mating call in the form of a metallic click with very indistinct pitch, or a brief series of such clicks, although one genus, *Pseudhymenochirus*, does have a more complicated call (Rabb, 1969). These characters may be truly primitive, but it would be difficult to prove in these completely aquatic frogs, which call under water and therefore have very little use for a typical vocal apparatus. The pipid voice could also be regarded as an adaptation to a specialized way of living.

In all the other families of Anura I feel that the voice has reached the level at which it best functions as an isolating mechanism. The voices of one family cannot be regarded as less developed than the voices of others.

CONCLUSIONS

The overwhelming purpose of the mating call in most species of Anura is to act as a premating isolating mechanism by being different from the voice of all other sympatric species and by being uniform within a species.

The morphology of the voice follows the taxonomy up to the genus level, but no further. Closely related allopatric forms normally have very similar or identical voices. Exceptions may be explained as the result of populations being subjected to different sound environments or as incidental by-products of other adaptive processes.

The important role of the voice as an isolating mechanism in some cases allows morphologically similar or even identical species to occur together, but in general the differences in voice among sympatric species are accompanied by differences in morphology. There seems to be a tendency toward similarity in voices of species from the same habitat in the tropical rain-forest.

Although the voices of anurans are very important characters in the evolutionary processes, they may be of little phylogenetic use because of their important role as isolating mechanisms in closely related sympatric species.

REFERENCES

Blair, W. F. 1955. Mating call and stage of speciation in the *Microhyla olivacea–M. carolinensis* complex. Evolution 9:469-489.

———. 1956. Call difference as an isolating mechanism in southwestern toads. Texas J. Sci. 8(1):87-106.

———. 1958. Mating call in the speciation of anuran amphibians. Am. Naturalist 92:27-51.

———. 1962. Non-morphological data in anuran classification. Systematic Zool. 11(2):72-84.

Bogert, C. M. 1960. The influence of sound on the behavior of amphibians and reptiles. *In* W. E. Lanyon and W. N. Tavolga, [ed.], Animal sounds and communication. Am. Inst. Biol. Sci. Publ. 7:137-320.

Brattstrom, B. H. 1962. Call order and social behavior in the foam-building frog, *Engystomops pustulosus*. Am. Zool. 2:50.

Capranica, R. R. 1965. The evoked vocal response of the bullfrog. M.I.T. Research Monograph 33, 106 p.

Duellman, W. E. 1967. Social organization in the mating calls of some Neotropical Anurans. Am. Midland Naturalist 77(1):156-163.

Frischkopf, L. S., and M. H. Goldstein. 1963. Response to acoustic stimuli from single units in the eighth nerve of the bullfrog. J. Acoust. Soc. America 35:1219-1228.

Heusser, H. 1968. Die Lebensweise der Erdkröte *Bufo bufo* (L.); Laichzeit: Umstimmung, Ovulation, Verhalten. Vierteljahres. Naturforsch. Ges. Zurich 113(3):257-289.

Inger, R. F. 1954. Systematics and zoogeography of Philippine amphibia. Fieldiana: Zool. 33:181-531.

Konishi, M. 1970. Evolution of design features in the coding of species-specificity. Am. Zool. 10:67-72.

Littlejohn, M. J. 1959. Call differentiation in a complex of seven species of *Crinia*. Evolution 13:452-468.

———. 1965. Premating isolating in the *Hyla ewingi* complex (Anura, Hylidae). Evolution 19(2):234-243.

———. 1968. The systematic significance of isolating mechanisms, p. 459-482. *In* Systematic biology. Nat. Acad. Sci., Washington, D.C.

Littlejohn, M. J., and T. C. Michaud. 1959. Mating call discrimination by females of *Pseudacris streckeri*. Texas J. Sci. 11(1):86-92.

Nevo, E. 1968. Discussion of the systematic significance of isolating mechanisms, p. 485-489. *In* Systematic biology. Nat. Acad. Sci., Washington, D.C.

Rabb, G. B. 1969. Fighting frogs. Brookfield Bandarlog. Chicago Zool. Soc. 37:4-5.

Savage, R. M. 1961. The ecology and life history of the common frog (*Rana temporaria temporaria*). Pitman & Sons, Ltd., London. 221 p.

Schiøtz, A. 1964. The voices of some West African amphibians. Vidensk. Medd. dansk. naturh. Foren., Kbh. 125:1-92.

———. 1967. The treefrogs (Rhacophoridae) of West Africa. Spolia Zool. Mus. haun. 25:1-346.

10

EVOLUTION OF ANURAN MATING CALLS BIOACOUSTICAL ASPECTS

Ian R. Straughan

The aim of this paper is to present, in a very preliminary way, the idea that mating calls and other acoustic behavior may be useful in establishing relationships of higher taxa within the Anura. I wish to acknowledge that my appreciation of the role of mating calls in anuran biology is the result not only of my own investigations but also of extremely helpful conversations with Murray Littlejohn and more recently with George Drewry.

* * * *

Studies of anuran mating calls have significantly increased our understanding of systematics and evolution within the order, particularly in defining species and their interrelationships. Electronic analyses and the subsequent precise descriptions of vocalizations have greatly enhanced their use as taxonomic tools. However, the use of mating calls in establishing taxonomic relationships above the level of genus has been suspect.

Lanyon (1969) has discussed the proper use of vocal characters in avian taxonomy and their inadequacy at higher taxonomic levels. With the exception of the problems resulting from learned vocalizations, his arguments are equally applicable to anuran mating calls. Given the usual approach of comparing detailed descriptions of vocalizations, their usefulness dilutes rapidly at higher taxonomic levels. However, as with any other attribute of taxonomic importance, relationships on a larger scale may be more readily derived from fundamental form, rather than from detailed structure. An examination of the evolution of mating calls themselves as functional units of a communications system may clarify some fundamental forms, and these may be useful in determining relationships of higher taxa.

Problems of parallelism and convergence exist for mating calls, as well as any other character, and homologies cannot be established from the calls alone because similar sounds may be produced by entirely different apparatuses or by divergent calls with identical vocal structures. As pointed out by Blair in the Introduction to this volume, all data must be considered together and weighed for each relationship before any realistic phylogenies can be drawn.

The mating call is part of a functional communication system, so to understand the factors that might influence its evolution we must analyze the function of that system. We assume, of course, that there is evolutionary development toward greater efficiency in all systems.

Cherry (1968) has elucidated the problems of communication in biological systems and provided the analogues from engineering systems for understanding and evaluating design efficiency in biological systems. Wilson and Bossert (1963) successfully predicted the design features of biochemical communications systems from physiochemical bases. The bases for such a study of sound communication are available in physics theory and engineering practice. Konishi (1970) discussed some of the design features of species-specific communications signals in general and sound signals in particular. In this functional context, anuran mating calls are the species-specific messages passed from the transmitter male to the receiver female. Fortunately, the calls can be sampled from the communications channel by recording without interfering with their functions. The function of this system presumably is attraction of females to calling, conspecific males. Until recently the definition of mating calls was largely contextual rather than functional, but evidence from call-discrimination tests is accumulating, which indicates that species-specific attraction of females is indeed a widespread phenomenon in anurans (see Littlejohn and Loftus Hills, 1968, for references to Hylidae; Awbrey, 1965, for Bufonidae; Littlejohn and Martin, 1969, for Leptodactylidae; and Straughan, 1966, for an extensible survey of call discrimination involving 34 species of 7 genera of Australian Leptodactylids and Hylids). Although Capranica (1965) demonstrated a species-specific evoked vocal response in males of *Rana catesbeiana* to what may be the mating call of that species, it did not demonstrate the attraction of females to that call. Not all calls defined as mating calls have been shown to have the function that the name implies, and they may indeed have other primary functions. Bogert (1960) has classified the functional types of calls used by anurans at least in a contextual sense. Some species have a repertoire of calls that are used in different contexts, and in referring to "mating calls" in these forms, one is no more narrowly defining the context than he is when he uses the term "breeding." Other call types, such as "territorial calls," may be important components in breeding assemblages, and care should be taken to ensure that the call in question is indeed a mating call.

On the other hand, secondary behavioral functions may be incorporated into the mating call system. Brattstrom and Yarnell (1968) and Heusser (1969) have attributed territorial and social heirarchical functions to operational mating calls. This dual role can be accomplished simply by adding a second set of receivers (other active males) to the transmitter (male)–receiver (female) communication system of the mating call. In such a system the two sets of receivers may not use the same clues from the call from specific recognition, and the signals (calls) need to carry relatively greater information, as shown in meadow larks by Lanyon (1966). Therefore, advanced mating calls (in this functional sense) are of necessity quite complex signals.

All mating calls can be defined by several descriptive parameters, which are basically related to intensity patterns in the two dimensions of frequency and time. Understanding the function of each of these call components in the communications system provides a basis for assessing the design features of the call in terms of its functional efficiency in its environmental situation. From this assessment, selective pressures and evolutionary trends may be postulated.

One of the most obvious components of mating calls is the dominant frequency, or pitch—the subjective measure of frequency to the human listener. Dominant frequency is nearly always incorporated in descriptions of mating calls, at first as pitch and later as an exactly measured parameter, but its functional role in communication has not been fully defined.

Recent studies of auditory function in anurans have shown that the ear behaves as a peripheral filter that is tuned to particular frequency bandwidths

(Capranica, 1965; Capranica and Frishkopf, 1966; Loftus Hills and Johnstone, 1970; Bishop, Straughan, and De Vorak, unpublished data). The ear is most sensitive to sounds of a rather restricted frequency bandwidth or a pair of frequency bandwidths, and in most cases (see Loftus Hills and Johnstone, 1970, for review) this "best frequency" matches the dominant frequency (or the frequency of maximum energy) of the mating call. This system is analogous to a radio transmitter broadcasting in a fixed frequency band, and the receiver being perpetually tuned to that frequency band or channel. Dominant frequency has a function similar to that in man-engineered communications systems in that it provides each species with its own delineated channel over which messages can be sent. Littlejohn and Martin (1969) found that the dominant frequencies in choruses of aggregated species of anurans were typically stratified across 4 KHz of the audio frequency spectrum, that is, the species had divided up the air waves into separate channels. Drewry and Rand (personal communication 1970) found very fine partitioning into frequency channels in multispecies assemblages of *Eleutherodactylus* in Puerto Rico. Even in choruses of more than 14 species that occur commonly in tropical Australia and Central America, where not enough channels are available at any one point in time, the dominant frequencies of species calling at that instant are similarly stratified (Straughan, unpublished data).

A basic design feature of anuran mating calls that would result in increased efficiency in terms of minimizing interference from other sound sources is a restricted frequency bandwidth. In more complex sound environments, selection for separate narrower bandwidths would lead to greater efficiency in communication for all species present.

If the dominant frequency of the call defines only the channel being used, the species-specific information must be carried by some other component of the call, unless channel separation is so effective that discrimination is affected by different species not being able to hear each other. The specific dominant-frequency component alone is sufficient to evoke calling responses in males of *Rana catesbeiana* (Capranica, 1965), but whether females are attracted specifically to this component has not been tested.

Littlejohn (1965*b*) and Straughan (1966) inferred from interactions between populations of species-pairs of the *Hyla ewingi* complex in southeastern Australia that the temporal pattern (specifically, the pulse repetition rate) encodes the species-specific information of the calls. The ability of females to distinguish between calls of *H. ewingi* and *H. verreauxi* on the basis of differences in only the pulse repetition rate has been established experimentally by Loftus Hills and Littlejohn (1971). Discrimination between the calls of *H. verreauxi* and another species of the *ewingi* complex (as yet unnamed) was also found to depend on temporal patterning (Straughan, 1966). In this case, note duration (or length of a pulse train), as well as the pulse repetition rate, could be used as discriminatory clues, which suggests that the two-phase temporal pattern contributed redundancy to the communcation system. Redundancy is another way in which communication systems can circumvent noise and may be indicative of more advanced systems.

Breaking up the call into discrete time units (pulses and notes, as used above) with a specific pattern in time is not the only way to encode species-specific information in the mating call, but this is the only system that has been verified in anurans to date.

Information can be carried by frequency modulations, amplitude modulations (other than discrete pulses) and frequency spectra. Mating calls show all these forms of modification, often in combinations, and these patterns undoubtedly contribute to the species specificity of calls.

A principle of communication systems is that the amount of information that can be carried in unit time is proportional to the frequency bandwidth. Therefore, as the bandwidth is narrowed the length of the signal must be extended proportionally to convey the same amount of information; that is, for mating calls to maintain their species specificity when selection is operating to narrow the bandwidth (for reduced interference), the call must become longer. This would tend to enhance the use of some form of temporal, rather than frequency, spectral coding. What form of encoding is being used by a particular species of frog cannot be determined without a detailed examination of the sound analyzing characteristics of its auditory system. Marler (1969) elucidated this problem in avian communication. It results from the fact that amplitude-modulated sounds of fixed frequency may be perceived either as sounds pulsed in time at a particular modulation frequency or as continuous raspy sounds with the fixed frequency clouded by side bands of frequencies that are separated from the original by the modulation frequency, which depends on the relative time- and frequency-resolving powers of the analyzing system. In some contexts it is possible to decide which would be the more appropriate way to hear (or analyze) sounds for efficient functioning. In Puerto Rico the total audio frequency range of the local species of *Eleutherodactylus* is consumed by narrow bandwidth calls that leave no open channels, but by adding side bands to its carrier frequency, one species penetrates this saturated sound environment with a loud, sharp click of wide frequency bandwidth (Drewry and Rand, personal communication 1970.) If the sound were perceived as a series of pulses of the dominant frequency there would exist the possibility of the call being obscured by other sounds in that frequency channel. Provided that species could hear over a wider frequency bandwidth, competition for a sound channel could be avoided.

Mating calls are broadcast to no particular receiver, and to be functional, the transmitter must be discoverable. Assuming that frogs locate the sound source by binaural intensity comparisons, the most appropriate sounds would be either high frequencies or sounds of wide frequency bandwidth. The dominant frequency of frog calls are, in general, inversely proportional to body size (Blair, 1964). The lower limit of size fixes the maximum frequency of sound produced in anura. Furthermore, Loftus Hills and Johnstone (1970) found that sensitivity of the anuran ear decreased markedly at high frequencies, restricting frogs to sounds of frequencies less than 5 KHz; therefore, sounds of wide frequency bandwidth are more appropriate to enhance discoverability in anuran communications systems. However, this design requirement is opposite to that for noise-free transmission, that is, narrow frequency bandwidth. The design of the call has to somehow compromise between the efficiencies for these two functions. This may be accomplished without loss of efficiency in either function by having a compound call that has a component carrying species-specific information, which may be in a narrow band, and a location component in the form of wider bandwidth clicks.

Another consideration in the design of this system is the broadcast distance, which is a function of carrier frequency, air density, humidity, temperature, sound absorption properties of vegetation, and other factors, and local abiotic noise. Schiøtz (this volume) raises the question of habitat and the type of call necessary for functioning efficiently in different environments, exclusive of other sound sources. Some degree of convergence in calls of species found in similar habitats would be expected.

Mating calls, as part of a functional communication system, are therefore under several adaptive influences, and can be modified. Modifications may be produced initially through changes in the central nervous system's motor control of the vocal apparatus, for example, on the number or rate of thoracic compressions or through active control of the vocal cords. More advanced forms may produce similar calls with less energy expenditure by evolving special anatomical modification, such as vocal sacs and additional laryngeal structures which, once set in motion, can resonate or oscillate passively to modify the call structure. By associating call types with the type of vocal apparatus, Martin (1972) has developed a phylogeny of calls in the genus *Bufo* based on paralleled development of call complexity and laryngeal modifications. When call components can be associated with a particular anatomical modification, relationships between species exhibiting particular call types are simplified, provided that the anatomical modifications are homologues. Without some anatomical base, homologues in call structure cannot be determined with certainty.

Although calls have generally been regarded as good species-specific characters, their origin has been disregarded as some mysterious by-product of speciation that developed *in vacuo* unless character displacement was noted. As pointed out by Littlejohn (1965*a*), calls obviously function in an environment of biotic (including all soniferous animals) and abiotic components that exert selective pressures on them, resulting in their evolution. The rates of evolution of mating calls may well be different in different groups and in different environments, and evidence of call evolution must be weighed against other evidence to develop meaningful phylogenetics. For example, members of the genus *Cyclorana,* which Lynch (this volume) concludes to be primitive leptodactylids, have very advanced complex mating call types (Straughan, 1966), while osteologically more advanced Australian leptodactylids have very simple and presumed less-advanced mating calls.

For a mating call to be functional it must have the design features necessary to broadcast in the particular physical environment, to avoid being obscured by local noise, to convey species-specific information, and to be discoverable. There is obviously more than one way to achieve this multiple function. It is reasonable to expect that closely related groups of frogs might use similar strategies for doing so, and by comparing their basic communication systems (extrapolated at this stage from the form of the mating call), we may be able to describe more broadly based relationships. Furthermore, calls are used in other behavior contexts, and if the total pattern of vocal behavior is included, more realistic relationships should be determinable.

The first anuran to use sound communication probably had discoverability as the only necessary design feature. The functional-looking ear of the fossil ascaphids described by Estes and Reig (this volume) and the reduced ear in modern ascaphids leads me to believe that the ancestors of modern families were soniferous and that *Ascaphus* and *Leiopelma* are secondarily mute. Pipid frogs were apparently present almost as early as more generalized frogs (Estes and Reig, this volume) and appear to be an early aquatic divergence from the main anuran line. Modern pipids have vocal behavior that is unlike that of any other group of frogs (Rabb and Rabb, 1963, and personal communication). No other modern family can be so clearly characterized by a fundamental call pattern. Within the family Bufonidae there is a great deal of similarity in the degree of tuning of the dominant frequency and in the basic way in which species-specific information is encoded in amplitude modulations. The form of amplitude modulation itself undergoes considerable development within the family (Martin, 1972), but the basic system is unique to this family and *Odontophrynus,* a South American leptodactylid. Whether this group represents the precursors of the Bufonidae cannot be decided on the similarity of the basic call pattern alone, but this evidence may provide a clue about where to look for the ancestors of the bufonids. Within the presently defined family Leptodactylidae, there is great similarity between the basic call structure of neotropical leptodactylids of the genus *Leptodactylus* and Australian leptodactylids of the genus *Limnodynastes*. It is necessary to examine the whole sound production and hearing systems of these two groups further to determine if these similarities are due to real homologues or to a convergence related to aquatic breeding habits. Within the family Hylidae also, neotropical and Australian hylas show an unusual similarity in the basic spectral structure and temporal pattern of their calls, and both are distinguishable from other hylids. The diversity of habitats occupied by these two groups argues against convergence, and the possibility of close affinity might be fruitfully investigated. The common North American species of *Rana* exhibit the most simple communication system associated with the so-called mating call that has yet been described. Even the most primitive procoela have a more advanced system based on the functional aspects of their calls—well-developed tuning and temporal or side band spectral patterning. Indeed, some of the presumed primitive procoela (Cycloraninae) have very advanced calls, as pointed out earlier. Simplication of a necessary functional system without passing through some peculiar adaptive zone during evolutionary development (e.g., as with vision in snakes) is not very probable. Thus, the ranids would have diverged from the procoela prior to the development of the most primitive call types found in that group today. Given the magnitude of divergence and diversification of calls within the procoela, evolution would need to have occurred very rapidly along many parallel lines if the ranids are to be considered a relatively late derivative of this group. Evidence from call patterns then would suggest placing the divergence of the ranid and procoela lines at a much earlier time.

So little is known about the actual functional systems associated with mating calls that any interpretation of their evolutionary implications at this stage is pure speculation. However, the few interpretive possibilities raised here lead me to believe that study of the mating call and other vocal behaviors in anurans will lead to clearer understanding of the evolution of major taxa within the order, as well as elucidate relationships at the species group and generic levels.

REFERENCES

AWBREY, F. T. 1965. An experimental investigation of the effectiveness of anuran mating calls as isolating mechanisms. Ph.D. Disseration, Univ. Texas, Austin.

BLAIR, W. F. 1964. Acoustic behavior of Amphibia, p. 694-708. *In* R. G. Bushnel, [ed.], Acoustic behaviour of animals. Elsevier Publ. Co., Amsterdam.

———. 1973. Major problems in anuran evolution. This volume.

BOGERT, C. M. 1960. The influence of sound on the behavior of amphibians and reptiles. *In* W. E. Lanyon and W. N. Tavolga, [ed.], Animal sounds and communication. Am. Inst. Biol. Sci. Publ. 7:137-320.

BRATTSTROM, B. H., AND R. M. YARNELL. 1968. Aggressive behavior in two species of leptodactylid frogs. Herpetologica 24(3):222-228.

CAPRANICA, R. R. 1965. The evoked vocal response of the bullfrog. M.I.T. Research Monograph 33, 106 p.

CAPRANICA, R. R., AND L. S. FRISHKOPF. 1966. Responses of auditory units in the medulla of the cricket frog. J. Acoust. Soc. America 40:1263(A).

CHERRY, C. 1968. On human communication: a review, a survey, and a criticism. 2nd ed. M.I.T. Press, Cambridge, Mass.

ESTES, R., AND O. A. REIG. 1973. The early fossil record of frogs. This volume.

HEUSSER, H. 1969. Der rudimentäre Ruf der männlichen Erdkröte *(Bufo bufo)*. Salamandra 5:46-56.

KONISHI, M. 1970. Evolution of design features in the coding of species-specificity. Am. Zool. 10:67-72.

LANYON, W. E. 1966. Hybridization in meadowlarks. Bull. Am. Mus. Nat. Hist. 134:1-25.

_______. 1969. Vocal characters and avian systematics, p. 291-310. *In* R. A. Hinde, [ed.], Bird vocalizations. Cambridge Univ. Press.

LITTLEJOHN, M. J. 1965*a*. Vocal communication in frogs. Australian Nat. Hist. 15:52-55.

_______. 1965*b*. Premating isolation in the *Hyla ewingi* complex (Anura, Hylidae). Evolution 19(2):234-243.

LITTLEJOHN, M. J., AND J. J. LOFTUS HILLS. 1968. An experimental evaluation of premating isolation in the *Hyla ewingi* complex (Anura, Hylidae). Evolution 22:659-663.

LITTLEJOHN, M. J., AND A. A. MARTIN. 1969. Acoustic interaction between two species of leptodactylid frogs. Animal Behaviour 17:785-791.

LOFTUS HILLS, J. J., AND B. M. JOHNSTONE. 1970. Auditory function, communication, and the brain–evoked response in anuran amphibians. J. Acoust. Soc. America 47:1131-1138.

LOFTUS HILLS, J. J., AND M. J. LITTLEJOHN. 1971. Pulse repetition rate as the basis for mating call discrimination by two sympatric species of *Hyla*. Copeia 1:154-156.

LYNCH, J. D. 1973. The transition from archaic to advanced frogs. This volume.

MARLER, P. 1969. Tonal quality of bird sounds, p. 5-18. *In* R. A. Hinde, [ed.], Bird vocalizations. Cambridge Univ. Press.

MARTIN, W. F. 1972. Evolution of vocalization in the genus *Bufo*, p. 279-309. *In* W. F. Blair, [ed.], Evolution in the genus *Bufo*. Univ. Texas Press, Austin.

RABB, G. B., AND M. S. RABB. 1963. On the behavior and breeding biology of the African pipid frog *Hymenochirus boettgeri*. Z. Tierpsychol. 20:215-241.

SCHIØTZ, A. 1973. Evolution of anuran mating calls: ecological aspects. This volume.

STRAUGHAN, I. R. 1966. An analysis of the mechanism of sex and species recognition and species isolation in certain Queensland frogs. Ph.D. Dissertation, Univ. Overland, Brisbane, Australia.

WILSON, E. O., AND W. H. BOSSERT. 1963. Chemical communication among animals. Recent Progr. Hormone Research 19:673-716.

11

ECOLOGICAL GENETICS OF ANURANS AS EXEMPLIFIED BY *RANA PIPIENS*

David J. Merrell

It seems wise to consider first the nature of ecological genetics and to compare and contrast it with population ecology and population genetics, which are its parents. A review of the literature reveals that the word has been used with somewhat different meanings by different authors (Ford, 1964; Sokal, 1965; Lerner, 1963; Sheppard, 1958; Wallace, 1968), while others who have engaged in what clearly seems to be research in ecological genetics have not described their research in these terms (Turesson, 1923; Clausen and Hiesey, 1958; E. Anderson, 1949; Stebbins, 1950, 1959; Dobzhansky, 1951; and others). In view of this diversity of opinion about the nature of ecological genetics, it seems appropriate to outline briefly my own views on the subject. Ecological genetics is essentially a union between population ecology and population genetics, combining certain aspects of each discipline but differing in certain respects from both.

POPULATION ECOLOGY

Population ecology deals with the kinds of organisms present in an area and with their distribution and numbers. In the case of animals, the population ecologist tries to determine the kinds of species present, their distribution, and their abundance and how these elements are influenced by the physical conditions in the environment, the vegetation, and the other animals present—members of the same species and of other species, which may be predators, prey, competitors, symbionts, or commensals. Population ecology may deal in statics, which involves an attempt to describe a population at one point in time, or it may deal in dynamics, which focuses on the physical or biological factors that cause changes in the kinds of species present or in their distribution and abundance.

However, even though the population ecologist may recognize that variation exists, the genotypes of the organisms are generally assumed to be similar and unchanging, as though the organisms were a group of identical black boxes in their response to the environment. In his research the population ecologist, through sampling, identifies the organisms present and determines their distribution. In order to ascertain their numbers, he may capture, mark, release, and recapture individuals in order to calculate the Lincoln-Petersen index, or he may use some other form of census. To study the behavior of individuals he may monitor them continuously, through the use of radioactive isotopes or radiotelemetry.

In order to study population trends he may attempt to estimate reproductive success by measuring fecundity, fertility, or the birth rate, and by measuring longevity, or death rates. Through such information he may be able to estimate the reproductive potential of the population or its natural rate of increase. As these data accumulate, he will be able to detect fluctuations and changes in the numbers, distribution, and kinds of organisms present in an area and to attempt to relate these changes to observed changes in the physical and biological environment.

Although some research in population ecology is carried on in the laboratory, much of it is conducted in the field.

POPULATION GENETICS

The unit of study in population genetics is the breeding population, which may be as small as a deme, a local breeding population, or as large as the entire species. It may, with introgression, be even larger than a single species.

Statics in population genetics usually involves the description of the balance between mutation and selection, or some type of balanced polymorphism, or some form of gene-frequency equilibrium, such as the Hardy-Weinberg equilibrium. Dynamics entails the effort to determine the causes of gene-frequency change due to the effects, alone or in combination, of mutation, selection, migration, and random genetic drift. In population genetics, research has tended to be theoretical and mathematical and has concentrated on the causes of changes in the frequency of genes and genotypes. A static concept of the environment has usually been employed to simplify the calculations. Experimental research has often been conducted in the laboratory, where the environmental conditions can be more rigorously controlled.

The major requirement for research in population genetics is the presence of detectable genetic variation, preferably with a known genetic basis, in the population under study. It may take the form of alternative alleles at a single locus (as in melanism in moths), different chromosome inversion types (as in many *Drosophila* species), different blood tvpes, differences in various proteins or enzymes, the presence of lethal genes, differences between ecotypes, and so on.

The parameters the population geneticist attempts to estimate are gene frequencies, mutation rates, selection coefficients, migration rates, and "effective" population size as the means of estimating the importance of genetic drift.

ECOLOGICAL GENETICS

Ecological genetics, in contrast to population ecology and population genetics, is the study of the adaptation of natural populations to their physical and biological environments, and the mechanisms by which they respond to environmental change. It requires the realization that populations are dynamic units that are very precisely adapted genetically to their environments and that are sensitive to, and within limits responsive to, any change in their environmental conditions. The ecological geneticist does not consider a population a set of identical black boxes that all respond in an identical fashion to their impinging environment (as often seems the case in population ecology), nor does he regard the environment as a fixed set of parameters that act uniformly on a variable population (as often seems the case in population genetics). Rather, the interplay, the interaction between the genetically variable population and its ever-changing environment, is the central focus of attention in ecological genetics. Thus, the ecological geneticist must be concerned not just with the kinds of organisms present and their distribution and numbers but also with the gene pool of the populations under study. Ecological genetics is really the direct study of evolution at the level at which it actually occurs. Such studies, which examine the ways in which natural populations are adapted to their environments and the mechanisms by which they respond to environmental change, hold promise for advances of considerable theoretical and practical significance in a world where rapid environmental change, due primarily to human activities, seems to be the order of the day.

Ecological genetics requires a combination of field and laboratory research. In the field, gene frequencies may be estimated in different populations in the same type of habitat, in different habitats, and under clinal conditions. In the same population, gene frequencies may be estimated seasonally or annually or in different age classes. Gene frequencies may also be determined in relation to particular environmental factors, such as temperature, moisture, and predation pressure.

Experimental field studies may involve the release of mutants into existing populations to study their fate—to estimate, for example, selection pressures or rates of dispersal. It is also possible to establish entirely new colonies containing both mutant and wild types and then follow the fate of the colony.

For organisms from different populations, a standard method for estimating the relative influence of genetic and environmental factors on the phenotype is reciprocal transplantation, rearing each in the other's environment, or else rearing them together in a common environment. This technique has been used particularly with plants but would no doubt be fruitful if used more often with animals.

The mode of inheritance of the genetic traits under study can best be determined by crosses carried out in the laboratory, although in some cases it can be inferred from the field data. Useful information on viability, fertility, longevity, and other traits can often be obtained by studies of the genetic variants under laboratory conditions. Moreover, the possible range of phenotypic expression controlled by a given genotype can best be ascertained by rearing the individuals under a variety of controlled environmental conditions in the laboratory. In addition, selection experiments in the laboratory—for example, on dominance modification—may lead to fruitful inferences about events in the field. Also, experimental laboratory populations simulating the natural populations but under controlled conditions may be helpful in the study of the effects of particular aspects of population dynamics, such as competition, selection, mutation, and drift. Thus, it can be seen that ecological genetics represents a fusion of the concepts and techniques of both genetics and ecology.

STUDIES WITH *RANA PIPIENS*

As an illustration of a study in ecological genetics, the work on the *burnsi* and *kandiyohi* polymorphisms in populations of leopard frogs *(Rana pipiens)* of the northern Middle West will be used. In 1922 Weed described two new species of frogs from Minnesota as *R. burnsi* and *R. kandiyohi,* but Moore (1942) showed *burnsi* to be a simple dominant, and Volpe (1955) later showed *kandiyohi* to be dominant also. S. C. Anderson and Volpe (1958) found *burnsi* and *kandiyohi* to be nonallelic, actually not even linked but on different chromosome pairs, for a cross of the double-dominant heterozygote (B/+; K/+) to wild type (+/+; +/+) gave equal numbers of BK, K, B, and wild type progeny.

As to distribution of the dominant genes, relatively little was known. Breckenridge's (1944) map was based on less than 250 frogs. For the most part, records of individuals were available, rather than samples from populations, so that gene-frequency data were virtually nonexistent.

Thus, the first step in the study was to get some idea of the distribution and frequency of the dominant *burnsi* and *kandiyohi* genes. Although the range of the *Rana pipiens* species complex covers most of North America, these genes are common in only a small part of the range in the northern Middle West, which nonetheless covers an area of some 100,000 square miles. Present information is based on data from more than 20,000 frogs (Merrell, 1965).

Burnsi frogs were found in Minnesota, western Wisconsin, northern Iowa, and the lake region in the eastern Dakotas. The maximum frequencies obtained were of the order of 10 per cent, with the gene frequency, therefore, about 5 per cent. The highest frequencies were found in the Anoka Sand Plain, which was formed by the damming of the Mississippi River by the Grantsburg sublobe of the Wisconsin ice. The distribution of *burnsi* cuts across the ecotone from coniferous forest to prairie, but its range is pretty much confined to the area from which the ice retreated only 10,000 years ago. Incidentally, despite distribution maps indicating subspecies differences between prairie and forested regions (e.g., Wright and Wright, 1949), no such differences were detected in this area. However, it was observed that the usual wild type frogs within the range of *burnsi* had fewer dorsal spots than wild type frogs collected either to the north or the south of the range of *burnsi*.

Frogs bearing the *kandiyohi* gene are also confined to the recently glaciated areas, but they seem to be further restricted to the prairie regions of western Minnesota and the eastern Dakotas. Again, the highest frequencies observed were about 10 per cent of the frogs, with the gene frequency 5 per cent. Thus, the ranges of *burnsi* and *kandiyohi* do overlap, but because both genes are rare, only one double-dominant heterozygote was found among the 20,000 frogs collected.

ADAPTIVE SIGNIFICANCE

The most commonly proffered explanation for the *burnsi* polymorphism was heterozygous advantage—that the *burnsi* heterozygote was superior in fitness to both homozygotes. However, the low frequency of the gene posed a problem, for a simple analysis (Merrell, 1969) showed that the difference in fitness between wild (+/+) and the *burnsi* heterozygotes (B/+) would have to be of the order of 0.01 or 0.001 to maintain the gene at this low frequency. Such a difference in fitness is virtually undetectable experimentally. Since other theoretical possibilities exist, there is no particular reason to favor heterozygous advantage over these other possibilities. Furthermore, among lab-reared frogs, there was no evidence for differences in viability among B/B, B/+, or +/+ individuals (Merrell, 1963, 1969, unpublished manuscript). There was no excess of B/+ frogs, and no deficiency of B/B frogs, which might be expected by analogy with the dominant polymorphisms in Nabours' grouse locusts (1929).

The finding that the highest *burnsi* frequencies are associated with the Anoka Sand Plain and the finding that dorsal-spot numbers are lower within the range of *burnsi* than beyond it suggest that selection favors a reduction in spot number in this region. Two genetic mechanisms must be at work, the *burnsi* gene and the genetic modifiers that reduce the amount of spotting produced by the wild type allele. Other crosses showed dorsal-spot number to be influenced by heredity (Merrell, 1965, unpublished manuscript). The unspotted pattern may be particularly advantageous during the winter, when the frogs rest on the sandy bottoms of the lakes and streams in this area.

Two types of evidence indicate that *burnsi* survives the winter better than wild type (Merrell and Rodell, 1968). Comparison of *burnsi* frequencies in eleven populations in the fall and the following spring indicated an increase in the relative frequency of *burnsi* during this period. Furthermore, in six populations suffering from winter kill, *burnsi* was found to be significantly more frequent among the frogs that survived. Therefore, *burnsi* may have a physiological adaptive advantage over wild type during the winter and be maintained by a form of seasonal selection.

If this is true, it should be possible to detect an advantage for the wild type gene during the warm season. However, an analysis of data from nine populations followed during three summers has not yet revealed any such superiority. A part of the problem is statistical, posed by the low frequency of *burnsi.*

As was the case with *burnsi,* no significant differences in viability were observed between *kandiyohi* heterozygotes (K/+) and wild homozygotes (+/+) reared together in the laboratory (See below). However, from records on age and size at metamorphosis for the laboratory crosses, an interesting and unexpected observation has emerged. The following data are from crosses with large numbers of progeny for two crosses of *kandiyohi* (K/+) and wild (+/+) and for one cross of *burnsi* (B/+; +/+) and *kandiyohi* (+/+; K/+). (The data for one of the K × P crosses were very kindly provided by C. F. Rodell.)

Progeny

Cross		P	K		Total	X^2_{2df}		
1. K × P		96	109		205	0.82	N.S.	
2. K × P (Rodell)		94	117		211	2.52	N.S.	
	Total	190	226		416	3.12	N.S.	
		P	K	B	BK	Total	X^2_{3df}	
3. B × K		86	72	75	72	305	1.75	N.S.

N.S. Not significant.

It can be seen that there is no significant difference in viability. However, the data on age at metamorphosis revealed that in Cross 1, 50 per cent of the *kandiyohi* tadpoles had completed metamorphosis by August 6, 1966, but that the 50 per cent level was not reached by the wild type tadpoles until August 19, 13 days later. In Cross 3, 50 per cent of the K tadpoles had metamorphosed by July 15, 1966, 50 per cent of both + and BK was reached on July 25, and 50 per cent of B on July 27. Thus, it appears that *kandiyohi* tadpoles tend to develop and undergo metamorphosis more rapidly than the other genotypes.

The problem of developing a suitable statistical test for these data remains. Something analogous to the probit analysis used in bioassay seemed appropriate (where the 50 per cent of a lethal dose is used as the most sensitive measure of differences in resistance) until the third set of data (Cross 2) was examined. In this cross, the 50 per cent level was reached on the same day by both K and P, but over the rest of the distribution, the rate of metamorphosis of K preceded that of P. It is hoped that a transformation that is suitable for the analysis of all of these data will be found.

These results suggest that the *kandiyohi* type tends to metamorphose significantly sooner than other genotypes reared in the same tank. Since *kandiyohi* is associated with the prairie habitat, the results also suggest that the adaptive significance of *kandiyohi* lies in its ability to carry development through to metamorphosis faster out on the prairie where, in times of drought, the breeding ponds may dry up before metamorphosis is completed. This finding parallels that of Zweifel (1968), who studied nine amphibian species in the arid Southwest and found "a clear correlation between the rate of development and the length of time the breeding sites may be expected to retain water." He argued that a rapid rate of development also protects against temperature extremes because the temperature tolerance of the tadpoles broadens as development proceeds. This sort of inference obviously needs further checking, both experimentally and in the field, but for the first time we appear to have a clue about the possible adaptive significance of the *kandiyohi* gene, which enables it to persist at a significant frequency in the prairie populations of the leopard frog.

In one study (Merrell, 1968) it was possible to compare population sizes in breeding ponds by using mark-recapture data, with the "effective size" of the breeding population estimated from counts of egg masses. The data showed that "effective sizes" were smaller—often much smaller—than estimates based on the mark-recapture data, which suggests that it may not be valid to estimate the effective size of the breeding population from simple census data. Furthermore, since the effective sizes were small enough for random genetic drift to cause gene-frequency differences between different local breeding populations, it raised the question of why such differences were not observed. The *burnsi* gene, for example, is widely and uniformly distributed at a low frequency of a few percentage points within its range.

Observations of the spring, the summer, and the fall migrations of the frogs, however, suggest that the extensive movements of the frogs during these migrations prevent local differences in gene frequency from becoming established (Merrell, 1970).

REFERENCES

Anderson, E. 1949. Introgressive hybridization. Wiley, New York. 109 p.

Anderson, S. C., and E. P. Volpe. 1958. *Burnsi* and *kandiyohi* genes in the leopard frog, *Rana pipiens*. Science 127:1048-1050.

Breckenridge, W. J. 1944. Reptiles and amphibians of Minnesota. Univ. Minnesota Press, Minneapolis. 202 p.

Clausen, J., and W. M. Hiesey. 1958. Experimental studies on the nature of species. IV. Genetic structure of ecological races. Carnegie Inst. Washington Publ. 615:1-312.

Dobzhansky, T. 1951. Genetics and the origin of species. 3rd ed. Columbia Univ. Press, New York. 363 p.

Ford, E. B. 1964. Ecological genetics. Methuen, London. 335 p.

Lerner, I. M. 1963. Ecological genetics: Synthesis. Proc. 11th Intern. Congr. Genet. 2:489-494.

Merrell, D. J. 1963. Rearing tadpoles of the leopard frog, *Rana pipiens*. Turtox News 41:263-265.

_______. 1965. The distribution of the dominant *burnsi* gene in the leopard frog, *Rana pipiens*. Evolution 19:69-85.

_______. 1968. A comparison of the estimated size and the "effective size" of breeding populations of the leopard frog, *Rana pipiens*. Evolution 22:274-283.

_______. 1969. Limits on heterozygous advantage as an explanation of polymorphism. J. Heredity 60:180-182.

_______. 1970. Migration and gene dispersal in *Rana pipiens*. Am. Zool. 10:47-52.

Merrell, D. J., and C. F. Rodell. 1968. Seasonal selection in the leopard frog, *Rana pipiens*. Evolution 22:284-288.

Moore, J. A. 1942. An embryological and genetical study of *Rana burnsi* Weed. Genetics 27:408-416.

Nabours, R. K. 1929. The genetics of the Tettigidae (Grouse locusts). Bibliogr. Genet. 5:27-104.

Sheppard, P. M. 1958. Natural selection and heredity. Hutchinson, London. 212 p.

Sokal, R. R. 1965. Ecological genetics. [Review of Ford.] Evolution 19:574-575.

STEBBINS, G. L. 1950. Variation and evolution in plants. Columbia Univ. Press, New York. 643 p.

———. 1959. The role of hybridization in evolution. Proc. Am. Phil. Soc. 103:231-251.

TURESSON, G. 1923. The scope and import of genecology. Hereditas 4:171-176.

VOLPE, E. P. 1955. A taxo-genetic analysis of the status of *Rana kandiyohi* Weed. Systematic Zool. 4:75-82.

WALLACE, B. 1968. Topics in population genetics. Norton, New York. 481 p.

WEED, A. C. 1922. New frogs from Minnesota. Proc. Biol. Soc. Washington 34:107-110.

WRIGHT, A. H., AND A. A. WRIGHT. 1949. Handbook of frogs and toads. 3rd ed. Comstock, Ithaca, N. Y. 640 p.

ZWEIFEL, R. G. 1968. Reproductive biology of anurans of the arid Southwest, with emphasis on adaptation of embryos to temperature. Bull. Am. Mus. Nat. Hist. 140:1-64.

12

EVOLUTION OF ANURAN KARYOTYPES

James P. Bogart

Until recently it was believed that anuran karyotypes were very stable and that each family typically had a similar karyotype. New information on many anuran species is not compatible with such theories, and in fact, anurans might well prove to be a very good group with which to demonstrate many aspects of chromosome mutation. Anuran choromosomes are large, easily obtained, typically few in number, and they demonstrate structural diversity within some groups of related species. Great care must be exercised, however, if karyotypic data are to be used as a taxonomic character. Each group of frogs must be individually studied with respect to its particular method of karyotype evolution and differentiation. When the karyotypes of a large number of related species are examined, patterns usually emerge. Thus, it is important to study closely related species before suggesting any meaningful comparisons among distantly related forms.

The karyotypes of six species, each representing a different genus of frogs, are presented in Figure 12–1. These species would not be considered closely related if almost any of the usual comparative characters were used. *Rhinoderma darwinii* from Chile has been placed in the families Bufonidae, Dendrobatidae, and Leptodactylidae; somctimes it is given its own family. *Glossostoma* is a genus of South American microhylid. *Phyllomedusa* or *Pithecopus palliata* is a hylid species from Amazonian Ecuador and Peru. *Eleutherodactylus portoricensis* is one member of this very large leptodactylid genus. *Heleophryne purcelli* is a debated species; this genus may be the only representative of the family Leptodactylidae in Africa. All of these species have 26 chromosomes, a number that has classically been associated with the family Ranidae. Although all have the same number of chromosomes, the karyotypes are not identical. The numbers of large and small chromosomes, secondary constrictions, centromeric positions, and percentage lengths must all be considered when a species' karyotype is being defined or compared. Conversely, the number of chromosomes among related species might be quite different. Centric fusions and fissions have obviously played a role in the karyotype evolution of the eleutherodactyline leptodactylid frogs. Chromosome numbers ranging from 18 to 36 are evident in some species of this lineage (Figure 12–2). Combining the telocentric chromosomes of *Eleutherodactylus ventrimarmoratus* ($2n = 36$) and *E. conspicillatus* ($2n = 34$) from Peru, *E. ricordi* ($2n = 32$) from Florida, *Syrrhophus marnockii* ($2n = 30$) from Texas, *S. leprus* ($2n = 26$) from Vera Cruz, Mexico, *E. melanostictus* ($2n = 22$) and *E. fleischmanni* ($2n = 20$) from Costa Rica always results in a karyotype with 18 chromosomes, such as that found in *E. podiciferus* from Costa Rica. This apparent multiformity of karyotypes is probably restricted to a certain line of related leptodactylid frogs. Karyotypically, the *marmoratus* group of the genus *Leptodactylus* appears to be more closely related to this line of leptodactylid frogs than to other members of the genus *Leptodactylus.* Other species or "lines" of *Eleutherodactylus* have karyotypes that are apparently fixed around the numbers 22 to 26.

Unlike the situation discovered in the genus *Eleutherodactylus,* there seems to have been only one chromosome number change in the genus *Bufo;* this was a reduction from 22 to 20 chromosomes in the fairly large African *regularis* species group (Bogart, 1968*a*). The similarity of most *Bufo* karyotypes is probably reflected by the ease with which species in this genus may be hybridized. These relationships have been well documented by W. F. Blair (1973, this volume). The major chromosomal variation that may be observed in the genus *Bufo* is in the position of secondary constrictions (Bogart, 1969). *Bufo*-like karyotypes have been found among other species of related genera, such as *Atelopus, Melanophryniscus,* or *Pedostibes,* and there has been very little morphologically detectable differentiation in the karyotypes of these related forms (Morescalchi and Gargiulo, 1968; Bogart, 1972, unpublished data). Thus, we have two very different approaches to karyotype evolution exemplified on the one hand by the family Bufonidae, with its stable number of chromosomes and subtle changes,

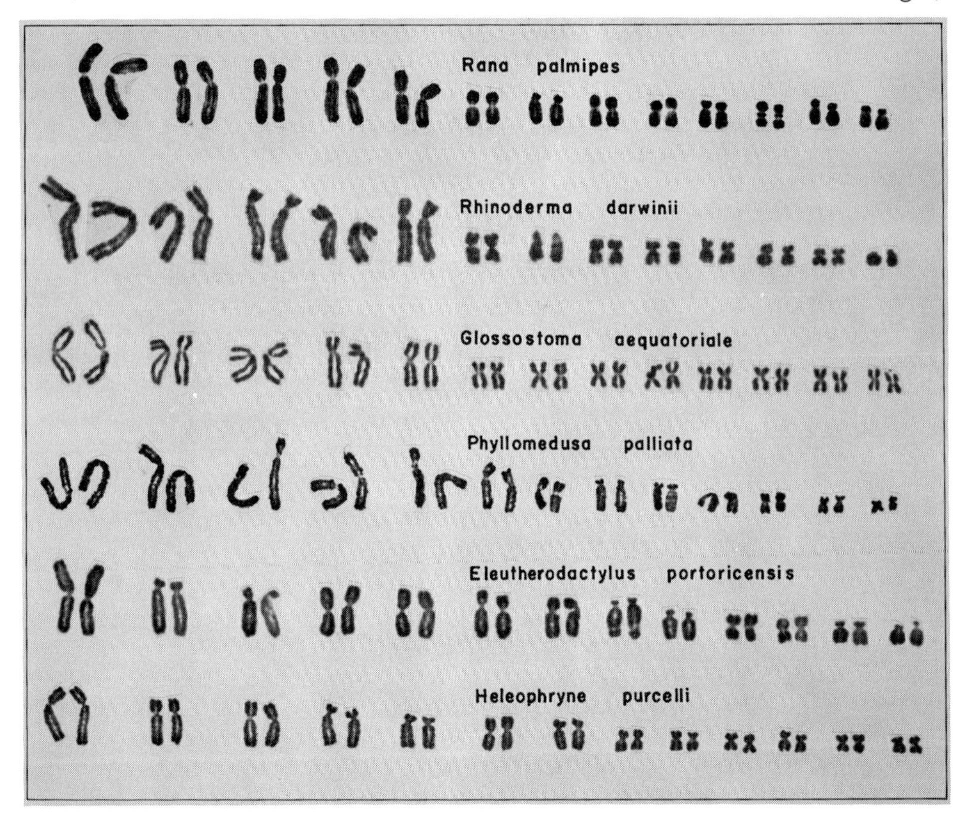

Figure 12–1. Series of karyotypes which demonstrate that distantly related species may have the same chromosome number. *Rana palmipes* from Colombia; *Rhinoderma darwinii* from Chile; *Glossostoma aequatoriale* from Ecuador; *Phyllomedusa palliata* from Peru; *Eleutherodactylus portoricensis* from Puerto Rico; and *Heleophryne purcelli* from South Africa.

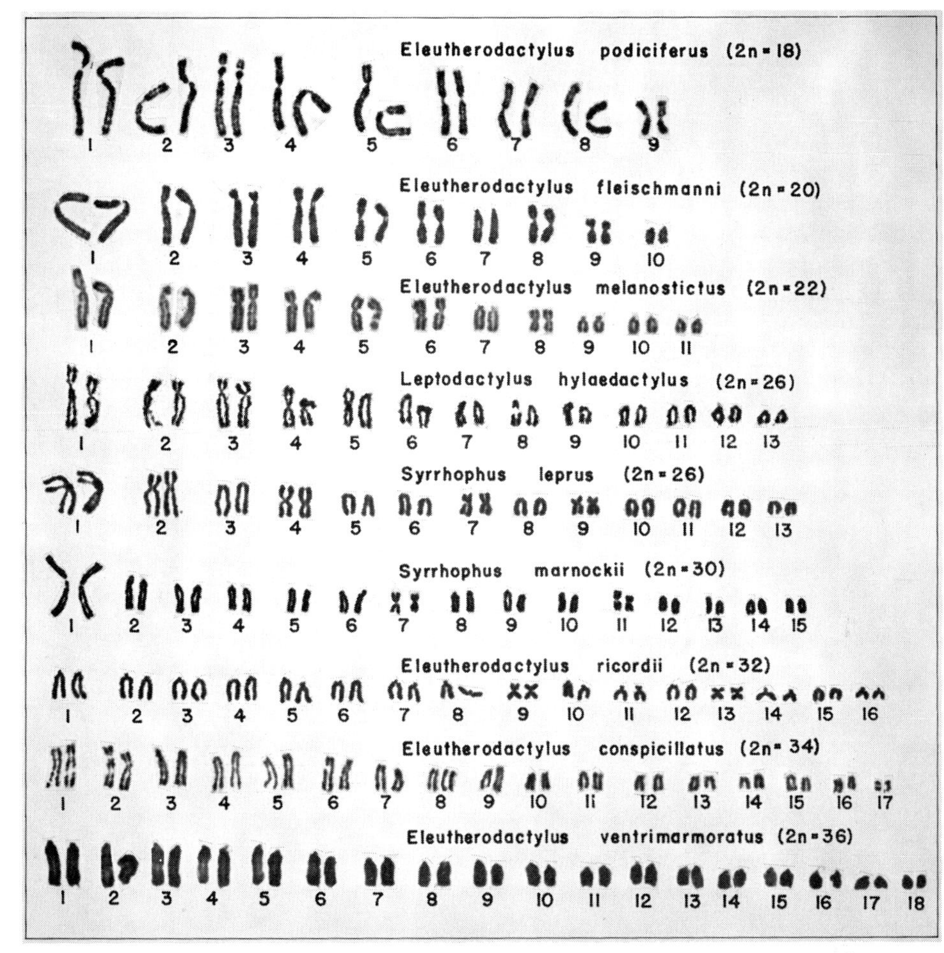

Figure 12–2. Series of karyotypes which demonstrate that chromosome numbers may vary considerably among closely related species. *Eleutherodactylus podiciferus, E. fleischmanni,* and *E. melanostictus* are from Costa Rica; *E. conspicillatus, E. ventrimarmoratus,* and *Leptodactylus hylaedactylus* are from Peru; *E. ricordii* is from Florida; *Syrrhophus marnockii* is from Texas; and *S. leprus* is from Vera Cruz state in Mexico.

and on the other hand by the genus *Eleutherodactylus,* which is characterized by considerable variability in chromosome numbers.

Some of the genera in the family Hylidae demonstrate karyotype differentiation that may prove to be quite meaningful in helping to understand the evolution of the family. Excepting the genus *Acris,* the holarctic hylid frogs have similar karyotypes (Figure 12–3) and hybrid combinations indicate that these species are comparatively closely related. *Phrynohyas venulosa* from Colombia and *H. sordida* from Panama are included with the holarctic hylids because their karyotypes are similar to the majority of holarctic hylids. These species may be an important link between the holarctic and neotropical hylid frogs. All of these species have 24 chromosomes and similar karyotypes. *Pseudacris clarki* and *P. streckeri* are externally somewhat specialized, but chromosomally and genetically they are more similar to North American *Hyla* species than would be indicated by their generic distinctness. This fact is borne out by the production of very viable, chromosomally normal hybrids between *Pseudacris* and some North American species of *Hyla*. *Hyla baudini* and *Hyla sordida* are sometimes considered to be generically distinct from *Hyla* (Duellman and Trueb, 1966; Goin, 1961; Starrett, 1960), but again viable, diploid hybrids are produced in cross combinations with *Hyla baudini* and *H. chrysoscelis*. Perhaps the *baudini* group is similar to the ancestral group that gave rise to the holarctic hylids. To give this group generic distinction would needlessly confuse the evolutionary picture.

Two large chromosomal groupings are found in neotropical members of the genus *Hyla:* those with karyotypes having 24 chromosomes and those possessing 30 chromosomes. The karotypes of frogs in the so-called *rubra* group of the genus *Hyla* (Figure 12–4) indicate that *H. rubra* from Peru and *H. rostrata* from Colombia are similar to each other and to the majority of 24-chromosome *Hyla*, whereas *H. brieni* and *H. catharinae*—both from the eastern coastal mountains of Brazil—have many more submetacentric chromosomes. *Hyla perpusilla,* also from southeastern Brazil, might be considered intermediate among these five species. It is notable that the 30-chromosome species of *Hyla* are, generally, relatively small in

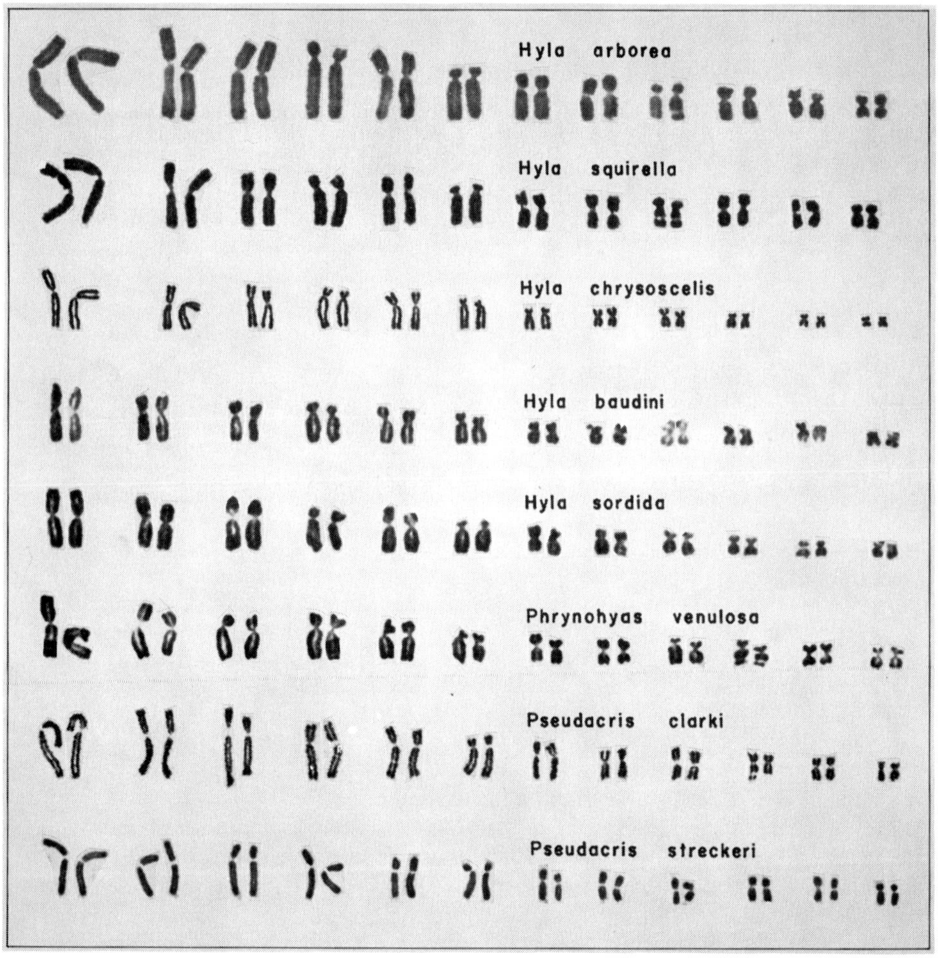

Figure 12–3. Karyotypes of some of the 24-chromosome hylid frogs. *Hyla arborea* from Israel; *H. squirella, H. chrysoscelis, Pseudacris clarki,* and *P. streckeri* from Texas; and *H. baudini* from Mexico are representatives of the holarctic hylid frogs. *Hyla sordida* from Panama and *Phrynohyas venulosa* from Colombia have very similar karyotypes to the holarctic *Hyla*.

body size. Some of the 30-chromosome species whose karyotypes are presented in Figure 12–5 are representatives of larger species groupings. The karyotypes of the 30-chromosome *Hyla* seem to fall into natural groupings that apparently reflect relationships between the various species and species groupings. Specimens of *Hyla minuta* from Peru exhibit karyotypes that are similar to karyotypes of *H. minuta* from far-off coastal Brazil. Later in this discussion we will see evidence of variation in chromosome number and structure within a single species. It will become evident that it is important to examine the chromosomes of more than one population of the same species, especially if these populations are isolated or if there is any doubt about the specific status of a population. *H. leali* may be grouped with *H. minuta* karyotypically, for neither of these species have any telocentric chromosomes and the karyotypes are similar. *H. microcephala, H. bipunctata, H. elongata, H. rhodopepla,* and *H. anceps* all have one and possibly the same pair of telocentric chromosomes, which could indicate that these species are somewhat related.

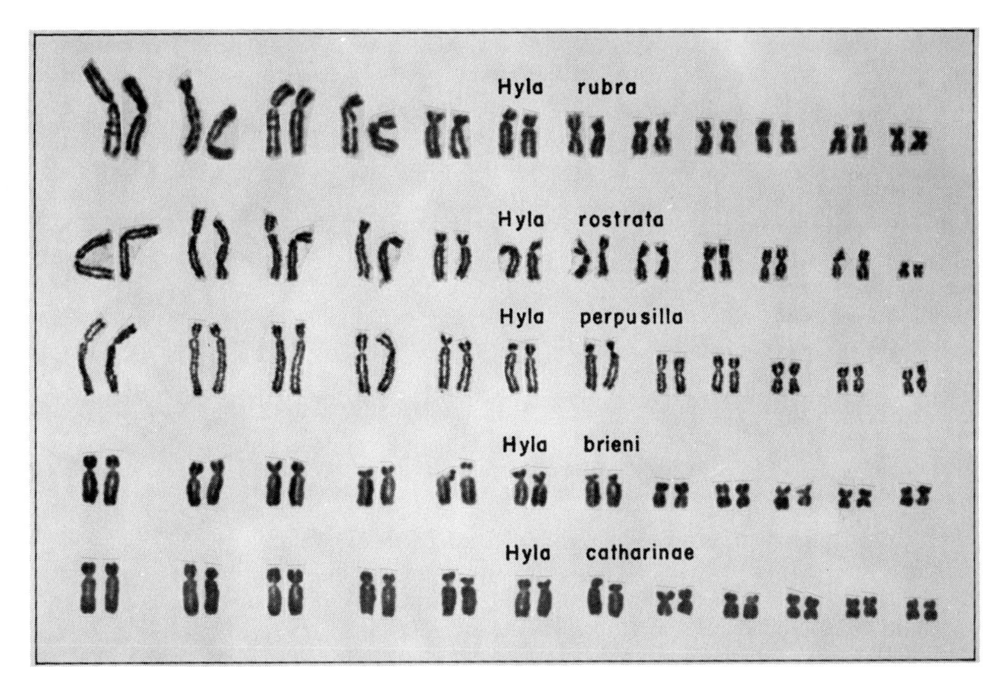

Figure 12–4. Karyotypes of some members of the 24-chromosome, neotropical, *rubra* group of *Hyla*. *H. rubra* is from Colombia; *H. rostrata* is from Peru; *H. perpusilla, H. brieni,* and *H. catharinae* are from southeastern Brazil.

The karyotypes of other 30-chromosome species of *Hyla* are presented in Figure 12–6. All these species have more than one pair of telocentric chromosomes, and there are some quite unexpected chromosomal similarities, such as the close karyotypic similarity between *H. labialis* and *H. marmorata.* Since the karyotypes between species believed to be related by other criteria are similar, it may be extrapolated that the species with morphologically similar karyotypes may be referred to a relatively recent common ancestor. *Hyla decipiens,* with two pairs of telocentric chromosomes, may be related to the *H. microcephala, H. bipunctata, H. elongata, H. rhodopepla, H. anceps* line, which has only one pair of telocentric chromosomes. *H. decipiens* might also be related to *H. nana,* for both

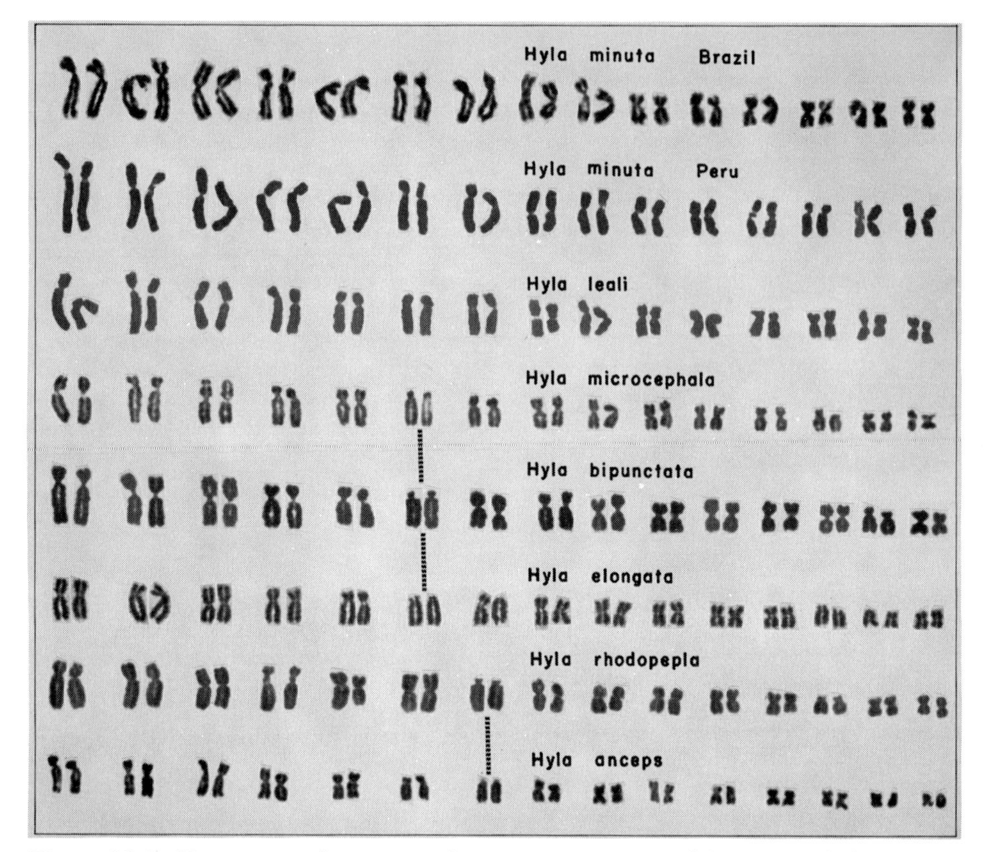

Figure 12–5. Karyotypes of some 30-chromosome neotropical hylas. *H. leali* and *H. rhodopepla* are from Peru; *H. microcephala* is from Colombia; *H. bipunctata, H. elongata,* and *H. anceps* are from southeastern Brazil; and karyotypes representing two populations of *H. minuta,* one from southeastern Brazil and the other from Peru.

these species have the same two pairs of large telocentric chromosomes. In order to make more meaningful comparisons among these 30-chromosome species of *Hyla,* idiograms must be constructed and compared from averaged measurements of many cells for each species—a time-consuming task that has not yet been completed.

Beçak (1968) published the karyotype of *H. albopunctata,* which has only 22 chromosomes. This is the type species of a group in which Cochran (1955) included *H. polytaenia* and *H. pulchella.* The fact that *H. albopunctata* has 22 chromosomes does not mean very much unless its karyotype is compared with those of closely related species. It is only through such studies that numerical changes may better our understanding of the karyotypic variation and, perhaps, the evolution of the group. In *Bufo,* a large species grouping can be derived from a single reductional event. Perhaps all the species related to *H. albopunctata* have only 22 chromosomes. *H. crepitans* and *H. pardalis* are members of the *faber* group of South American *Hyla.* The karyotypes (Figure 12–7) indicate that *H. polytaenia, H. pulchella, H. crepitans,* and *H. pardalis* all have 24 chromosomes. The morphology of their chromosomes is similar to that of the rest of the 24-chromosome *Hyla,* with the exception of *H. brieni* and *H. catharinae* of the *rubra* group. *Hyla albopunctata* apparently has a specialized karyotype that is different in many ways from the karyotypes of *H. polytaenia* and *H. pulchella.* Evidently

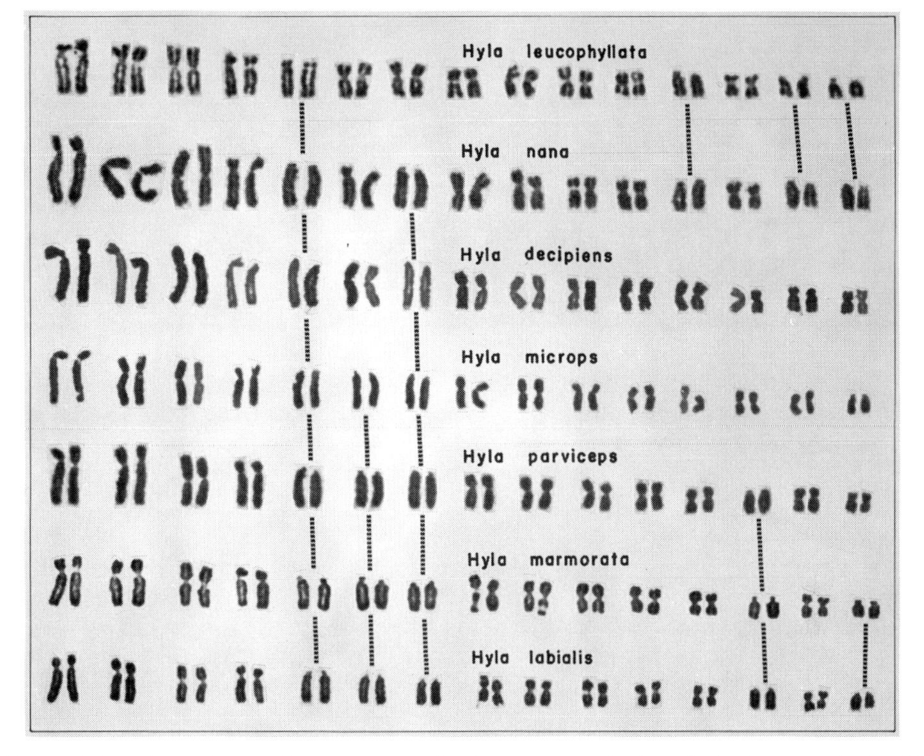

Figure 12–6. Karyotypes of neotropical 30-chromosome Hylas. *Hyla leucophyllata, H. parviceps,* and *H. marmorata* are from Peru; *H. nana* is from northern Argentina; *H. decipiens* and *H. microps* are from southeastern Brazil; and *H. labialis* is from Colombia.

there was a chromosomal reduction that involved one of the small pairs of chromosomes. No telocentric chromosomes are apparent in other members of this group that have been studied, and therefore there is no evidence for centric fusion. Perhaps the reduction could have been brought about by the end-to-end fusion of two chromosomes or the translocation of the genetic material to one or more chromosomes from a small pair that became a heteromorphic, or "B," chromosome and was eventually lost. *Hyla polytaenia*'s last pair of chromosomes consistently appears to be less heavily stained than the other chromosomes, which could indicate that this pair of chromosomes has less DNA and is in fact approaching a heterochromatic composition.

An even more striking case of chromosome reduction in the genus *Hyla* is presented in the green, *albomarginata* group of South American *Hyla* (Figure 12–8). All these *Hyla* species appear to be related by call, color, white peritonium, cloritic blood, and habits. *Centrolenella* is very specialized but has many similarities with this group of green hylas. They are often found sympatrically. The populations of *Hyla albofrenata* that I studied are morphologically distinct and have different chromosome numbers, so it is most probable that we are dealing with two species, not one. The same holds true for *H. albosignata. H. punctata* and *H. granosa* from Peru and *H. albofrenata* from Tijuca forest area in Rio de Janeiro all have the expected 24 chromosomes and thus are similar in morphology to many of the North and South American 24-chromosome *Hyla. H. albofrenata* from outside of São Paulo at Boracéa has only 22 chromosomes. *H. albosignata* from Boracéa has 20 chromosomes, and *H. albosignata* from Ter-

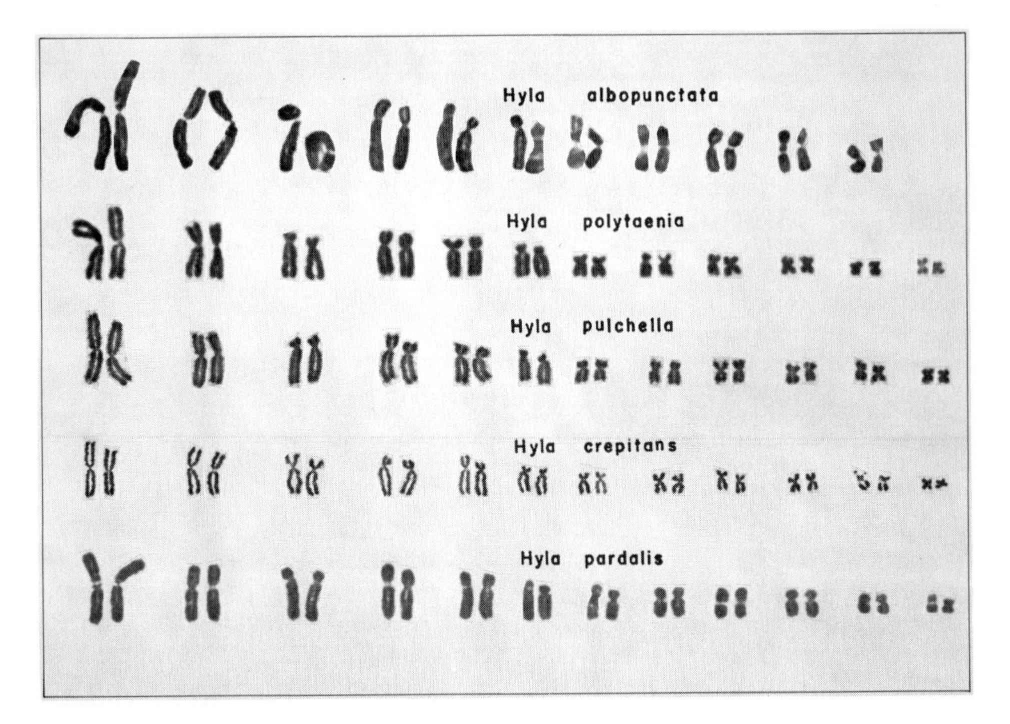

Figure 12–7. Karyotypes of some 24-chromosome neotropical hylas and the 22-chromosome *Hyla albopunctata*. *H. albopunctata*, *H. polytaenia*, and *H. pardalis* are from southeastern Brazil; *H. pulchella* is from Argentina; and *H. crepitans* is from Colombia.

esopolis outside of Rio de Janeiro has only 18 chromosomes. Again, the smallest chromosomes seem to be involved in the reduction, but the actual mechanism responsible for the reduction is still uncertain. Once again there are no telocentric chromosomes evident in this group. *Centrolenella eurygnatha* and other members of this genus have only 20 chromosomes, and the chromosome configurations of species in this genus are like those of the 20-chromosome *Hyla albosignata* from Boracéa. Chromosomal, morphological, and ecological similarities may indicate some phylogenetic relationship, rather than represent merely a remarkable case of convergent evolution. So far, work with North and South American hylas has revealed two main groupings. These are the 24-chromosome group and the 30-chromosome group. The few species with 22, 20, and 18 chromosomes are individual cases, seemingly derived from related species having 24 chromosomes. No species of South American *Hyla* has yet been found with 26 or 28 chromosomes; such a species would bridge the gap between these two major groups. If the number change resulted from centric fission, we might expect to find some species with intermediate numbers, as well as 30-chromosome species having six pairs of telocentric chromosomes derived through the centric dissociation of three metacentric pairs to produce 30-chromosome karyotypes from an original 24. It is possible that the chromosomal mutations took place a very long time ago and the intermediate steps are not now to be found. We must also bear in mind the many species that have not been sampled. There is some evidence, however, that the 30-chromosome and 24-chromosome *Hyla* are derived from two entirely different lines of hylid frogs and that the apparent morphological similarity is a product of incomplete study and convergence. The actual transition from one number to the other (i.e., from 24 to 30) may not have taken place at all; it may have involved a common ancestral karyotype that had 26 chromosomes. The karyotypes of some of the species of hylid frogs that are relevant to the evolution of karyotypes in the family Hylidae are presented in Figure 12–9. *Gastrotheca gracilis* has a karyotype quite similar to some of the so-called primitive genera of leptodactylid frogs, such as *Telmatobius, Calyptocephalella (=Caudiverbera), Thoropa, Batrachyla, Cyclorhamphus, Lepidobatrachus, Ceratophrys,* and others (Bogart, 1967, 1970, unpublished data). Most workers agree that the Hylidae broke off from the Leptodactylidae, and if this is the case, a primitive karyotype for the hylid frogs would most likely have 26 chromosomes. A 24-chromosome karyotype is not common in the family Leptodactylidae. *Gastrotheca, Phyllomedusa, Fritziana goeldii,* and at least some of the Australian *Hyla* still maintain a fairly unspecialized leptodactylid-like karyotype, and it may well be that these groups are living representatives of some of the most primitive hylid frogs. *Fritziana ohausi,* with 28 chromosomes, and *Flectonotus fissilis,* with 30, could reflect the mechanism that gave rise to the 30-chromosome *Hyla.* From the examples provided by *Fritziana* and *Flectonotus,* centric dissociation is responsible for the increase in number. If we assume that the 30-chromosome *Hyla* had a 26-chromosome ancestor, we would expect to find some species with four pairs of telocentric chromosomes, such as is found in *H. microps* and *H. parviceps.* Evidently, pericentric inversions have shifted the position of the centromeres in many cases from a telocentric to a more metacentric position. Chromosomally, the genus *Hyla* appears to be a very artificial assemblage. The Australian species in this genus are probably primitive and more closely related to *Gastrotheca, Phyllomedusa,* or some species included in the family Leptodactylidae. The 30-chromosome *Hyla* and the 24-chromosome *Hyla* were probably independently derived from a 26-chromosome ancestor.

Other hylid genera besides *Hyla* have 24 chromosomes. Recently Barrio (1970; Barrio and de Rubel, 1970) has studied the chromosomes and calls of members of the family Pseudidae, which have karyotypes very similar to those of some 24-chromosome hylas. *Trachycephalus* and *Corythomantis (=Aparasphenodon)* also have 24 chromosomes. *Corythomantis brunoi* is one of the cask-headed Brazilian hylids, *Pseudis paradoxa* from Argentina is a questioned hylid, and *Acris* is a North American genus in the family Hylidae. The karyotypes of these three species are presented in Figure 12–10. *Corythomantis brunoi* and *Pseudis paradoxa* have a 24-chromosome karyotype that is similar to many of the 24-chromosome *Hyla. Acris crepitans* has only 22 chromosomes, and the karyotype is very similar to some species of the genus *Leptodactylus.* Chromosomally, at least, it is very possible that *Acris* could be a leptodactylid, rather than a hylid, genus.

So far, two independent occurrences of polyploidy have been discovered in the Hylidae. Wasserman (1970) discovered that *Hyla versicolor* is a tetraploid species ($2n = 4X = 48$), and Beçak et al. (1970) discovered that *Phyllomedusa burmeisteri* in Brazil is also a tetraploid, with $2n = 4X = 52$. Speciation by means of polyploidy in anurans would seem to offer an exciting new area of study.

Morescalchi (1968*c*) has compared the chromosomes of some of the so-called primitive species of anurans. He has concluded that reduction in chromosome number is a characteristic of the higher anurans and that the primitive species all have a large number of chromosomes. Unfortunately, in many cases there are not enough related species of the "primitive" anurans to tell anything about their mechanism of chromosomal variation. For example, *Ascaphus truei* has been found to have 44 chromosomes (Morescalchi, 1967) or 46 chromosomes (Bogart, 1968*b*). Two female specimens of the New Zealand species *Leiopelma hochstet-*

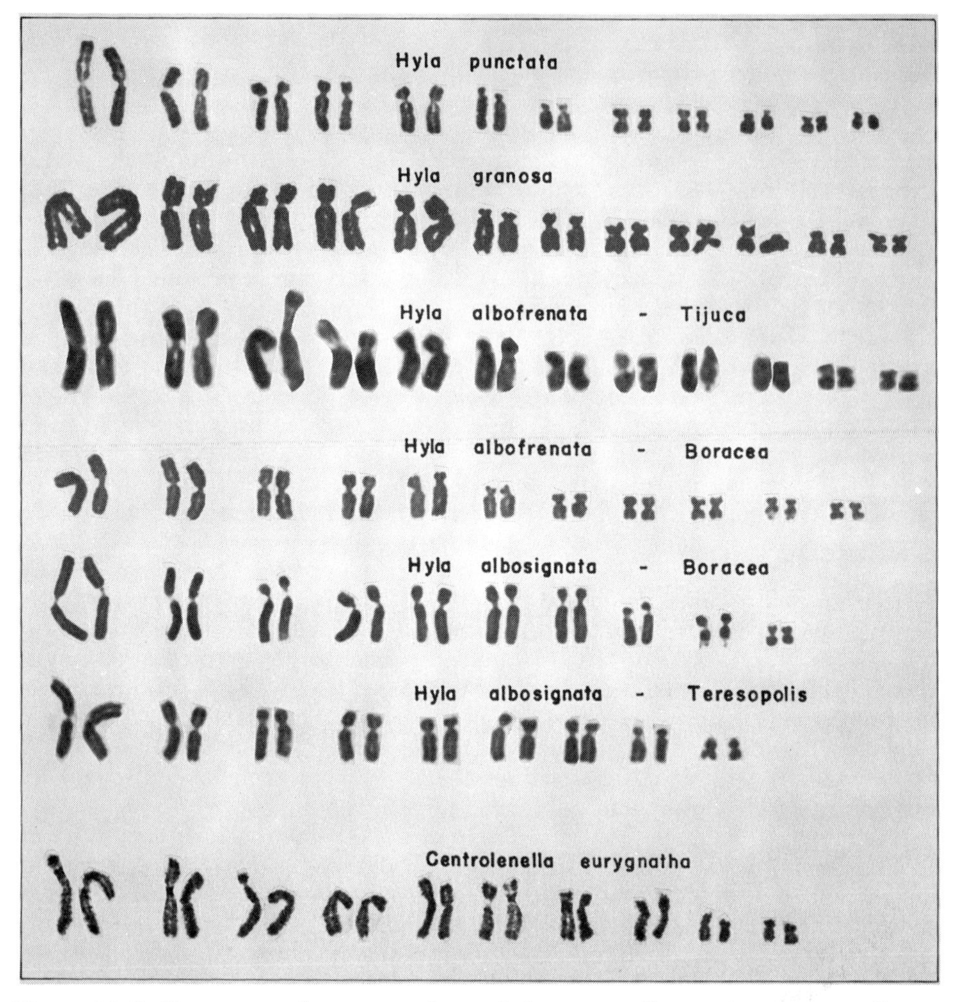

Figure 12–8. Karyotypes of some members of the green *albomarginata* group of neotropical hylas. *Hylas punctata* and *H. granosa* are from Peru. *Hyla albofrenata* (2n = 24) from Tijuca forest in Rio de Janeiro, Brazil; *H. albofrenata* (2n = 22) and *H. albosignata* (2n = 20) are from Boracéa, northeast of São Paulo, Brazil; and *H. albosignata* (2n = 18) is from Teresopolis, west of Rio de Janeiro in Brazil.

teri were found to have 34 and 23 chromosomes by Morescalchi (1968*a*). Thus, in the most primitive family, Ascaphidae, three chromosome numbers have been found among two species in the family, making it difficult to compare the karyotypes of these species with any other anuran. In the Pipidae, *Pipa parva* was found to have 30 telocentric chromosomes (Morescalchi, 1968*b*), and its karyotype is quite different from that of other members of the family that have been studied. Rhinophrynidae is supposedly a very primitive family, but *Rhinophrynus dorsalis* has a karyotype very similar to some of the genera of microhylid frogs and has a karyotype consisting of 22 chromosomes (Bogart, unpublished data).

Although it is much too premature to come to any definite conclusions concerning the evolution of karyotypes among all Anura, some speculation is warranted, based on the evidence available for some of the families. Beginning from

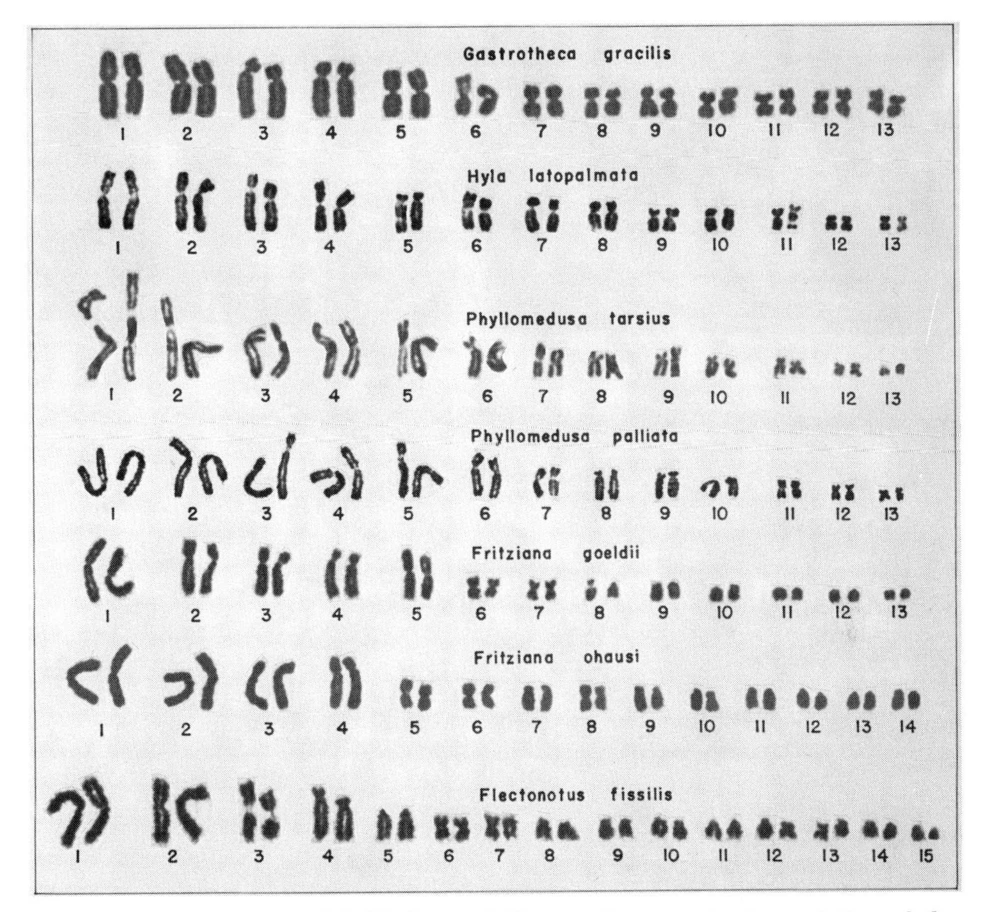

Figure 12–9. Karyotypes of hylid frogs which may demonstrate the evolution of the 30-chromosome *Hyla* species. *Gastrotheca gracilis* is from northern Argentina; *Hyla latopalmata* is from eastern Australia; *Phyllomedusa tarsius* and *P. palliata* are from Peru; and *Fritziana goeldii, Fritziana ohausi,* and *Flectonotus fissilis* are from southeastern Brazil.

a 26-chromosome pelobatid-like karyotype, it is possible to derive all the karyotypes of the so-called higher families of anurans. This pelobatid-like karyotype is present in such leptodactylid genera as *Calyptocephala*(?), *Telmatobius, Cyclorhamphus, Thoropa, Ceratophrys,* and *Lepidobatrachus.* Some of these genera are considered to be quite primitive leptodactylids; the fossil record for *Calyptocephala*(?) goes back to the lower Oligocene (Schaeffer, 1949). Some Australian leptodactylids, such as *Mixophrys*, have 24-chromosome karyotypes that are like those found in some South American leptodactylids (Bogart and Morescalchi, unpublished data). A 26-chromosome karyotype could be the ancestral type for the family Hylidae, as was speculated above. A reduction in chromosome number in the family Leptodactylidae must have produced at an early time 22- and 24-chromosome karyotypes, which represents major dichotomies within the family Leptodactylidae. Leptodactylid genera with 22 chromosomes include all the *Eupemphix*-like genera, such as *Eupemphix, Edalorhina, Pleurodema,* and *Engystomops,* as well as the genus *Leptodactylus,* with the exception of the *marmoratus* group. Also having similar 22-chromosome karyotypes are the chromosomally

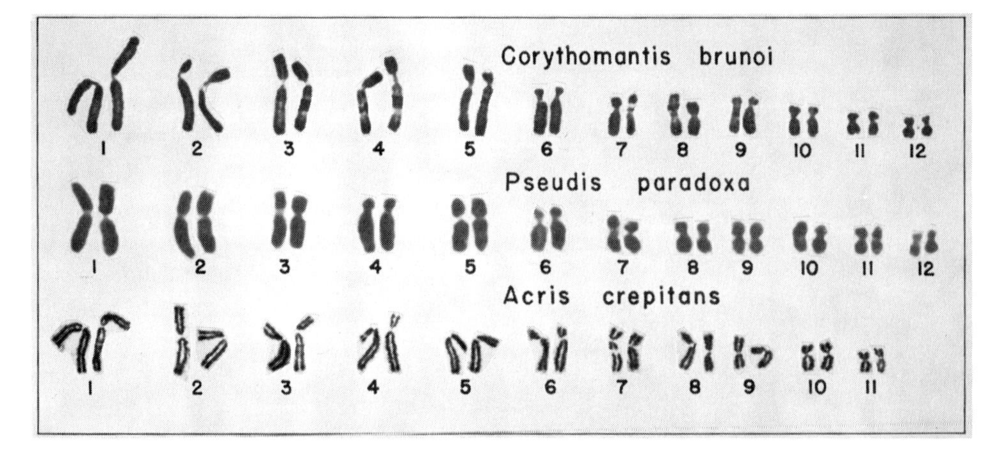

Figure 12–10. Karyotypes of the casque-headed hylid, *Corythomantis (=Aparasphenodon) brunoi* from southeastern Brazil; *Pseudis paradoxa* of the family Pseudidae from southeastern Brazil; and *Acris crepitans* from Texas.

prebufonid leptodactylid genera, such as *Odontophrynus, Melanophryniscus, Dendrophryniscus, Macrogenioglottus,* and *Stombus.* The family Ranidae may also be derived from an early 26-chromosome lineage. Recent results from study of some African species indicate that there is considerable variation in karyotypes among some of the ranid frogs—if indeed they really are ranid frogs—living in that continent. The family Microhylidae apparently has karyotypic roots among the two major lineages (i.e., 22 chromosome and 26 chromosome). It is very possible that this family is not monophyletic. The family Dendrobatidae may be derived chromosomally from a 26-chromosome ancestor, such as the leptodactylid *Elosia.* Dichotomies in this family are demonstrated by chromosomal reductions. Several species of *Phyllobates* (including *Colostethus* and *Hylaxis*) have 24 chromosomes, and *Dendrobates* species have 20 and 18 chromosomes.

Comparative material from Africa and Australia should help to supply the karyotypic information in the areas of doubt, and we should be able to discover other mechanisms of chromosomal mutation that will help us to interpret the differences between various karyotypes.

Acknowledgments

First and foremost, I would like to thank the principal investigator of this study, Dr. W. F. Blair of the University of Texas, who provided encouragement, equipment, and financial assistance. My friends and fellow students William Birkhead, William Martin, Terry Matthews, Craig Nelson, Richard Newcomer, Dennis Ralin, and Richard Sage all assisted in some of the field work. Professional assistance was kindly offered by and gratefully accepted from the following people: Dr. Avelino Barrio, Instituto Nacional de Microbiologia, Buenos Aires; Dr. Flavio Bazán, former Director General, Servicio Forestal y de la Caza, Lima, Peru; Drs. Maria Luiza and Willy Beçak, Instituto Butantan, São Paulo, Brazil; Dr. Henrique Rodrigues da Costa, Universidade Federal do Rio de Janeiro; Dr. Eugenio Izecksohn, Universidade Federal Rural do Rio de Janeiro; Dr. Raymond F. Laurent, Instituto Miguel Lillo, Tucumán, Argentina; Dra. Bertha Lutz, Museu Nacional, Rio de Janeiro; Dr. P. E. Vanzolini, Director de Museu de Zoologia, São Paulo, Brazil; and Dr. Luiz Dino Vizotto, Departamento de Zoologia, Faculdade de Filosofia e Letras, São Jose do Rio Prêto, Brazil. Living material was kindly sent from Australia by Dr. Jeanette Covacevich, from Puerto Rico by Dr. George Drewry, from Florida by Dr. Roy McDiarmid, and from Israel by Dr. Eviatar Nevo. I extend my thanks to all these associates. My greatest debt of gratitude is to my companion and wife, Jo Ellen, who was a constant source of encouragement, a very able translator, and a dependable field and laboratory assistant.

REFERENCES

BARRIO, A. 1970. Caracteres del canto nupcial de los pseudidos (Amphibia, Anura). Physis 29:511-515.

BARRIO, A., AND DELIA PISTOL DE RUBEL. 1970. Caracteristicas del cariotipo de los pseudidos (Amphibia, Anura). Physis 29:505-510.

BEÇAK, M. L. 1968. Chromosomal analysis of eighteen species of anura. Caryologia 21:191-208.

BEÇAK, M. L., L. DENARO, AND W. BEÇAK. 1970. Polyploidy and mechanisms of karyotypic diversification in amphibia. Cytogenetics 9:225-238.

BLAIR, W. F. 1972. Evidence from hybridization, p. 196-232. *In* W. F. Blair, [ed.], Evolution in the genus *Bufo*. Univ. Texas Press, Austin.

BOGART, J. P. 1967. Chromosomes of the South American amphibian family Ceratophridae with a reconsideration of the taxonomic status of *Odontophrynus americanus*. Canadian J. Gen. Cytol. 9:531-542.

_______. 1968*a*. Chromosome number difference in the amphibian genus *Bufo:* the *Bufo regularis* species group. Evolution 22(1):42-45.

_______. 1968*b*. Los cromosomas de anfibios anuros del género *Eleutherodactylus*. Act. IV Congr. Latinoam. Zool.

_______. 1969. Chromosomal evidence for evolution in the genus *Bufo*. Ph.D. Dissertation, Univ. Texas, Austin.

_______. 1970. Systematic problems in the amphibian family Leptodactylidae (Anura) as indicated by karyotypic analysis. Cytogenetics 9:369-383.

_______. 1972. Karyotypes, p. 171-195. *In* W. F. Blair, [ed.], Evolution in the genus *Bufo*. Univ. Texas Press, Austin.

COCHRAN, D. M. 1955. Frogs of southeastern Brazil. U.S. Natl. Mus. Bull. 206, 409 p.

DUELLMAN, W. E., AND L. TRUEB. 1966. Neotropical hylid frogs of the genus *Smilisca*. Univ. Kansas Publ. Mus. Nat. Hist. 17(7):281-375.

GOIN, C. J. 1961. Synopsis of the genera of hylid frogs. Ann. Carnegie Mus. 36(2):5-18.

MORESCALCHI, A. 1967. Note citotassonomiche su *Ascaphus truei* Stejn. (Amphibia Salientia). Atti. Soc. Pelorit. Sci. Fis. Mat. Nat. 13:23-30.

_______. 1968*a*. The karyotypes of two specimens of *Leiopelma hochstetteri* Fitz. (Amphibia, Salientia). Caryologia 21:37-46.

_______. 1968*b*. I cromosomi di alcuni Pipidae (Amphibia, Salientia). Experientia 24:81-82.

_______. 1968*c*. Hypotheses on the phylogeny of the salientia, based on karyological data. Experientia 24:964-966.

MORESCALCHI, A., AND G. GARGIULO. 1968. Su alcune relazioni cariologiche del genere *Bufo* (Amphibia, Salientia). Estr. Rend. Accad. Sci. Fis. Mat. Soc. Naz. Sci., Lett., Arti, Napoli, 4, 35:117-120.

SCHAEFFER, B. 1949. Anurans from the early tertiary of Patagonia. Bull. Am. Mus. Nat. Hist. 93:45-68.

STARRETT, P. 1960. Descriptions of tadpoles of middle american frogs. Misc. Publ. Mus. Zool. Univ. Michigan 110:1-37.

WASSERMAN, A. O. 1970. Polyploidy in the common tree toad *Hyla versicolor* LeConte. Science 167:385-386.

13

THE GEOGRAPHIC DISTRIBUTION OF FROGS: PATTERNS AND PREDICTIONS

Jay M. Savage

INTRODUCTION

Some fifteen years ago, I presented my views of the significance of the distribution of frogs in a symposium entitled Problems in the Phylogeny of the Salientia (*Copeia,* 1955, 4:319). A number of papers from that symposium were published subsequently (Brattstrom, 1957; Jameson, 1957; Orton, 1957). My contribution, "Distributional Patterns and Salientian Evolution," was based upon the same kinds of data and working principles used in preparing the present paper. The report, although written before the present resurgence of interest and evidence for continental drift (Blackett et al., 1965; King, 1967; Runcorn, 1962) and seeming in retrospect a reasonable interpretation for its day, left me dissatisfied with my conclusions, and for this reason it was never published. Perhaps it is just as well that it was not. Like most other biogeographers in the 1940's and 1950's, I was influenced significantly by Matthew's (1915) and Simpson's (1953) ideas in my evaluation of distributional evidence. Their views were based primarily upon the Cenozoic history of mammal groups, which fit a hypothesis of permanency of the ocean basins and relative continental stability. A major feature of their explanation of mammal history involved the concept of the origins of major groups on the northern land masses, with southward migration and invasion of more southern areas. The extensive fossil record of mammals supported these ideas. Nevertheless, the peculiar physiology of mammals (endothermic) and the fact that most extant families arose from the Miocene onward bars the use of what may be a perfectly valid formulation of biogeographic events for mammals in the Cenozoic as a model to explain the history of frogs (ectothermic and with most extant families present by the Cretaceous). It is now becoming increasingly apparent that the Matthew-Simpson theory does not form an adequate basis for explaining the Mesozoic origins and subsequent radiation of fresh-water fishes, amphibians, or reptiles and probably does not apply to Mesozoic origins of mammals as well. Thus, much that I had to say fifteen years ago about the history of the frogs and their distribution now appears to be questionable, as does the interpretation of Darlington (1957).

For the present paper, I began anew to evaluate the distributional history of the Anura. As the study proceeded, I realized that a completely different approach to interpretation was essential. It is now obvious that the key to understanding present patterns of frog biogeography lies in the Jurassic to early Cretaceous history of the two major world continents, Laurasia and Gondwanaland,

and their subsequent fragmentation and metamorphosis from mid-Cretaceous onward.

The aim of this paper is to relate the known facts of frog distribution to the known facts of present and past climate and geography. Using the theory of continental drift as an acceptable description of broad geodynamic events in earth history, I have developed a biogeographic theory that integrates distributions, climatic associations, major geologic changes, and drift. This theory seems to explain more adequately the current frog distribution patterns than those theories based upon other models of world geologic history (Wallace, 1876; Darlington, 1957). The theory, as presented, is founded upon the methods and principles outlined in earlier reports (Savage, 1960, 1963, 1966, 1973). Because of the complexity of the subject, I have outlined only the major patterns and directly related historic events. Detailed analysis of phylogenetic histories, key faunal areas, and new fossil discoveries, as well as critical evaluation of my ideas by those familiar with frogs, surely will modify and shape the theory. I hope that the more controversial points and deficiencies in interpretation will challenge others to attack the questions raised by my formulations.

THE MAJOR GROUPS OF ANURANS

All attempts to group living frogs into major phyletic lines based upon adult morphology (Cope, 1865; Nicholls, 1916; Noble, 1931; Griffiths, 1963; Inger, 1967; Kluge and Farris, 1969) reveal serious weaknesses when paleontological, biogeographic, and developmental features are considered. As pointed out by Hecht (1963), the concept of four major evolutionary lines within the Anura that is based upon larval features (Orton, 1953, 1957) and further substantiated by larval developmental characteristics and larval and adult jaw muscle evolution (Starrett, 1968, this volume) conforms more closely to the probable history than Noble's (1931) system, which was adopted and modified slightly by most other authors. A review of all the arguments leads me to support the general pattern espoused by Hecht (1963), although my views differ from it in some ways, as noted below.

Principal Phyletic Lines

Essentially the frogs appear to be formed of three major more-or-less independent lines that were established by early Jurassic times, approximately 180 million years ago. The first line is the suborder Xenoanura of Starrett (this volume), represented by living forms in which the larvae lack an oral disk, beaks, and denticles and have paired spiracles and well-developed sensory barbels (Orton's Type I larvae). Representatives of this stock are known from the upper Jurassic. A second line, the suborder Scoptanura (Orton's Type II larvae), has larvae that lack an oral disk, beaks, denticles, and sensory barbels and have a single median spiracle located far posterior on the underside of the body near the anal tube, although the spiracle may be rotated laterally in association with the tube in some specialized genera. The earliest incontestible fossil of this line is from the Miocene, but given the difficulty of identifying fossil frog bone fragments with living families, many of which have not had the skeleton intensively studied, earlier finds may be anticipated. The radiation and extreme diversity of external morphology, skeleton (Parker, 1934), and musculature (Starrett, 1968) in the sole nominal family of this group (Microhylidae) transcend the limits of any other frog family. In the case of jaw muscles, the family includes conditions transcending the limits for the tetrapod classes. These features are further strong evidence for the ancient establishment and diversification of the stock. A third stock, known from the upper Jurassic, is represented by living groups in which the larvae have an oral disk, horny beaks on the jaws, horny denticles, and a single spiracle and lack sensory barbels. Some specialized forms within this group have larvae in which the denticles are lost, but they always retain the oral-disk structure and beaks and are obviously derived from forms having the basic oral disk, with beaks and denticles. In a number of genera and families in this large group, the larval stage may be modified or absent, with some form of direct development present (Starrett, this volume). Two distinctive subgroups are seen within this lineage: those with larvae having a median ventral spiracle in the brachial region and double denticle rows (Orton's Type III)—the suborder Lemmanura—and those with larvae having a sinistral brachial spiracle and denticles in a single row (Orton's Type IV)—the suborder Acosmanura. Definitive fossils belonging to these two lines, based upon concordance in adult structures, are recorded from the upper Jurassic (Hecht, 1963).

One family of recent frogs does not fall readily into any of the suborders. This group, the Leiopelmatidae, usually has been included with the Ascaphidae as a single family within Type III. Living leiopelmatids lack a larval stage and undergo a form of direct development, so that no data are available on how they might resemble the larvae of *Ascaphus* and the discoglossids. Jurassic fossils from Argentina (Hecht, 1963; Estes and Reig, this volume) are extremely close to *Leiopelma* in adult skeletal features. It seems possible that the features shared by *Ascaphus* and *Leiopelma*, particularly the perforate amphicoelous vertebrae, are a cluster of primitive characters retained in two independent evolutionary lines. Most of these common features are conditions expected in ancestral groups that might have given rise to rhinophrynids, pipids, microhylids, discoglossids, and these families. Perhaps further study will establish the Leiopelmatidae in a separate suborder, but for the present, I regard its status as uncertain.

Classification of Frogs

For purposes of further analysis all subfamilies and families without subfamilial divisions are treated as equal units. This procedure allows a maximum number of natural clusters of genera to be analyzed without distorting the results by hiding distinctive lines within a single undivided family. The controversies about family limits are minimized as a result, since each unit is more or less equivalent. Forty-four familial units of anurans are recognized in the classification adopted below. This number includes 1 extinct and 43 living groups and approaches the maximum number of families that might be accepted by herpetologists.

Superorder SALIENTIA

Order PROANURA
- Family Triadobatrachidae

Order ANURA

Suborder I (Xenoanura)
- Family Rhinophrynidae
- Family Pipidae
  - Subfamily Pipinae
  - Subfamily Xenopinae
- Family Paleobatrachidae

Suborder II (Scoptanura)
- Family Microhylidae
  - Subfamily Dyscophinae

Subfamily Cophylinae
Subfamily Asterophryinae (includes Sphenophryninae, *Calluella,* and *Colpoglossus*)
Subfamily Microhylinae (includes *Melanobatrachus*)
Subfamily Brevicipitinae
Subfamily Hoplophryninae
Subfamily Phrynomerinae

Suborder III (Lemmanura)
Family Discoglossidae
Family Ascaphidae

Suborder IV (Acosmanura)
Family Pelobatidae
Subfamily Megophryinae
Subfamily Pelobatinae
Subfamily Pelodytinae
Family Myobatrachidae
Subfamily Myobatrachinae
Subfamily Cycloraninae
Family Pelodryadidae
Family Heleophrynidae
Family Leptodactylidae
Subfamily Ceratophryinae
Subfamily Hylodinae
Subfamily Leptodactylinae
Subfamily Telmatobiinae
Family Centrolenidae
Family Bufonidae
Family Brachycephalidae
Family Allophrynidae
Family Pseudidae
Family Rhinodermatidae
Family Hylidae
Family Dendrobatidae
Family Sooglossidae
Family Ranidae
Subfamily Raninae (includes *Hylarana* in part)
Subfamily Platymantinae (includes *Hylarana* in part)
Subfamily Phrynobatrachinae
Subfamily Mantellinae
Subfamily Rhacophorinae
Subfamily Arthroleptinae
Subfamily Hemisinae
Family Hyperoliidae
Subfamily Astylosterninae
Subfamily Hyperoliinae
Subfamily Scaphiophryninae

Suborder uncertain
Family Leiopelmatidae

Within the acosmanurans several clusters of related families are recognizable. The two principal groups are referred to throughout the remainder of the paper by informal collective terms as leptodactyloids (families Myobatrachidae, Heleophrynidae, Leptodactylidae, Rhinodermatidae, Centrolenidae, Brachycephalidae, Allophrynidae, Pseudidae, and Dendrobatidae) and ranoids (families Ranidae and Hyperoliidae).

Several traditional concepts of various units have had to be changed, and these changes should be noted. Probably the most difficult and diverse of frog groups are the microhylids. Parker's (1934) world-wide revision placed great emphasis on the morphology of vertebrae, pectoral girdles, and palates as a basis for suprageneric classification. Accumulating evidence (Carvalho, 1954; Starrett, 1968) indicates that Parker's subfamilies are probably arbitrary morphological groupings that do not necessarily represent clusters of closely related genera. It seems clear that the distinction between the Asterophryinae and Sphenophryninae is invalid. The two differ only in that the former usually have an amphicoelous vertebra just anterior to the sacrum and the other presacral vertebrae procoelous (diplasiocoelous), while the latter have all presacral vertebrae procoelous. One genus, *Genyophryne,* referred to the Asterophryinae by Parker (1934) has all procoelous vertebrae. Elsewhere in the family we now know that the first presacral vertebra may be variously procoelous or amphicoelous in closely related genera (Carvalho, 1954). Both of Parker's nominal subfamilies are centered in New Guinea, and the genera of both have direct development. Inclusion together as the Asterophryinae seems completely appropriate.

I have also placed the two nominal Indo-Malayan genera *Calluella* and *Colpoglossus* (Inger, 1966, synonymized them) in the Asterophryinae. Parker regarded them as members of the same subfamily as *Dyscophus,* a Madagascar genus. *Calluella* appears to be a primitive genus from which the more highly evolved asterophryines may have developed by further expansion of the vomerines (seen in *Colpoglossus*), loss of teeth, and elimination of a larval stage. If this interpretation is correct, the resemblances between *Calluella* and *Dyscophus* derive from their similarities in primitive microhylid features, not from a close relationship.

Parker placed all New World and most Asian genera in the single subfamily Microhylinae, probably because he regarded the Old World *Microhyla* and New World *Gastrophryne* as congeneric, a view refuted by Carvalho (1954). Although I do not treat them as separate taxonomic categories, my impression is that the two clusters of genera form distinct units that resemble one another because of parallel reduction specializations. The genus *Melanobatrachus* of southern India is included with the Asian microhylines; it differs from them only in lacking the external and middle-ear apparatus—features that have been lost repeatedly in frog history in diverse groups—and in having fusion of the ethmoid and parasphenoid, which also occurs in a New World genus (*Synapturanus*). In terms of pectoral-girdle and palatal structure, *Melanobatrachus* is similar to any proposed ancestor for Old World microhylines.

Parker included *Melanobatrachus* and the two endemic montane eastern African genera *Hoplophryne* and *Parahoplophryne* within a distinct subfamily, principally because of the loss of the external and middle-ear apparatus and the ethmoid-parasphenoid fusion in the three genera. The African genera differ from all other Afro-Malagasian microhylids in many features, but they resemble the phrynomerines in several ways. They do not seem closely allied to *Melanobatrachus,* and I regard them as a specialized relict endemic African subfamily. Both features of ear reduction and ethmoid-parasphenoid fusion are examples of end-point specializations that may be expected in unrelated genera.

In dealing with the complexities of ranoid classification, I have followed the opinions of Poynton (1964) in grouping ranid, rhacophorine, and hyperoliid genera. The characters of the Arthroleptinae and Hemisinae suggest that they are probably not ranids or hyperoliids, and I suspect that they have closer relationships with the Sooglossidae. Contrary to Poynton, I regard the Hyperoliidae as a family distinct from the other ranoids. The distinction between the Raninae and Platymantinae is tenuous (Laurent, 1951) but makes some morphologic and geographic sense if the Indian and Malayan *Hylarana* are placed in the latter group. The *Hylarana* of Africa do not seem to be closely allied to those of Asia, and the two may represent parallel adaptive trends. Liem (1969) placed the Mantellinae and Rhacophoridae together as a family—the Rhacophoridae—and restricted the name Hyperoliidae to the hyperoliines. All of the other units that Poynton (1964) recognized are placed by Liem in the Ranidae. Liem suggested that his Rhacophoridae arose from the Phrynobatrachinae and the Hyperoliidae from the Astylosterninae.

The arciferal procoelous tree frogs of Australia and New Guinea have usually been regarded as confamilial with the Hylidae. I can find no evidence that supports this conclusion, and the recent studies by Tyler (1971) on superficial throat musculature refute the association. The Australasia tree frogs do not appear to be closely allied to any Australian genus of Myobatrachidae, although they must have arisen from a common leptodactyloid stock. For these reasons, I regard the Australasian tree frogs as a distinct family, for which the first available name is Pelodryadidae (Günther, 1858). The family name is based upon the distinctive green tree-frog genus *Pelodryas* Günther, 1858 (monotype: *Rana caerulea* White, 1790).

DISTRIBUTION OF FROG FAMILIAL UNITS

At the present time, frogs have the most extensive geographic and ecologic range of the three living orders of amphibians. The limbless, blind, burrowing caecilians (Order Gymnophiona) have a circumtropical distribution (Figure 13–1). The salamanders (Order Caudata) are essentially restricted to the northern extratropical regions, except for an extensive radiation in tropical America (Figure 13–2). Frogs and toads (Order Anura) are found almost everywhere that salamanders and caecilians occur, except in areas having extremely cold winter conditions in the extreme north or on a few high peaks, and are the only amphibians on Madagascar, the West Indies, New Guinea, Australia, New Zealand, and the islands of the western-to-southwestern Pacific.

The distribution of frogs is analyzed in three ways for this study: general ecologic limits (zonation), geographic distribution (regionality), and fossil localities (history). In order to ascertain the general ecologic limits for each familial unit, an attempt was made to plot the distribution of each unit against parameters of major bioclimatic significance. The parameters were established according to Holdridge's (1964) system, as amplified by Tosi (1964) and Holdridge (1967). The limits of climatic data available for each group made it impossible to obtain a clear picture of humidity limits for many of the units, so a more general approach was applied. The approximate limits for temperature-precipitation and correlated evapotranspiration requirements for each of the three orders of amphibians (Figure 13–3) indicate that within each major temperature zone, frogs tend to be distributed widely with relation to available water. Consequently, a map (Figure 13–4) was prepared indicating the approximate latitudinal regions of the earth as defined by mean annual biotemperatures, reduced to sea level, according to the Holdridge diagram (Figure 13–3). Biotemperature is defined as equal to the mean unit period in Centigrade with substitution of 0° for all values less than 0° C and 30° for all values in excess of 30° C. The number of frog familial units on major land areas in each latitudinal zone is indicated on this map.

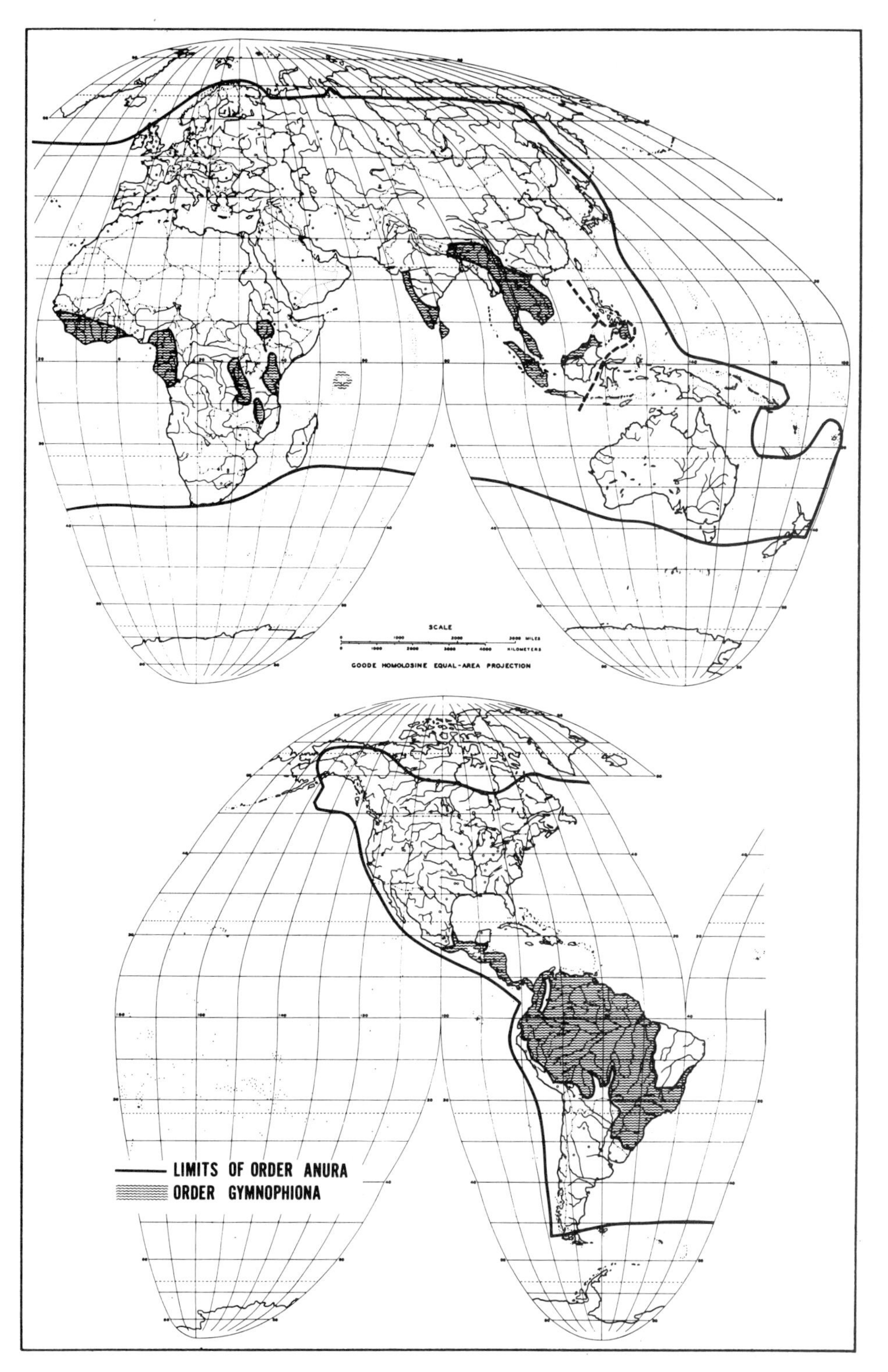

Figure 13–1. Comparison of distribution of frogs (Anura) and caecilians (Gymnophiona).

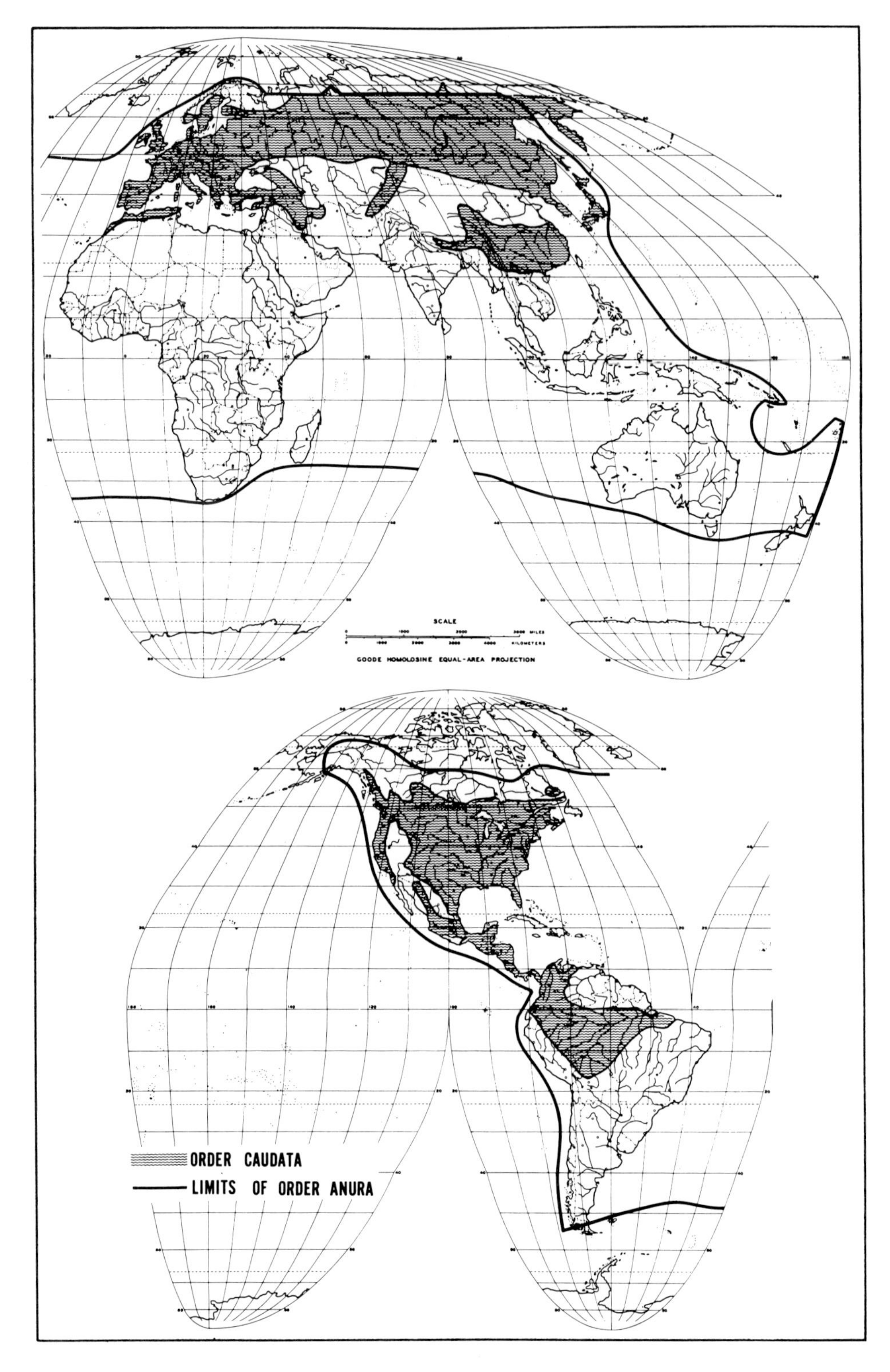

Figure 13–2. Comparison of distribution of salamanders (Caudata) and frogs (Anura).

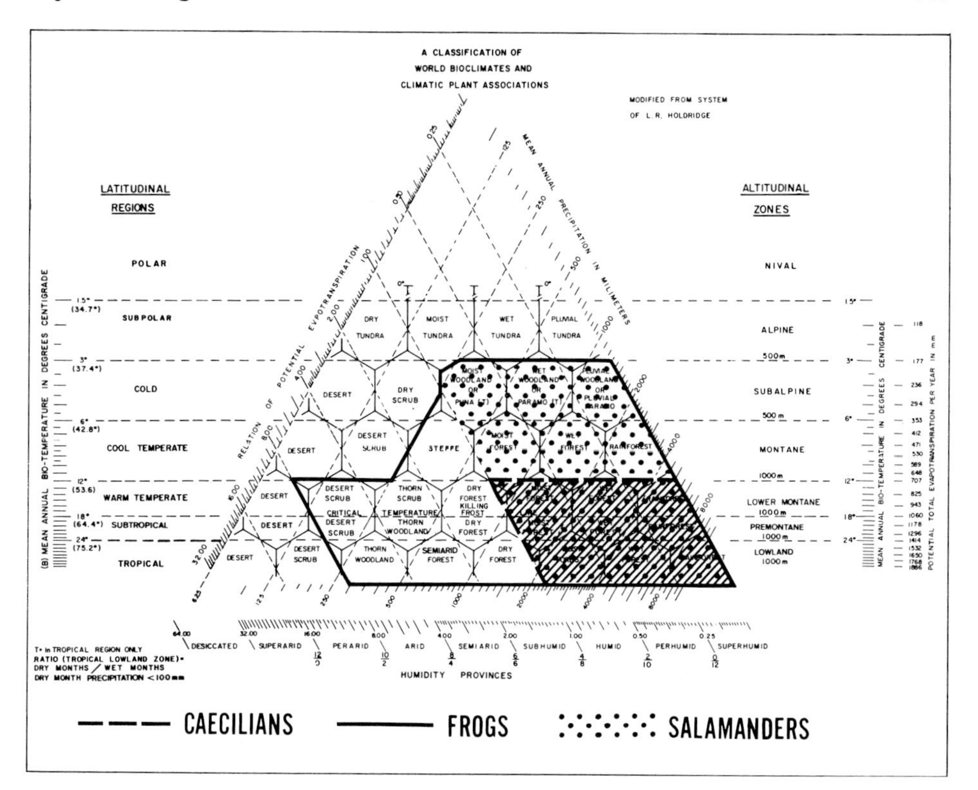

Figure 13–3. Ecologic distribution of living amphibians on world bioclimatic diagram. See text for explanation.

A series of maps (Figures 13–5–13–29) illustrate the known geographic distribution for each familial unit as accurately as data and map scale permit. All maps are standardized for easy comparison to one another and to the map of climatic latitudinal biotemperature regions (Figure 13–4).

Finally, the known localities for fossil representatives of each group are included on the maps of recent distribution. Not all fossil records are indicated. All localities of fossils of an age earlier than Miocene are shown for each unit, with the following abbreviations indicating geologic age: T, Triassic; LJ, lower Jurassic; MJ, middle Jurassic; UJ, upper Jurassic; C, Cretaceous; E, Eocene; O, Oligocene. Only records outside the present distribution of a family are plotted for the Miocene and Pliocene (M, Miocene; P, Pliocene).

General Ecologic Distribution

Living amphibians occur in Tropical to Cold Latitudinal Regions (Figures 13–3, 13–4). Caecilians are restricted to the Tropical Region, where they occur in Lowland, Premontane, and Lower Montane Altitudinal Zones. Salamanders and anurans occur in Tropical, Subtropical, Warm Temperate, Cool Temperate, and Cold Latitudinal Regions and all altitudinal zones in each region. The ecologic limits of salamanders and caecilians are encompassed entirely within the tolerances of the frogs. Caecilians occur nowhere without frog or salamander associates, and their ecologic limits are included within the tolerances of both groups. The portion of the model of major world bioclimates (Figure 13–3) marked with dashed lines indicates the amount of overlap for all three groups and corresponds

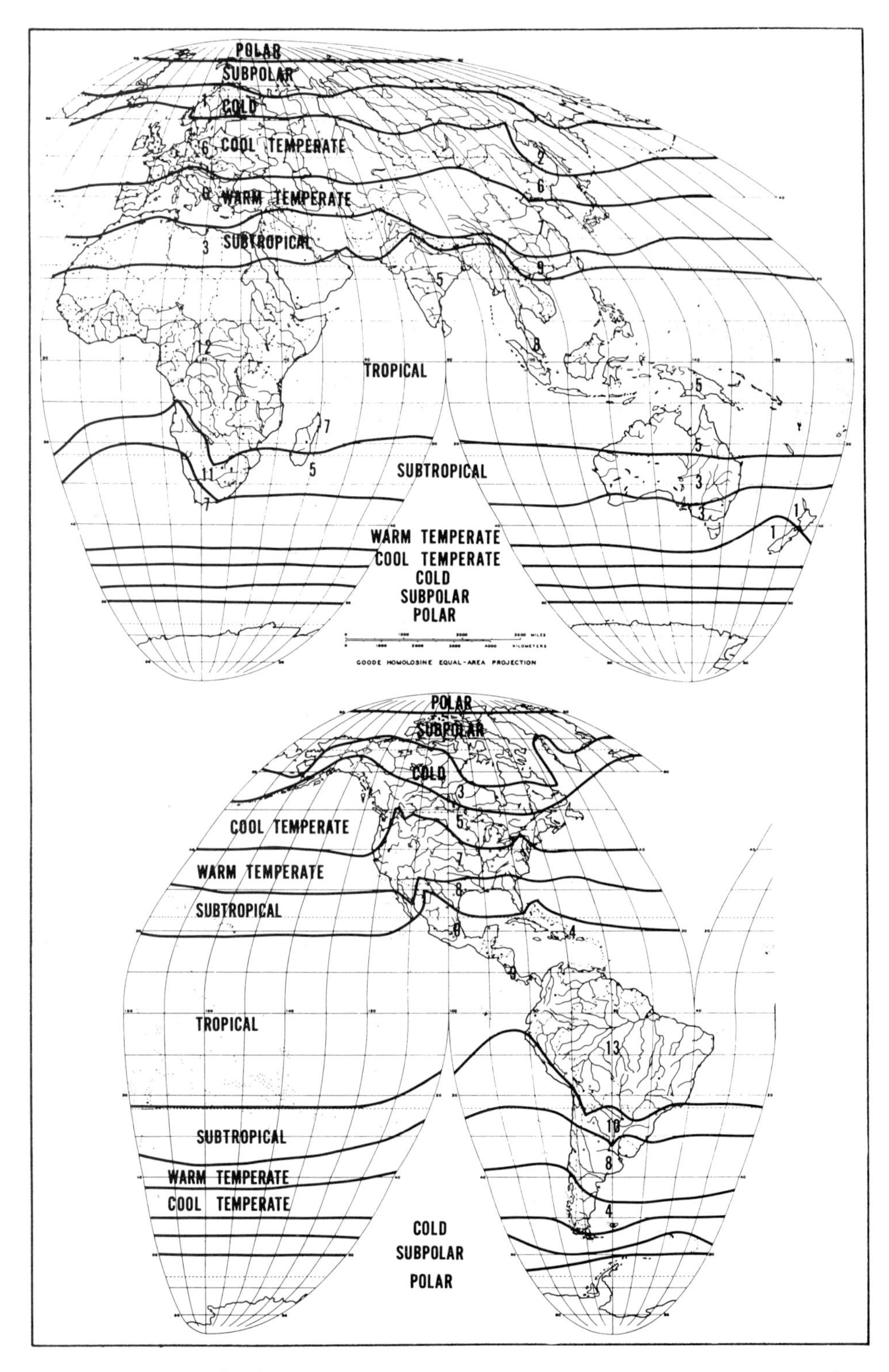

Figure 13–4. Latitudinal zonation of living frog groups. Numbers indicate the number of familial level units found in the latitudinal zone on a particular land mass.

to the ecologic distribution for caecilians. Within the Anura 37 (86 per cent) of the familial units are found in the Tropical Region; 32 (74 per cent) occur in Subtropical Regions; 25 (58 per cent) occur in Warm Temperate Regions; 11 (25 per cent) range into Cool Temperate Regions; and 3 (7 per cent) are found in Cold Regions. The number of units by latitudinal region for each major geographic area is indicated on the map (Figure 13–4). If distributions are analyzed with the northern and southern extratropical regions taken separately, the following pattern emerges:

North Temperate (Cold, Cool, Warm)	11 groups	25%
North Subtropical	13 groups	30%
South Subtropical	24 groups	56%
South Temperate (Cool, Warm)	19 groups	37%

Further analysis of distribution patterns shows that modern frogs have centers of diversity and evolution associated with one of three primary latitudinal zones: Northern Extratropical (Subtropical Warm Temperate to Cold Regions), Tropical, or Southern Extratropical. The familial units by zone and geographic center or centers of diversity, with the number of latitudinal biotemperature regions in which the units occur, are:

Northern Extratropical (4 units)
- Ascaphidae (North America)–2
- Discoglossidae (Eurasia)–4
- Pelobatinae (Eurasia and North America)–4
- Pelodytinae (western Eurasia)–2

Tropical (33 units)
- Rhinophrynidae (Middle America)–2
- Pipinae (South America)–1
- Xenopinae (Africa)–3
- Dyscophinae (Madagascar)–2
- Cophylinae (Madagascar)–2
- Asterophryinae (Indo-Malaya)–2
- Microhylinae (southern Asia and South America)–4
- Hoplophryninae (Africa)–1
- Phrynomerinae (Africa)–3
- Megophryinae (southeastern Asia)–3
- Pelodryadidae (New Guinea)–3
- Ceratophryinae (South America)–3
- Hylodinae (South America)–2
- Leptodactylinae (South America)–3
- Telmatobiinae (South America)–3
- Centrolenidae (South America)–1
- Bufonidae (South America)–5
- Brachycephalidae (South America)–1
- Allophrynidae (South America)–1
- Pseudidae (South America)–3
- Hylidae (South America)–5
- Dendrobatidae (South America)–1
- Sooglossidae (Seychelles)–1
- Raninae (Africa)–5
- Platymantinae (southeastern Asia)–2
- Phrynobatrachinae (Africa)–3

Mantellinae (Madagascar)–2
Rhacophorinae (Africa)–4
Arthroleptinae (Africa)–2
Hemisinae (Africa)–2
Astylosterninae (Africa)–1
Hyperoliinae (Africa)–3
Scaphiophryninae (Madagascar)–1

Southern Extratropical (6 units)
Brevicepitinae (Africa)–3
Leiopelmatidae (New Zealand)–2
Myobatrachinae (Australia)–3
Cycloraninae (Australia)–3
Heleophrynidae (Africa)–2
Rhinodermatidae (South America)–2

These data show that 77 per cent of the units have tropical centers of distribution, 9 per cent are Northern Extratropical, and 14 per cent are Southern Extratropical groups.

Geographic Distribution of Frogs

The present distribution of frog familial units (Figures 13–5–13–29, Tables 13–1–13–2) fall into five major distribution patterns:

Northern Extratropical—distribution both recent and fossil, associated with temperate and cold latitudinal regions of Eurasia and North America

Cosmopolitan Tropical—distribution includes most of major land masses, but greatest diversity in tropical areas

Widespread Tropical—wide-ranging groups with greatest diversity in tropical areas

Tropical—groups essentially limited to tropical areas

Southern Extratropical—groups with highest diversity in southern temperate regions

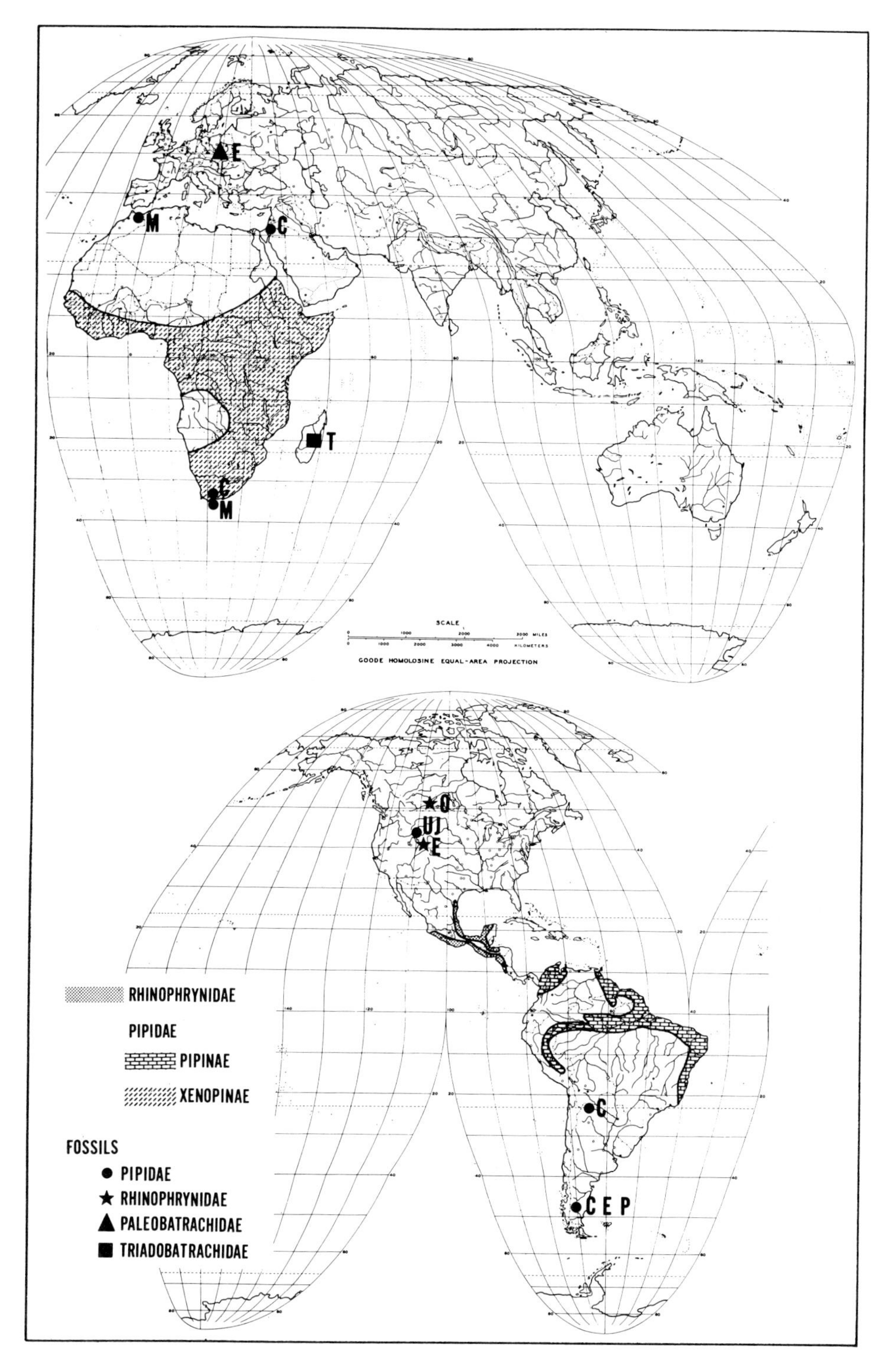

Figure 13–5. Distribution of proanuran amphibians and xenoanuran frog families. Letters indicate fossil ages as identified in text.

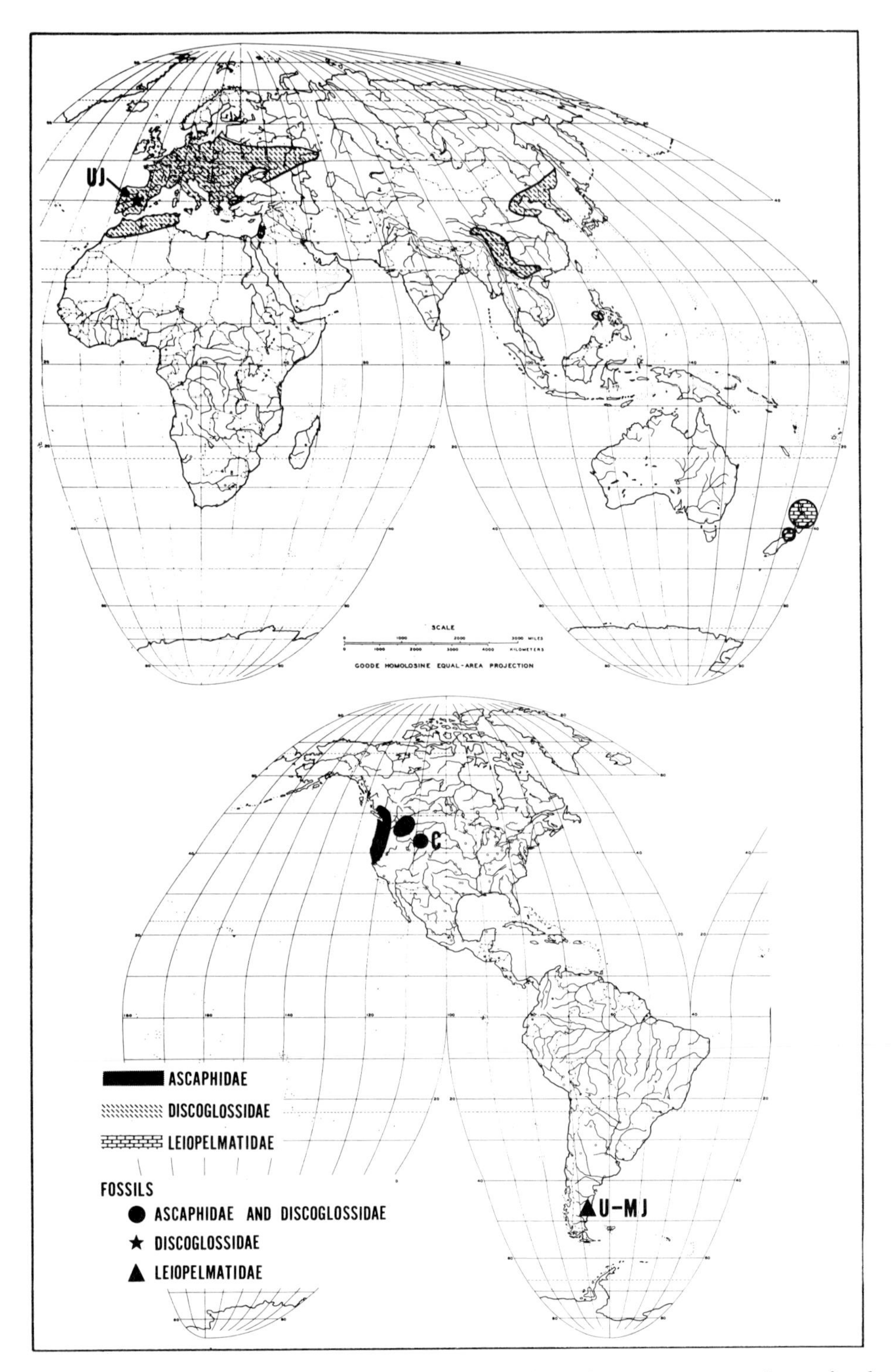

Figure 13–6. Distribution of lemmanuran and leiopelmatid frogs. Letters indicate fossil ages as discussed in text.

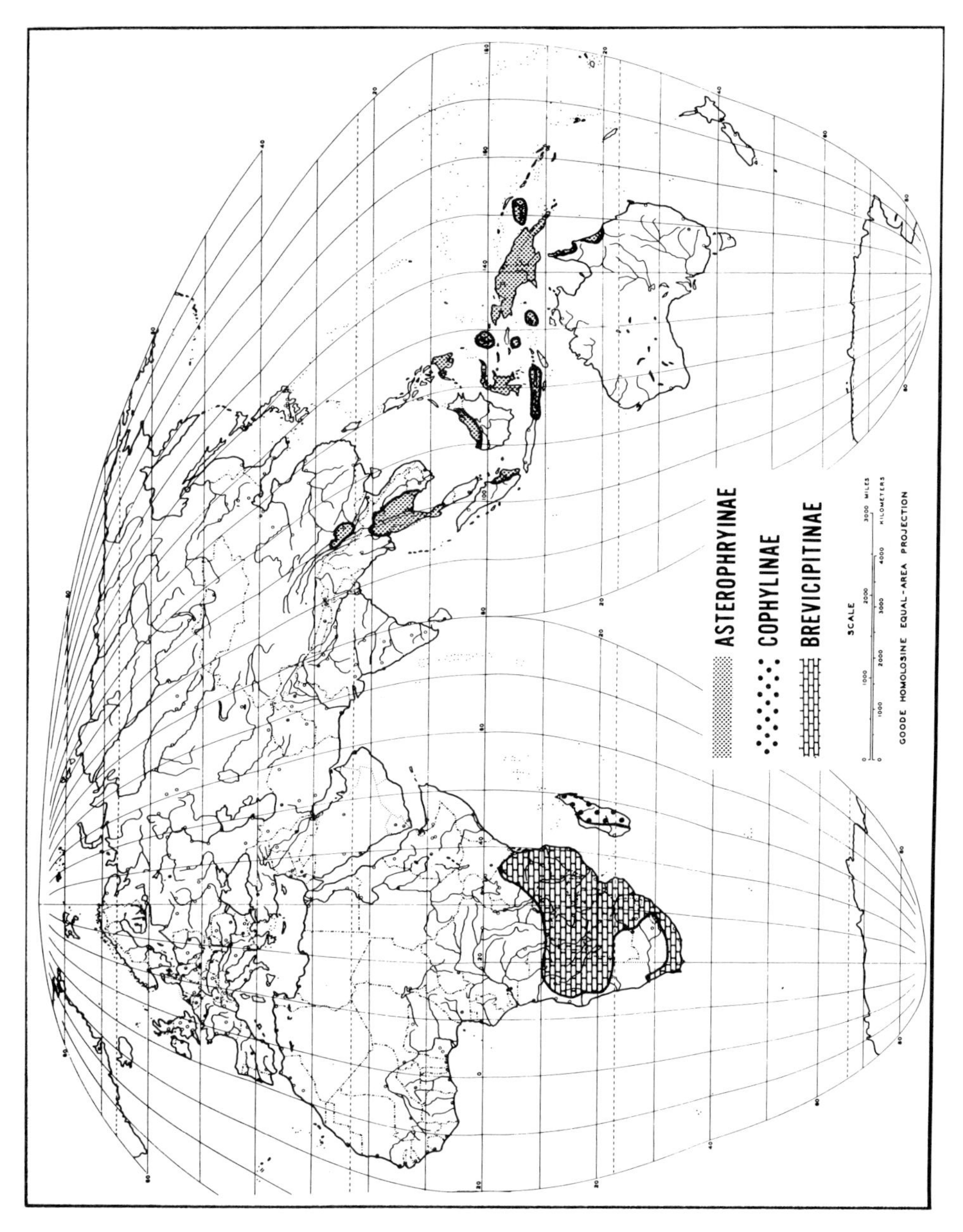

Figure 13–7. Distribution of three subfamilies of scoptanuran frogs.

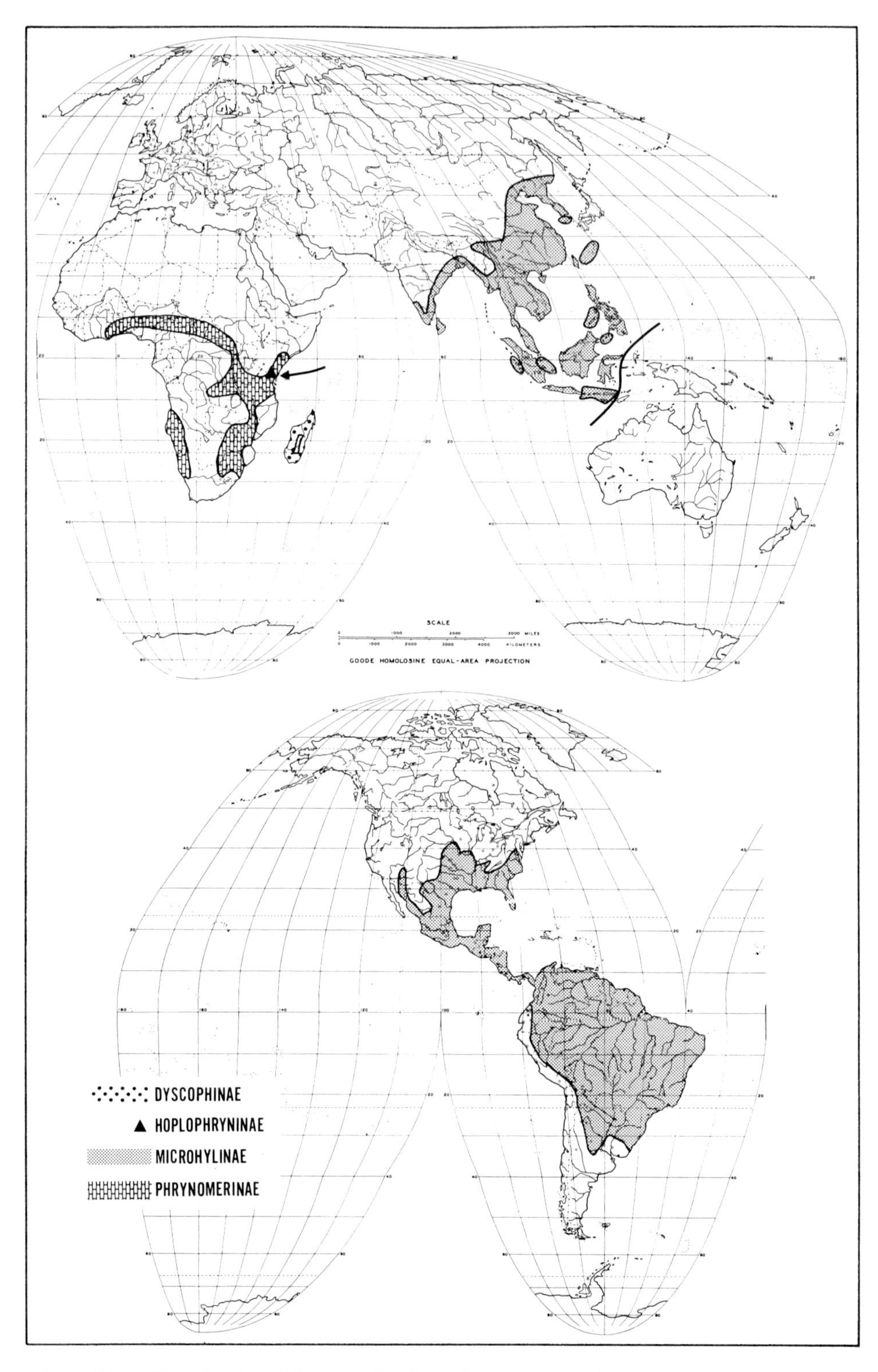

Figure 13–8. Distribution of four subfamilies of scoptanuran frogs.

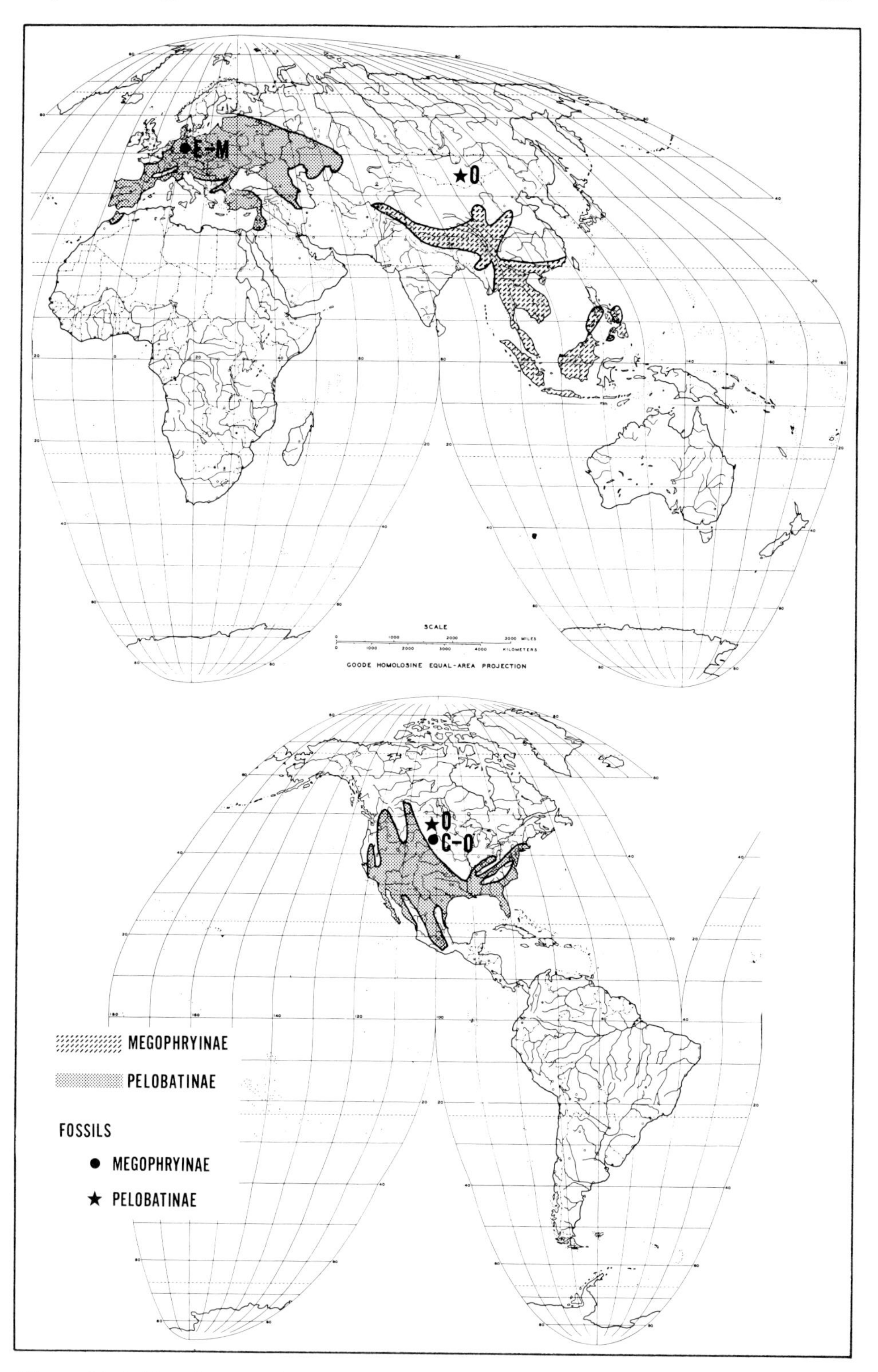

Figure 13–9. Distribution of two subfamilies of the frog family Pelobatidae. Letters indicate fossil ages as identified in text.

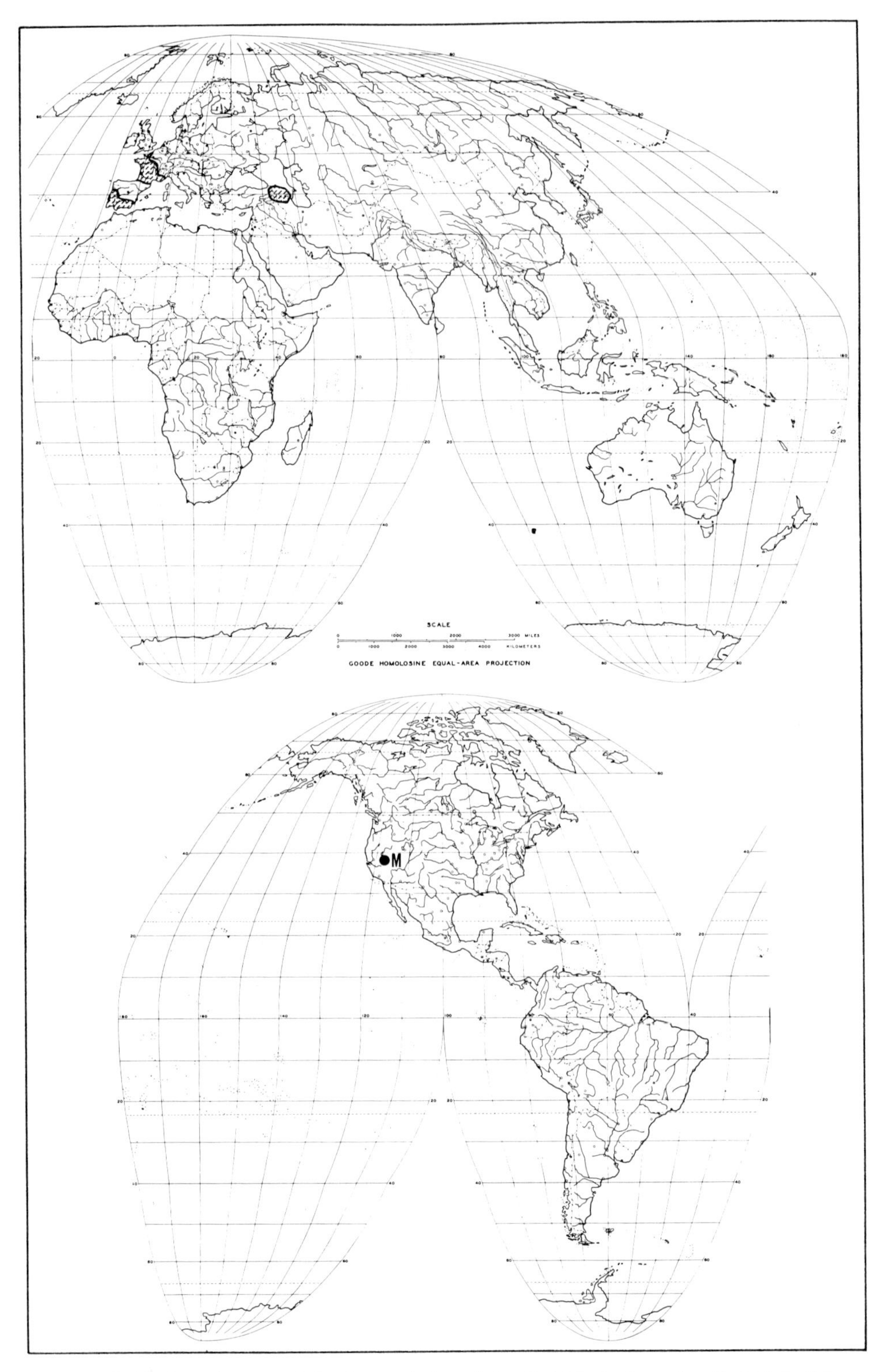

Figure 13–10. Distribution of the frogs of the subfamily Pelodytinae, family Pelobatidae. The single fossil is from Miocene.

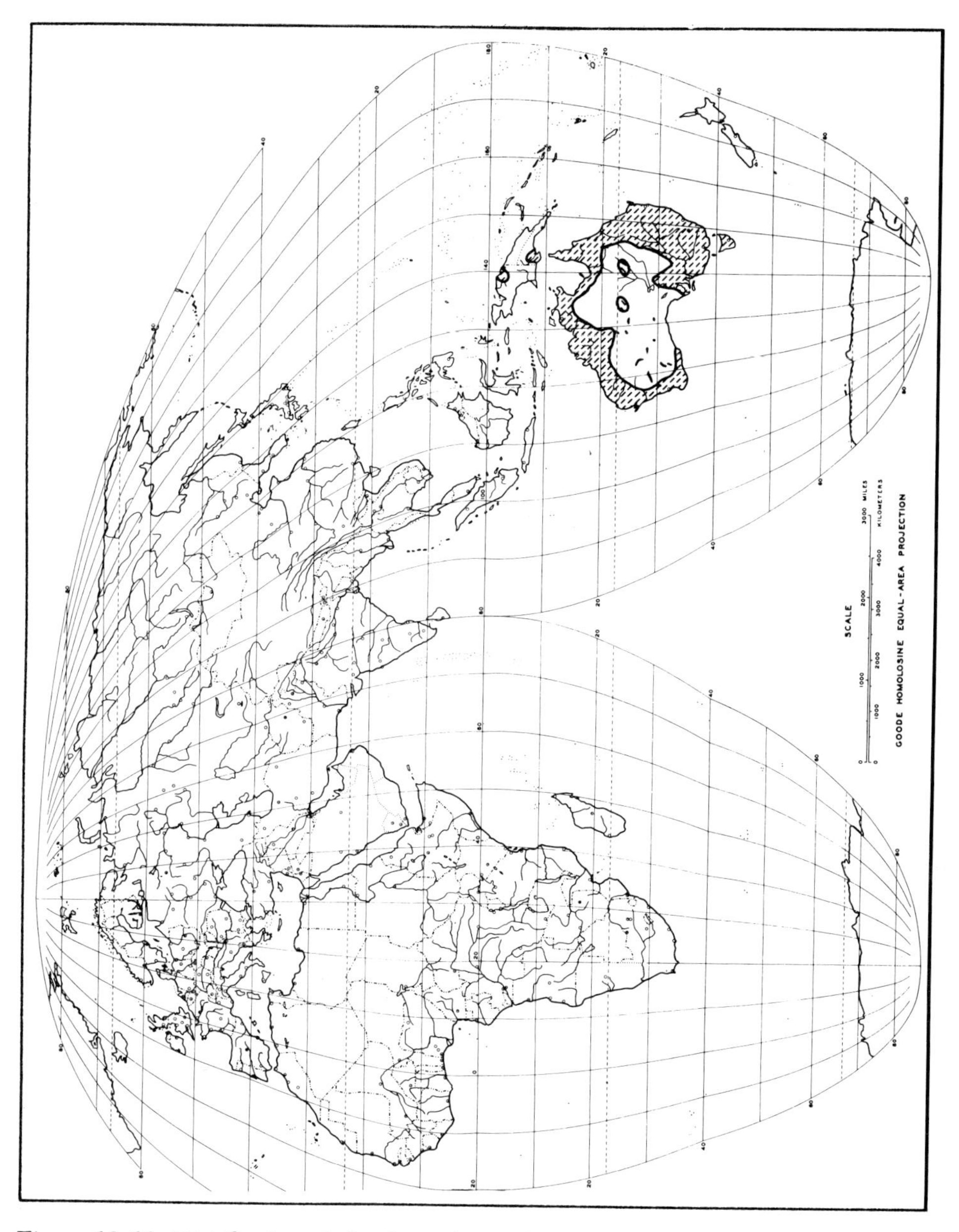

Figure 13–11. Distribution of the Australasian frog subfamily Myobatrachinae, family Myobatrachidae.

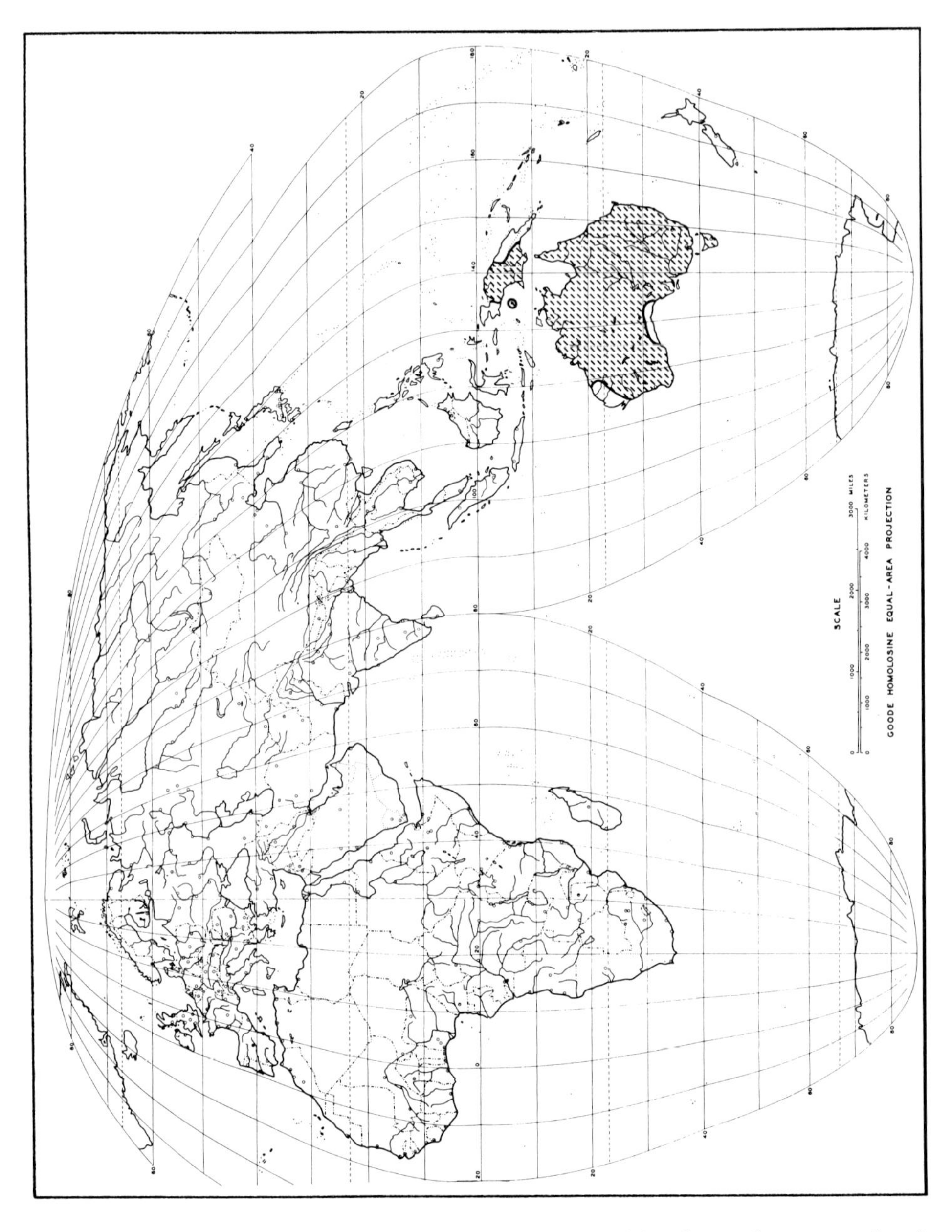

Figure 13–12. Distribution of the Australasian frog subfamily Cycloraninae, family Myobatrachidae.

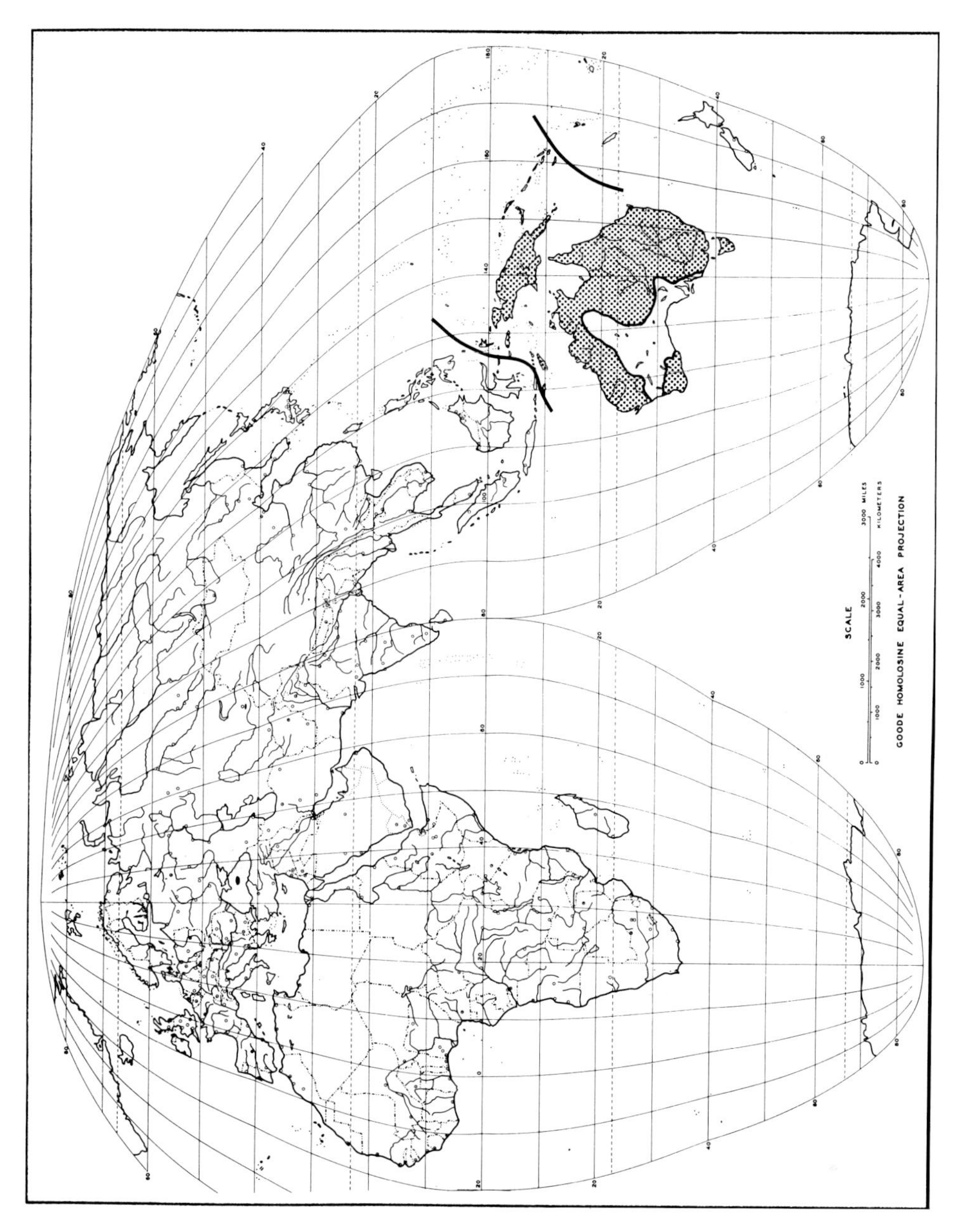

Figure 13–13. Distribution of the Australasia tree frogs, family Pelodryadidae.

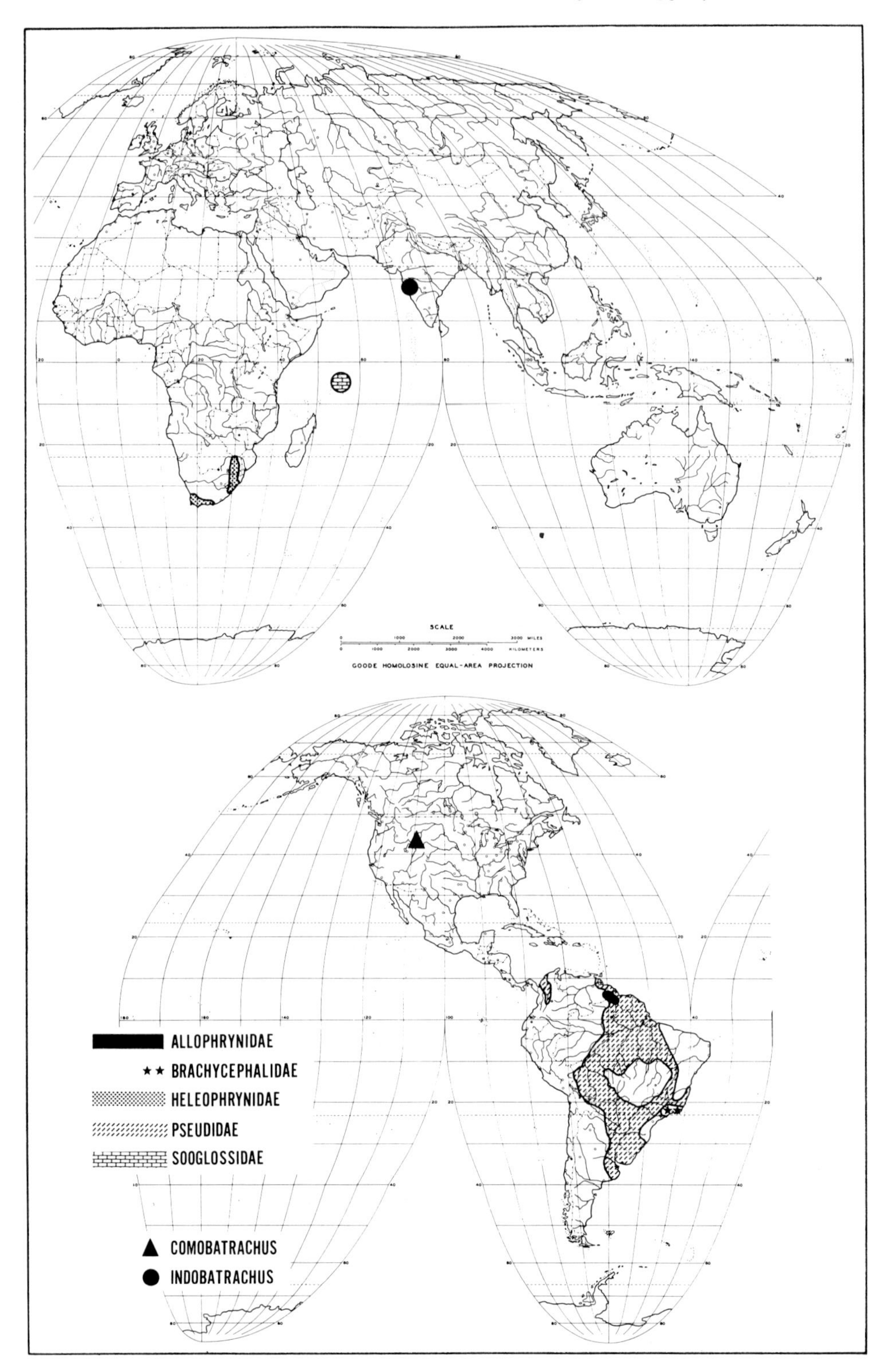

Figure 13–14. Distribution of diverse acosmanuran frog families and the fossil genera *Comobatrachus* and *Indobatrachus*.

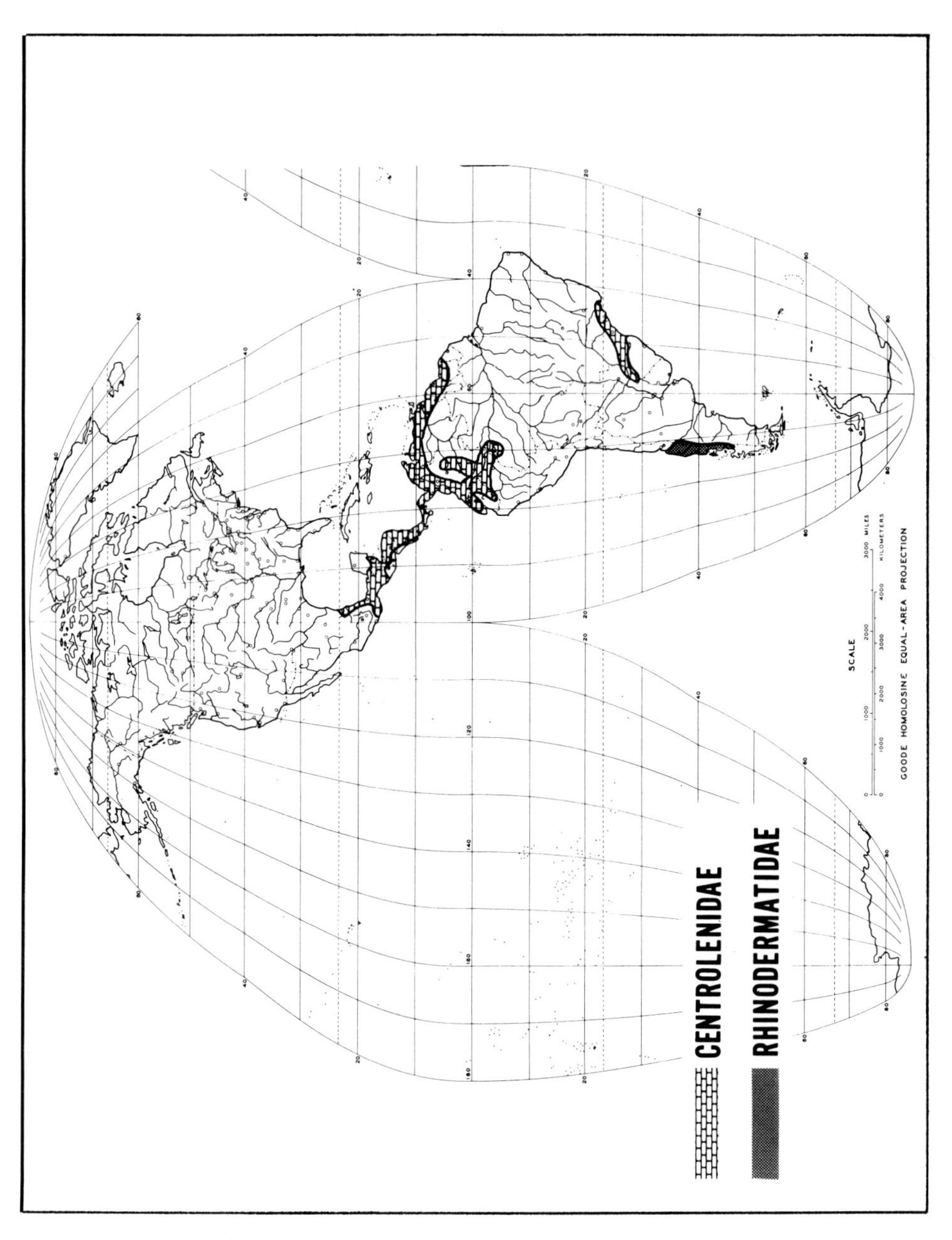

Figure 13–15. Distribution of two New World lepodactyloid derivative stocks.

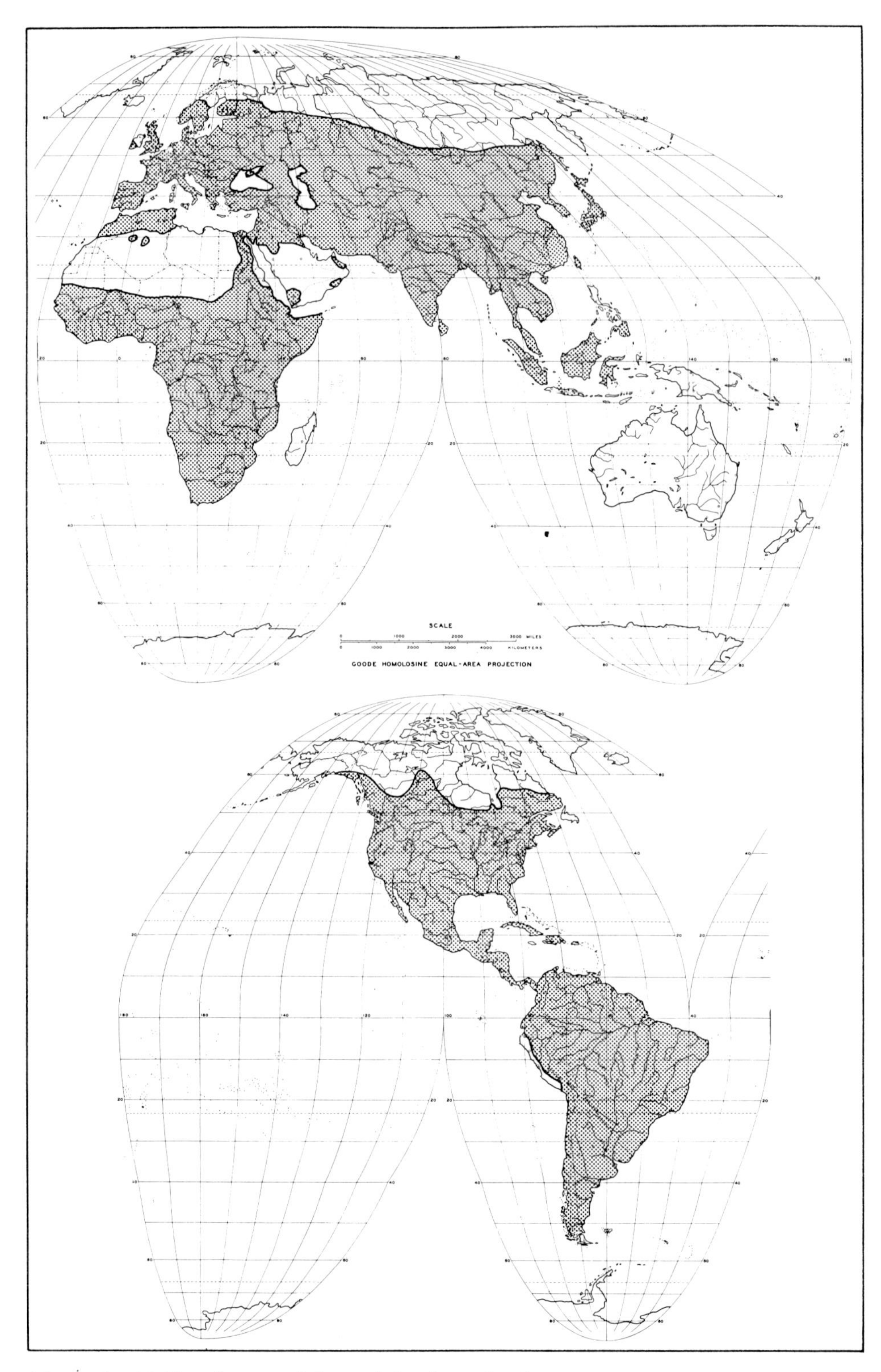

Figure 13–16. Distribution of the toad family Bufonidae.

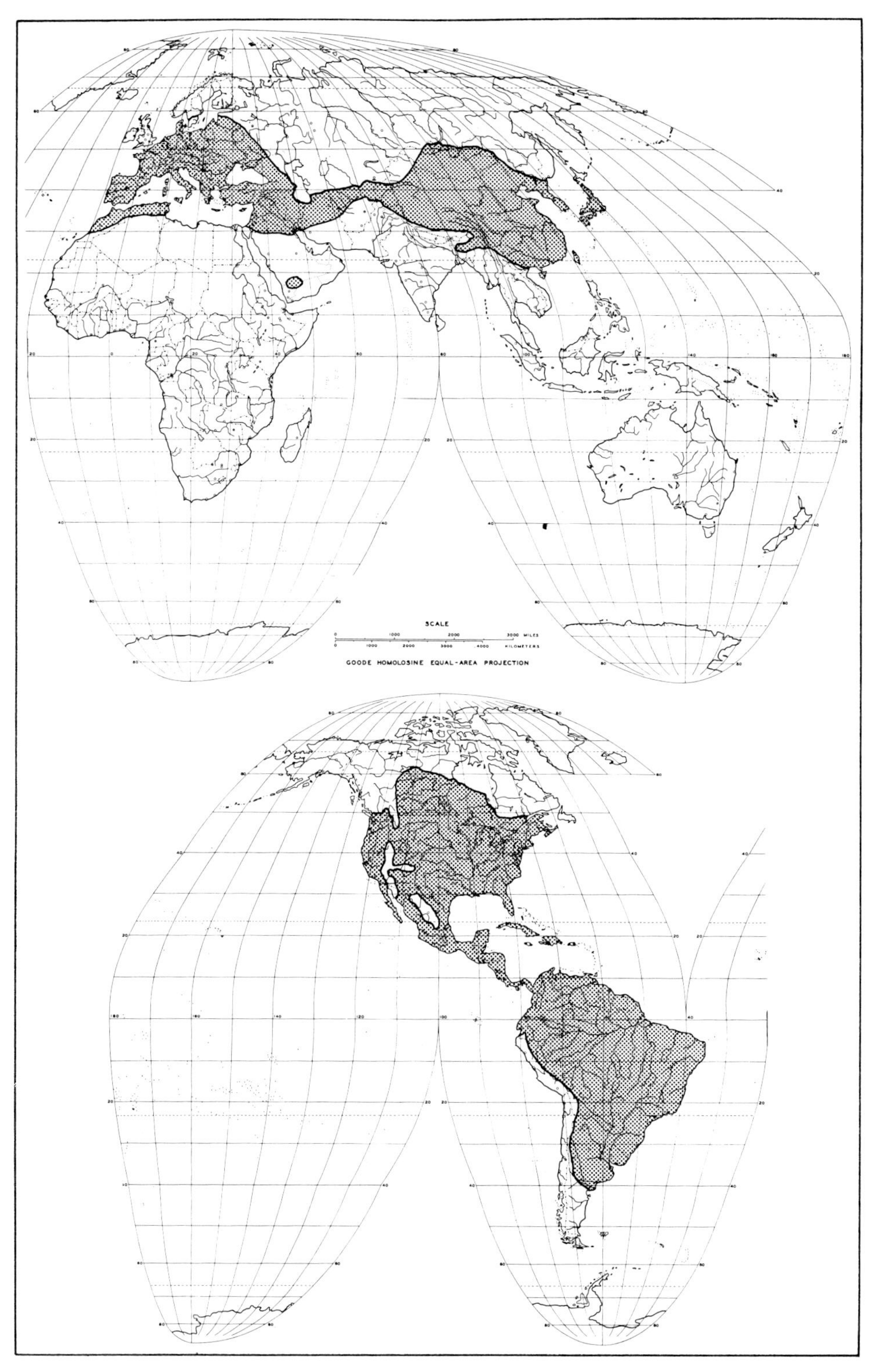

Figure 13–17. Distribution of the tree-frog family Hylidae.

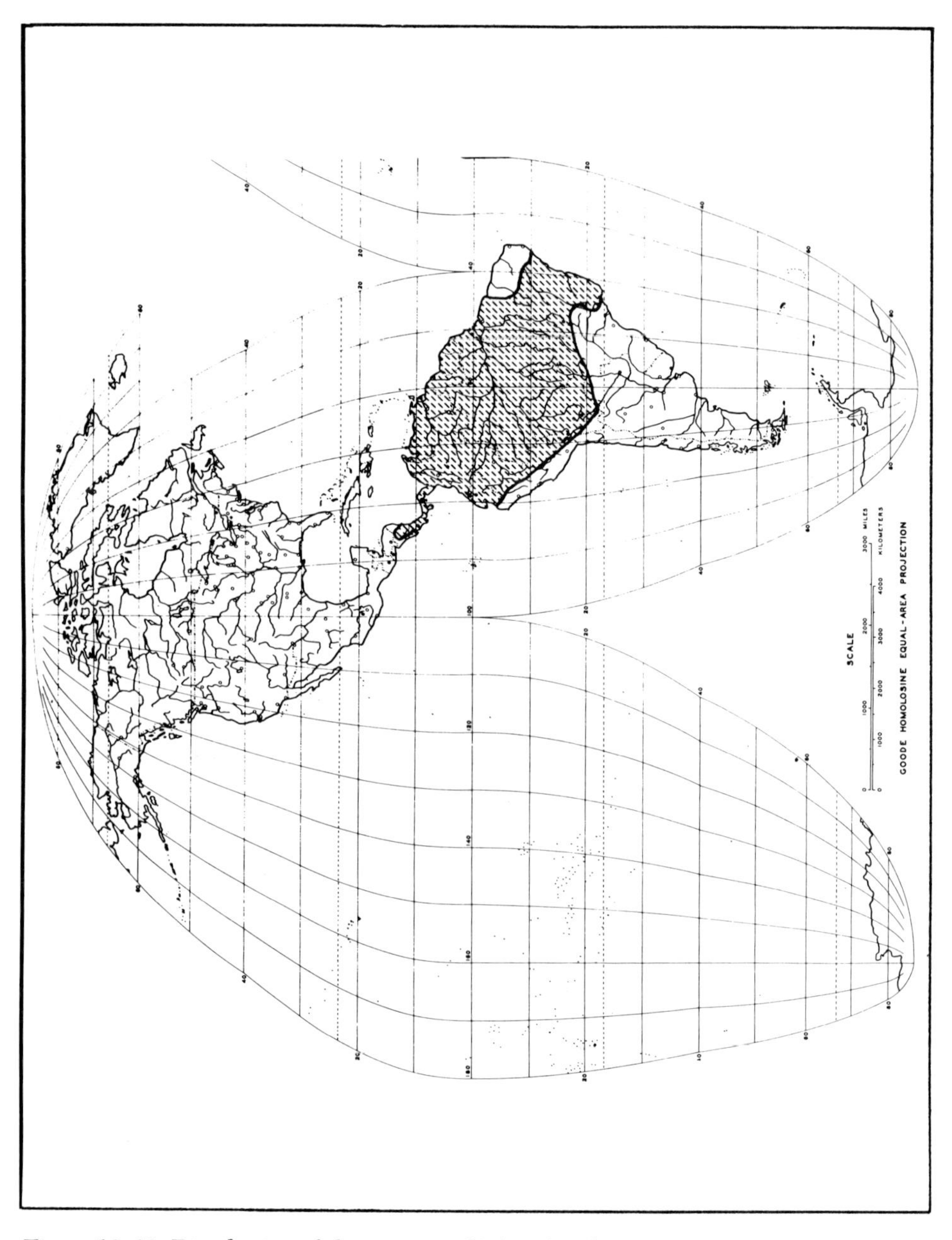

Figure 13–18. Distribution of the New World frog family Dendrobatidae.

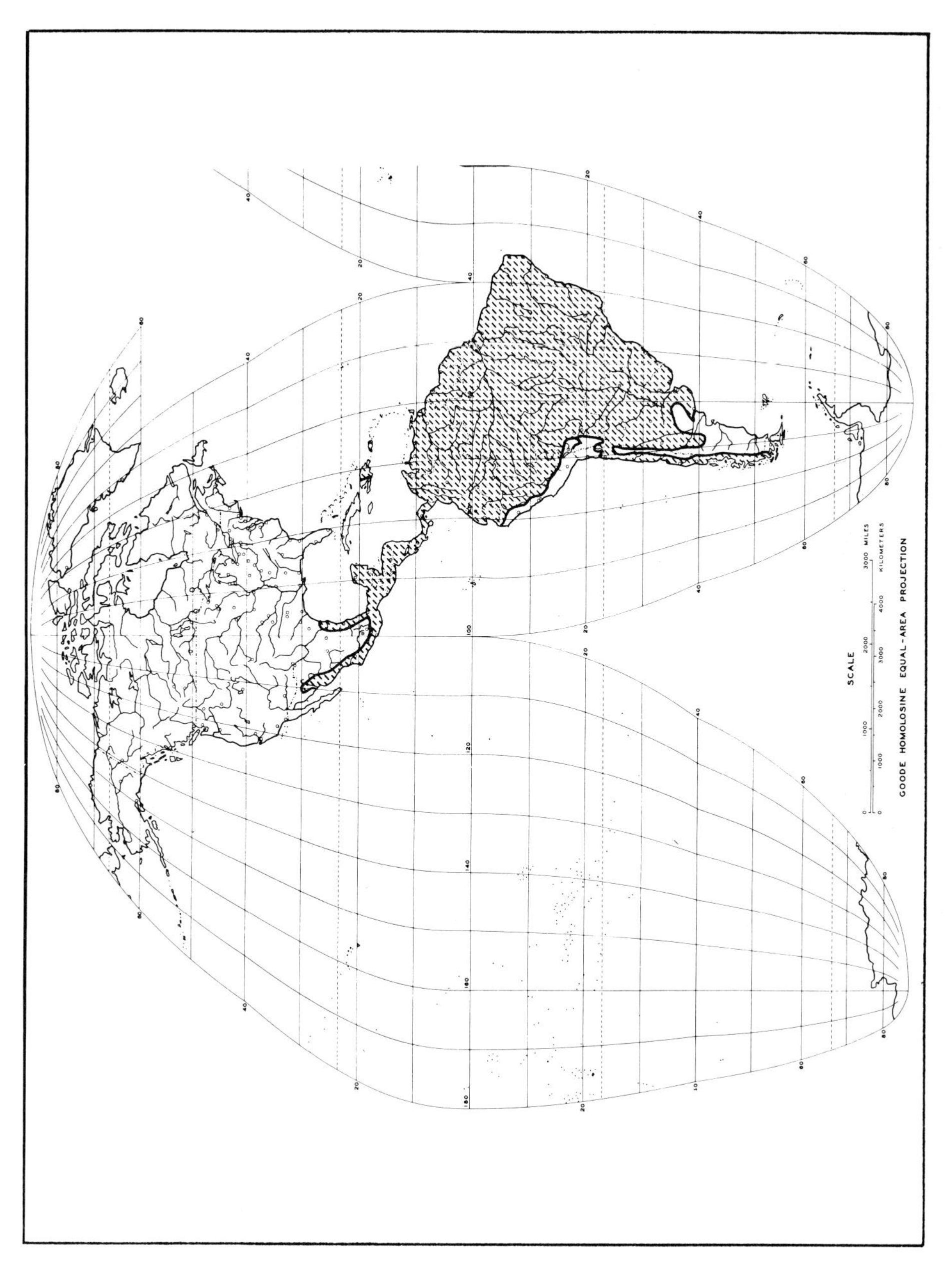

Figure 13–19. Distribution of the New World frog subfamily Leptodactylinae, family Leptodactylidae.

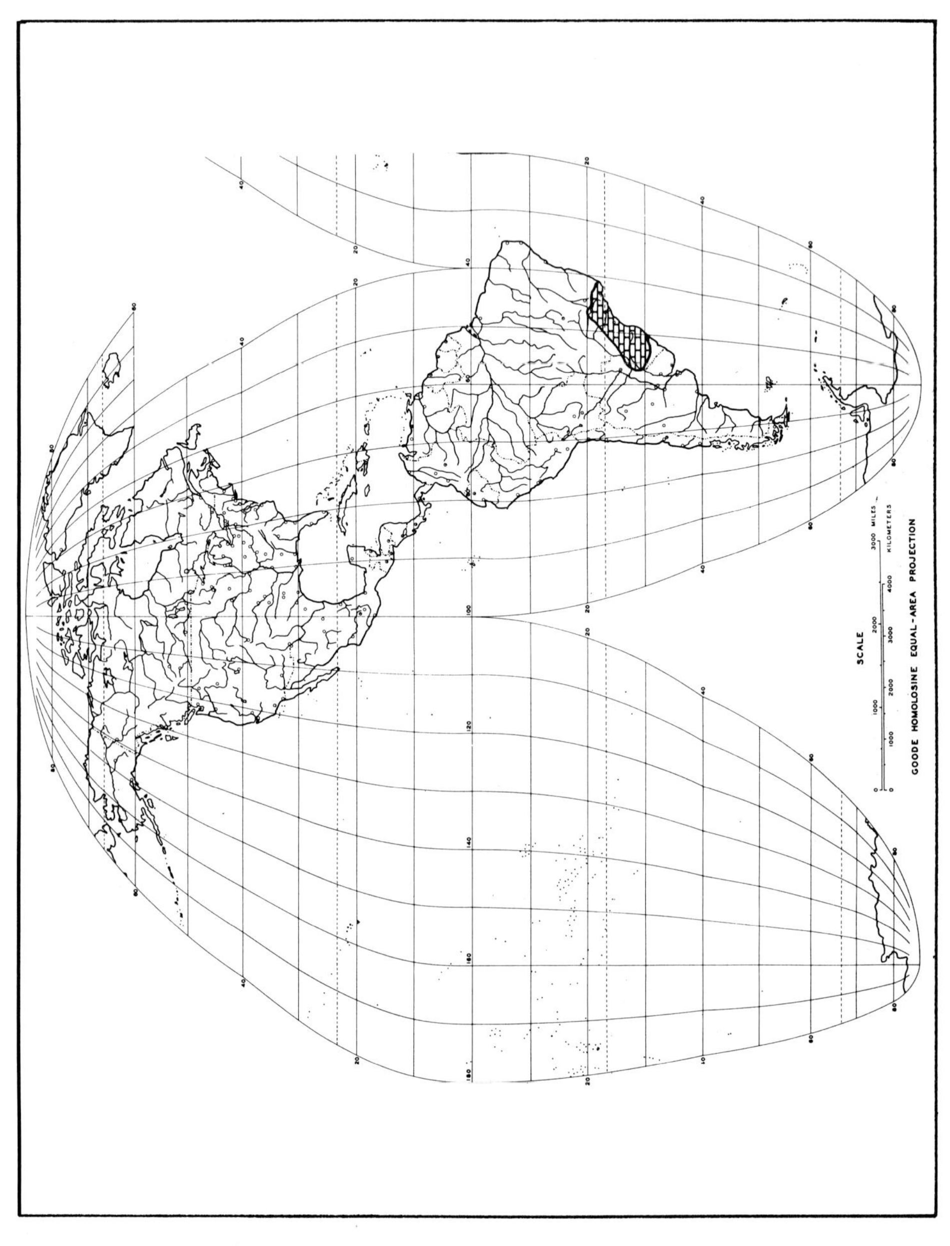

Figure 13–20. Distribution of the South American frog subfamily Hylodinae, family Leptodactylidae.

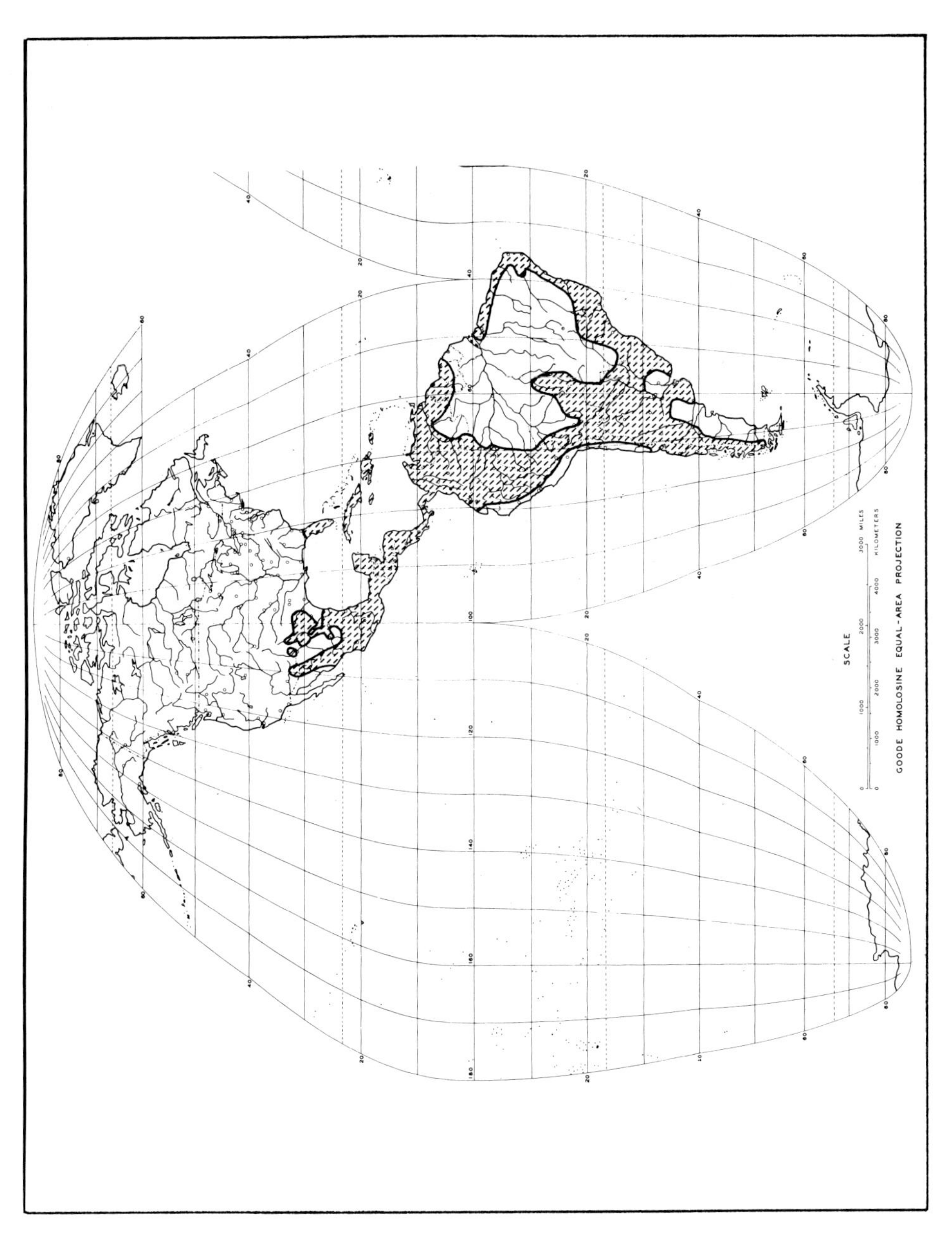

Figure 13–21. Distribution of the New World frog subfamily Telmatobiinae, family Leptodactylidae.

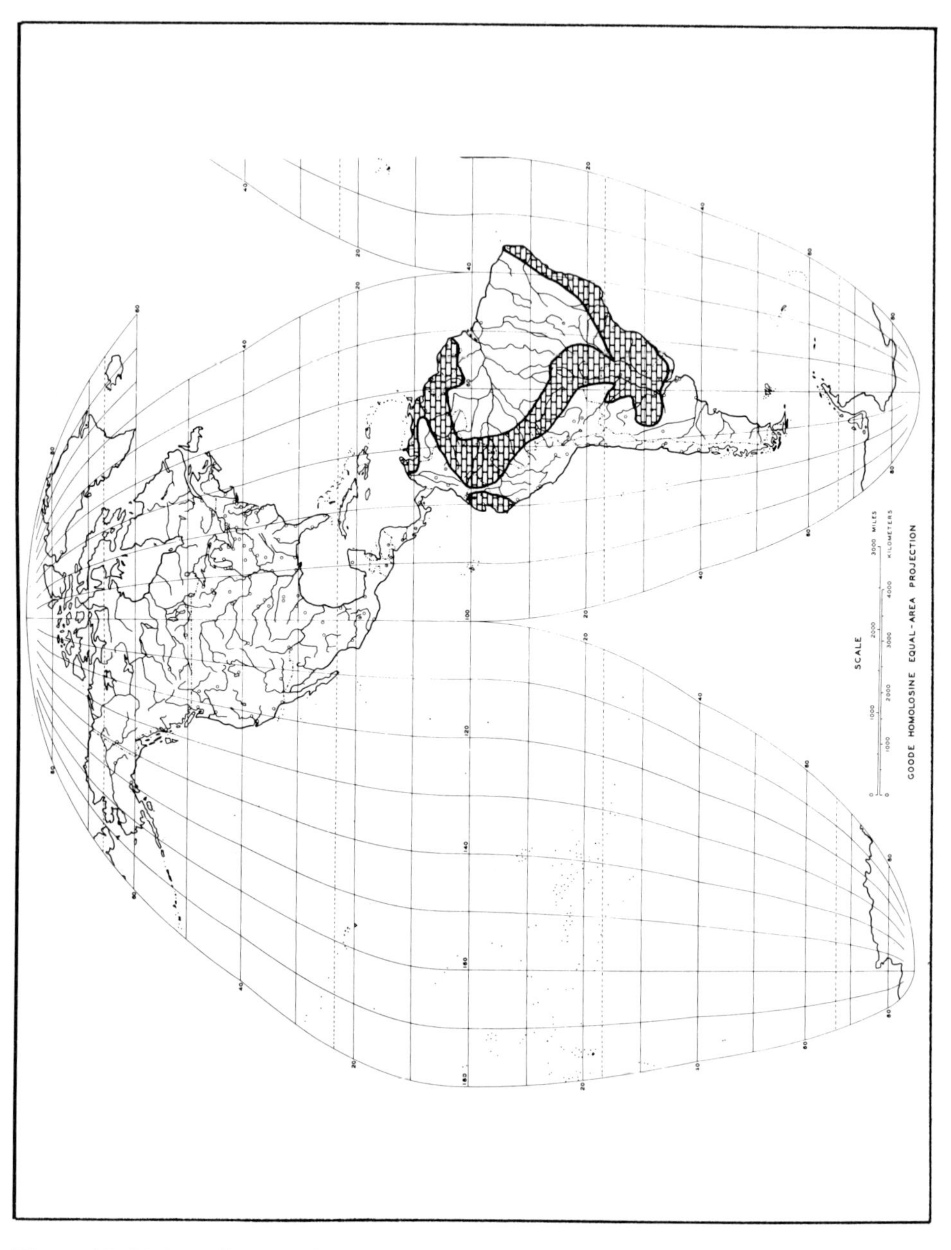

Figure 13–22. Distribution of the South American frog subfamily Ceratophryinae, family Leptodactylidae.

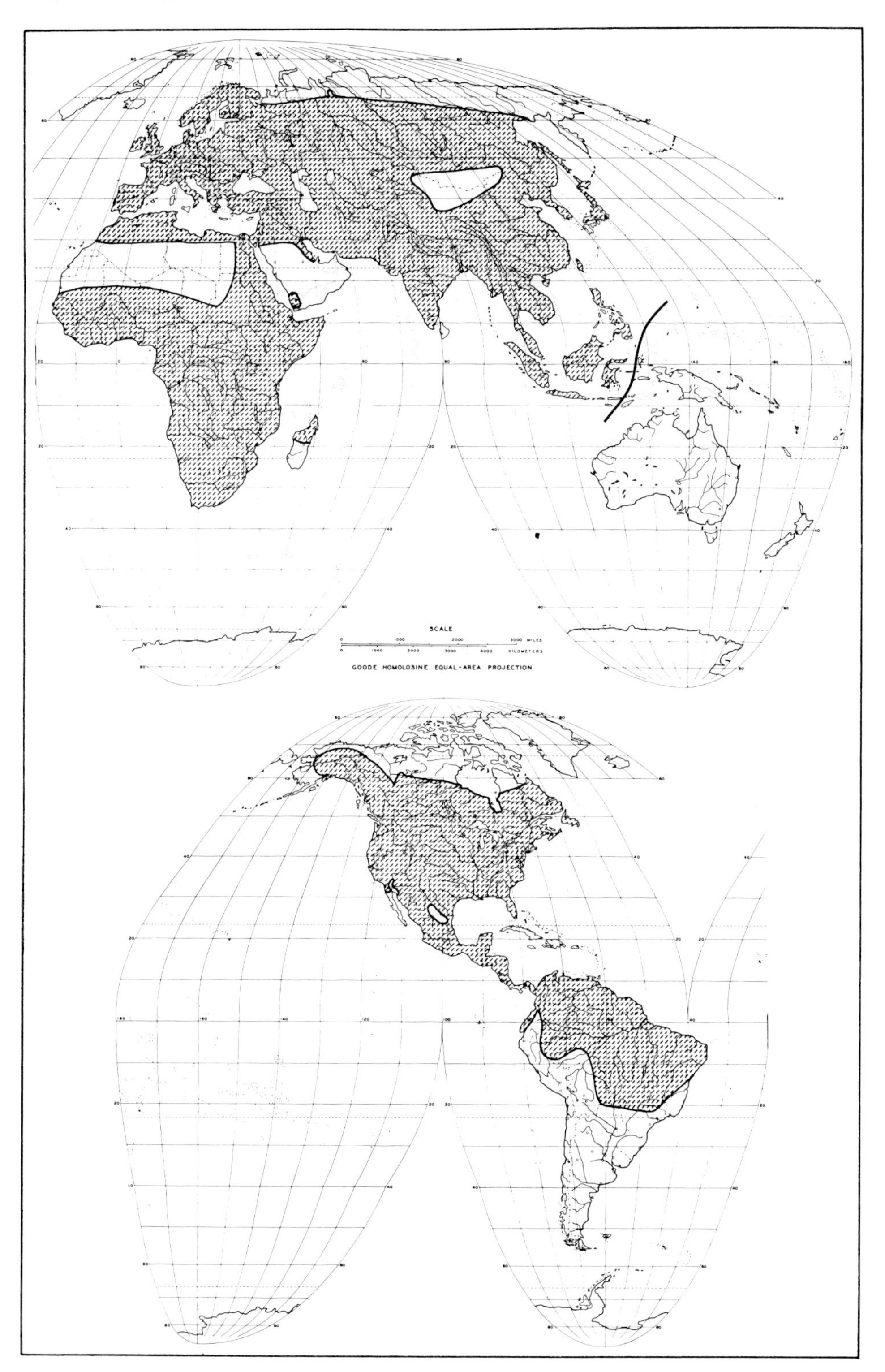

Figure 13–23. Distribution of the cosmopolitan frog subfamily Raninae, family Ranidae.

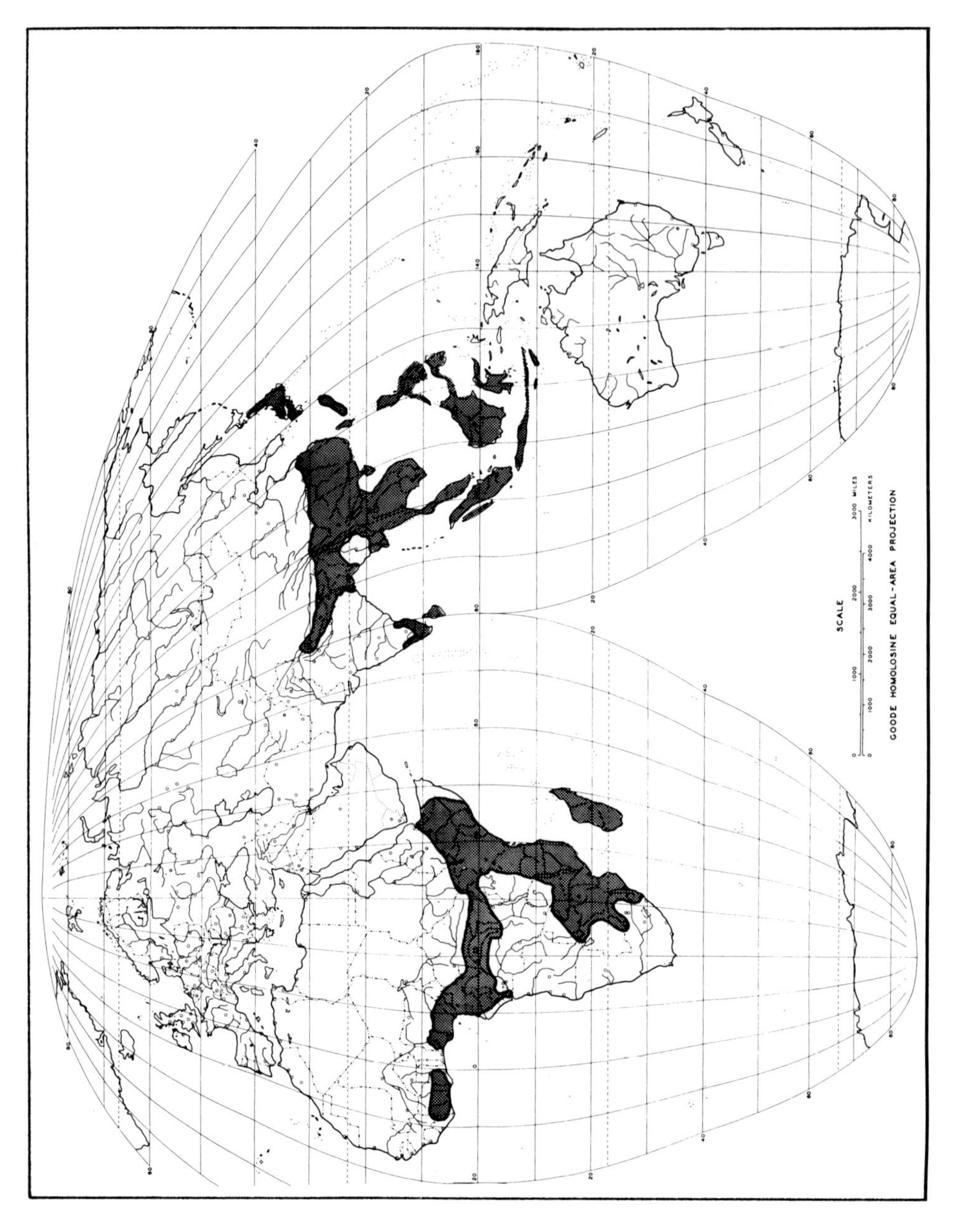

Figure 13–24. Distribution of the Old World tree-frog subfamily Rhacophorinae, family Ranidae.

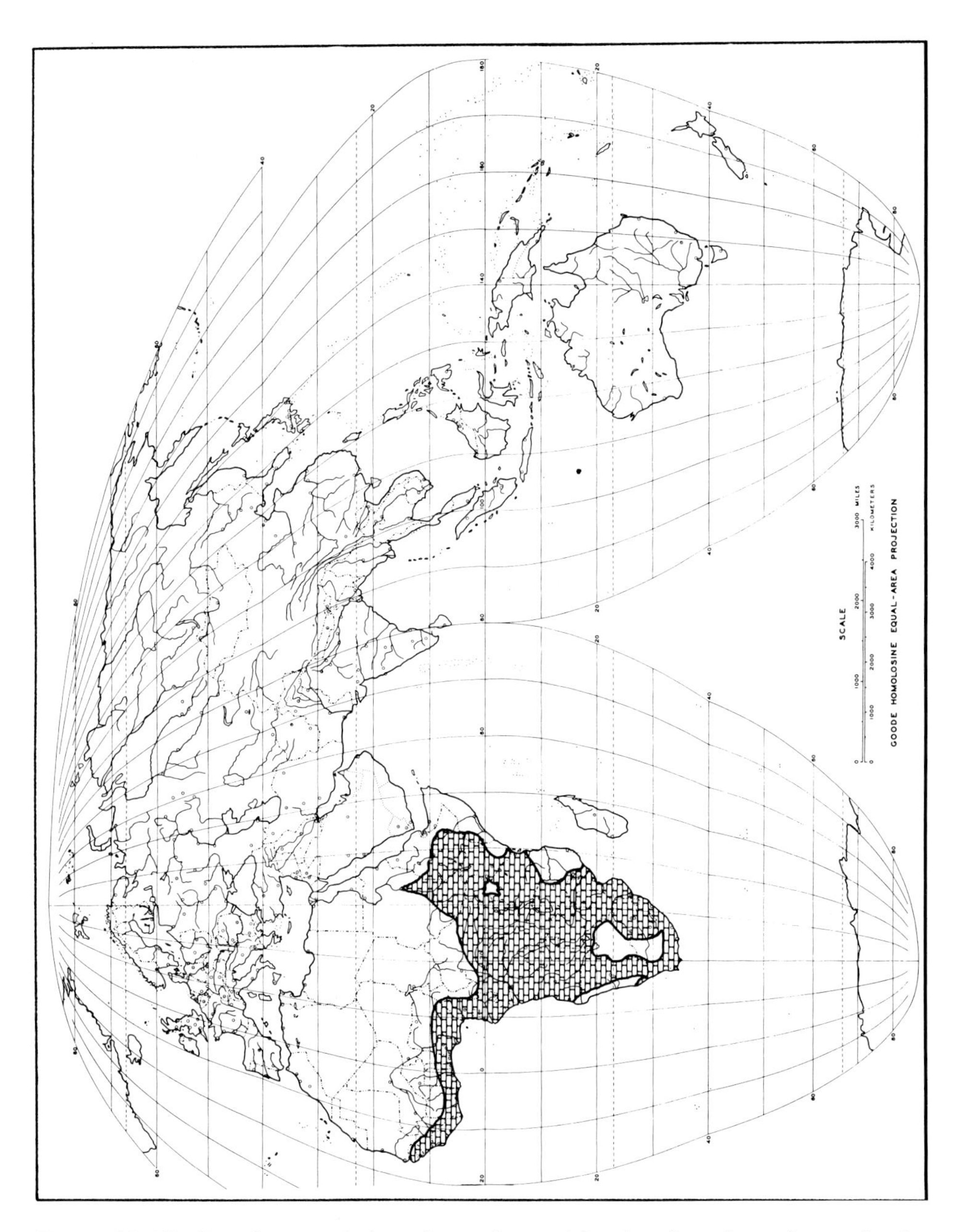

Figure 13–25. Distribution of the African frog subfamily Phrynobatrachinae, family Ranidae.

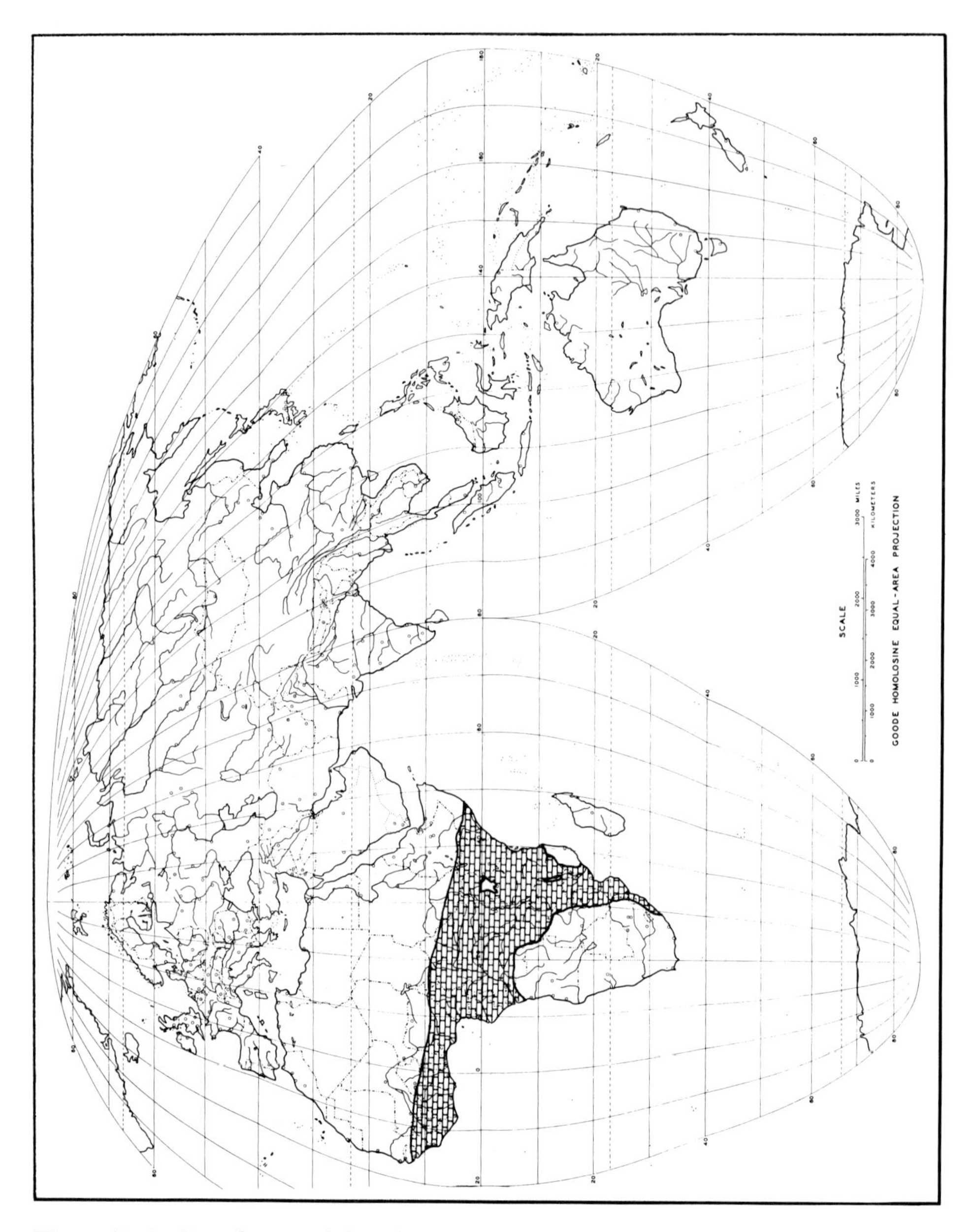

Figure 13–26. Distribution of the African frog subfamily Arthroleptinae, family Ranidae.

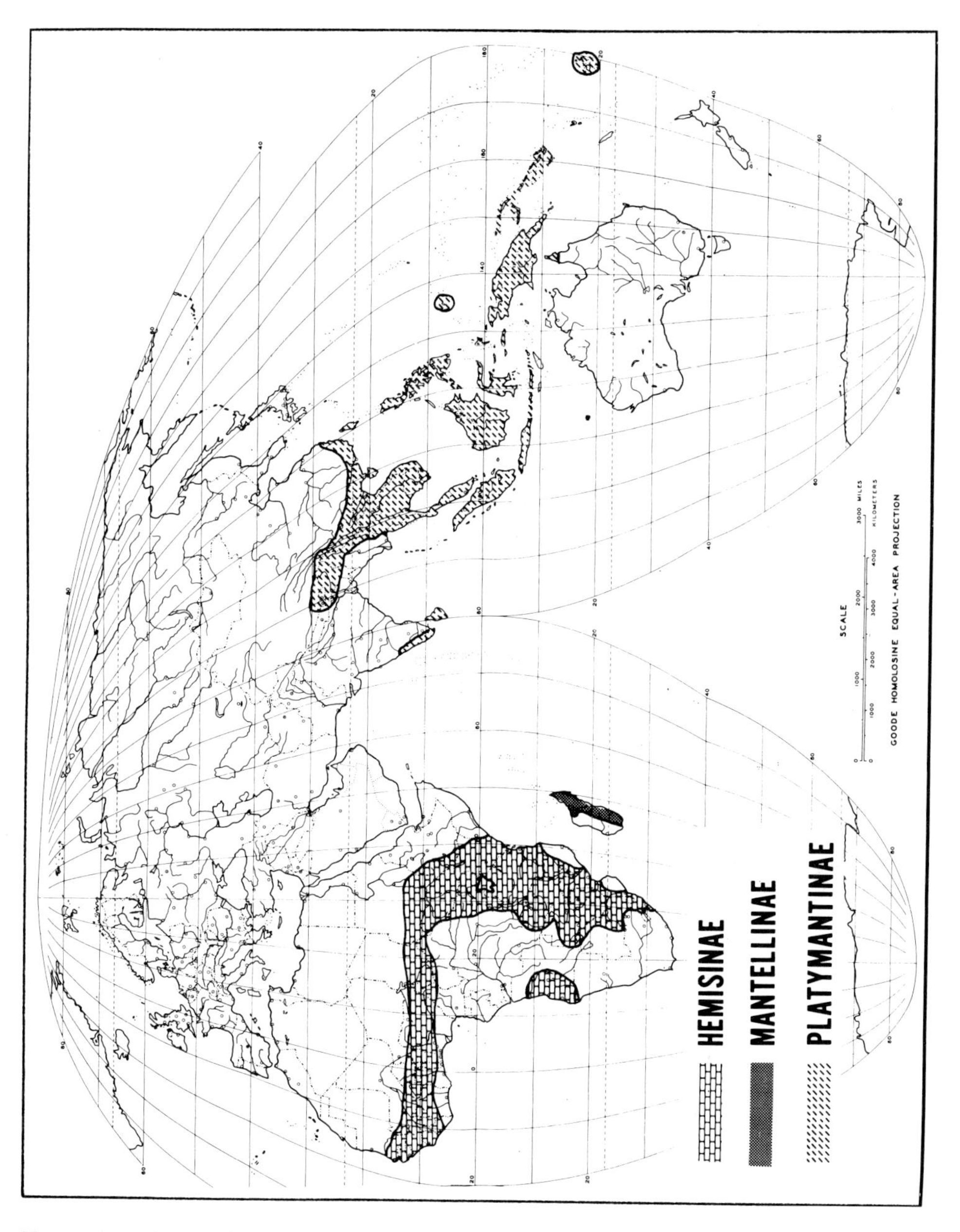

Figure 13–27. Distribution of three subfamilies of Old World frog family Ranidae.

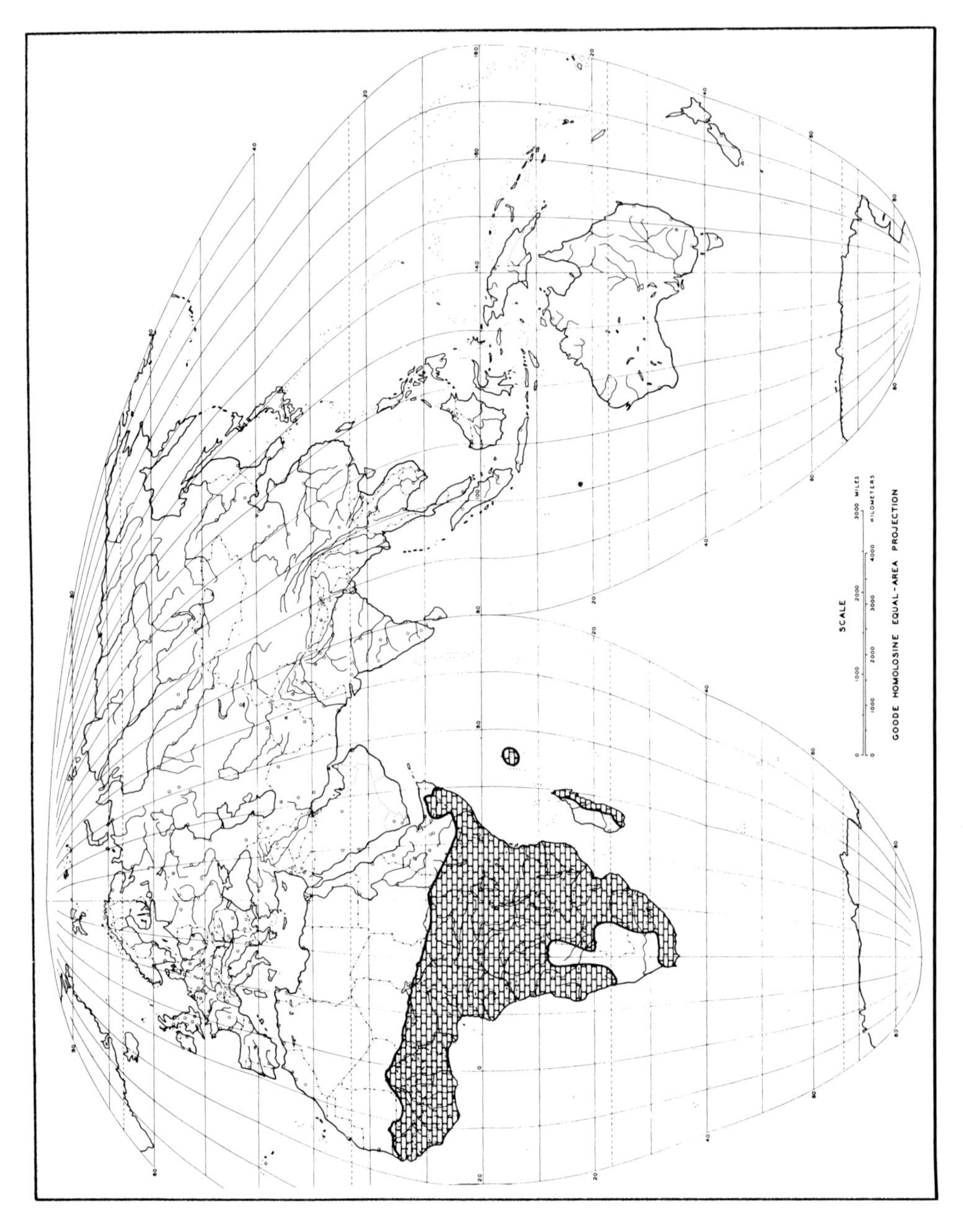

Figure 13–28. Distribution of the African and Malagasian tree frogs, subfamily Hyperoliinae, family Hyperoliidae.

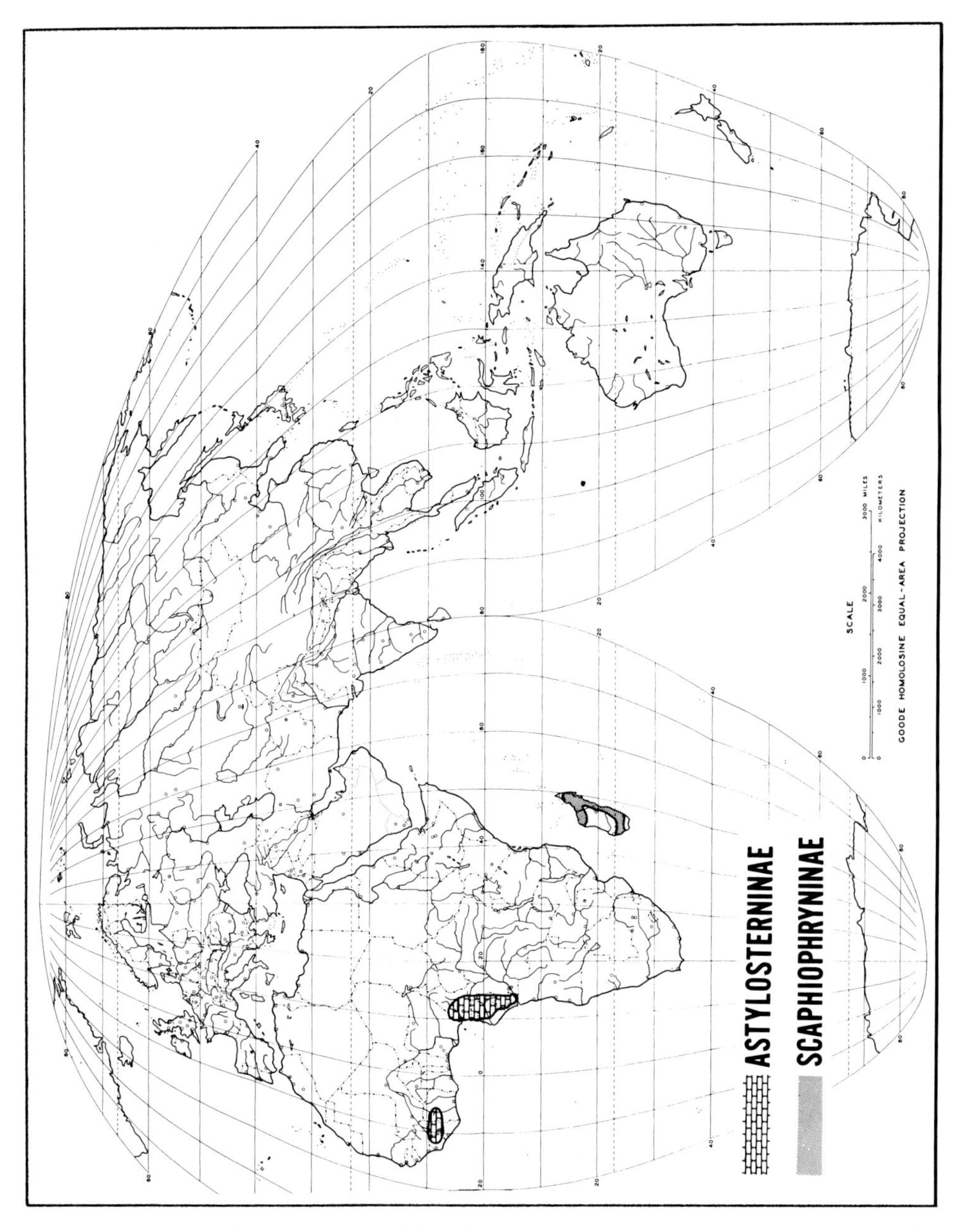

Figure 13–29. Distribution of two subfamilies of the African and Malagasian frog family Hyperoliidae.

A number of distinctive distribution patterns within the broad groupings provide a basis for further evaluation of the world frog faunas as follows (the number of units within each pattern is indicated in parentheses):

1. Northern Extratropical (4)
A. Temperate Regions of Europe, North Africa, northeastern Asia, and western China; one genus in Philippines (1)
Discoglossidae
B. Temperate to Subtropical Regions of Europe, North Africa, and North America into deserts of Mexico (1)
Pelobatinae
C. Temperate Europe (1)
Pelodytinae
D. Temperate western North America (1)
Ascaphidae

2. Essentially Cosmopolitan, with Tropical Center (2)
A. Cold to Tropical Regions on all continental masses, except Madagascar, New Guinea, and Australia (1)
Bufonidae
B. Cold to Tropical Regions on all land masses, except New Guinea and Australia (1)
Raninae

3. Widespread, with Tropical Center (3)
A. Cool Temperate to Tropical Regions of eastern Asia and Indo-Malayan area; Warm Temperate, Subtropical, and Tropical Regions of North, Central, and South America (1)
Microhylinae
B. Cold to Tropical Regions of North America and Tropical to Warm Temperate Regions of Central and South America; Cool Temperate to Subtropical Regions of Eurasia and North Africa (1)
Hylidae
C. Tropical Africa, Madagascar, and Indo-Malaya to Warm and Cool Temperate eastern Asia and Japan (1)
Rhacophorinae

4. Tropical (27)
A. African and Malagasian endemics
i. Tropical to southern Warm Temperate Africa, Madagascar, and Seychelles (1)
Hyperoliinae
ii. Tropical to southern Warm Temperate Africa (2)
Xenopinae
Phrynobatrachinae
iii. Tropical to southern Subtropical Regions of Africa (4)
Phrynomerinae
Arthroleptinae
Hemisinae
Astylosterninae
iv. Malagasian endemics (4)
Dyscophinae
Cophylinae
Mantellinae
Scaphiophryninae

v. Tropical Montane East African endemic (1)
Hoplophryninae
vi. Seychelles endemic (1)
Sooglossidae
B. Asian–New Guinean endemics (3)
i. Tropical, Subtropical, and Temperate southeastern and western China to Afghanistan; southeast Asia, exclusive of India (1)
Megophryinae
ii. Warm Temperate western China through Tropical Malaya, Indo-Malayan islands, New Guinea to northern Australia and the Solomon, Palau, and Fiji islands (1)
Platymantinae
iii. Subtropical western China south through Tropical Indo-Malaya to northern Australia (1)
Asterophryinae
C. New Guinean–Australian endemic (1)
Pelodryadidae
D. American endemics
i. Tropical North, Central, and South America to Subtropical Region in the north and Cool Temperate Region in the south (2)
Leptodactylinae
Telmatobiinae
ii. Tropical Middle and South America into Subtropical Region in the south (1)
Centrolenidae
iii. Tropical South America into southern Central America (2)
Dendrobatidae
Pipinae
iv. Tropical South America into southern Warm Temperate Region (2)
Ceratophryinae
Pseudidae
v. Tropical endemics (3)
Hylodinae
Brachycephalidae
Allophrynidae
vi. Lowland Tropical Middle American endemic (1)
Rhinophrynidae

5. Southern Extratropical
A. South African endemics (2)
Brevicepitinae
Heleophrynidae
B. Australian–New Guinean endemics (2)
Myobatrachinae
Cycloraninae
C. New Zealand endemic (1)
Leiopelmatidae
D. South American endemic (1)
Rhinodermatidae

A combination of data for geographic regions (Table 13–1) and latitudinal biotemperature regions (Table 13–2) provides a basis for recognizing five major recent world frog faunas:

Holarctic–distributed throughout the Northern Hemisphere, with three centers of further differentiation: western Eurasia and northwestern Africa, eastern and southern Asia, and North America

African—found throughout Africa, except in the northwest, with a depauperate sample in the Sahara Desert

Malagasian—on the island of Madagascar

Australasian—distributed over Australia, New Guinea, and islands immediately to the west, with a depauperate sample extending eastward through the western Pacific to Fiji

South American—distributed over tropical America from Mexico south, on the West Indies, and throughout South America

The single genus *Leiopelma,* with three species, is restricted to New Zealand. No other native frogs occur there, and it does not seem appropriate to group the islands with any other major faunal unit. On the other hand, they do not qualify, on the basis of their one relict, as a major frog fauna.

The frogs of the Seychelle Islands in the Indian Ocean are another problem. The fauna is comprised of two genera and three species of the peculiar endemic Sooglossidae, a *Rana,* and an endemic hyperoliine *(Megalixalus).* The latter two are definitely African in origin, but the endemic family, although it is probably a relict that formerly occurred in tropical Africa, makes the fauna distinctive. For these reasons, I do not group frogs of these islands with the African fauna, nor do I regard them as a major frog faunal unit.

The traditional views of frog distribution (Noble, 1931; Darlington, 1957) disagree with these findings in a notable way. The older views regard tropical Asia, from India through the Sunda chains to the Philippines, as an area supporting a major frog fauna. My analysis indicates that the Oriental or Indo-Malayan frog fauna is a complex mixture of familial-level units that arrived from at least three primary sources during middle to late Cenozoic. At this level the Indo-Malayan region appears to have an accumulation of groups from three sources: (a) subtropical to tropical lines from eastern Asia that are being displaced southward by cooling and drying trends (a discoglossid, several megophryines, and bufonids), (b) derivatives from an Indian source (microhylines, tropical ranines, and rhacophorines), and (c) invaders from New Guinea (asterophryines). No endemic familial-level units occur in the area or have a center of distribution there. The microhylines, ranines, and rhacophorines range far north into temperate China, and the ranine derivative platymantines reach New Guinea, northern Australia, and as far as Fiji. At present, there is a gradual transition from cool temperate to tropical conditions in humid eastern Asia, and no sharp demarcation divides the frog groups between the Burma-Thailand-Indochina area and eastern China. The faunas of tropical and temperate Asia differ strikingly in species and generic composition (Inger, 1966), but the clusters do not represent the same magnitude of differentiation noted between major frog faunas elsewhere at suprageneric levels. Only one temperate Asian familial-level unit, the Hylidae, fails to be represented well inside the tropical region. This family barely reaches Tonkin, North Vietnam, from the north. The New Guinean microhylid group Asterophryinae extends through the region into western China. In short, the southeastern to eastern Asian region forms a complex filter, barrier and bridge, as well as a relict refuge, for frogs from New Guinea, India, and temperate China. It is parallel to the situation that exists in Central America (Savage, 1966) but without a continuous land connection from the south.

Most workers regard the Malagasian frog fauna as a part of the African (Ethiopian) unit. Madagascar has been continuously separated from Africa by an ever-widening marine barrier of great depth since Jurassic times. Four of the

Table 13–1. Geographical Distribution of Frogs

Familial groups or units	Europe, northwest Africa	Africa	Madagascar	Seychelles	Central eastern Asia	Southern Asia	New Guinea–Australia	Pacific Islands	New Zealand	North America	Central America	South America
Rhinophrynidae										X	X	
Pipinae												X
Xenopinae		X										
Dyscophinae			X									
Cophylinae			X									
Asterophryinae					X	X	X					
Microhylinae					X	X				X	X	X
Brevicepitinae		X										
Hoplophryninae		X										
Phrynomerinae		X										
Leiopelmatidae									X			
Ascaphidae										X		
Discoglossidae	X				X	X						
Megophryinae					X	X						
Pelobatinae	X									X		
Pelodytinae	X											
Myobatrachinae							X					
Cycloraninae							X					
Pelodryadidae							X					
Heleophrynidae		X										
Ceratophryinae												X
Hylodinae												X
Leptodactylinae										X	X	X
Telmatobiinae										X	X	X
Centrolenidae										X	X	X
Bufonidae	X	X			X	X				X	X	X
Brachycephalidae												X
Allophrynidae												X
Pseudidae												X
Rhinodermatidae												X
Hylidae	X				X					X	X	X
Dendrobatidae											X	X
Sooglossidae				X								
Raninae	X	X	X	X	X	X				X	X	X
Platymantinae						X	X	X				
Phrynobatrachinae		X										
Mantellinae			X									
Rhacophorinae		X	X		X	X						
Arthroleptinae		X										
Hemisinae		X										
Astylosterninae		X										
Hyperoliinae		X	X	X								
Scaphiophryninae			X									
Totals												
Number of Units 43	6	13	7	3	8	8	5	1	1	10	9	15
Percentage 100	14%	30%	16%	7%	19%	19%	11%	2%	2%	23%	21%	35%

Table 13–2. Distribution of Frogs by Land Mass and Latitudinal Biotemperature Regions

Eurasia Familial groups or units	Tropical	Subtropical	Warm temperate	Cold temperate	Cold
Asterophryinae	X	X			
Microhylinae	X	X	X	X	
Discoglossidae	X	X	X	X	
Megophryinae	X	X	X		
Pelobatinae		X	X	X	
Pelodytinae			X	X	
Bufonidae	X	X	X	X	X
Hylidae		X	X	X	
Raninae	X	X	X	X	X
Platymantinae	X	X			
Rhacophorinae	X	X	X	X	
Total 11	8	10	9	8	2
Extratropical Asia: Total 9		(9)	7	6	2
Tropical Asia: Total 8	8	(9)			

Africa (except northwest) Familial groups or units	Tropical	Subtropical	Warm temperate
Xenopinae	X	X	X
Brevicipitinae	X	X	X
Hoplophryninae	X		
Phrynomerinae	X	X	
Heleophrynidae		X	X
Bufonidae	X	X	X
Raninae	X	X	X
Phrynobatrachinae	X	X	X
Rhacophorinae	X	X	
Arthroleptinae	X	X	
Hemisinae	X	X	
Astylosterninae	X		
Hyperoliinae	X	X	X
Total 13	12	11	7

Madagascar Familial groups or units	Tropical	Subtropical
Dyscophinae	X	X
Cophylinae	X	X
Raninae	X	
Mantellinae	X	X
Rhacophorinae	X	X
Hyperoliinae	X	X
Scaphiophryninae	X	
Total 7	7	5

Australia–New Guinea Familial groups or units	Tropical	Subtropical	Warm temperate
Asterophryinae	X		
Myobatrachinae	X	X	X
Cycloraninae	X	X	X
Pelodryadidae	X	X	X
Platymantinae	X		
Total 5	5	3	3

North America

Familial groups or units	Tropical	Sub-tropical	Warm temperate	Cold temperate	Cold
Rhinophrynidae	X	X			
Microhylinae	X	X	X		
Ascaphidae			X	X	
Pelobatinae	X	X	X	X	
Leptodactylinae	X	X			
Telmatobiinae	X	X	X		
Centrolenidae	X				
Bufonidae	X	X	X	X	X
Hylidae	X	X	X	X	X
Raninae	X	X	X	X	X
Total 10	9	8	7	5	3

Central America

Familial groups or units	Tropical
Rhinophrynidae	X
Microhylinae	X
Leptodactylinae	X
Telmatobiinae	X
Centrolenidae	X
Bufonidae	X
Hylidae	X
Dendrobatidae	X
Raninae	X
Total 9	9

South America

Familial groups or units	Tropical	Subtropical	Warm temperate	Cold temperate
Pipinae	X			
Microhylinae	X	X	X	
Ceratophryinae	X	X	X	
Hylodinae	X	X		
Leptodactylinae	X	X	X	X
Telmatobiinae	X	X	X	X
Brachycephalidae		X		
Centrolenidae	X	X		
Bufonidae	X	X	X	X
Allophrynidae	X			
Pseudidae	X	X	X	
Rhinodermatidae			X	X
Hylidae	X	X	X	
Dendrobatidae	X			
Raninae	X			
Total 15	13	10	8	4

West Indies

Familial groups or units	Tropical
Leptodactylinae	X
Telmatobiinae	X
Bufonidae	X
Hylidae	X
Total 4	4

seven familial-level units on the island are endemic. The others seem to be Cenozoic arrivals from Africa.

Table 13–3. Comparison of Major World Frog Faunas

Fauna	Holarctic	African	Malagasian	Indo-Malayan	Australasian	South American
Holarctic	**14**[a]	3[b]	1	7	2	7
African	*78*[c]	**13**	3	3	0	2
Malagasian	*93*	*77*	**7**	2	0	1
Indo-Malayan	*50*	*77*	*75*	**8**	4	3
Australasian	*76*	*100*	*100*	*50*	**5**	0
South American	*56*	*88*	*93*	*81*	*100*	**16**
Number of endemics	3	10	4	0	3	9
Percentage of world fauna	*7%*	*23%*	*9%*	*0%*	*7%*	*21%*
Percentage of area fauna	**21%**	**77%**	**57%**	**0%**	**60%**	**56%**

[a]Numbers in boldface type indicate number of groups occurring in region.
[b]Numbers in roman type indicate number of groups in common.
[c]Numbers in italic type indicate coefficient of differences.

Comparisons of the major world frog faunas are summarized (Table 13–3), with data on the number of units within each fauna, the number of units shared between faunas, and the degree of endemism. The coefficient of difference was calculated as a measure of distinctness between faunas from the formula:

$$CD = 1 - C/N_2 \times 100.$$

C is the number of units shared by the two compared faunas; N_2 is the number of units in the larger of the two faunas. Data for the Indo-Malayan area are included for comparison. A more detailed analysis of the transition between the Holarctic fauna of India and Indo-Malaya and that of the Australasian fauna is also provided (Table 13–4).

On the basis of mean CD values (the mean of the CD values for comparisons with each of the four other faunas), the Australasian and Malagasian faunas are the most distinctive, with means of 94 per cent and 91 per cent respectively. Africa (84 per cent) and South America (86 per cent) are equally distinctive and the Holarctic (76 per cent) shares the most units with other faunas.

Distribution of Fossil Frogs and Paleoclimatic Associations

The significance of the fossil record to the evolution and radiation of frogs was reviewed by Hecht (1963) and is discussed fully elsewhere in this volume (Estes and Reig). My comments are limited to consideration of particularly crucial fossil finds and the distribution of fossil records. In addition, the relationship between the general paleoclimatic associations for fossils and the distribution of Cenozoic geofloras is given, based on the ideas of Axelrod (1960).

Table 13–4. Distribution Patterns of Frogs in Tropical Asia–Australia

Familial groups	India-Ceylon	Western China	Southeastern Asia	Indo-Malaya	Austro-Malaya	New Guinea	Northern Australia
Asterophryinae		X	X	X	X	X	X
Microhylinae	X	X	X	X			
Discoglossidae		X	X	X			
Megophryinae		X	X	X			
Myobatrachinae						X	X
Cycloraninae						X	X
Pelodryadidae					X	X	X
Bufonidae	X	X	X	X	X		
Raninae	X	X	X	X			
Rhacophorinae	X	X	X	X	X		
Platymantinae	X	X	X	X	X	X	X
Total—11	5	8	8	8	5	5	5
Composition	5As	7As	7As	7As	3As	1As	1As
	0Aus	1Aus	1Aus	1Aus	2Aus	4Aus	4Aus
Total composition	7 As groups, 4 Aus groups						

X Presence.
As Asian.
Aus Australian.

Within the xenoanurans, the pipids have a long fossil history (Figure 13–5). All records (Nevo, 1968) are from localities with climates corresponding to tropical or subtropical conditions today and are from areas in which the family still persists. The South American group is associated with present-day derivatives from Neotropical-Tertiary geoflora, and the African line is associated with Paleotropical-Tertiary derivatives. The allied extinct Paleobatrachidae are known from Eocene tropical deposits in Europe. The family Rhinophrynidae occurred in tropical to subtropical paleoclimates associated with marginal Neotropical-Tertiary vegetation in Eocene-Oligocene North America.

The fossil record for the scoptanurans is limited. The single family (Microhylidae) is questionably reported from Miocene deposits in Europe and Africa and from Miocene onward in North America. The upper Jurassic frog *Comobatrachus* from North America (Hecht and Estes, 1960) may represent this family or some primitive group of this suborder.

Lemmanurans are known from Cretaceous North America (Ascaphidae, Figure 13–6), Jurassic Spain, and Cretaceous North America (Discoglossidae, Figure 13–6), with several Oligocene-Miocene genera of discoglossids in Europe. The records of Ascaphidae and Discoglossidae are associated with warm temperate components of the Arcto-Tertiary geoflora.

The leiopelmatids have been recorded from lower and middle Jurassic finds in Argentina (Figure 13–6). The probable Cenozoic association of the group is with Antarcto-Tertiary geofloral components.

Among acosmanurans, pelobatid frogs have an extensive fossil record (Zweifel, 1956; Kluge, 1966; Estes, 1970). The most primitive group (Megophryinae) is definitely known from Cretaceous to Miocene deposits in North America and from the Eocene to the Miocene in Europe. The more advanced subfamily Pelobatinae is known from the Oligocene onward in North America and Eurasia. The Pelodytinae are known from the Miocene in North America. These fossils are from warm temperate to subtropical situations and seem to represent an association with marginal warm temperate components of Arcto-Tertiary vegetation.

Leptodactyloid frogs are definitely known as fossils from the Eocene onward in South America. Records for North America prior to the Pleistocene are questionable (Lynch, 1969). The presumed leptodactyloid genus *Indobatrachus* (Noble, 1930; Chiplonker, 1941; Verma, 1965) from the Eocene of southern India can only be placed within that stock with extreme question. Nothing in the descriptions of the three nominal species of the genus permit family identification. *Indobatrachus* could be a megophryine pelobatid, a myobatrachid, a sooglossid, an arthroleptine, or an extinct stock not closely related to any of these (Noble, 1930; Griffiths, 1959; Hecht, 1963; Lynch, 1969). Until the available material is critically re-examined, inclusion of *Indobatrachus* with any family seems unwise.

The fossil record for the remaining groups of higher frogs offers no surprises. Bufonids and ranids are known from areas within their present geographic limits from Miocene onwards in North America and Eurasia, well within present limits of distribution, and the former is now known from Paleocene Brazil (Estes and Reig, this volume).

HISTORICAL SOURCES OF THE FROG FAUNAS

The composition of the frog fauna in any existing world land area is determined in complex fashion by the interaction of present and past ecology, geographic accessibility, long-term physiographic events, and the evolutionary history of the familial units. A review of the distributional data, major bioclimatic and vegetational associations, and the fossil record indicates five distinctive patterns, with several subpatterns that suggest common histories.

The first pattern involves groups of frogs that have primitive larvae (Types I and II) and that are essentially tropical in distribution throughout time. Representatives are found today in continental areas that have had continuous tropical conditions since Jurassic times, although some recent genera also range into extratropical areas.

The second pattern includes groups of frogs that have larvae (Type III) with advanced mouth parts and that have their distribution centered on extratropical northern land masses from the Jurassic onward.

The third pattern is represented by the distribution of a single group (Leiopelmatidae) that is known from Jurassic South America and recent New Zealand. The association with southern extratropical climate seems consistent through time.

The fourth pattern consists of a series of extremely successful modern frogs with advanced larval mouth parts (Type IV), all having distributions centered on tropical Africa or America. The earliest known fossils for members of this pattern are Paleocene South America and Eocene in areas now extratropical, but the groups probably go back at least to the Cretaceous in tropical regions.

The fifth pattern involves groups of frogs with advanced larvae (Type IV). Frogs in this pattern have basically southern centers of distribution, but they are apparently successfully invading tropical areas in America and New Guinea. The earliest age for known fossils of this group is Eocene, but probably the basic stocks go back in southern extratropical areas to the Jurassic.

These basic distribution patterns have been recognized by previous students of biogeography, most importantly Darlington (1957). He saw clearly that the model proposed for mammal dispersal by Matthew (1915) and developed further by Simpson (1953, 1966) did not correspond to the facts of amphibian distribution. Even Darlington did not fully appreciate the significance of the patterns, and he attempted an explanation of them that cannot be substantiated in the light of the early fossil history of frogs (Hecht, 1963; Estes and Reig, this volume).

The most striking and consistent features of these patterns, which confound any attempt to explain frog distribution from a Cenozoic mammal model (Simpson, 1953) or from a radiation in successive waves from tropical southeast Asia to other areas (Darlington, 1957), are the continuous long-term association of the ancient major phyletic lines from the Jurassic to the present, with the same general bioclimatic and vegetational associations; the obvious independence of Africa and South America as major sources of evolution and centers of dispersal of modern frogs; the unimpressive, nondistinctive composition of the fauna of southeast tropical Asia, which is incompatible with a presumed source region; and the unique features of the southern extratropical frog faunas.

Darlington was unaware that frogs evolved at the same time as salamanders and caecilians and that the main phyletic lines were well established by the middle Jurassic, with most families present prior to the Tertiary. He, of course, did not accept the concept of continental drift as a factor in frog dispersal and later explicitly denied its relevance (Darlington, 1965) to modern biogeographic patterns. In actuality, the present patterns of frog distribution can only be explained adequately by reference to drift theory. Any other approach, in addition to being contrary to the accumulating geodynamic evidence, effects unexplainable models of frog evolution and dispersal.

It is not my intention to attempt to review or develop the current status of continental-drift theory, which now is becoming accepted widely and investigated by geologists, paleontologists, and geophysicists. It does, however, form the foundation for my explanation of present distribution patterns. Many paleogeographic facets of continental history have not been fully worked out by drift analysis, so that details and time sequences for many events remain equivocal. The broad features of Mesozoic and Cenozoic land masses and their history are known and provide the basis for my theory. Although detailed studies of short time-sequences or particular geographic areas require more intensive analysis than is possible in a paper of this scope, I have attempted to explain the major features of frog evolution and distribution in broad terms in the hope that critical readers may be motivated to test my ideas against other interpretations or independent evidence from other sources. Some of these testings will surely provide further support for the patterns I discern, and of course, my proposals will be modified as new evidence accumulates and herpetologists and biogeographers with different or original approaches compare my hypotheses with their views of reality.

The Geologic and Climatic Matrix

As I have pointed out elsewhere (1960, 1966), the development of a biogeographic theory requires an interpretation based upon present and fossil distribution patterns, basic ecologic associations—particularly with climate and vegetation—and evaluation of similarities and differences in faunal composition set within a framework of known geodynamic and paleoecologic events. Crucial to the elucidation of a comprehensive theory of frog evolutionary biogeography is an understanding of the broad features of land mass organization and world climate for Mesozoic times. World patterns of Cenozoic change are less critical because basic patterns of land mass distribution were set by late Cretaceous and the principal features of Cenozoic climatic change are well known (Axelrod, 1960). The latter involved restriction of tropical vegetation to an ever-narrower low latitude region and the development of subhumid to arid climates at mid-latitudes on the western margins of the continents, as cooling effects—most markedly from the Oligocene onwards—expanded extratropical climates toward the equator.

At the time of the establishment of the three basic frog phyletic lines (Types I, II, and III) in late Triassic, the two supercontinents, Laurasia in the north and Gondwanaland to the south, formed the principal world land masses and were separated by the great tropical Tethys Sea (Figure 13–30). Subsequent geodynamic events ultimately led to the fragmentation and drift of Laurasia to form North America, Greenland, and Eurasia, while Gondwanaland became South America, Africa, Antarctica, India, Madagascar, Australia and New Guinea, and probably New Zealand (Blackett et al., 1965). The greater portion of the masses had a warm hot climate, comparable to tropical regions today. Only on the extreme north and south were warm temperate areas present.

The two supercontinents began to break up beginning in the Triassic, but the rifts do not seem to have formed significant barriers to terrestrial faunal exchanges until in the Jurassic. Gondwanaland was split into two halves during the middle Jurassic by a disruption that began much earlier and ran from north to south along what is now the coast of eastern Africa. The separation created two land masses—western Gondwanaland, which became South America, Africa, and Madagascar, and eastern Gondwanaland, which later fragmented into India, Antarctica, and Australia. Eastern and western Gondwanaland drifted apart during the remainder of the Jurassic, the former to the east and the latter westward (Figure 13–31). During early Cretaceous both portions of Gondwanaland fragmented again. By the middle Cretaceous the six major southern land masses existed as separate entities, and Gondwanaland was no more. In the north, Laurasia split into two major masses during early Cretaceous, more or less along the same zone that separated South America from Africa.

The Cretaceous period was the most critical time in frog evolution, for it affected present distribution patterns. Ancestors of all modern families must have been present by late Jurassic, approximately 150 million years ago, on one or more of the three major land masses: Laurasia, eastern Gondwanaland, or western Gondwanaland. The disruption of these masses during the early Cretaceous set the stage for rapid differentiation in the faunas isolated on the fragments. The extent of the Cretaceous, a period of time (75 million years) exceeding all of the Cenozoic (65 million years), certainly was sufficient for extensive evolutionary radiation. In general, tropical conditions prevailed over most of the world during the Cretaceous, and maximum marine embayments over the developing continents were characteristic. Antecedents of modern extratropical vegetation were present under mild warm temperate conditions in the far north and south (Figures 13–33, 13–34).

By the Paleocene the modern continents were separated but still drifting toward their current positions. The primary geofloras contributing to modern vegetation also were established by the Paleocene. The temperate Arcto-Tertiary components were present in the far north, and temperate Antarcto-Tertiary components in the extreme south. Much of the land area supported tropical vegetation with Paleotropical-Tertiary components over much of Eurasia, Africa, India, and Australia, and Neotropical-Tertiary components in the Americas. The semiarid to arid vegetation that came to dominate on the western continental margins

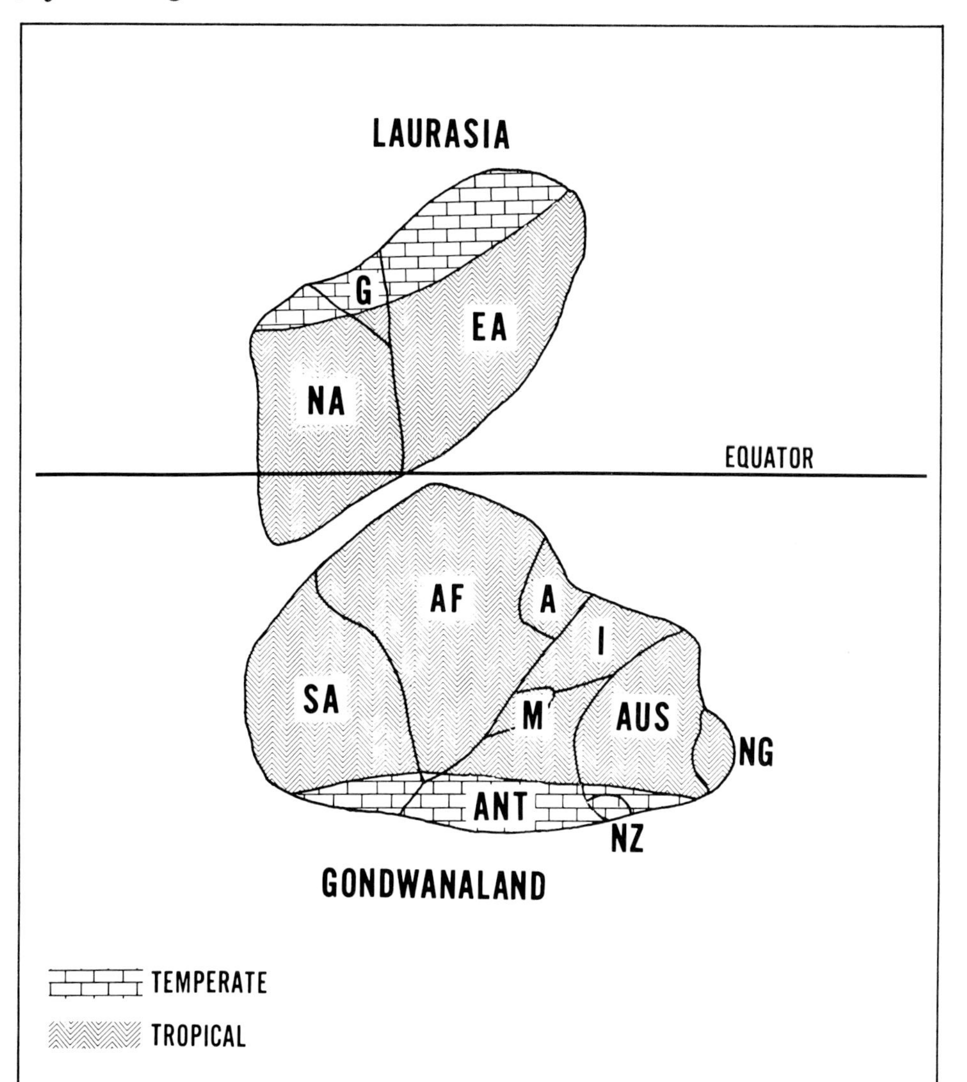

Figure 13–30. Generalized distribution of world land masses derived from Laurasia and Gondwanaland, and climates in Triassic Period: EA, Eurasia; G, Greenland; NA, North America; A, Arabia; AF, Africa; ANT, Antarctica; AUS, Australia; I, India; M, Madagascar; NG, New Guinea; NZ, New Zealand; SA, South America.

at temperate-subtropical latitudes in late Tertiary was represented at this time by tropical ancestors (Figure 13–35).

The essential features of paleogeography early in the Tertiary as they affected frog distribution were as follows:

1. broad connection via the Bering land bridge between North America and eastern Eurasia
2. isolation of western Eurasia
3. isolation of Africa as an island or several large islands
4. isolation of India as an island
5. isolation of Madagascar as an island

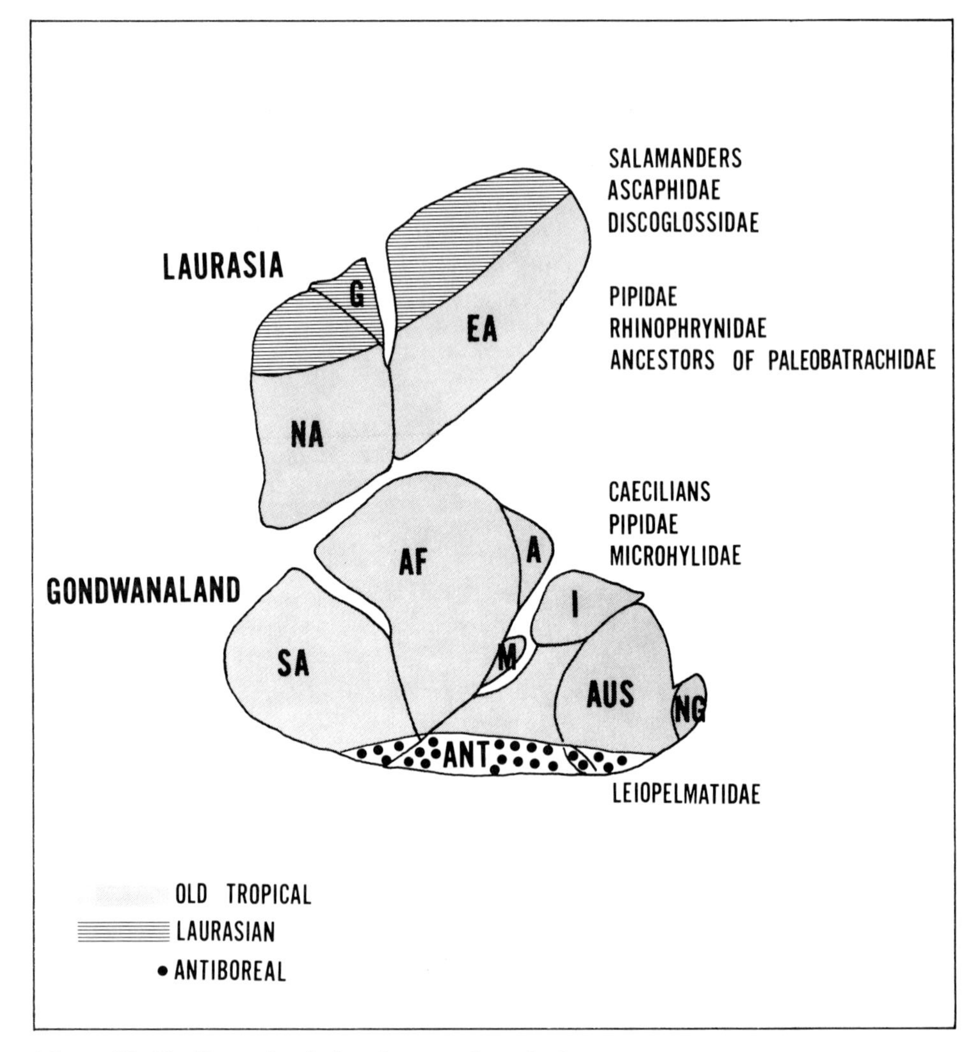

Figure 13–31. Generalized distribution of world land masses and frog source units in early Jurassic. For abbreviations, see Figure 13–30.

6. isolation of Australia as an island
7. connection of North and South America via the Panamanian land bridge

This situation allowed for differentiation in isolation on each isolated land mass, interchange between North America and eastern Eurasia by the Bering land bridge, and interchange between North America and South America through the Isthmian Link.

The principal changes in land mass relationships and connections during the Cenozoic that significantly affected frog distributions are as follows:

1. submergence of the Isthmian Link region across the Panamanian portal from late Paleocene to late Miocene, which effectively reisolated South America as an island but allowed overland exchange with North America on the newly uplifted Isthmian Link from early Pliocene onwards (Savage, 1966, 1973).

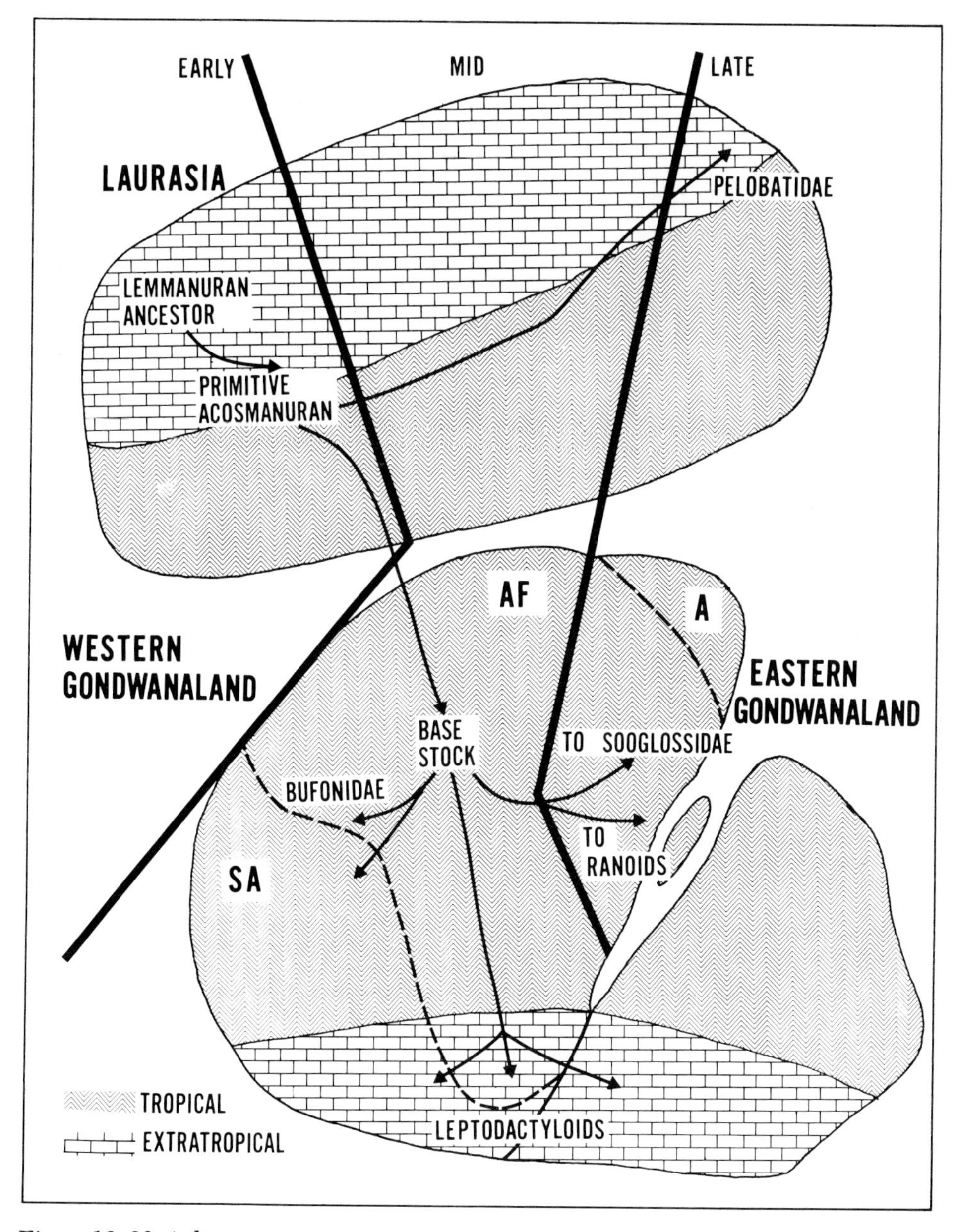

Figure 13–32. A diagrammatic representation of the phylogeny of the principal acosmanuran frog lines, superimposed upon Laurasia, western Gondwanaland, and eastern Gondwanaland for Jurassic times; approximate time of origin is indicated by time divisions. The Pelobatidae probably originated in North America in middle-to-late Jurassic, rather than in Eurasia, as the diagram may suggest. For abbreviations, see Figure 13–30.

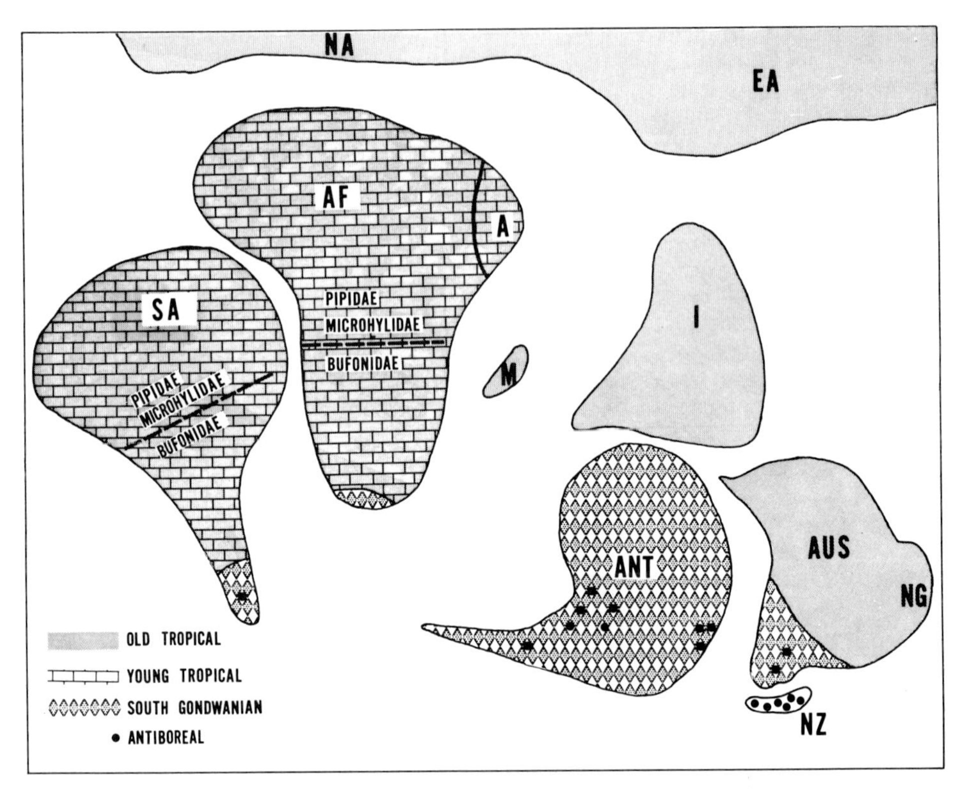

Figure 13–33. Generalized distribution of southern land masses and frog source units in early Cretaceous. South America and Africa are recently separated and India, Antarctica, and Australia have begun drifting apart. For abbreviations, see Figure 13–30.

2. intermittent land-positive areas continuously present in the Indo-Malayan archipelago region from Paleocene to present (Holmes, 1965), but relatively little stability until Eocene; alternating cycles of uplift and submergence from Eocene onwards from Sumatra to the Philippines, with land areas larger than currently present, but never a continuous land mass.

3. connection of the Indian island to the Eurasian land mass in Oligocene-Miocene (Holmes, 1965; King, 1967); during a considerable period of later Cenozoic, isolation of the Indian peninsula from southeast Asia by an extensive embayment covering Assam and Burma (King, 1967).

4. uplift of the high mountains and northern portions of New Guinea to present levels from Miocene onwards (King, 1967; Laseron, 1969); the southern portions of the present island seem to be part of the original Gondwanaland, represented by the Australian shield; the older regions appear to have been land positive and continuous with northern Australia at various times in late Cenozoic and to have suffered the same vicissitudes as Arnhiem Land, the Gulf of Carpentaria, and the continental shelf between these areas and New Guinea (Sahul shelf), including late Cenozoic subsidence.

5. restricted connection of Africa to Eurasia across the Red Sea and Arabia to Persia and Baluchistan from late Miocene onward, although the connection was variously modified and today is restricted to the Sinai area (Moreau, 1966); northwest Africa does not seem to have ever been connected to Europe but rep-

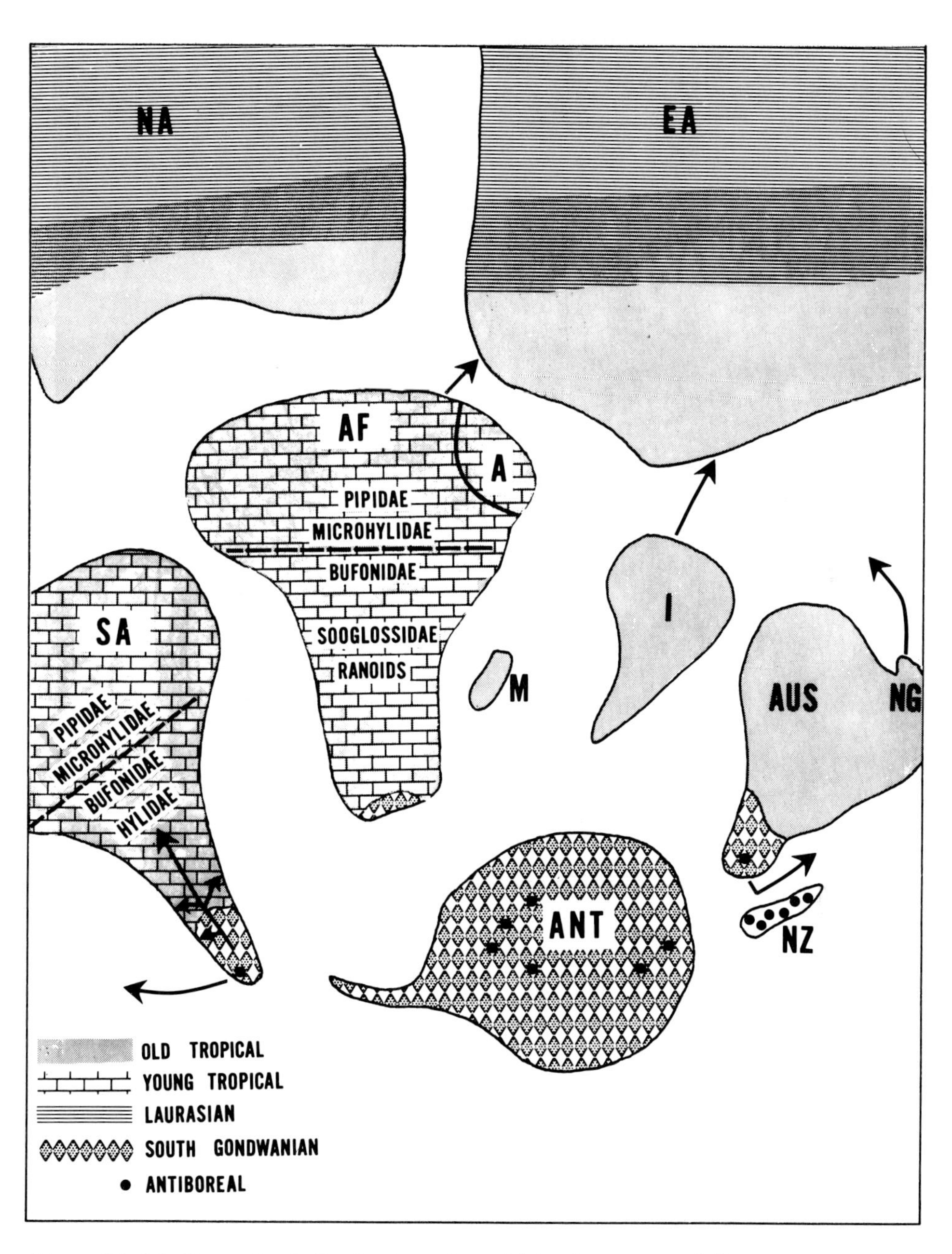

Figure 13–34. Generalized distribution of world land masses and frog source units in middle Cretaceous. For abbreviations, see Figure 13–30.

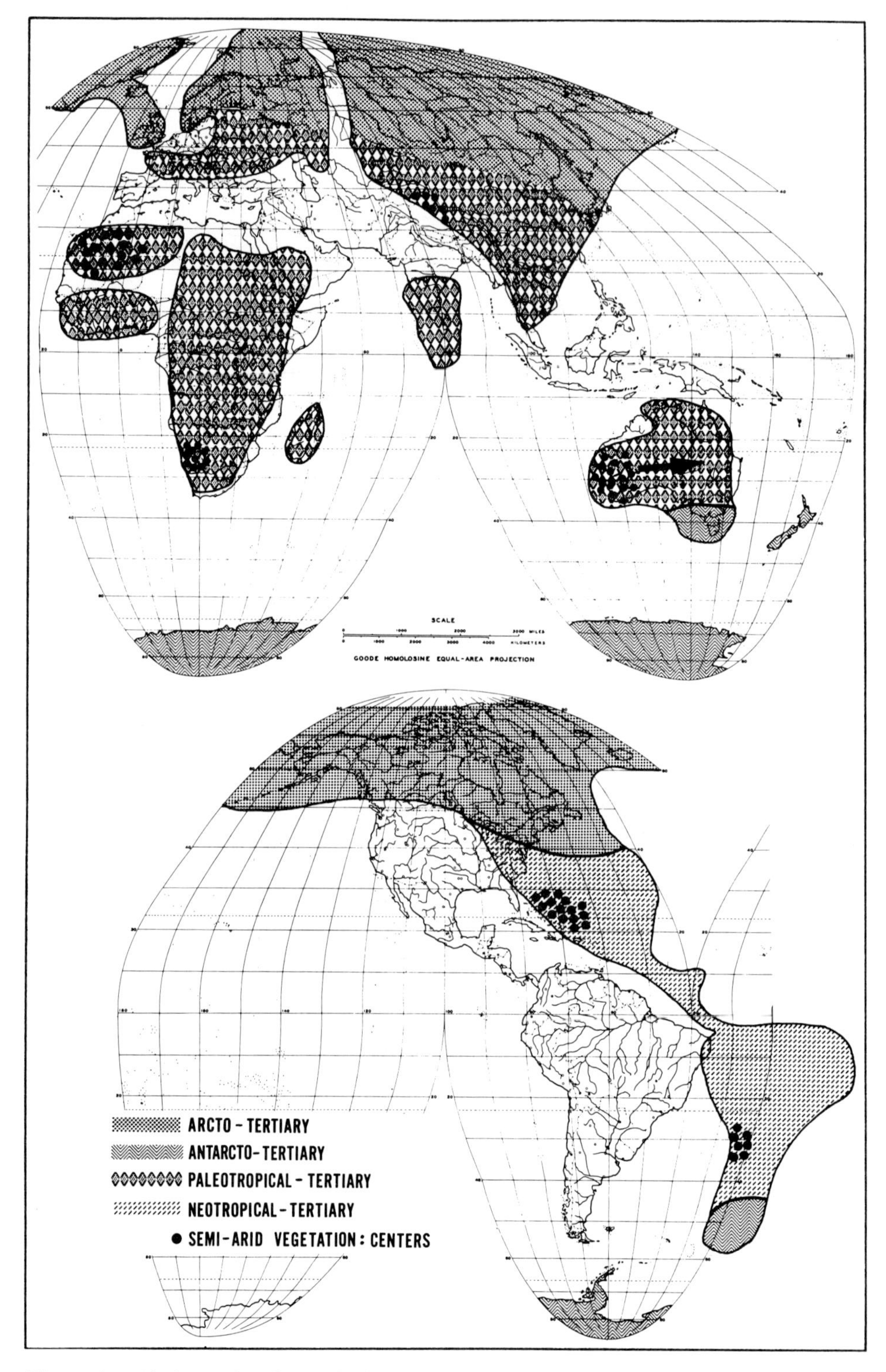

Figure 13–35. Generalized distribution of world land masses, climates, and major geofloral units in early Cenozoic (Paleocene).

resents an Oligocene uplift that added to Africa after major patterns of geography and distribution were established to the south and east on older land-positive areas of the continent (King, 1967).

6. the intermittent disappearance of the Bering land bridge between North America and Eurasia in Tertiary, with final submergence in late Pliocene (Simpson, 1947).

The primary world climatic changes of note during the Cenozoic are as follows (Axelrod, 1960):

1. the gradual elimination of tropical conditions and vegetation in Europe and North America, with tropical groups becoming restricted to the southern paleopeninsulas of southeast Asia (Thailand-Indochina-Malaya) and Middle America, respectively, by Miocene.

2. expansion of temperate conditions southward over the North American and Eurasian masses.

3. elimination of most of terrestrial life from Antarctica.

4. development of semiarid to arid climate and vegetation over most of Australia by Miocene.

5. expansion of semiarid to arid climates and vegetation from Miocene onwards over western North America, western Asia, western south Africa, and western South America and across northern Africa.

Historical Source Units

The five major patterns of frog distribution previously described suggest comparable histories for groups conforming to a particular pattern. On the basis of recent and fossil distributions, the known broad patterns of paleoclimates and their associated vegetation, combined with the evidence for geodynamic change related to continental drift and fragmentation, the families of frogs sharing a similar history may be grouped together as primary historical source units. The five historical source units are briefly defined below, and their composition, geographic relations, and Cenozoic vegetational associations are summarized (Table 13–5).

Old Tropical—derivatives of groups distributed in tropical regions of Laurasia and Gondwanaland by early Jurassic, with a disjunct circumtropical distribution today; associated with tropical vegetation throughout history; distributional history similar to that of caecilians.

Laurasian—derivatives of stocks established in Laurasia by Jurassic and associated with North America and Eurasia during entire history; probably circumpolar in distribution during most of Tertiary, but ranges fragmented by cooling and drying in late Cenozoic; associated with extratropical climates and vegetation through Cenozoic, but two groups have invaded tropical Asia; distributional history is comparable to that of salamanders.

Antiboreal—a group of frogs known from Jurassic onward in extreme southern remnants of Gondwanaland; probably widely spread in Mesozoic but represented today only in New Zealand; associated with southern extratropical conditions throughout time.

Young Tropical—derivatives of groups that evolved in tropical areas on the western Gondwanaland continent in late Jurassic to early Cretaceous; in Cenozoic these groups developed extensive tropical radiations and expanded on to North

America and Eurasia; associated originally with tropical vegetation but invading extratropical areas as well.

South Gondwanian—descendants of a southern extratropical group that became fragmented by continental drift on several land masses derived from Gondwanaland; the basic stock of this unit is associated with temperate climates and vegetation, but since Cretaceous has invaded and diversified in tropical areas.

EVOLUTION OF THE PATTERNS

The earliest fossil frogs are from 170 million years ago in the Jurassic. A probable frog precursor, *Triadobatrachus* (Hecht, 1962), is known from lower Triassic approximately 200 million years ago. By the end of Jurassic at least four

Table 13–5. Historical Units of Recent Frog Faunas

Units and familial groups	Geographic source	Basic vegetation association	Parallel pattern
I. Old Tropical	Tropical Gondwanaland	Tropical-Tertiary	Caecilians
Rhinophrynidae	Laurasia	Neotropical-Tertiary	
Pipidae	Gondwanaland	Tropical-Tertiary	
Microhylidae	Gondwanaland	Tropical-Tertiary	
II. Laurasian	Laurasia	Arcto-Tertiary	Salamanders
Ascaphidae			
Discoglossidae			
Pelobatidae			
III. Antiboreal	Temperate Gondwanaland	Antarcto-Tertiary	
Leiopelmatidae			
IV. Young Tropical	Tropical Africa and America	Tropical-Tertiary	
Bufonidae	Tropical Africa and America	Tropical-Tertiary	
Sooglossidae	Tropical Africa	Paleotropical-Tertiary	
Ranidae	Tropical Africa	Paleotropical-Tertiary	
Hyperoliidae	Tropical Africa	Paleotropical-Tertiary	
V. South Gondwanian	Temperate Gondwanaland	Antarcto-Tertiary	
A. Basic Temperate stocks			
Primitive Myobatrachidae	Australia	Antarcto-Tertiary	
Heleophrynidae	South Africa	Antarcto-Tertiary	
Primitive Leptodactylidae	South America	Antarcto-Tertiary	
B. Extratropical Derivative	South America	Antarcto-Tertiary	
Rhinodermatidae			
C. Tropical Derivatives			
Pelodryadidae	North Australia	Paleotropical-Tertiary	
Ceratophryinae	South America	Neotropical-Tertiary	
Hylodinae	South America	Neotropical-Tertiary	
Brachycephalidae	South America	Neotropical-Tertiary	
Allophrynidae	South America	Neotropical-Tertiary	
Pseudidae	South America	Neotropical-Tertiary	
Centrolenidae	South America	Neotropical-Tertiary	
Hylidae	South America	Neotropical-Tertiary	
Dendrobatidae	South America	Neotropical-Tertiary	

of the living groups are definitely known as fossils (the xenoanuran Pipidae, the lemmanuran Discoglossidae, and the Leiopelmatidae). All four major lineages were probably present by that time, although in addition to xenoanurans, lemmanurans, and leiopelmatids, only the fossil *Comobatrachus,* which is either a scoptanuran or acosmanuran, is known prior to the Cretaceous. Descendent stocks of all four lineages persist today, and all evidence suggests that almost all families of frogs had appeared by the end of the Cretaceous.

The most striking feature of past and present distributions is the fundamental and consistent association of distributions of each group with either Laurasia or Gondwanaland, and those supercontinents' Cretaceous fragments that by continental drift have come to form today's major land areas. These associations are as follows: In Laurasia, Rhinophrynidae, Paleobatrachidae, Ascaphidae, Discoglossidae, Pelobatidae, and salamanders; in Gondwanaland, Pipidae, Microhylidae, Leiopelmatidae, Leptodactyloidea, Pelodryadidae, Bufonidae, Hylidae, Sooglossidae, and caecilians.

The only family not conforming to this basic pattern is the most primitive frog group, the Pipidae. The pipids probably originated in Gondwanaland but occur as fossils on both supercontinents. Today they are restricted to the African and South American fragments of western Gondwanaland.

In the following sections, I outline my views of the biogeographic history of frogs in the form of a general theory. The temptation is very strong in such a circumstance either to sketch the conclusions without supportive data or to lead the reader through all the logical steps (some would say contortions) that support each statement that is made. The genesis of ideas seems to be based upon a simultaneous unconscious testing and experimental association of a vast amount of data, feelings, impressions, and experiences that come into consciousness as insights of apparent patterns of relationships (Areiti, 1967), and only after these processes occur do we use the ponderous, narrow, and linear conscious logical process to support our discoveries. The theory I have developed is my best approximation of what seems to have occurred in frog history. Every statement made in the explanation could be preceded by precautionary modifiers ("apparently," "probably," "presumably," or "possibly"), but since I regard the function of the theoretical biologist to be parallel to that of his colleagues in physics, that is, to develop explanatory hypotheses to be tested by future studies, I have reduced these modifiers to a minimum; they are to be understood. Similarly, I have attempted to present detailed logical arguments in support of only the most crucial parts of the theory, leaving the reader an opportunity to challenge many statements that are apparent to me but need further investigation. By such processes the insight and imagination of the theoretician are tested against others' perceptions of reality.

Mid-Mesozoic Patterns

The presence of unquestioned lemmanurans (Discoglossidae) in the upper Jurassic strongly suggests that the origin of the three basic lineages of frog evolution (Types I, II, and III) and the leiopelmatids occurred in the Triassic and that they had already undergone substantial diversification prior to the Jurassic. Although the fossil evidence is meager, it indicates that the basic pattern of distribution for all groups of amphibians, except that of the acosmanurans, had been set by that time (Figure 13–31). These early patterns parallel one another in terms of continental and ecologic zonation to the extent that they probably reflect a real situation (an asterisk indicates a Jurassic fossil).

Laurasia		
	Warm Temperate	*Tropical*
	Salamanders* . .	ancestors of Rhinophrynidae
	Ascaphidae	Pipidae*
	Discoglossidae*	ancestors of Paleobatrachidae
Gondwanaland		
	Warm Temperate	*Tropical*
	Leiopelmatidae*	Caecilians
		Pipidae
		Microhylidae

My interpretation suggests that the primitive groups (the xenoanurans and scoptanurans) evolved in warm moist (tropical) areas of Gondwanaland. The filter-feeding larvae of these stocks, among other features, apparently limit them to regions where high productivity in ponds or still waters provides the tadpoles with organic detritus or microorganisms in abundance. Although the scoptanuran larva is a marked departure from the more primitive xenoanuran type, the major evolutionary shift in frogs lies in the development of larvae with an oral disk, beaks, and denticles. By the early Jurassic this highly advanced condition must have been present in larvae of the early lemmanurans (Ascaphidae and Discoglossidae). Available data indicate that in the Jurassic, frogs with filter-feeding larvae were tropical in distribution. Those with advanced mouth parts in the larvae were northern, probably in association with warm temperate conditions.

The basic structure of oral disk, beaks, and denticles in the larvae of lemmanuran and acosmanuran frogs unquestionably is designed for grazing on attached algae (Starrett, this volume). The concomitant changes in the larval head, skeleton, and musculature, in the branchial region, and in the elongation and coiling of the intestinal tract are similarly associated with algae-grazing habits. The filter-feeding larvae of the more primitive lines (the xenoanurans and scoptanurans) cannot remove filamentous or attached algae from the surfaces of rocks, submerged vegetation, or the bottom. The tadpoles of living lemmanurans apparently are restricted to the basic algae-grazing adaptive zone. In one genus *(Ascaphus)* the larvae are specialized algae-grazers with numerous denticle rows, which allow them effectively to scrape algae off exposed rocky surfaces in the spray zone of fast-moving brooks, and with the oral disk enlarged and modified into an adhesive organ, which prevents the tadpoles from being swept downstream by the unruly waters.

After the evolutionary breakthrough provided by the modification of the mouthparts, branchial cavity, digestive tract, and associated structures for algae grazing, previously unrealized opportunities for adaptive radiation in larval structure and function developed among many parallel descendent lines within the acosmanurans. All are based upon this single major ecologic-evolutionary shift in larval organization. The radiation is expressed in many ways. Algae-grazing larvae provided a means for invasion of many habitats not available to filter-feeders, which are essentially restricted to ponds. Algae-grazing acosmanuran larvae are found in ponds, rivers, fast-moving streams, tree holes, and water trapped in epiphytes. Many genera found in fast-moving brooks have the basic algae-grazing oral disk modified as an adhesive organ, as in *Ascaphus*. The mouthparts essential for algae-grazing, particularly the beaks and denticles, provided a basis for a radiation of advanced tadpoles that feed on one or more of the following: small crustaceans; insect larvae; insects; and larger prey, including frogs' eggs and other tadpoles. In several genera the beaks and denticles are greatly enlarged for grasping large prey, while others take large bites from the leaves and stems of aquatic higher plants. Reduction in denticle rows and specialization of the oral disk into a dorsally directable funnel for surface feeding also occur in several genera. These and many other derivative modifications (Orton, 1953; Starrett, 1968, this volume) have developed independently in diverse acosmanuran lines, but each is founded upon the basic algae-grazing larval organization that is the common denominator of evolution in higher frogs.

There seems little question that the earliest frogs evolved in tropical situations. The Proanura are from the Triassic of Madagascar, and all families with filter-feeding larvae (Pipidae, Rhinophrynidae, Microhylidae and presumably Paleobatrachidae) seem to have been, and still are, tropical. Where did frog groups with advanced algae-grazing larvae originate? The most ancient of these groups, today represented by the Ascaphidae and Discoglossidae, are basically extratropical in distribution and seem to have been so through much of geologic time. It appears likely that they developed from ancestors with filter-feeding larvae that invaded peripheral tropical to warm temperate regions, where selection pressures required a more efficient larval feeding mechanism. Under warm moist tropical conditions the amount of organic materials available to filter-feeders is more than adequate. In less equable conditions in which the amount of organic material suspended in the water might be reduced seasonally or for other reasons, development of an oral disk, beaks, and denticles for grazing on attached algae might be favored. It seems probable that this development—the key that opened the way for the extensive radiation of lemmanuran and acosmanuran frogs—originated in north temperate areas.

The status of the Leiopelmatidae in this context remains a problem. The sole living genus has a form of direct development, so we have no evidence about the kind of tadpole that may have been present in its ancestor. Frogs referred to this family are known from the lower and middle Jurassic of Argentina. All recent workers have assumed that the leiopelmatids are closely allied to the Ascaphidae and by logical extension that early leiopelmatids had larvae of Type II. I am extremely doubtful that the leiopelmatids can be placed with the ascaphids. The two families share many primitive features (Ritland, 1955*a*, 1955*b*; E. M. T. Stephenson, 1951, 1952, 1955, 1960; N. G. Stephenson, 1951*a*, 1951*b*, 1955), but these may well be retained from an earlier stage in frog evolution.

As an alternate hypothesis I suggest that lemmanurans and leiopelmatids are independent derivatives of a primitive frog stock with filter-feeding larvae. In Laurasia this stock gave risk to the lemmanuran line, with algae-grazing larvae. In the far south the line evolved into a group, the leiopelmatids, with direct development. This feature of the life history may represent another way for filter-feeding larval lines to take advantage of temperate or at least nontropical lowland conditions, since many montane microhylids have also developed this advance. As indicated by the embryologic studies of the Stephensons cited above, the leiopelmatids may be relicts, rather than derivatives of the primitive frog stock that gave rise to the lemmanurans. My hypothesis requires that xenoanurans, scoptanurans, and the ancestors of the leiopelmatids must have been present in tropical regions in the Triassic. The derivatives on Laurasia (lemmmanurans) had the same general geographic relations in the Jurassic as they do today. It seems likely that lemmanurans have never occurred on the southern continents.

Nothing permits further examination of the historical origin of this ancient persistent geographic pattern; it can only be accepted as the situation in the early Jurassic. This pattern sets the stage for further frog evolution as the events of later Mesozoic fragmented the world land masses. It is certain that three of the historical source units of the modern frog faunas were already clearly defined in early Jurassic. Old Tropical groups (Pipidae and Microhylidae) were probably widespread over tropical areas. Within this unit three distinctive patterns are suggested: Tropical Laurasia and Gondwanaland, Pipidae; Tropical Laurasia, ancestors of Rhinophrynidae (North America) and ancestors of Paleobatrachidae (Eurasia); Tropical Gondwanaland, Microhylidae. The pipids subsequently became restricted to western Gondwanaland, the Rhinophrynidae persists to this day in tropical North America, and the Paleobatrachidae is known only from the Eocene of Europe. The microhylids even now have only a few genera that occur outside the limits of fragments of Gondwanaland. The pattern for this family corresponds very closely to that for caecilians, which also may be considered an Old Tropical (Gondwanaland) group.

The Laurasian source unit was present in peripheral tropical to warm temperate regions on Laurasia. Members of the unit include the Ascaphidae and Discoglossidae, which from earliest times have been northern Laurasian extratropical lines. Salamanders, of course, also conform to this pattern. The Antiboreal unit (Leiopelmatidae) was restricted to warm temperate regions of southern Gondwanaland, and its restriction to New Zealand today seems to be as a relict. King (1967) indicated that New Zealand probably was part of Australia in the Triassic-Jurassic and became fragmented prior to the Cretaceous. Leiopelmatids appear to persist there because no other frogs, until recent introductions, were in competition with them, while they disappeared in South America, Antarctica, and Australia.

A major difficulty in discussing later events in frog history and evolution remains the lack of evidence for the origins of the modern frogs (acosmanurans). Present distribution patterns and the single fossil *Comobatrachus* from the upper Jurassic confirm that this stock probably arose from an ancestor allied to the lemmanurans in late Triassic or early Jurassic. The next-oldest identifiable acosmanuran is *Eopelobates* from the Cretaceous of North America, referrable to the megophryine line of the Pelobatidae but probably ancestral to it (Estes, 1970). Given the present distribution and diversity of acosmanurans, it is obvious that none of the living families is closely allied to the ancestor of this line. Of the living acosmanurans, the most primitive family (Pelobatidae) apparently arose in the Jurassic from an ancestor related to the lemmanurans. The primitive subfamily Megophryinae is known from the Cretaceous, Eocene, and Oligocene of North America and Eocene to Miocene in Europe in the form of *Eopelobates*, and the more highly advanced modern megophryines occur in temperate and tropical southeastern Asia. Estes (1970) concluded that the latter group and *Eopelobates* diverged no later than the Cretaceous. The other two stocks, the Pelobatinae and Pelodytinae, probably arose in the Eocene from the *Eopelobates* line. The pelobatids are not close to any existing group of frogs and seem to have had a long independent history on the northern land masses, where all genera occur today. For this reason, I regard them as members of the Laurasian source unit.

All evidence supports the view that modern frogs (acosmanurans) were derived from lemmanuran antecedents. None of the living acosmanurans are closely related to known lemmanurans, which differ strikingly in tadpole structure and adult structure. The ancestors of the acosmanuran line must have evolved from a northern lemmanuran stock in the Jurassic, if not earlier. Such an ancestor very likely was associated with warm temperate conditions on Laurasia, with one derivative line remaining there (that lineage giving rise to the Pelobatidae) and with a second major derivative line invading and radiating in Gondwanaland. In my view, the pelobatids are a relict-derived group of acosmanurans of substantial age but not directly derivable from lemmanurans or ancestral to any other modern frog family.

Lynch (1969) has argued that megophryine pelobatids are the direct ancestors of early leptodactyloids, primarily because some leptodactyloids (Myobatrachinae) share with the former the condition of amphicoelous vertebrae and free intervertebral bodies. Even this condition is a questionable similarity since, as pointed out by Noble (1930) and Parker (1940) and confirmed in material I have examined, the intervertebral body is partially fused to the anterior part of the centrum in most myobatracines and thus produces a semiopisthocoelous condition comparable to that in the Rhinophrynidae (Walker, 1938). In megophryines the intervertebral bodies are free and ossified and the vertebral centra are nonperforate amphicoelous. Although Lynch fails to define the Leptodactylidae *(sensu latu)*, none of his subgroups approaches the pelobatid megophryines closely, unless a very careful selection of comparative characters is made, as in his table (1969, Table 5). Because this point is critical, I have summarized the principal differences between leptodactyloids and megophryines (Table 13–6). These features confirm the distinctness of the two groups and do not support Lynch's contentions that megophryines gave rise to leptodactyloids. As we shall see later, such a view is also contrary to the biogeographic evidence. In most cases megophryine characters tend to show advancement over primitive leptodactyloid features so that some case might be for megophryine derivation from leptodactyloids —a proposal that does not merit any more serious consideration than the reverse.

Table 13–6. Anatomical Differences Among Megophryine, Primitive Leptodactyloid, and Some Advanced Leptodactyloid Frogs

Anatomical feature	Megophryines	Primitive leptodactyloids	Some advanced leptodactyloids
Sacral diapophyses	Expanded	Cylindrical	Dilated
Cricoid ring	Divided dorsally	Complete	Divided ventrally
Sacral-coccygeal articulation	Single condyle	Double condyle	Fused
Semitendinosus-sartorius muscles	Single	Separate	Separate
Depressor mandibulae	Single anterior slip on dorsal fascia	Three slips on dorsal fascia, squamosal, and tympanic ring	Various, some approach megophryines
Adductor mandibulae	Subexternus only	Both subexternus and externus	Either subexternus or externus
Larval denticle rows	More or less than 2/3 (0;4/5–8/8)	2/3	More or less than 2/3

According to my view a primitive basal stock of acosmanurans developed by early Jurassic. By middle Jurassic a tropical derivative lineage was established in Gondwanaland, where it underwent significant evolutionary radiation. Arising from the central tropical acosmanurans were three lines leading to extant families. Each of these major evolutionary lines—the leptodactyloids, the bufonids, and the ranoids—have a distinctive morphology and distribution. They seem to have diverged from the original tropical acosmanurans independently and at slightly different times. The primitive leptodactyloids seem to have diverged at the earliest time and to have become associated with southern extratropical vegetation prior to the initial separation of Gondwanaland into eastern and western land masses in middle Jurassic. Their present distribution, with primitive genera in temperate South America, subtropical to temperate South Africa, and temperate Australia and extensive radiation on the former and latter continents, supports this concept. They do not appear to have ever occurred on the northern continents until a few genera became established in Middle America and North America across the Isthmian Link with South America in the Cenozoic. Lynch (1969) reached a similar conclusion—that basic leptodactyloid stocks originated in southern temperate areas—but his concept is confused by his acceptance of megophryine pelobatids as direct ancestors to leptodactyloids and his failure to realize the differences in bufonid and leptodactyloid history.

The bufonids appear to have arisen in tropical areas on western Gondwanaland after the separation of Gondwanaland into eastern and western fragments. During the fragmentation of western Gondwanaland in late Jurassic—early Cretaceous, bufonids that were almost certainly of the genus *Bufo* became isolated on the two island continents of Africa and South America. The difference in chromosome complements between the toads of the strictly African *Bufo regularis* group ($2n = 20$) and all other *Bufo* ($2n = 22$) may be a reflection of the ancient split. The fact that all advanced bufonid genera including the "atelopodids" are apparently derived from *Bufo* (Tihen, 1960; McDiarmid, 1969), further confirms the long history of this genus in Africa and South America.

The ranoids appear to be tropical African in origin and advanced over an ancestral stock that may be represented today by the Sooglossidae, Hemisinae, and Arthroleptinae, or these latter three may be independent derivatives of the same primitive line that gave rise to the ranoids. Origin of the central ranoid line must have been in the Cretaceous, since Africa was completely isolated from the rest of Gondwanaland by that time and ranids have only recently barely reached South America and northern Australia.

The views regarding the origin of acosmanurans and of the three major lineages of modern frogs expressed in the last several paragraphs form a hypothesis that agrees with climatic and geodynamic events, as well as modern distributions. It cannot be fully confirmed until additional fossils and other evidence have tested its basic outlines. One point is clear: if the hypothesis is approximately correct, a basic stock of acosmanurans must have been present in tropical Gondwanaland by middle Jurassic, for leptodactyloid ancestors were almost certainly present in southern temperate Gondwanaland by the time of fragmentation into eastern and western land masses. The ancestors of the bufonids also must have been present in tropical portions of the western Gondwanaland continent prior to its disjunction into African and South American island continents in early Cretaceous. Derivatives of this same basic tropical acosmanuran stock must have given rise to the ranoid lineage in Africa following the isolation of that continent in early Cretaceous. No existing family appears to be a satisfactory stem group for this radiation, but no evidence of the nature or occurrence of such a basal stock is available, at least in South America and Africa.

Is there any suggestion that such a stock existed and has been replaced by the radiation of ranoids in the tropics of the Old World? The sole possibility at present is the enigmatic *Indobatrachus* from Eocene Deccan (India). According to the hypothesis developed here, the eastern Gondwanaland mass separated from the western sector before the appearance of bufonid or ranoid frogs. In addition, if the restriction of primitive leptodactyloids to temperate areas held true, none of these frogs could be expected in India when it became fragmented off from the other land masses in earliest Cretaceous. The anuran fauna of the Indian island continent must have contained microhylids. All other recent families found there today seem to be post-Eocene invaders from the northwest (ranoids) or northeast (bufonids). It seems probable that in addition to microhylids, the basic acosmanuran tropical stock might also have been represented in

India during Cretaceous to Oligocene, prior to its connection with Eurasia. The status of *Indobatrachus* thus becomes of special importance. Lynch (1969), who did not examine the several available specimens, accepts and paraphrases Noble's (1930) conclusions in every detail, even to relating the fossil to Australian myobatrachines. The vertebrae of *Indobatrachus* seem to be nonperforate amphicoelous, with free and possibly ossified intervertebral bodies, dilated sacral diapophyses, and a bicondylar sacral-coccygeal articulation. Osteologically, the fossil appears to be at an early stage of acosmanuran evolution. It may represent such a stage or it may be an early leptodactyloid that became isolated on India in the Cretaceous and drifted northward with that island. In either event, it provides no evidence that favors a pelobatid origin for leptodactyloids or the northern origin and occurrence of the latter. If anything, *Indobatrachus*—whether a preleptodactyloid stock or a leptodactyloid—conclusively eliminates any idea of the presence of leptodactyloids on Laurasian areas prior to Cenozoic invasions. The relatively abundant materials of this frog need to be re-examined by workers who are alert to these possibilities and familiar with the osteology of modern frogs and the limits of fossil interpretation.

Other remnants of the early acosmanurans of tropical Gondwanaland are to be expected in fossil finds on the southern continents. Certainly the rather late appearance of definitive microhylids (Miocene), bufonids (Eocene), leptodactyloids (late Cenozoic), hylids (Miocene), and ranoids (Eocene) as fossils on the northern continents suggests a long history to the south. Although well advanced toward ranid features, the family Sooglossidae and its possible allies the Hemisinae and Arthroleptinae seem to represent a transition between early acosmanurans and ranids (Griffiths, 1959). Perhaps *Indobatrachus* has affinities with this stock, but the three groups are advanced in most discernible features over the fossil.

The grand events of frog evolution in Jurassic must remain conjectural. My interpretation is summarized (Figure 13–32) by relating geography to a proposed phylogeny of acosmanurans. Under this interpretation, by the late Jurassic the main groups of frog evolution were present and, as Gondwanaland became completely broken up in the early Cretaceous, in position to affect modern distribution patterns. The probable occurrences of modern families or their precursors are summarized (Table 13–7) for late Jurassic–early Cretaceous. As the developments during Cretaceous and Cenozoic are described, each group will be followed in relation to its geography. Basically, the situation at this point was as follows:

Laurasia
  Temperate—Laurasian Unit
  Tropical—Old Tropical Unit
Gondwanaland
  Tropical—Old Tropical Unit; Young Tropical Unit (South America and Africa)
  Temperate—Antiboreal Unit; South Gondwanian Unit

Cretaceous Patterns and Events

During the early Cretaceous (Figure 13–33) the drifting continents became widely separated, and each fragment underwent physiographic evolution independently. Events in Laurasia during the Cretaceous seem not to have markedly affected frog evolution, although rhinophrynid and paleobatrachid differentiation in North America and Europe may have been proceeding independently. The Laurasian Unit families were probably circumpolar in distribution on both land masses at this time.

The effects of fragmentation and drift on the southern masses were much more profound, for the acosmanurans had become established and were to undergo major radiation on at least three of the continents. Perhaps the best way to examine what probably transpired during the 75 million years of Cretaceous time is to consider each mass individually.

SOUTH AMERICA. At the beginning of Cretaceous (Figure 13–33, Table 13–7), the fauna must have been broadly zoned: The tropical zone must have supported anurans of the Old Tropical Unit and Bufonidae of Young Tropical Unit; the temperate zone, the Antiboreal Unit and South Gondwanian basic stock. The principal events of Cretaceous involved radiation within the bufonids and invasion and diversification of leptodactyloid derivatives in subtropical to tropical areas. Sometime during the Cretaceous isolation of the continent, the tropical New World family of tree frogs (Hylidae) arose from some leptodactyloid stock. Several lines of evidence point to the validity of this interpretation. By far the most convincing is the chronology provided by the land bridge present during the Paleocene between North and South America across the Isthmian Link region (Savage, 1966, 1973). This bridge was inundated by the sea during Eocene to early Pliocene times, so that South America was an island during much of the Cenozoic. Microhylids, bufonids, and hylids all have distributional and evolutionary patterns indicating that they first reached North America from the south during the Paleocene. It is clear under these circumstances that they must have been in South America as differentiated groups in the Cretaceous.

Carvalho's (1954) study of the Microhylidae indicates that the two Middle American endemic genera arose from a common ancestor that also developed into a South American radiation of twelve genera. The ancestral stock was in both Middle and South America in the Paleocene but underwent independent evolution north and south of the submerged link during middle Cenozoic. Since the Pliocene, three genera of the South American radiation have moved into Panama over the re-established isthmus. No other conclusion is possible but that microhylids were in South America prior to the Paleocene (Savage, 1966). Contrary to the views that I have previously expressed, bufonids must have been in South America in early Cenozoic. Three lines of argument support this concept: As shown by McDiarmid (1969), the radiation of bufonids in South America leading to the distinct endemic genera *Oreophrynella, Atelopus, Dendrophryniscus,* and *Melanophryniscus* must have occurred during the Eocene-Pliocene from an ancestor essentially like *Bufo;* Blair (1970) and Tihen (1962) have presented evidence that the radiation of *Bufo* in South America is ancient and extensive and certainly not a post-Pliocene affair; and Estes and Wassersug (1963) have established the presence of a modern *Bufo* species in Miocene South America prior to the Pliocene land reconnection. Estes and Reig's (this volume) report of *Bufo* from the Paleocene of Brazil clinches this argument. In addition, it is now quite clear that the tropical North American and Holarctic radiations of toads arose from invaders from South America that became isolated and differentiated north of the Panamanian portal. The situation with the tree frogs, Hylidae, corresponds to that for the Bufonidae. South America is a center of speciation and high group diversity that must have developed during the Eocene-Miocene isolation. Nevertheless the family also is known during this time in North America (Tihen, 1964), and one group, *Hyla* (*sensu strictu,* Gaudin, 1969), has invaded Eurasia.

The leptodactylid situation is less clear. Only one large group (usually placed in a single genus *Eleutherodactylus*) seems to have been north of the Panamanian portal during mid-Cenozoic. Perhaps other genera became extinct there, but the fantastic radiation south of the portal area into four subfamilies, some forty essentially endemic genera, and five endemic derivative families,

Table 13–7. Probable Distribution of Families or Ancestral Stocks by Late Jurassic–Early Cretaceous

Laurasia:							
	North America				*Eurasia*		
Temperate	Salamanders				Salamanders		
	Ascaphidae						
	Discoglossidae				Discoglossidae		
	Pelobatidae				Pelobatidae		
Tropical	Pipidae				Pipidae		
	Rhinophrynidae				Paleobatrachidae		
Tethys Sea							
Gondwanaland:							
	South America	*Africa*	*Madagascar*	*India*	*Australia*	*New Zealand*	*Antarctica*
Tropical	Caecilians	Caecilians		Caecilians			
	Pipidae	Pipidae					
	Microhylidae	Microhylidae	Microhylidae	Microhylidae	Microhylidae		
	Bufonidae	Bufonidae		*Indobatrachus*			
		Sooglossidae					
		Ranoids					
Temperate	Leiopelmatidae				Leiopelmatidae	Leiopelmatidae	Leiopelmatidae
	Leptodactyloids	Leptodactyloids			Leptodacytyloids		Leptodactyloids

leaves no doubt of the long history of the group in South America (Lynch, 1969).

The broad pattern sketched here is one that has been repeated, with variations, on other southern continents. At the beginning of the Cretaceous the dominant terrestrial frogs in tropical South America must have been the scoptanuran microhylines. Also present were bufonids, probably of the genus *Bufo.* This genus is marvelously adapted for a generalized terrestrial ecologic role. Early in its history, selection pressures molded the stock into a relatively narrow range of morphologic variation and larval structure. The result has been radiation and wide distribution into almost every major climatic zone, without significant change in the basic conservative toad morphology. Except for minor radiations of specialized derivative genera in tropical Africa (Tihen, 1960), Indo-Asia (Inger, 1966), and the neotropics (Savage and Kluge, 1961; McDiarmid, 1969), toads from all over the world retain a basic common morph and show general adaptation, rather than special modifications. It is almost as though evolution discovered a very adaptable toad morph, and once it was developed, no major improvements could be devised.

In temperate areas of South America, the basic leptodactyloid stock was differentiating during the Cretaceous. As Lynch (1969) has pointed out, the primitive base genera of both the leptodactyline and telmatobiine lines of evolution are even today associated with extratropical regions of this continent. The basic algae-grazing tadpoles of the leptodactylids apparently had the potential for diversification and radiation into a great many habitats and ecologic niches not utilized by the filter-feeding larvae of microhylids or the conservative pond larvae of *Bufo.* However, the leptodactylids' fundamental adaptation to temperate conditions seems to have delayed their expansion northward into tropical South America. Almost certainly the first major derivative of the original leptodactyloid stock to penetrate and radiate in tropical South America were the tree frogs, Hylidae. The advantages of their algae-grazing larvae and an initial acquisition of a general adaptation for arboreal adult life made them extremely suitable noncompetitive invaders into the tropical fauna composed of aquatic frogs (pipids), terrestrial forms with filter-feeding larvae (microhylines), and the terrestrial toads *(Bufo).* Substantial radiation must have taken place in the Cretaceous, since such diverse hylids as *Smilisca, Agalychnis,* and the ancestors of *Hyla (sensu strictu)* invaded North America in the Paleocene. Later in the Cretaceous, antecedents of the tropical radiations of leptodactyloids must have moved into subtropical and tropical southern South America. Clearly they had advantages for diversification not available to the microhylids, with their filter-feeding larvae, and to the toads. Nevertheless, none of the leptodactylids or their specialized derivative families seem to have been able to pass through the tropical regions of South America and invade Middle America by the end of the Paleocene, except possibly *Eleutherodactylus.* I can only conclude that the main pulse of tropical leptodactyloid radiation and diversification is an early Cenozoic phenomenon. The southern extratropical leptodactyloid derivative family Rhinodermatidae probably differentiated in the Cretaceous and may represent a specialized stock that retains many primitive basic leptodactyloid features. A summary of the situation in mid-Cretaceous is diagramed (Figure 13–34). The primary Cretaceous event in South America was the origin and differentiation of the tree frogs.

AFRICA. At the beginning of Cretaceous the fauna was probably broadly zoned as follows: The tropical zone must have supported the Old Tropical Unit and the Bufonidae and ancestors of ranoids of the Young Tropical Unit; the temperate zone, the South Gondwanian Unit.

The principal events of the Cretaceous involved radiation within the bufonids and the origin and radiation of ranoids in tropical latitudes. The pattern suggested here differs from that in South America in that tropically adapted acosmanurans ancestral to ranoids developed to completely dominate the tropical fauna. As a consequence the temperate leptodactyloids remained restricted to a few areas in southern Africa and failed to compete with the ranoid stocks. By the Cenozoic, Africa must also have had its distinctive fauna of relict Old Tropical groups, the Young Tropical bufonids and the beginning of the fantastic ranid-hyperoliid radiation into seven major subfamilies and fifty genera. In South America, where no ranoids existed, the leptodactyloids evolved into tree frogs (Hylidae and Centrolenidae) and myriad other adaptive types; in Africa the ranoids exploded in a similar manner into tree frogs (Hyperoliidae and Rhacophorinae) and a vast spectrum of other adaptive types.

MADAGASCAR. This subcontinent seems to have inherited only scoptanurans from the original Gondwanaland mass. Its separation from Africa and eastern Gondwanaland in the Jurassic occurred before the origin of bufonids or ranoids, and its geographic position apparently precluded the temperate African leptodactyloids from reaching the island. In the Cretaceous the fauna probably consisted of microhylids, which are now differentiated into two distinctive endemic subfamilies. Pipids may also have occurred on the island and may have become extinct there during later times.

INDIA. The Indian Cretaceous fauna must have been similar to that of Madagascar, except for the possible presence of the ancestors of *Indobatrachus*. It is possible that pipid frogs occurred on these island subcontinents, but if so they became extinct before leaving any record. The microhylid fauna of India is diverse and suggests a long history for the family on the subcontinent (Parker, 1934).

ANTARCTICA. If any of my major premises regarding Mesozoic distributions are correct, the Antarctic continent must have been inhabited by leiopelmatids and leptodactyloids during the Cretaceous; the recent discoveries by Colbert (1970) of lower Triassic reptiles and amphibians on the continent that have relationships to other Gondwanaland faunas certainly supports an Africa–South America–Australia–New Zealand–Antarctic connection and corroborates the concepts developed in this paper.

AUSTRALIA. In early Cretaceous the frog fauna was probably broadly zoned, with the Old Tropical Unit represented by microhylids in tropical areas and Antiboreal (Leiopelmatidae) and South Gondwanian (basic leptodactyloids) groups in extratropical situations. By the end of the period the tree-frog line (Pelodryadidae) may have begun to differentiate in tropical areas, but it seems likely that this was a Cenozoic event.

Cenozoic Developments

The distribution of land and major geofloral components in the Paleocene is mapped in broad terms on the accompanying figure (Figure 13–35) and provides a generalized view of the extent of tropical and extratropical climatic conditions. Perhaps the best way to trace further developments in frog distribution patterns is to relate the probable history for the major groups as they are associated with land and climatic change from the Paleocene onward.

Old Tropical Unit

RHINOPHRYNIDAE. This group has been restricted to the North American tropics for all of its history; it did not enter South America across the Paleocene Isthmian

Link and gradually became restricted southward in the late Cenozoic by cooling temperatures in North America. Its association with marginal tropic–warm temperate and semiarid conditions may explain its failure to reach South America through the probably humid evergreen forests of the Paleocene isthmian connection.

PIPIDAE. Known after the Jurassic only from derivative areas of the western Gondwanaland continent, this group experienced a substantial radiation on South America (Pipinae) and Africa (Xenopinae) but now remains as depauperate relict groups that survive because of strictly aquatic specializations. The family may have been present in Cretaceous Madagascar, India, and Australia but is now extinct in these regions.

PALAEOBATRACHIDAE. This is an extinct tropical European radiation known only from Eocene.

MICROHYLIDAE. By far this is the most successful of ancient stocks but shows by distribution and history both its age and the effects of competition with acosmanurans. At the time of fragmentation of Gondwanaland in the Cretaceous, microhylid stocks must have been present in tropical portions of the principal southern land masses (South America, Africa, Madagascar, India, and Australia). On each mass the family's subsequent history appears to have been unique. Unfortunately, the understanding of the evolution within the family is inadequate. Parker's (1934) excellent morphologic work, with main phyletic lines based upon vertebral, palatal, and pectoral girdle characters, will surely prove to be an oversimplification, as Carvalho (1954) and the studies of musculature (Starrett, 1968; Tyler, 1971) now suggest. Distributionally, five major areas of radiation and evolution are found: South America, with twelve extremely diverse and specialized genera associated for the most part with the ancient Brazilian and Guianan geologic shields; Africa, with three diverse and distinctive subfamilies (Phrynomerinae, Brevicipitinae, and Hoplophryninae) and eight genera; Madagascar, with two endemic subfamilies (Dyscophinae and Cophylinae) and eleven genera; India, with a major radiation of about 10 genera that have invaded the rest of tropical and eastern Asia; New Guinea, the center of radiation for the Asterophryinae (about twelve genera) that also occur in extreme northern Australia and reach western China to the north.

During the Cenozoic the following events seem to have affected microhylid evolution (Figure 13–36):

1. Development of the ranoid radiation in Africa led to the elimination of almost all scoptanurans except for specialized forms. The phrynomerines (five species) and brevicipitines (seventeen species) are highly specialized stocks, and the hoplophrynines (three species) are specialized montane relicts. Except for *Phrynomerus*, all of the African microhylid genera were eliminated essentially from tropical lowland situations. I can only interpret this situation as an indication of an expansion of ranids and hyperoliids into the habitats and niches of generalized microhylids through the Cenozoic. The limits placed on microhylid evolution by their larvae already has been pointed out.

2. On Madagascar, primitive generalized microhylids survived and radiated, but the number of forms (about thirty species) now may be reduced, since ranoids invaded the island in the late Cenozoic.

3. Microhylids survived and evolved in India (four genera, ten species), with an invasion of the Indo-Malayan region by descendent stocks after India became connected to Asia; a secondary Indo-Mayalan radiation produced seven endemic genera, two of which range into the temperate regions of China.

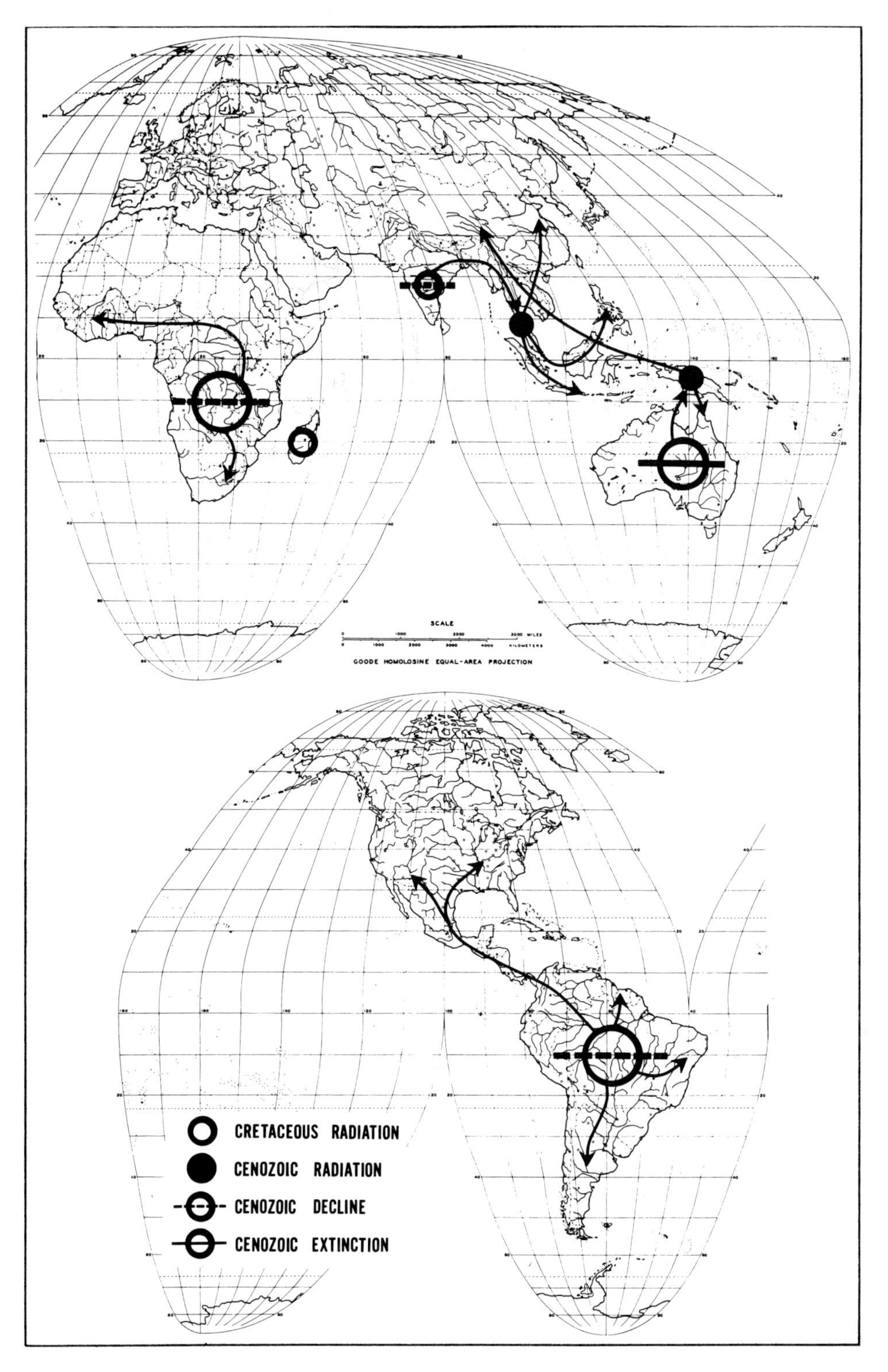

Figure 13–36. Mesozoic and Cenozoic radiations and Cenozoic decline of microhylid frogs.

4. The family was gradually eliminated from Australia as tropical areas on that continent became arid and remained so from the middle Cenozoic onward; the remaining stock became established and radiated in New Guinea in middle Cenozoic, with one line moving northward through Indonesia to China; primitive members of this line had larvae, but the speciose genera in New Guinea all have direct development.

5. There was a substantial radiation in South America, with one stock crossing into Middle America across the Isthmian Link in Paleocene and evolving into two genera; these genera experienced separate evolution during Eocene to Pliocene, with both lines adapted to fossorial ant-eating or termite-eating habits. Except the Middle American genera and two in South America, most are monotypic and relict in distribution.

Antiboreal Unit

This group (family Leiopelmatidae) became extinct in South America, Antarctica, and Australia during the Cenozoic. It survives only in New Zealand.

Laurasian Unit

The Laurasian families appear to have had similar histories during most of the Cenozoic. Ascaphids, discoglossids, and pelobatids are all known from Cretaceous North America. Their association with Arcto-Tertiary or marginal Arcto-Tertiary to tropical associations corresponds with present distribution patterns. Generally these groups have been restricted southward by decreasing temperatures in late Cenozoic (Figure 13–37). The family histories may be summarized as follows:

ASCAPHIDAE. As far as known, they have been always associated with extratropical coniferous forests in western North America, but they were possibly widespread in early Cenozoic.

DISCOGLOSSIDAE. The family probably had a circumpolar distribution in early Cenozoic, but the range became fragmented by cooling trends in the north and by increasing aridity at middle latitudes in late Cenozoic. One line represented by present-day *Barbourula* seems to have become associated with tropical conditions in Asia and, as tropical habitats became restricted southward from the latitude of central China from the Oligocene onward, came to invade the Indo-Malayan paleopeninsula and today remains as a hanging relict near the furthest extreme of the insular chain in the Philippines. The distribution of discoglossids parallels so closely that of the deciduous forest derivatives of the Arcto-Tertiary geoflora and the salamandrids that a common historical association seems certain. Discoglossids have disappeared in North America and reached tropical Asia; salamandrids still occur in North America and across central Asia. The history of both groups correlates closely with the development of Arcto-Tertiary vegetation (Axelrod, 1960).

PELOBATIDAE. The primitive members of this family are known from the Cretaceous onwards from localities near the boundary between warm temperate and tropical situations. During late Cenozoic cooling and drying, the distribution became fragmented as isotherms were displaced southward. Two genera invaded the Indo-Malaya paleopeninsula with some success, while other remnants are associated with derivatives of Arcto-Tertiary geofloras, and one line *(Scaphiopus)* has invaded the semiarid Madro-Tertiary associations of western North America.

The accompanying table (Table 13–8) summarizes the probable Cenozoic history of the Laurasian Unit.

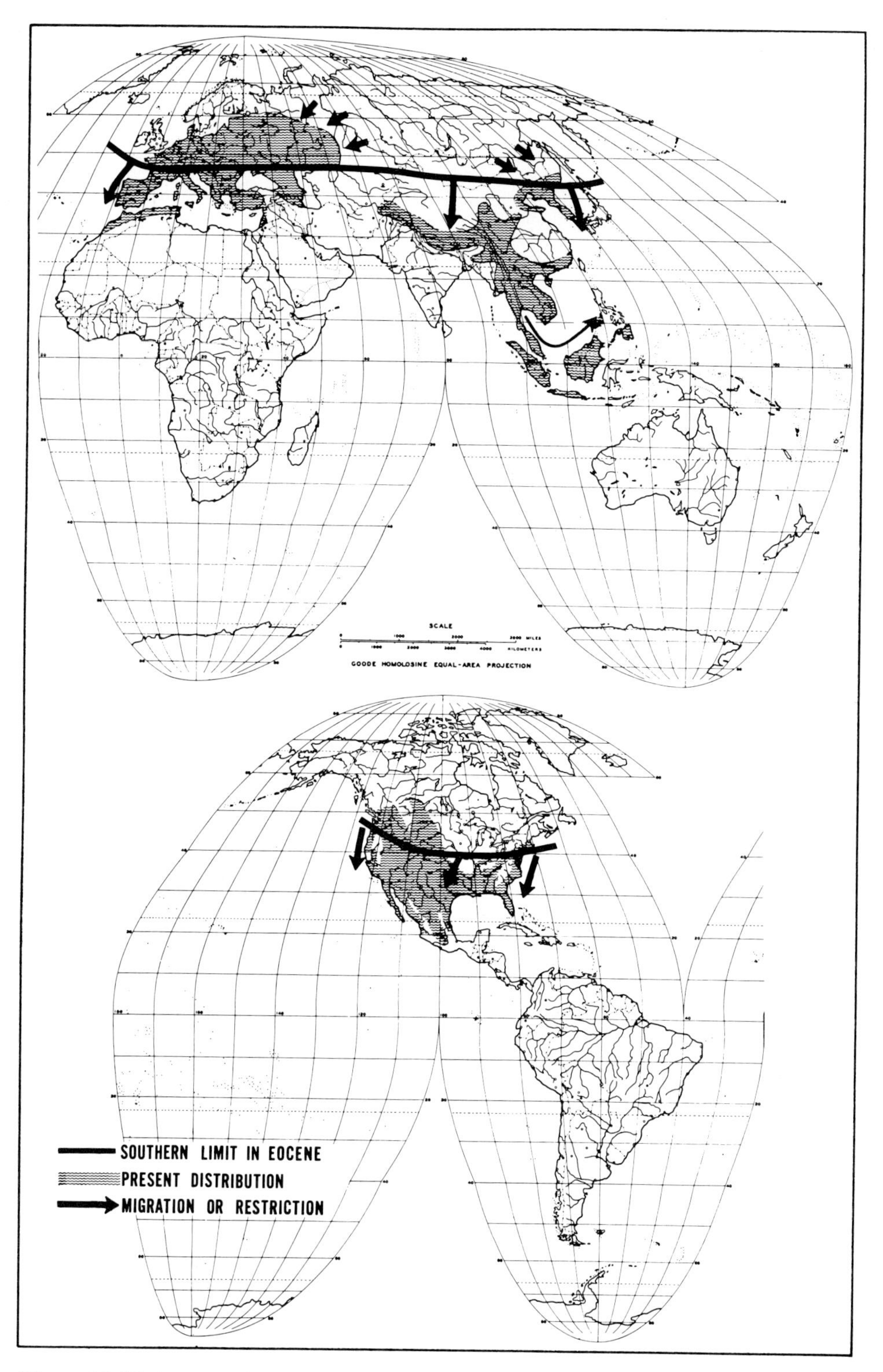

Figure 13-37. Cenozoic history of the Laurasian source unit.

Table 13–8. Probable Cenozoic History of Laurasian Unit

	North America	Western Eurasia	Eastern Eurasia	Indo-Malaya
Initial Geofloral Association	Arcto-Tertiary	Arcto-Tertiary	Arcto-Tertiary	
Early Cenozoic	Circumpolar in extratropical and tropical transitional areas			Absent
Oligocene	Extinction of discoglossids and megophryines	Extinction of ascaphids and paleobatrachids	Extinction of ascaphids	Absent
Miocene	Invasion of dry southern areas	Southward displacement Extinction of megophryines	Southward displacement	Invasion from north
Late Cenozoic	Extinction of pelodytines	Contraction of range west and south	Extinction of pelodytines Contraction of range south and east	Hanging relicts
Recent Geofloral Association	Arcto- and Madro-Tertiary	Arcto-Tertiary	Arcto-Tertiary	Paleotropical-Tertiary

Young Tropical Unit

BUFONIDAE. As previously indicated, the toads had two centers of Cretaceous development as the result of the fragmentation of western Gondwanaland. One stock appears to have been restricted to Africa, where it has undergone substantial radiation throughout the Cenozoic. Several phyletic lines are represented by this radiation, but one major stock is composed of *Bufo regularis* and its allies, with a diploid chromosome complement of $2n = 20$ (Bogart, 1968). All other toads, as far as is known, have $2n = 22$. The few South African toads (*carens, gariepensis,* and *rosei*) and other toad species with 22 chromosomes elsewhere on the continent may be relicts of the original African stock that gave rise to the 20 chromosome line. Without question *Bufo* and its several diverse and specialized derivative genera in Africa (Tihen, 1960) form a series of ancient and distinctive evolutionary units endemic in origin and not immigrating out of the continent. Blair's (1970) recent proposal that *Bufo* entered Africa in Miocene fails to account for these radiations and does not fit the distributional facts.

The second major center of bufonid evolution is in South America. The evidence for the presence of *Bufo* in that area in the Cretaceous has been reviewed above. Both Blair (1963) and I (1966) previously have suggested, mistakenly, that bufonids entered South America in the middle Tertiary. It is now clear that the bufonids moved northward across the Isthmian Link into tropical North America in the Paleocene. After the link subsided by the Eocene, further evolution and diversification seem to have taken place in three centers: (a) continued radiation on the South American island continent (*arenarum, granulosus, haematiticus, marinus, spinulosus,* and *typhonius* groups and the atelopodine genera); (b) differentiation of groups in tropical North America (*valliceps* and allied groups and *Crepidophryne*); (c) invasion of temperate Arcto-Tertiary areas of North America in Eocene, with isolation from tropical ancestral stock by changes leading to mid-latitude aridity; evolution in association with temperate climates and vegetation during the rest of the Cenozoic (*boreas* and *americanus* groups and derivatives).

Apparently in the Eocene, derivatives of the tropical North American group crossed the Bering land bridge and established another secondary center of radiation in tropical Asia. This stock gradually was displaced southward into Indo-Malaya by cooling trends at northern latitudes in later Cenozoic, as in the case of the Laurasian unit. A second invasion of Eurasia in the Oligocene by temperate North American bufonids probably led to the development of the recent species of the northern Eurasian *(viridis)* group. India has been invaded from the east by members of the Indo-Malayan line and from the northeast and northwest by northern forms. The lack of significant differentiation in the subcontinent is strong evidence for its isolation during most of the Cenozoic. Northwest Africa has been recently invaded by members of the *viridis* group.

The views expressed here differ from those presented by Blair (1970) in his discussion of bufonid history. While he pointed out the unique long-term consistency of the *Bufo* adaptation and its ecologic flexibility without major morphologic change, he underestimated its probable age by at least 100 million years. The nature of the African and South American radiations of the family make it evident that the basic stock was in western Gondwanaland in the Cretaceous, almost 120 million years ago. Blair's idea that bufonids originally entered Africa in the Miocene has been questioned earlier in this paper. He also has failed to explain the occurrence of *Bufo* both north and south of the Panamanian portal in the Miocene, although it is clear that in order for toads to be on both American land masses at that time, they must have crossed the Isthmian Link in the Paleocene.

Blair's (1970) concept of only two major adaptive lines in *Bufo*—a broad-skulled (or thick-skulled) group and a narrow-skulled (or thin-skulled) group—is questionable. The former group was regarded as tropical to subtropical in distribution and adaptation, and the latter extratropical or montane. These groups correspond to his (Blair, 1963) southern line (from a *melanostictus*-like ancestor) and northern line (from a *calamita*-like ancestor). Several species with intermediate skulls were recognized under his more recent arrangement. I strongly suspect that primitive *Bufo* (as Blair also suggests, 1970) had intermediate skulls and that narrow-skulled and broad-skulled forms evolved several times and in parallel. In tropical Asia, toads of each of the three types appear to be represented. Without any other features to support this division, it has questionable significance. Nevertheless, I agree completely with Blair that *Bufo* and allied derivative genera in Indo-Malaya came from a stock like the *valliceps* group in tropical North America and that temperate Eurasian bufonids (*viridis* group) belong to the same lineage as those predominating in temperate North America (for example, the *boreas* and *americanus* groups). A summary of my ideas on Cenozoic bufonid dispersal and evolution is provided on the accompanying map (Figure 13–38).

SOOGLOSSIDAE. This family is restricted to the Seychelle Islands, an ancient fragment (600 million years old) of Gondwanaland. The group is almost certainly a relict of the African line that gave rise to the ranoids. Sooglossids probably occurred in Africa in the Cretaceous but have been replaced there by more ad-

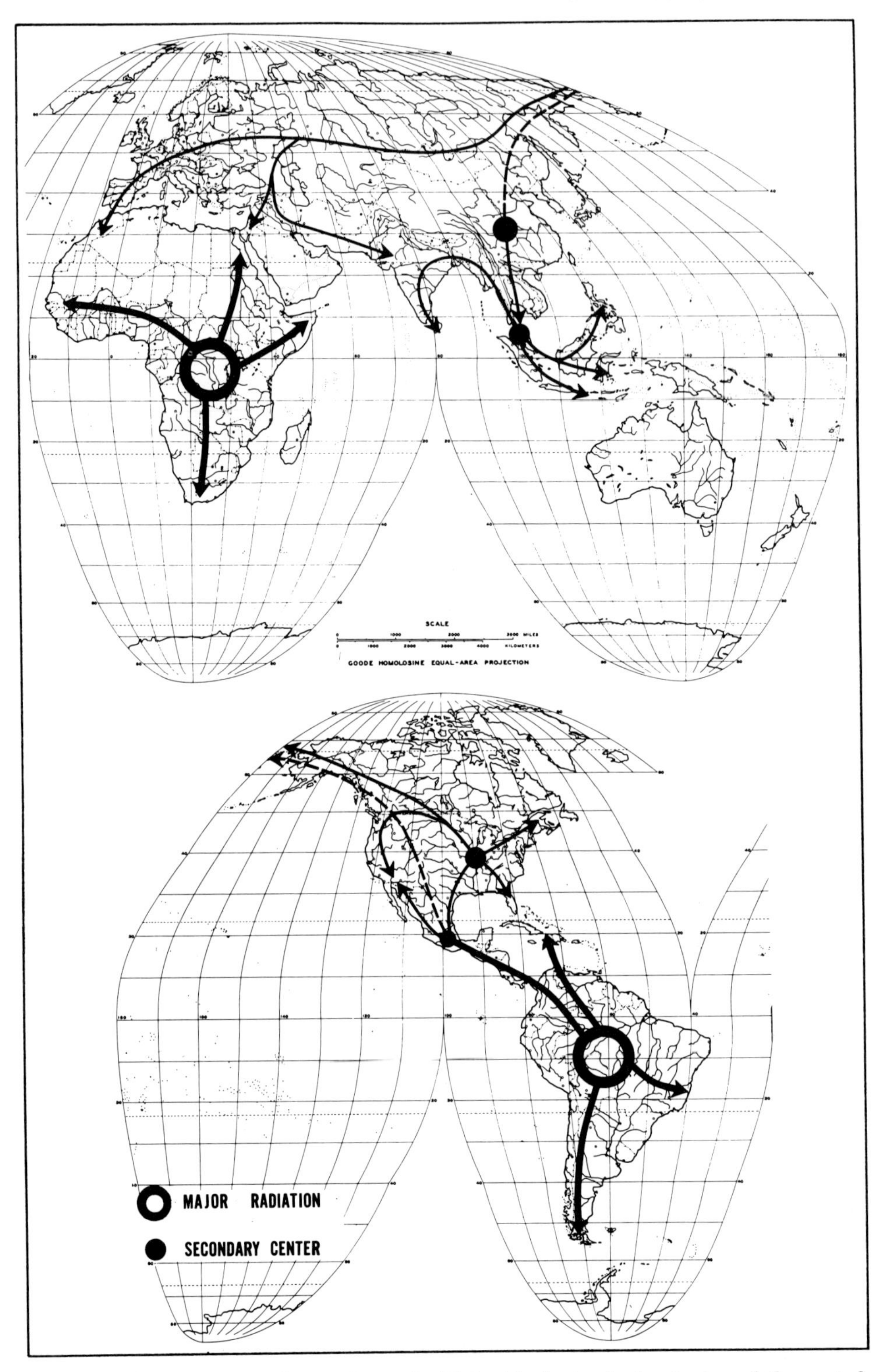

Figure 13–38. Cenozoic radiation in toads (Bufonidae). At the beginning of the period two separate stocks probably separate since early Cretaceous, had been established—one in tropical South America and the other in tropical Africa.

vanced groups, unless the peculiar Arthroleptinae and Hemisinae are also remnants of this early stock.

RANIDAE. This family is so clearly African in origin, diversification, and radiation that no one has seriously questioned these propositions, although Darlington (1957) prefers a less-precise center, the Old World tropics. In the Cenozoic, extensive radiation took place in this family in Africa among the endemic phrynobatrachines, arthroleptines, hemisines, and the more wide-ranging ranines. Two ranid stocks apparently dispersed from the continent into tropical Eurasia by the Eocene. The first of these, the Raninae, are diversified highly in Africa (at least 9 genera and 150 species) and appear in the Eocene of Europe and Miocene of North America. Apparently tropical representatives of this line invaded Eurasia in early Tertiary. From the original non-African ranines, two major evolutionary lines developed. One retained tropical associations, invaded the Indian subcontinent in Oligocene-Miocene, and has undergone divergence in tropical Asia. A closely allied derivative group apparently arose early in the history of tropical Asian ranids to become the Platymantinae. They were successful invaders of the Indo-Malayan chain and diversified and immigrated to reach New Guinea, northern Australia, and Fiji, probably in post-Miocene times. They seem to have been followed down the island chain by a later wave of ranids from the original tropical ranine stock. The second major ranine line became associated with temperate Arcto-Tertiary vegetation, expanded across Eurasia, diversified into several distinct lines, and invaded North America by the Eocene. This stock invaded South America across the Isthmian Link from the Pliocene onward. I have sketched elsewhere (1966) the history of *Rana* in Middle America from Eocene to Pliocene. The members of this radiation include the Old World forms (*esculenta* and *temporaria* groups) and the several New World groups.

The second African ranid line that invaded Eurasia is the tree-frog subfamily Rhacophorinae. This stock is represented by one African genus *Chiromantis* (12 species), six Asian genera (about 50 species), and a group of Madagascar forms (*Boophis*, with 50 species). The primitive members of this line were probably somewhat like *Rhacophorus-Boophis* and may have originated in Africa and immigrated to Asia along with the tropical ranines in early Cenozoic. The Madagascar genus seems to be derived from the primitive African line. Evolution seems to parallel closely that of ranines in India and Indo-Malaya (Figure 13–40). Apparently the basic stock in Africa has become extinct, but the specialized derivative genus *Chiromantis* persists there. The elimination of rhacophorines in Africa is correlated apparently with the rise of the hyperoliid radiation of tree frogs. Fairly early in the Cenozoic another stock of primitive ranids must have reached Madagascar, where they differentiated into the endemic Mantellinae (4 genera, 50 species). The Cenozoic radiation of ranids is summarized on the accompanying maps (Figures 13–39, 13–40). For an alternate view of rhacophorine and mantelline relations, see Liem (1969).

HYPEROLIIDAE. This group is a Cenozoic radiation of major proportions in Africa of two subfamilies, 18 genera, and approximately 300 species. The arid barrier across north Africa, Arabia, and Baluchistan seems to have prevented expansion into Asia over the late Miocene to recent land connections. Evolution of diversity in this family corresponds to the elimination of the older tree-frog line—the ranid rhacophorines—in Africa, until only one specialized genus of the latter remains. One hyperoliid derivative stock has developed on Madagascar, where its ancestors must have arrived in the early Cenozoic by overwater immigration (the Scaphiophryninae, with two genera and 10 species). A more recent overwater immigration included the ancestors of *Heterixalus* (two species) and *Hyperolius* (9–10 forms) of the modern African group of genera (Figure 13–40).

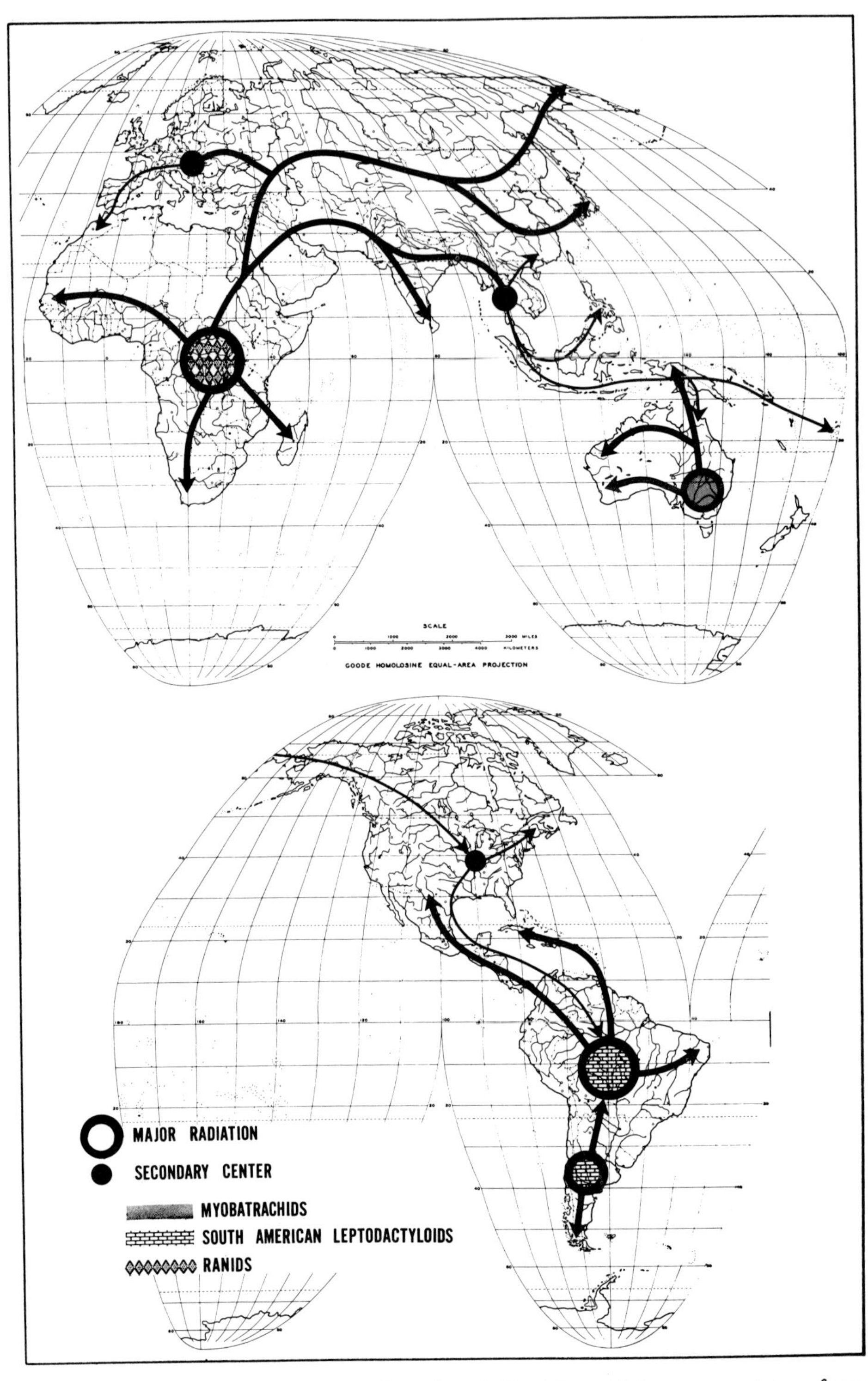

Figure 13–39. Cenozoic radiation in leptodactyloid and ranid frogs, except tree frogs.

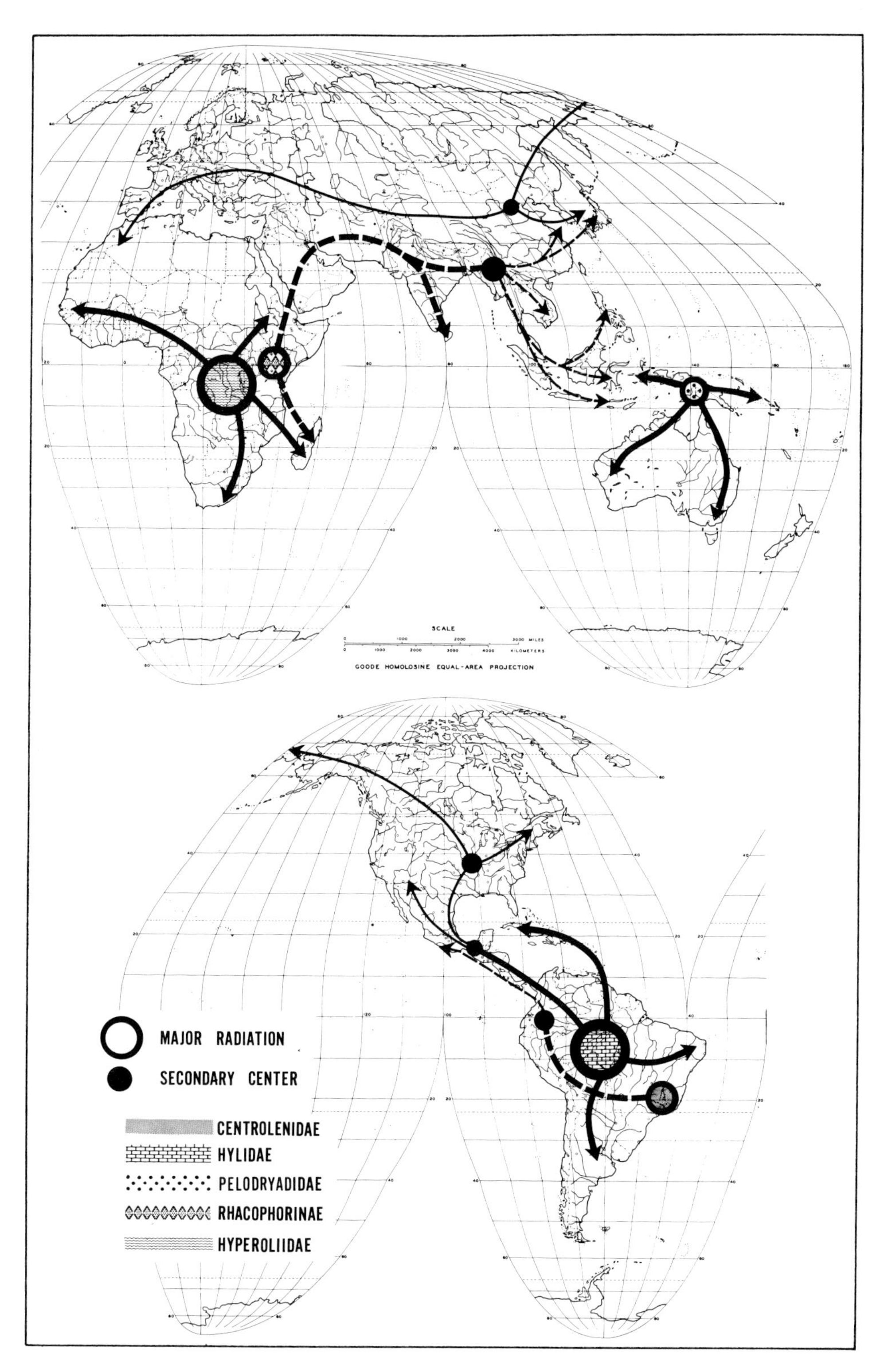

Figure 13–40. Cenozoic radiations in tree frogs.

South Gondwanian Unit

The various member groups of this unit are each derived ultimately from a southern extratropical leptodactyloid stock. It is more instructive to follow the history of these derivatives by continental centers than to study these groups by family.

AFRICA. No major radiation of the family Heleophrynidae seems to have occurred. Since the basic leptodactyloid stock was adapted to temperate latitudes, the heleophrynid line seems to have remained in this zone and probably was blocked from invasion and radiation in tropical Africa by the explosive evolution of the tropical-adapted ranoids in the Cretaceous and early Tertiary.

SOUTH AMERICA. The earliest tropical derivatives of the temperate leptodactyloid stock, as previously discussed for the Cretaceous, were the tree frogs of the family Hylidae. When the Paleocene land bridge was established, hylids moved northward into tropical North America. After the subsidence of the land connection, evolution proceeded in three centers: (a) radiation of a host of genera and species in the South American continental island (e.g., *rubra, leucophyllata, maxima,* and *raddiana* groups and *Gastrotheca, Phyllomedusa*); (b) evolution in tropical North America of several lines (e.g., *uranochroa* and *bistincta* groups and *Anotheca, Agalychnis,* and *Smilisca*); (c) invasion of temperate Arcto-Tertiary areas of North America in Eocene, with isolation from tropical ancestral groups by the changes leading to midlatitude aridity.

The stock from this last group *(Hyla sensu strictu)* developed into the several distinctive extratropical species groups (e.g., *versicolor, eximia,* and *cinerea* groups; *Acris, Pseudacris,* and *Limnaeodus*). In later Cenozoic this stock crossed the Bering land bridge into Eurasia from which it reached Europe and differentiated into seven or eight closely related species (Gaudin, 1969). It will be noted that this pattern closely parallels that for the evolution of New World and temperate Old World bufonids. In the Pliocene, when the Panamanian portal became emergent, some exchange between tropical Middle America and South America took place (Savage, 1966) and continues until the present. The Cenozoic history of the hylids is summarized on the accompanying map (Figure 13–40).

The major features of leptodactyloid evolution in the New World seem centered on tropical South America from the Eocene onward. It is questionable whether leptodactylids entered North America along with bufonids and hylids across the Paleocene Isthmian Link. Of the members of the family found in Central America today, all species, with the sole exception of *Eleutherodactylus* and its allies, represent South American recent arrivals that used the Pliocene-to-Recent isthmian connection. The latter genus may have arrived in the Paleocene and differentiated north and south of the portal until the Pliocene, as I have previously suggested (1966). It possibly arrived by overwater transport (all members of the genus have incapsulated, land-laid eggs and no tadpole stage) from the south in later Tertiary.

The fossil record includes Eocene-to-Recent records of the family in South America. Its radiation there is further substantiated by the seven familial units endemic to the region and the fact that only since the Pliocene have two of these entered Middle America. One North American Eocene fossil has been attributed to the family *Eorubeta* (Hecht, 1960). Lynch (1969) concluded, after an exhaustive comparison with the skeletons of most New World frogs, that the specimen is not a leptodactylid. The accompanying map (Figure 13–39) summarizes the history of the leptodactyloid radiation in the New World. It seems probable that much of the evolutionary radiation of tropical leptodactyloids occurred in the Cenozoic.

AUSTRALIA. The original leptodactyloid stock in this region was associated with temperate conditions and its derivatives; the family Myobatrachidae (2 subfamilies and 17 genera) retain this association today. The leptodactyloids of Australia have their present distribution centered in warm temperate, southeastern Australia, although they occupy a variety of arid, semiarid, and tropical habitats as well. The allied but distinctive tree frogs (Pelodryadidae), on the other hand, have a tropical New Guinea (3 genera, 50 species) center of distribution and seem to have invaded Australia (4 genera, 30 species) from the north in the middle Cenozoic.

During most of the Cenozoic, temperate, humid environments were restricted to southern areas. The climates of these areas were warmer, more humid, and milder in early Cenozoic than now. The originally tropical regions of central Australia became dominated by semiarid to arid subtropical-temperate climate and vegetation early in history so that during middle to late Cenozoic the continent consisted of three major climatic zones: a northern tropical humid zone, more or less the region north of present 20° S latitude; a central semiarid to desert belt; and a southern warm temperate belt, more or less south of present 32° S latitude. Myobatrachids evolved and radiated in the southern temperate zone and gradually invaded the subtropical-tropical dry regions during middle Cenozoic. Apparently, the ancestors of the tree frogs reached the northern tropical area prior to the establishment of strong climatic zonation. This ancestor must have been a leptodactyloid, but none of the existing Australian genera show close relationships to the tree frogs. This stock developed and radiated in northern Australia and probably invaded New Guinea across the generally uplifted straits in the Miocene. As the vicissitudes of later Cenozoic aridity virtually eliminated tropical forests from Australia, pelodryadid evolution was centered in New Guinea. The present tree-frog fauna (*Litoria*, *Pelodryas*, and allied genera) of Australia appears to be derived by a series of invasions from New Guinea, principally across the Torres Strait region with further southward movement through a humid east-coast corridor connecting the scattered tropical evergreen forests of northern Queensland through a series of evergreen deciduous and eucalypt associations to the lowland temperate associations of southeastern Australia. This concept is diagramed in summary form on the accompanying maps (Figures 13–39, 13–40).

Competition, Dominance, Coexistence, and Replacement

Darlington (1957) developed a hypothesis that competition and replacement were the processes responsible in large part for present patterns of amphibian distribution. His concept depended upon the assumptions that frog evolution was centered on the Old World tropics, with radial immigration in successive waves to other areas of the world, and that each advanced level of frog organization came to dominance through competition and eventually replaced the more primitive stocks in a region. The evidence most strongly supporting the hypothesis, according to Darlington (1957, Figure 26), is the complementarity of ranid and leptodactylid distribution and also that of the tree-frog groups, the rhacophorids (which are equal to the Hyperoliidae and the ranid Rhacophorinae of this paper), and the hylids (which are equal to the Hylidae, Centrolenidae, and Pelodryadidae of this paper). The essentially allopatric distribution of these paired groups led him to conclude that ranids had replaced leptodactylids by competition and that the rhacophorids superceded the hylids in a similar manner. Confronted with his assumptions of the permanence of ocean basins and continental masses during the history of frog evolution and of the Old World tropical origin for all frog groups, as well as his emphasis on dominance and competition, Darlington had no choice but to generalize a similar replacement pattern as the explanation for the history of frog dispersal. In his view the most primitive families were eliminated by competition in Old World tropical regions by the more advanced families as they appeared and radiated at later times. He believed that all families of higher frogs arose in this fashion and, except for rhacophorids, invaded the Americas across the Bering land bridge. He further concluded that pipids reached South America from the north by a similar but earlier process, that salamanders, discoglossids, and pelobatids are retreating out of the tropics northward, and that caecilians, and microhylids, followed a pattern similar to that for the higher frogs. His theory of amphibian dispersal is neatly summarized by a diagram (Darlington, 1957, Figure 27) indicating that the Ascaphidae-Leiopelmatidae line originated in the Old World tropics and radiated north and south, that subsequently the leptodactyloid line originated in the same area, radiated, and in the process (competitive dominance) replaced the earlier stock, except in North America and New Zealand, and finally that the ranids followed the same path and have eliminated the leptodactyloids over most of Africa, Eurasia, and North America. The essence of this explanation lies in the concept of competitive extinction, a view that does not appear to be valid in the light of frog distributional data.

An examination of the data indicates that each major frog fauna contains a mosaic combination of various groups (Table 13–1) derived from several historical sources and that in each fauna different dominance relations exist:

Eurasia: bufonids and ranids, about equally represented
Africa: ranid and hyperoliid radiations
Madagascar: microhylids (most diversity), ranids (most species), and hyperoliids (intermediate diversity and number of species)
Indo-Malaya: ranids (most species), microhylids and bufonids (most diversity)
Australia–New Guinea: leptodactyloids and pelodryadids, but microhylids highly diverse in New Guinea
North America: bufonids, hylids, and ranids in about equal numbers
Middle America: hylids (most diversity and numbers of species) and leptodactyloids (almost as many species)
South America: leptodactyloids and hylids

The most important aspects of these patterns are as follows: (a) acosmanurans predominate in all regions, except possibly Madagascar; (b) toads, family Bufonidae, are well represented in tropical areas, except Australasia and Madagascar, and predominate in northern extratropical regions; (c) a major adaptive type in all tropical regions and dominant in North America are tree frogs families Hylidae, Pelodryadidae, Centrolenidae, Ranidae (Rhacophorinae), and Hyperoliidae; and (d) ranoids predominate in Old World tropical areas and areas on the northern land masses, while leptodactyloids and their derivatives predominate in tropical America and in Australasia. In short, the acosmanuran radiations of Cenozoic have come to overshadow the Mesozoic radiations of Laurasian and Antiboreal source units in northern and southern extratropical areas and of Old Tropical groups in tropical areas. These Mesozoic radiations do not seem to have been eliminated by competition with more advanced frogs because Laurasian and Young Tropical and one South Gondwanian family (Hylidae) occur together throughout much of the Holarctic region and Young Tropical families and South Gondwanian groups coexist with Old Tropical stocks in tropical areas.

The history of frog radiation and dispersal does not appear to be based primarily upon competitive replacement, but upon pre-emptive ecologic occupancy of broad adaptive zones by primitive stocks, which denied more advanced lines ecologic accessibility to the zones. As a result the advanced lines have undergone substantial radiations into ecologic adaptive zones not available to the primitive stocks but have failed to replace them.

The fauna of northern regions has a base Laurasian layer, with bufonid, hylid, and ranid additions. The frog fauna of tropical areas has a base of Old Tropical families, to which ranoids (Old World) or leptodactyloids and derivatives (New World, Australasia) were added. These patterns suggest that replacement through competition is reduced or absent between acosmanurans and older groups to the point that it is not an important regulator of frog distribution. In this regard, dominance by frog groups does not reflect evolutionary success through competitive elimination of other frogs.

Why is replacement through competition relatively unimportant in the case of frog distributions—or, for that matter, in the distribution of amphibians? Perhaps it has been overemphasized as a feature in the evolution in all major groups. In the case of frogs an answer seems to lie in the general adaptations of the three primitive stocks of frogs present by the Jurassic. The most fundamental of these adaptations involves the basic features of tadpole life, which have been shown (Starrett, this volume) to be the key to understanding the evolution of frogs. Of secondary significance are the basic climatic associations of the groups and the ecologic roles of the adults:

xenoanurans—	filter-feeders midwaters	tropical
scoptanurans—	filter-feeders infusoria	tropical
lemmanurans—	algae-grazers	extratropical

In tropical regions xenoanurans and scoptanurans had radiated and occupied most larval filter-feeding niches long before the arrival of acosmanurans in these areas. Acosmanurans have radiated into a host of tropical adaptive larval types (Orton, 1953, 1957; Starrett, this volume). Because of long-term pre-emptive occupancy of filter-feeding niches by larval rhinophrynids, pipids, and microhylids, the recent arrivals have not been able to invade these adaptive subzones. Although the acosmanuran larva is an advance over the more primitive larvae, as I have pointed out elsewhere (1969), the presence of a specially adapted primitive stock prevents invasion of its adaptive zone by a presumably more advanced line. The acosmanurans were able to radiate into unoccupied adaptive subzones that could not be utilized by the ancient lines (whose tadpoles lacked beaks and denticles) but could not displace the primitive stocks in their role as tropical filter-feeders.

The situation is less clear when lemmanurans are compared to acosmanurans. Both groups occur sympatrically in northern temperate areas. Perhaps a detailed study will show that as in the case of Old Tropical groups, the larvae of the Ascaphidae and Discoglossidae occupy niches that have not been invaded in the northern areas by pelobatids, bufonids, hylids, and ranids. These latter groups do not seem to have eliminated the Ascaphidae and Discoglossidae by competitive replacement, although the superior general adaptation of the acosmanurans may have encouraged maintenance or perfection of special adaptation in the older line.

Within acosmanurans pre-emptive occupancy of adaptive zones by groups having the earliest ecologic accessibility appears to have blocked occupancy by more recent arrivals on the scene. Pre-emption appears to involve both geographic (ecologic) opportunities and total ecologic adaptation (adult and larval) in these cases and supplies the basis for understanding patterns of complementarity, without competitive dominance and replacement. Darlington's (1957) position is that complementarity results from the elimination of an early stock by competitive replacement of a more advanced stock. My idea is that complementarity results from independent geographic origin and radiation of various stocks so that more recent arrivals in a particular area cannot move into the adaptive zones pre-empted by the older native stocks. In this concept the competitive advantage lies with the stock having the longest history in the area. In effect, the older stock has occupied the ecologic subzones available to it and has a competitive edge because of many generations of selective molding, and the more recent arrival is prevented from becoming established or is forced to invade and develop in ecologic subzones unoccupied by the older native group.

In the higher frogs the two most striking cases of complementarity involve the leptodactyloid-ranid situation (Figure 13–39) and the tree frogs (Figure 13–40). According to the ideas developed in this paper, leptodactyloids have had a long history of evolution centered in South America and Australia, while ranids radiated from tropical Africa. In two cases ranids have arrived from the north very late in leptodactyloid history in South America and northern Australia. These arrivals were probably 5 million to 10 million years ago. In neither case have the ranids been able to diversify substantially, and certainly no indication of ranid replacement of leptodactyloids is suggested. On both continents the long-term effects of leptodactyloid radiation seem to have led to marked diversification and perfection of adaptations for many ecologic subzones. The pre-emptive occupancy of the subzones by leptodactyloids effectively prevents potential ranid expansion and radiation by blocking ecologic accessibility to the subzones. The competitive advantages lie in these cases with the older, resident stocks.

Almost the opposite situation is present in Africa. As asserted earlier in this paper, the basic stock of leptodactyloid evolution was probably temperate adapted; in South America and Australia, the stock had only filter-feeding larval groups (Pipidae and Microhylidae) to contend with in tropical regions. In these areas many ecologic opportunities for stocks with the basic algae-grazing larvae existed, and over the millions of subsequent years, these opportunities were exploited. This was not the case in Africa, for shortly after that continent became an island in the Cretaceous, the ancestral ranoid stock underwent rapid radiation in the tropics, diversified, and filled most subzones open to frogs with the basic algae-grazing larvae. The ranoid expansion effectively pre-empted the available tropic subzones, and the African temperate-adapted leptodactyloids did not have an open arena for diversification. Ecologic accessibility was denied the leptodactyloids by the ranoid radiation, just as it was denied to the ranids later in time by the tropical leptodactyloid radiations in South America and Australia.

The distribution of tree frogs also emphasizes the role of pre-emptive occupancy of adaptive subzones as an evolutionary regulator. Of the primitive frog Types I, II, and III, only a few microhylids are arboreal, but four major and one minor arboreal radiations occurred in the acosmanurans. These radiations show strong complementarity and some coexistence, but in one case competitive reduction of an early radiation by a later one is suggested. The five tree-frog families are Hylidae, Pelodryadidae, Centrolenidae, Ranidae (Rhacophorinae), and Hyperoliidae. As may be seen from the accompanying map (Figure 13–40), three major radiations (Hylidae, Pelodryadidae, and Hyperoliidae) do not overlap in distribution. The rhacophorines occur with hyperoliids in Africa and on Madagascar but have a complementary distribution to hylids and pelodryadids in the Eurasian, Indo-Malayan, and Australian regions. Centrolenids are sympatric with hylids throughout their range.

Only the reduced number of rhacophorines in Africa suggests possible competitive reduction. This group is represented by one genus in Africa and one on Madagascar, both sympatric with the presumably advanced hyperoliids. As previously mentioned, the hyperoliids seem to have occupied most arboreal subzones in Africa at the expense of the older rhacophorines, although it remains possible that the latter never had a major radiation on the continent. Rhacophorine and hylid stocks overlap to some degree in temperate east Asia; rhacophorines and pelodryadids overlap slightly in the Lesser Sundas. The overlap of both stocks in eastern Asia involves peripheral portions of the groups' ranges where only very few forms occur, and both seem to have different adaptive advantages that allow coexistence. In the Sundas a similar balance at the peripheries of the ranges is in effect. It appears that rhacophorines have pre-empted tropical arboreal subzones in Asia and denied ecologic accessibility to a hylid invasion from the north and pelodryadid invasion from the east into Indo-Malaya. Pelodryadids have pre-empted tropical arboreal subzones in the New Guinea–Australia region and denied ecologic accessibility in the area to the rhacophorines. In each case the most ancient group in a geographic area is able to maintain itself against potential invaders by pre-emptive occupancy and adaptation to the local arboreal environments.

The centrolenids have a fragmented range in South America and Middle America (Figure 13–15) but hylids coexist everywhere with them. The centrolenids invariably lay their eggs on vegetation overhanging fast-moving streams. The larvae are adapted for stream conditions (Starrett, 1960) and fall into the water after considerable development. The combination of these adaptations suggests that they reduce competition with other tree frogs and probably accounts for the centrolenids' evolution in regions sympatric with the hylids.

The tropical regions of the world are very nicely divided among the hyperoliids (Africa), rhacophorines (Asia), and hylids (America). A dynamic tension seems to exist where tree-frog groups are in ecologic contact to prevent impingement of the most-recent arrival into the ecologic space of the older stock. Contrary to Darlington's (1957) concept, this equilibrium prevents or substantially reduces the potential for radiation and competitive replacement by immigrants.

The distribution of the remaining major group of acosmanurans, the toads (family Bufonidae), has its own distinctive pattern relating to this problem (Figure 13–38). As discussed previously, toads have an ancient but generally adapted terrestrial anuran organization. Among the basic adaptations for this mode of life are a generalized slow-moving water or pond larva (denticle rows 2/3), integumentary and physiologic adaptations to reduce water loss, hopping locomotion, and highly toxic skin secretions. The modal toad adaptation apparently conferred immediate evolutionary advantage, but as it became perfected by natural selection, the stock lost the genetic potential for expansion into other ecologic roles. Toads of the genus *Bufo* are highly successful components of the modern frog fauna because their specializations for terrestrial life, including their wide range of ecologic tolerances, are unique, but they are also restricted by them to a single ecologic adaptive zone. Only in a few instances have derivatives of *Bufo* invaded other ecologic subzones and then only to a limited extent. Toads are able to coexist with other groups because they occupy an ecologic zone that has not been entered by other stocks. Their special adaptations and wide range of tolerances have enabled them to pre-empt this zone everywhere but on Madagascar and in the Australasian area. In South America (McDiarmid, 1969) and Indo-Malaya (Inger, 1966), where leptodactyloids and ranoids, respectively, reached tropical areas later in time than the toads, minor radiations and evolution of specialized bufonid genera have taken place. Since toads preceded ranids in Africa, a similar pattern may be responsible for the several peculiar *Bufo* allies there as well (Tihen, 1960).

The principal themes of frog distributional history involve the following:

1. the origin of a core new general adaptation (e.g., the algae-grazing larva)

2. the invasion of geographically and ecologically accessible areas (e.g., spread of basic leptodactyloids through temperate Gondwanaland)

3. the radiation into available ecologic (adaptive) subzones (e.g., radiation of leptodactyloids and derivatives in South America)

4. the pre-emptive occupancy of accessible ecologic (adaptive) subzones by the older stock in the area so that recent arrivals radiate into subzones not available to older resident stocks (e.g. South American pipids and microhylids retain pre-eminence in filter-feeding larval niches; pipids in the aquatic adult subzone, and microhylids as adult fossorial ant and termite eaters; leptodactyloids and derivatives moved into many subzones, but not these)

5. the continued coexistence with, rather than replacement of, older stocks with a different core of general adaptation (e.g., leptodactyloids and derivatives with pipids and microhylids in South America)

6. the pre-emptive exclusion of the more recently arrived stock with a similar core of general adaptation (e.g., peripheral ecologic role of recently arrived ranids in South America)

Patterns of dominance arise as the result of themes 1 through 5; those of complementarity, from 4 and 6. Competitive dominance and replacement seem to have little to do with the origin and dispersal of frog groups. The composition of the world frog faunas is determined by time of origin and ecologic accessibility to a geographic area for a particular group. The faunas are limited not by competitive replacement but by pre-emptive occupancy of ecologic (adaptive) zones. Generally speaking, frog faunas are characterized by growth through the addition of new groups with different cores of general adaptation to a geographic region, followed by ecologic radiation in coexistence with earlier native groups.

ORIGINS OF THE MAJOR WORLD FROG FAUNAS

A SUMMARY

In previous sections, I have discussed the Mesozoic origins and Cenozoic developments for each family and faunal source through time. The following review is concentrated upon Cenozoic events for each major geographic region as these contributed to the patterns of present-day frog distribution.

At the beginning of the Cenozoic the original northern land mass was fragmented (Figure 13–35), but it supported an essentially circumpolar temperate frog fauna comprised of Laurasian source unit groups (Table 13–8). The composition of the tropical fauna on the northern land masses remains a mystery, although we know that it included Rhinophrynidae (North America), Paleobatrachidae (Europe), and megophryine pelobatids in both North America and Eurasia. Darlington (1957) believed that salamanders, ascaphids, discoglossids, and pelobatids were of tropical origin and that they withdrew northward and were replaced in the tropics through competition by advanced dominant modern frogs. My interpretation of the available evidence indicates that the Laurasian groups always have had basically northern distributions, with southern limits at latitudes with mild, warm subtropical to temperate conditions like those in the southeastern United States today. In the Cretaceous and early Tertiary these climates extended much farther north than they now do, and tropical conditions affected much of what is now the northern extratropical region. The zonation of climatic regions was much less strongly marked at that time so that an extensive ecotone between mild humid subtropical to warm temperate conditions and moist tropical climates was present (Axelrod, 1960). Although some genera of the northern lines were associated with the tropical-extratropical interface, none seemed to have a primarily tropical radiation. As a matter of fact, the distribution of the living tropical discoglossids and megophryines and the subtropical pelobatines seems to represent southward displacement from more northern latitudes. A similar southward displacement during the Cenozoic is seen in many salamander stocks now restricted to southeastern China and the southeastern United States.

Among the most significant geologic events of the Cenozoic was the establishment of the Isthmian Link between North America and South America. The connection provided a route for the first migrations of Gondwanaland frogs, particularly bufonids and hylid tree-frogs onto the Laurasian land masses, with profound results. Because of the difference in middle to late Cenozoic history, each of the three main Holarctic centers of frog diversity are considered separately.

NORTH AMERICA. A warm temperate (Arcto-Tertiary geofloral association) fauna of ascaphids and discoglossids, warm temperate–subtropical megophryines, and a tropical fauna of rhinophrynids and possibly megophryines was enriched by invaders from the south in the Paleocene. The immigrants from tropical South America included microhylids, bufonids, hylids, and possibly leptodactylids. Stocks of these southern invaders differentiated in tropical North America in isolation from South America during the Eocene through Miocene submergence of the Isthmian Link. By late Eocene one toad lineage and a stock of tree frogs became associated with warm temperate conditions in which Arcto-Tertiary vegetation predominated. The general cooling and drying trends of middle to late Cenozoic caused a gradual displacement southward of tropical conditions and associated biota to the Middle American paleopeninsula that existed until the reconnection of Central America and South America in the early Pliocene. To the north, over broad areas of formerly tropical climates in western North America and northern Mexico, temperate semiarid to desert conditions developed from the Oligocene onward. The expanding xeric regions became an effective barrier to interchange between the extratropical northern areas, with their Arcto-Tertiary floral components, and the tropical regions. Concurrent with these events discoglossids and megophryines became extinct in the New World, and pelobatines and pelodytines appear to have become associated with the developing xeric habitats. It also appears that in the Eocene, Eurasian temperate ranid stocks crossed the Bering land bridge onto the continent. By the Miocene the basic modern fauna of extratropical and tropical North America was established:

	From South	North American	Moving South	From Eurasia
Extratropical				
Arcto-Tertiary Formations: Conifer and broad-leaf forests	bufonids hylids	ascaphids	pelobatines	ranids
Madro-Tertiary Formations: Xeric vegetation			pelobatines pelodytines	
Tropical	microhylids bufonids hylids leptodactylids	rhinophrynids	ranids	

By this time tropical habitats were restricted to southern Mexico and Central America. When the latter was reconnected to South America in earliest Pliocene, several leptodactyloid lines, including leptodactylines, telmatobiines, centrolenids, and dendrobatids, moved northward into the Middle American tropics to a greater or lesser extent. Relatively late in the Cenozoic several Middle American tropical groups (microhylids and bufonids) invaded the central and eastern United States to complete the faunal mixing. The pelodytines are not known in the New World after the Miocene. Details of the exchange through Middle America have been described in detail in an earlier paper (Savage, 1966).

EASTERN ASIA. In early Cenozoic the amphibian fauna associated with temperate Arcto-Tertiary geofloral components consisted of the circumpolar salamanders, ascaphids, discoglossids, and pelobatid megophryines. Representatives of the latter two groups occur in tropical Indo-Malaya today and may be remnants of the early Tertiary tropical fauna of Eurasia. In the Eocene, bufonids derived from tropical North American stocks became associated with the tropics of eastern Asia by immigration across the Bering land bridge. Ranines that adapted to temperate areas also contributed to the developing frog fauna by immigration from western Eurasia. By the middle Tertiary, cooling and drying trends caused a southward displacement of isotherms and fragmentation of Arcto-Tertiary forests and their associated fauna and thus reduced or eliminated these biotas from central Asia. Correlated with these events was the development of pelobatine and pelodytine groups from ancestors that probably arrived in the region from North America in Eocene (Estes, 1970). The Indian continental island was added to the Eurasian land mass by the middle Tertiary, and the Indo-Malayan archipelago was well developed by this time. Shortly thereafter, the endemic Indian and New Guinea microhylid stocks invaded southeast Asia from the west and southeast, respectively. Rhacophorines from southeast Asia and the microhylids moved northward to varying extents into temperate eastern Asia during the rest of the Cenozoic. The events of the Tertiary are summarized below:

	From West	In Eurasia	From South	From North America
Temperate	ranids (E)	ascaphids discoglossids		pelobatines (E-O) pelodytines (E-O) bufonids (O) hylids (O)
Tropical		discoglossids megophryines	microhylids (M) rhacophorines (M) ranids (P)	bufonids (E)

E, Eocene; O, Oligocene; M, Miocene; P, Pliocene.

Among the most recent arrivals are temperate-adapted bufonids and hylids that reached the region from North America across the Bering land bridge in Oligocene. Tropical-adapted ranids (Raninae and Platymantinae) range into southern China today, but they probably represent rather recent immigrants from Indo-Malayan centers.

WESTERN EURASIA. A basic Cretaceous amphibian fauna of temperate (Arcto-Tertiary geofloral association) salamanders, discoglossids, megophryines, and probably ascaphids, as well as a tropical group probably composed of paleobatrachids and megophryines, was enriched by an invasion of southern tropical ranids of African origins in the Paleocene. The latter probably arrived from the southeast, possibly by overwater dispersal from northeast Africa-Arabia, since no land

bridge connected Africa and Europe in any region until Miocene. By Eocene at least one ranid stock had become associated with warm temperate conditions in areas of Eurasia dominated by Arcto-Tertiary floral components. Fairly early in the Cenozoic, at least by the Oligocene, ascaphids and paleobatrachids had become extinct. Cooling and particularly drying trends gradually eliminated tropical climates and vegetation from this area, and semiarid to desert temperate and subtropical conditions came to dominate southern Europe and areas at mid-latitudes to the east during the rest of the Cenozoic. In general, from the Oligocene onwards isotherms were displaced southward across all of Eurasia—a situation parallel to what occurred at the same time in North America. Concomitant with temperature depression in the west, drying trends affected Eurasia, and the distinctive semi-arid Mediterrano-Tertiary geoflora came to dominate what were formerly tropical areas. Cooling trends also fragmented the formerly circumpolar Arcto-Tertiary forests and associated faunas. As cooling intensified from the Miocene onwards, central Asia became uninhabitable to all but a few temperate-adapted amphibians. Southward displacement of isotherms forced the tropical floras and faunas of eastern Asia southward, but those of central Asia became extinct because they could not pass over the Himalayan barrier into tropical India or through the developing semiarid to desert conditions to the southwest. Similarly, the tropical groups of western Eurasia became extinct through the combination of cooling and drying, the barrier to southward immigration provided by the Tethys Sea remnants, and the Sahara Arabian–Baluchistan desert belt. These events, well documented by fossil floral history (Axelrod, 1960), correlate closely with frog distributions.

The onset of marked drying effects was in the Oligocene, and it was at this time that pelobatine and pelodytine lines reached Eurasia across the Bering land bridge and underwent evolution across Eurasia in marginal temperate-subtropic areas (Estes, 1970). At about this same time, temperate-adapted lines of bufonids and hylids also invaded Eurasia from North America and became an integral part of the temperate Eurasian fauna. By mid-Miocene, as tropical habitats were replaced by temperate semiarid conditions, megophryine pelobatids became extinct in western Eurasia. Pelobatines and pelodytines were associated with the latter climatic conditions in this region from the Miocene onwards. The cooling trends of later Cenozoic led to the reduction and restriction of the ranges of discoglossids, pelobatines, and pelodytines to the milder temperate climates of Europe and to their disappearance over most of Asia. Discoglossids also occurred as relicts of this formerly continuous distribution in temperate eastern Asia. Only a few species of salamanders, bufonids, hylids, and ranids have been able to maintain continuous distributions across central Asia during the stringent vicissitudes of Pleistocene cooling. In the Eurasian region west of the Urals, the present faunal composition was derived from the following sources (Tables 13–7, 13–8):

	Group	Source
Present from Paleocene on:	discoglossids	*in situ*
Eocene arrivals:	ranids	from southeast; ultimately Africa
Oligocene arrivals:	pelobatids bufonids hylids	from east; ultimately North America
Extinctions:	ascaphids and palaeobatrachids (Oligocene) megophryines (Miocene)	

The amphibian fauna of northwest Africa is derived completely from southern Europe. All of its members belong to European species or species groups; none are from tropical Africa. A similar situation is seen in the amphibian fauna of the Near East. Whatever frogs occurred in the tropical portions of western Eurasia in the early Cenozoic disappeared as semiarid to desert habitats replaced the earlier mesic tropical associations.

The history of the frog faunas of the southern continents requires that each land mass receive individual attention. At the end of the Cretaceous each southern continent formed an isolated island with basic faunas as previously described (Figure 13–34).

AFRICA. The families of amphibians on this continent were all established there prior to the Cenozoic. No additional major groups have been added to the fauna during the Cenozoic, although adaptive radiation occurred along many lines. Conversely, two groups of ranids (Raninae and Rhacophorinae) immigrated out of tropical Africa into Eurasia during this period. Since the Cretaceous, bufonids, ranids, and hyperoliids have come to predominate all of Africa. The ranoid radiations have enriched the fauna substantially and have had two other effects: (a) by pre-emptive occupancy of most tropical adaptive subzones, the ranoids have prevented any major radiation by the basic temperate leptodactyloids into tropical Africa, which is the converse of the situation for South America and Australia, and (b) by their presence, they have restricted further evolution by microhylids.

MADAGASCAR. The basic frog fauna of Madagascar consists of two groups of endemic microhylids that are probably descended from Triassic ancestors. Added to this fauna have been a ranid (Mantellinae) and a hyperoliid (Scaphiophryinae) endemic stock, each developed from overwater invaders of early Tertiary. Later Tertiary invaders include the ancestors of *Boophis* (Rhacophorinae), *Heterixalus* and *Hyperolius* (Hyperoliidae), and the very recent arrival of one species of *Rana.*

AUSTRALIA. The most significant feature of frog evolution in Australasia is the radiation of the temperate-adapted leptodactyloids. The general division of the continent into three broad transcontinental climatic zones—a northern tropical humid zone, a semiarid to desert central zone, and a southern moist temperate zone—during the middle and late Cenozoic is of particular significance. The strong zonation led to restriction of the tropical microhylids to northern areas and temperate myobatrachids to the south. One early group of leptodactyloids apparently became associated with the tropical zone before full development of the xeric barrier between north and south and gave rise to the tropical tree-frogs (Pelodryadidae). Their history seems closely associated with that of the microhylids during most of the Cenozoic.

In essence, at the beginning of the Tertiary, Australia was divided into a southern warm temperate belt, a northern tropical zone, and an ecotone of subtropical character with a mixture of geofloral components. In the south central areas semiarid components were beginning to differentiate (Figure 13–35). Myobatrachids dominated the southern area, and microhylids and the ancestors of the pelodryadids dominated the northern tropical region. Although all three groups may have occurred in the intermediate zone, the cooling and drying trends of Oligocene to Recent eliminated the associations in that region and replaced them with the semiarid to desert communities of today. Cooling allowed for expansion northward of warm temperate conditions and myobatrachids, and restricted the microhylid and pelodryadid lines and associated tropical conditions to smaller and smaller northern areas. At the same time, drying trends eliminated the environments and frogs of the intermediate subtropical areas so that by the Miocene, microhylids and pelodryadids were centered to the north and myobatrachids to the south of an extensive semiarid to desert region.

When New Guinea became emergent in the Miocene, microhylids and pelodryadids invaded overland from northern Australia and underwent radiation there. Continuing drying trends eliminated most moist tropical habitats in northern Australia from late Miocene onward, and the land connection between New Guinea and Australia was submerged. The distributional evidence strongly suggests that the Australian tree frogs and their derivatives (at least 4 genera, but only about 30 species) invaded the continent from New Guinea on several different occasions in post-Miocene times. Some of the pelodryadids have become adapted to nonarboreal conditions, and others to life in the subtropic to temperate scrub, woodland, and forest habitats peculiar to Australia. The microhylids have a similar history of re-establishment in Australia but presently have barely reached northern Queensland, where they are confined to isolated patches of humid tropical forest. In other words, the Australian microhylid and pelodryadid fauna is derived completely from New Guinea, since tropical humid conditions and associated frogs were eliminated from northern Australia, probably in early Pliocene, and only later became re-established.

The basic temperate leptodactyloid stock (Myobatrachidae) diversified in southern Australia, primarily in association with Antarcto-Tertiary geofloral components. After the original frog stocks that were adapted to subtropical but humid conditions were eliminated from central Australia, myobatrachids invaded semiarid to desert situations from the south. Several stocks became specialized to life in these environments and have undergone speciation in them. Very likely the ancestors of these lines were undergoing differentiation in the semiarid communities of western Australia in early Cenozoic and spread with them across the continent.

The cooling trends of late Cenozoic encouraged a general northward flow of myobatrachids, but two other factors seem more important in explaining modern distributions. First, as semiarid to desert conditions developed in south central Australia, drying trends in late Cenozoic fragmented the range of temperate forest and woodland situations and their leptodactyloids into southwestern and southeastern centers. Secondly, the uplift of the Great Dividing Range during the Pliocene, apparently associated with increased rainfall along its eastern slope, established a humid corridor from north Australia (tropical) through subtropical formations to southeastern Australia (temperate). This filter corridor has been broken from time to time by alterations in aridity and humidity during the Pliocene to Recent, but it has allowed a northward flow of myobatrachids and a southward movement of pelodryadids that continues to the present. The opening of this corridor explains the rapid southward immigation and differentiation of pelodryadids in eastern Australia. Because of their extremely high moisture needs, the microhylids apparently could not take the same advantage of the corridor that myobatrachids (northward flow) and pelodryadids (southward flow) did, for all known New Guinea and Australian microhylids lay their eggs on land and undergo direct development.

SOUTH AMERICA. In the early Tertiary the fauna consisted of pipids, microhylids, bufonids, hylids, and leptodactylids, with the first four groups well distributed in tropical areas and the latter beginning to radiate out of the temperate environments into southern subtropical to tropical regions. All of these groups must date from the Cretaceous on the continent, and several invaded tropical North America across the Isthmian Link in the Paleocene. As far as is known, no new familial-level groups invaded South America from the north during the period of land connection. During most of the Cenozoic (Eocene through Miocene), the continent was an island, and it was during this episode that hylids and leptodactylids underwent an extensive radiation, principally in tropic areas. At the same time, pipids and microhylids maintained themselves or perhaps became reduced in numbers, and bufonids developed several distinct lines, with some *Bufo* invading montane and temperate areas. After the reconnection of North America and South America in the early Pliocene, ranids moved into South America for the first time to complete the faunal picture. The key events in Cenozoic history in South America are summarized below. The groups indicated by the letter (A) are the portion of the lineage that underwent evolution in isolation in tropical North America during the Eocene through Miocene. The letter (B) demarks that portion undergoing independent diversification in South America during the same time interval.

Paleocene

Groups in South America:	Groups moving into North America:
pipids	microhylids (A)
microhylids	bufonids (A)
bufonids	hylids (A)
hylids	leptodactylids (A)*
leptodactylids	

Eocene-Miocene

Groups evolving in isolation in South America:	Groups originating in South America:	Groups evolving in isolation in Middle America:
pipids	brachycephalids	
microhylids (B)	allophrynids	microhylids (A)
bufonids (B)	centrolenids	bufonids (A)
hylids (B)	pseudids	hylids (A)
leptodactylids (B)	rhinodermatids	leptodactylids (A)*
	dendrobatids	

Pliocene-Recent

Groups invading South America from north:	Groups invading North America from south:
ranids	pipids
bufonids (A)	microhylids (B)
hylids (A)	bufonids (B)
leptodactylids (A)*	hylids (B)
	leptodactylids (B)
	centrolenids
	dendrobatids

The only South American leptodactylid (*) in Central America at the time of the Pliocene reconnection was the diverse genus *Eleutherodactylus* and several nominal genera that were clearly derived from it. This complex has encapsulated eggs and direct development and may have reached Central America by overwater transport across the Panamanian portal in the Eocene or Oligocene. Details of the complex faunal relations and interchanges between Middle America and South America from the Pliocene onward have been fully discussed elsewhere (Savage, 1966).

Two other southern areas warrant brief comment. The frog fauna of Antarctica almost certainly included leiopelmatids and leptodactyloids in the Cretaceous. World cooling trends in the Cenozoic completely eliminated them from that continent. The fauna of New Zealand contains the only living leiopelmatids (three species). They are relicts, and it seems likely that no other native frogs have occurred on the islands during the Cenozoic.

Finally, special note must be taken of the Indo-Malayan region, which has a complex history (Table 13–4, Figure 13–41). Three different source areas have contributed to the frog fauna of the region to differing degrees: India has been the western source for microhylines, rhacophorines, and tropical ranids; middle Tertiary tropical eastern Asia was the northern source area for discoglossids, megophryines, and bufonids; and New Guinea was the southern source for Indo-Malayan asterophryines.

The original Cenozoic fauna appears to have been a southern extension of the tropical east Asian fauna of early Tertiary that probably occurred over much of what is now south central China. Two remnants of this fauna are the tropical members of the Discoglossidae *(Barbourula)* and Pelobatidae (*Megophrys* and *Lepidobatrachium*). By the Eocene, tropical toads derived from North America became associated with this unit. Since areas of southeast Asia and the Indo-Malayan chain have been land positive since the Eocene, invasion from the north allowed formation of the diverse assemblage of bufonids (*Ansonia, Cacophryne, Pedostibes, Pelophryne, Pseudobufo* and *Ophryophryne*) and speciation in the pelobatids. The cooling and drying trends of the late Tertiary eliminated tropical habitats in mid-latitude Asia, and the Himalayas prevented tropical forms from reaching India from the north. The Assam-Burma marine embayment restricted opportunities for the penetration of Indian groups from the west. In essence, Thailand-Indochina, together with the Indo-Malayan chain, formed a paleopeninsula into which tropical Asian groups were forced by changing climates and restrictive physiographic barriers from the Miocene onward. The groups from India moved into the region after the Assam-Burma area was uplifted in the Oligocene. The asterophryines have filtered northward through the archipelago, beginning at least in the Miocene. The fauna of the Indo-Malayan chain is a composite derived from these three sources and parallels the situation in tropical Middle America during the Eocene-Miocene, when the Panamanian marine portal was open.

IN CONCLUSION. The fundamental pattern of amphibian distribution is derived from two primary sources: the ancient establishment, by the early Jurassic, of two major centers of evolution, one on the northern Laurasian land mass (salamanders, some xenoanurans, lemmanurans, and pelobatids) and another on the southern Gondwanaland supercontinent (caecilians, pipids, microhylids, and advanced acosmanurans); and the fragmentation and evolution in isolation during the Cretaceous of caecilians and frogs on the six land areas derived from Gondwanaland and of salamanders and northern frogs on Laurasia. Contrary to the ideas of Darlington (1957), frog evolution has not been centered on the Old World tropics. Substantial distinctive radiations of microhylids and acosmanurans have taken place in each tropical area, except on the Indian continental island, where caecilians and microhylids developed in a unique fashion. In Africa the radiations involved caecilians, microhylids, bufonids, and ranoids; in Australasia, microhylids, myobatrachids, and pelodryadids; and in South America, microhylids, bufonids, hylids, and leptodactyloids. The basic stocks for the radiations of leptodactyloids and their derivatives in Australasia and South America were

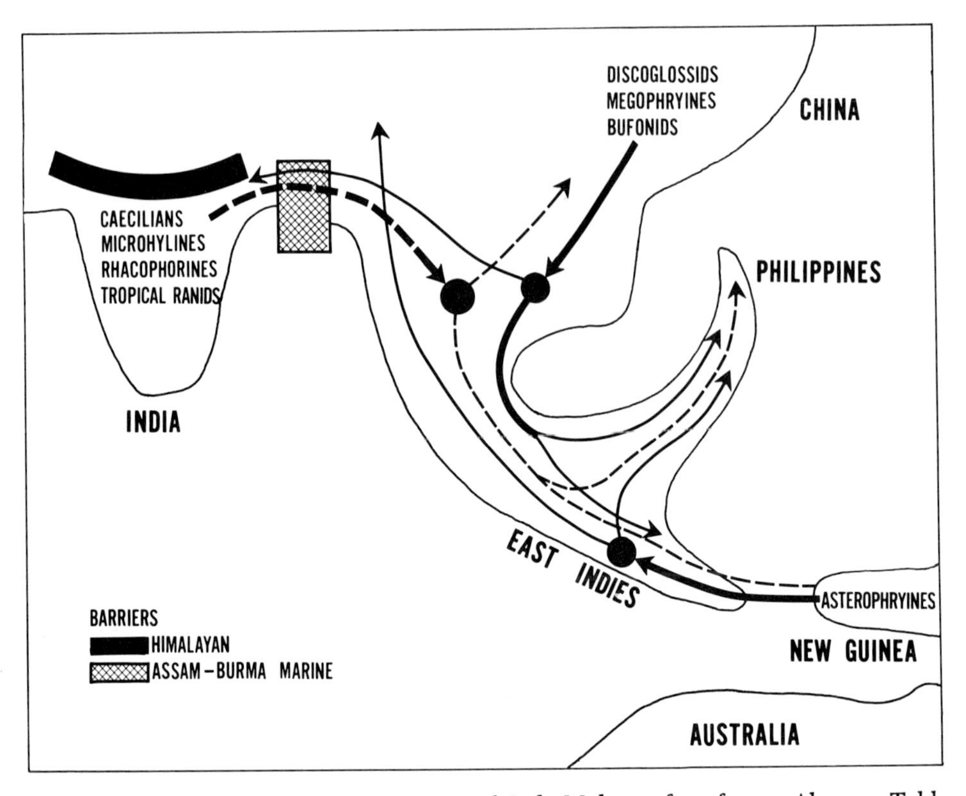

Figure 13–41. Diagram indicating sources of Indo-Malayan frog fauna. Also see Table 13–4.

temperate-adapted groups that invaded subtropic and tropical regions from the south. The tropical areas of India and southeast Asia have never been a major source region for frog evolution and radiation but have served as a refuge for tropical Asian groups displaced southward by late Cenozoic cooling trends and for the ancient Indian caecilian and microhylid stocks that developed while the subcontinent was a Cretaceous and early Tertiary island. In the Cenozoic the connections established between several southern land masses and northern regions allowed invasion of North America and Eurasia by southern tropical lines. The most successful radiations were by bufonids and hylids from South America and by ranids from Africa. These groups have followed the northern land masses as immigration routes, the former two families into Eurasia from North America and the latter into North America from Eurasia. Bufonids subsequently have moved southward from northern Asia into India and Indo-Malaya, and ranids have now reached South America. Remnants of the ancient fauna of Laurasia still persist in Eurasia (salamanders, discoglossids, and pelobatids) and North America (salamanders, ascaphids, and rhinophrynids). The peculiar endemic genus *Leiopelma* of New Zealand is a relict of an ancient temperate Gondwanaland radiation. The most consistent themes of amphibian evolution are the unique quality of radiations on each Gondwanaland derivative land mass; the role of pre-emptive occupancy of ecologic (adaptive) zones, rather than competitive replacement, as limiting faunal composition and faunal growth; and increasing diversity through addition of new groups that coexist with earlier stocks.

Acknowledgments

In a study of this nature it is impossible to acknowledge everyone who has provided comments, impressions, data, and hints over the many years that I have been interested in the problems of frog phylogeny, evolution, ecology, and distribution; nevertheless, whoever and wherever you are, I thank you. In the direct preparation of the paper I have benefited greatly from the encouragement of Dr. James L. Vial, the organizer of this symposium, who stimulated me to put my ideas into form; from the knowledge and ideas of Dr. Priscilla H. Starrett, whose brilliant insights into the significance of basic tadpole structure and function in frog evolution promises to revolutionize most of our concepts of anuran biology; from the contribution of Dr. Ian R. Straughan to my previously limited appreciation of frog evolution in Australia; and from the evaluation of the significance of recent cytogenetic evidence by Dr. John W. Wright. Their aid is appreciated greatly.

Finally, I wish to thank Ronald T. Harris for his care and excellence in executing the many figures that form the backbone of evidence and interpretation for this paper.

REFERENCES

Areiti, S. 1967. The intrapsychic self. Basic Books, New York. 487 p.

Axelrod, D. I. 1960. The evolution of flowering plants, p. 227-305. *In* Sol Tax, [ed.], Evolution after Darwin, vol. 1. Univ. Chicago Press, Chicago.

Blackett, P. M. S., E. Bullard, and S. K. Runcorn. 1965. A symposium on continental drift. Phil. Trans. Roy. Soc. London, A, 258:1-322.

Blair, W. F. 1963. Evolutionary relationships of North American toads of the genus *Bufo:* a progress report. Evolution 17(1):1-16.

———. 1970. Genetically fixed characters and evolutionary divergence. Am. Zool. 10(1):41-46.

Bogart, J. P. 1968. Chromosome number difference in the amphibian genus *Bufo:* the *Bufo regularis* species group. Evolution 22(1):42-45.

Brattstrom, B. H. 1957. The phylogeny of the salientia based upon skeletal morphology. Systematic Zool. 6(2):70-74.

Carvalho, A. L. de. 1954. A preliminary synopsis of the genera of American microhylid frogs. Occ. Pap. Mus. Zool., Univ. Michigan 555:1-19.

Chiplonker, G. W. 1941. A new species of fossil frog from the Intertrappean beds of Worli Hill, Bombay. J. Bombay Nat. Hist. Soc. 41(4):799-804.

Colbert, E. H. 1970. The fossil tetrapods of Coalsack Bluff. Antarctic J. 5(3):57-61.

Cope, E. D. 1865. Sketch of the primary groups of batrachia salientia. Nat. Hist. Rev. 5:97-120.

Darlington, P. J. 1957. Zoogeography: the geographical distribution of animals. John Wiley and Sons, New York. 675 p.

———. 1965. Biogeography of the southern end of the world. Harvard Univ. Press, Cambridge, Mass. 236 p.

Estes, R. 1970. New fossil pelobatid frogs and a review of the genus *Eopelobates.* Bull. Mus. Comp. Zool., Harvard Univ. 139(6):293-339.

Estes, R., and O. A. Reig. 1973. The fossil record of frogs: a review of the evidence. This volume.

Estes, R., and R. Wassersug. 1963. A Miocene toad from Colombia, South America. Breviora, Mus. Comp. Zool., Harvard Univ. 193:1-13.

Gaudin, A. J. 1969. A comparative study of the osteology and evolution of the holarctic tree frogs: *Hyla, Pseudacris, Acris* and *Limnaeodus.* Ph.D. Dissertation, Univ. So. California.

Griffiths, I. 1959. The phylogenetic status of the sooglossinae. Ann. Mag. Nat. Hist., Ser. 13, 2:626-640.

_______. 1963. The phylogeny of the salientia. Biol. Rev. 38(2):241-292.
GÜNTHER, A. C. L. G. 1858. Catalogue of the batrachia salientia in the collection of the British Museum. Taylor and Francis, London. 155 p.
HECHT, M. K. 1960. A new frog from an eocene oil-well core in Nevada. Am. Mus. Novitates 2006:1-14.
_______. 1962. A reevaluation of the early history of the frogs. Part I. Systematic Zool. 11(1):30-44.
_______. 1963. A reevaluation of the early history of the frogs. Part II. Systematic Zool. 12(1):20-35.
HECHT, M. K., AND R. ESTES. 1960. Fossil amphibians from Quarry Nine. Postilla, Peabody Mus. Nat. Hist., Yale Univ. 46:1-19.
HOLDRIDGE, LESLIE R. 1964. Life zone ecology. Trop. Sci. Center, San Jose, Costa Rica. 124 p.
_______. 1967. Life zone ecology. 2nd ed. Trop. Sci. Center, San Jose, Costa Rica. 206 p.
HOLMES, A. 1965. Principles of physical geology. 2nd ed. Ronald Press, New York. 1288 p.
INGER, R. F. 1966. The systematics and zoogeography of the amphibia of Borneo. Fieldiana: Zool. 52:1-402.
_______. 1967. The development of a phylogeny of frogs. Evolution 21(2):369-384.
JAMESON, D. L. 1957. Life history and phylogeny in the salientians. Systematic Zool. 6(2):75-78.
KING, L. C. 1967. The morphology of the earth. 2nd ed. Oliver & Boyd, London. 726 p.
KLUGE, A. G. 1966. A new pelobatine frog from the lower miocene of South Dakota with a discussion of the evolution of the *Scaphiopus-Spea* complex. Los Angeles Co. Mus. Contrib. Sci. 113:1-26.
KLUGE, A. G., AND J. S. FARRIS. 1969. Quantitative phyletics and the evolution of anurans. Systematic Zool. 18(1):1-32.
LASERON, C. F. 1969. Ancient Australia. Taplinger Pub., New York. 253 p.
LAURENT, R. F. 1951. Sur la nécessité de supprimer la famille des rhacophoridae mais de créer celle des Hyperoliidae. Rev. Zool. Botan. africaines 45(1):116-122.
LIEM, S. S. 1969. The morphology, systematics, and evolution of the old world tree-frogs (Rhacophoridae and Hyperoliidae). Ph.D. Dissertation, Univ. Illinois.
LYNCH, J. D. 1969. Evolutionary relationships and osteology of the frog family Leptodactylidae. Ph.D. Dissertation, Univ. Kansas.
McDIARMID, R. W. 1969. Comparative morphology and evolution of the neotropical frog genera *Atelopus, Dendrophryniscus, Melanophryniscus, Oreophrynella,* and *Brachycephalus*. Ph.D. Dissertation, Univ. So. California.
MATTHEW, W. D. 1915. Climate and evolution. Ann. New York Acad. Sci. 24:171-318.
MOREAU, R. E. 1966. The bird faunas of Africa and its islands. Academic Press, London. 424 p.
NEVO, E. 1968. Pipid frogs from the early cretaceous of Israel and pipid evolution. Bull. Mus. Comp. Zool., Harvard Univ. 136(8):255-318.
NICHOLLS, G. C. 1916. The structure of the vertebral column in the anura phanero-glossa and its importance as a basis of classification. Proc. Linn. Soc. London, Z, 128:80-92.
NOBLE, G. K. 1930. The fossil frogs of the Intertrappean beds of Bombay, India. Am. Mus. Novitates 401:1-13.
_______. 1931. The biology of the amphibia. McGraw-Hill Book Co., New York. 577 p.
ORTON, G. L. 1953. The systematics of vertebrate larvae. Systematic Zool. 2(2):63-75.
_______. 1957. The bearing of larval evolution on some problems in frog classification. Systematic Zool. 6(2):79-86.
PARKER, H. W. 1934. A monograph of the frogs of the family Microhylidae. British Museum (Nat. Hist.), London. 208 p.
_______. 1940. The Australasian frogs of the family Leptodactylidae. Novitates Zool. 42(1):1-106.
POYNTON, J. C. 1964. The amphibia of southern Africa. Ann. Natal Mus. 17:1-334.
RITLAND, R. M. 1955*a*. Studies on the post-cranial morphology of *Ascaphus truei*. I. Skeleton and spinal nerves. J. Morphol. 97(1):119-178.

———. 1955*b*. Studies on the post-cranial morphology of *Ascaphus truei*. II. Myology. J. Morphol. 97(2):215-282.
RUNCORN, S. K., ed. 1962. Continental drift. Academic Press, London. 388 p.
SAVAGE, J. M. 1960. Evolution of a peninsular herpetofauna. Systematic Zool. 9(3):184-212.
———. 1963. Development of the herpetofauna. Proc. 16th Intern. Zool. Congr. 4:12-14.
———. 1966. The origins and history of the Central American herpetofauna. Copeia 4:719-766.
———. 1969. Evolution. 2nd ed. Holt, Rinehart, and Winston, New York. 152 p.
———. 1973. The Isthmian link and the evolution of neotropical mammals. MS.
SAVAGE, J. M., AND A. G. KLUGE. 1961. Rediscovery of the strange Costa Rica toad, *Crepidius epioticus* Cope. Rev. Biol. Trop. 9(1):39-51.
SIMPSON, G. G. 1947. Holarctic mammalian faunas and continental relationships during the Cenozoic. Bull. Geol. Soc. America 58:613-688.
———. 1953. Evolution and geography. Condon Lecture Oregon State Syst. Higher Educ., Eugene.
———. 1966. Mammalian evolution on the southern continents. Neues Jahrb. Geol. Paleontol., Monatsh. 125:1-18.
STARRETT, P. 1960. Descriptions of tadpoles of middle american frogs. Misc. Publ. Mus. Zool. Univ. Michigan 110:1-37.
———. 1968. The phylogenetic significance of the jaw musculature in anuran amphibians. Ph.D. Dissertation, Univ. Michigan.
———. 1973. Evolutionary patterns in larval morphology. This volume.
STEPHENSON, E. M. T. 1951. The anatomy of the head of the New Zealand frog, *Leiopelma*. Trans. Zool. Soc. London 27(2):255-305.
———. 1952. The vertebral column and appendicular skeleton of *Leiopelma hochstetteri* Fitzinger. Trans. Roy. Soc. New Zealand 79(3):601-613.
———. 1955. The head of the frog *Leiopelma hamiltoni* McCulloch. Proc. Zool. Soc. London 124(4):791-801.
———. 1960. The skeletal characters of *Leiopelma hamiltoni* McCulloch, with particular reference to the effects of heterochrony on the genus. Trans. Roy. Soc. New Zealand 88(3):473-488.
STEPHENSON, N. G. 1951*a*. On the development of the chondrocranium and visceral arches of *Leiopelma archeyi*. Trans. Zool. Soc. London 27(2):203-252.
———. 1951*b*. Observations on the development of the amphicoelous frogs *Leiopelma* and *Ascaphus*. J. Linn. Soc. London 42(283):18-28.
———. 1955. On the development of the frog *Leiopelma hochstetteri* Fitzinger. Proc. Zool. Soc. London 124(4):785-795.
TIHEN, J. A. 1960. Two new genera of african bufonids, with remarks on the phylogeny of related genera. Copeia 3:225-233.
———. 1962. Osteological observations on new world *Bufo*. Am. Midland Naturalist 67(1):157-183.
———. 1964. Tertiary changes in the herpetofaunas of temperate North America. Senck. biol. 45(3/5):265-269.
TOSI, J. A. 1964. Climatic control of terrestrial ecosystems: a report on the Holdridge model. Econ. Geogr. 40(2):173-181.
TYLER, M. J. 1971. The superficial throat musculature of the Anura. MS.
VERMA, K. K. 1965. Note on a new species of fossil frog from the Intertrappean beds of Malabar Hill, Bombay. Curr. Sci. 34(6):182-183.
WALKER, C. F. 1938. The structure and systematic relationships of the genus *Rhinophrynus*. Occ. Pap. Mus. Zool., Univ. Michigan 372:1-11.
WALLACE, A. R. 1876. The geographical distribution of animals. Harpers, New York. 607 p.
WHITE, J. 1790. Journal of a voyage to New South Wales, with sixty-five plates of non-descript animals, birds, lizards, serpents, curious cones of trees and other natural productions. Bebrett, London. 299 p.
ZWEIFEL, R. G. 1956. Two pelobatid frogs from the tertiary of north america and their relationships to fossil and recent forms. Am. Mus. Novitates 1762:1-49.

PART III DISCUSSION

Metter

Since yesterday afternoon we've had fourteen papers on various individual subjects concerning the evolution of the anurans. Each one of those papers has given us a new insight into anuran groups; however, I'm sure that all of you would agree that if we took those fourteen papers and the discussions that followed and put all the information into a computer, it would spit out the material in such a way that we might not have any better picture of the evolution of anurans than we had before.

One of the things that has struck me is that practically every paper has shown that each of the individual workers was confronted with things that he didn't know what to do with. Dr. Savage doesn't know what to do with the leiopelmids, Dr. Straughan doesn't know what to do with the pipids, and so on down the line. Each one of them finds things that do not quite seem to fit the picture. I would suggest that in the future, as people are working on the evolution of the anurans, we should not only restress Dr. Blair's ideas that we need to consider all aspects, but we also should decide what we must get rid of and not consider at all, such as Dr. Straughan's pipids or Dr. Savage's leiopelmids.

For the start of the discussion this afternoon, I had originally planned to throw a little barb, but as I look around the audience, I see many people sharpening axes, and so I'm just going to let anyone from the audience start.

Thomas Fritts, University of Kansas

I would like to make clear one point of which I'm sure Dr. Bogart is aware. Chromosomal morphology is usually altered in taxa through one of several methods, that is, duplication, deletion, translocation, inversion, centric fusion or fission, or chromosomal breakage with simultaneous centromere development. The first of these mutations can result in sterility between individuals or taxa, with modification from those of the original. Simultaneously, any of these could exist but be microscopically undetectable, except in pachytene studies. Thus, karyotypes that seem identical can be quite different in terms of reproductive compatibility of the taxa, while those grossly different can be quite reproductively compatible. Such a condition frequently precludes any karyotypic studies in large, diverse groups without simultaneous hybridization and study of meiotic material, especially pachytene, which therefore precludes speculation about evolutionary distances between related taxa.

I was wondering if perhaps your [Bogart's] comments about the *Leptodactylus marmoratus* group and a few of the other taxa might have been based more on speculation than on your own data.

Bogart

That was a nice speech, and I do agree that we do have to be very careful, because some of these things may look very similar. But I started out my talk by saying that we have to understand the actual mechanisms exhibited in each closely related group before we can say anything about more distantly related animals. Where the chromosomes are very similar, hybridization experiments, especially in *Bufo,* have proved very useful. Anurans aren't very good material to study pachytene stages in hybrids, as several people have found out. In male hybrids, when they get to the stage of producing gametes, the chromosomes usually all clump or else the division is abnormal, which results in polyploid or aneuploid sperm. Now I don't really think I have to defend many of my arguments. In a gene system, if the chromosomes are exceedingly similar in species that are known by many other criteria to be fairly closely related, I don't think there's any problem. With some species, like *Acris,* or with the *marmoratus* group of the genus *Leptodactylus,* there are several other things besides chromosomes that lead me to believe that these species aren't very similar. I know that Ron Heyer has some very good information on the *marmoratus* group of *Leptodactylus,* but as far as putting these species in the genus *Eleutherodactylus,* I don't propose anything like that. I would say they are not very close to many *Leptodactylus,* and this is in terms of chromosomes. The chromosomes are very different, and it's about all I can go into here. Recently we've done a lot of blood work at Texas, and the enzyme systems fall into the same pattern. Until we can actually work on the gene patterns themselves, which is going to be very interesting in the future, we can't definitely say anything, but we can bring out these similarities and differences, which may have evolutionary significance.

Wassersug

I'd like to direct myself to Dr. Bogart's last comment. It seems to me that we've talked primarily at two levels of this symposium. One is problems dealing with species, and the other is with problems of higher taxa. Concerning problems of higher taxa, we get back to one that affects everyone, as do all problems in higher taxa, and that is the value of certain characteristics. For instance, some people put a great deal of emphasis on larval characteristics, while other people have totally ignored them here, and we find that their phylogenies differ radically. Now you've just said, Dr. Bogart, that you didn't feel that you needed to defend some of your ideas here. I think that defense is very important if we're going to find out whether the evolution of karyotypes has anything to do with the evolution of frogs. You'll have to show us why you feel any weight should be given to karyotypes.

Let me ask a specific question to demonstrate this point. You said, without telling us what the karyotype was for *Rhinophrynus,* that it had a karyotype much like the microhylids, although shortly after that, you said the microhylids had a fair range and gave an example. Were you trying to imply that, perhaps in support of what Savage and Starrett have said, *Rhinophrynus,* which I think everyone has agreed is pipoid in character, is really closely related to the microhylids?

Bogart

I'm sorry, I did skip over a section in my paper, which is a problem when trying to get things straight. *Rhinophrynus dorsalis* from around Tapachula, Mexico, has a karyotype very similar to the 22-chromosome microhylids, such as *Gastrophryne,* not the 26-chromosome microhylids. Recently I've been corresponding with Morescalchi. I sent him my karyotypes of *Rhinophrynus,* and he has some species of pipids with 22 chromosomes also. So, it will be a little while before we can get this all worked out. I really resent everybody saying that you can't put the leptodactylids here when you're including so much in the families Leptodactylidae and Hylidae. I don't think that these are set, definite, monophyletic families, as everybody here, I think, is agreed.

Wassersug

I'm not clear whether my question was directed to that at all. I'm trying to ask you what was the value, if any, of the use of karyotypes in higher taxa. I was talking with Dr. Etheridge before. Perhaps he would like to rephrase the question; maybe the way I phrased it is difficult.

Richard Etheridge, San Diego State College

I was very interested by a comment at the close of Ian's [Straughan] excellent talk. He felt that the evolution of vocal behavior may very well be independent of the evolution of other systems in anurans. Aside from the fact that changes in form and structure of karyotypes may be important in evolution of reproductive isolation, I would like to ask Dr. Bogart, first, if he believes that the modifications in the visible structure and number of chromosomes is in any way correlated or related to the evolutionary changes in other systems, particularly anatomy, or if he believes that evolutionary changes, in particular morphology and number of chromosomes, is totally independent of evolutionary changes in anatomy? What is your evidence—aside, of course, from the fact that on occasion the conclusions you come to on the basis of your chromosomal data are in agreement with the conclusions reached by morphologists?

Bogart

Well, I have to admit that a chromosome is just one morphological character. In some cases, where you don't have a lot of closely related animals to work with, you're getting into a field that you shouldn't be saying a whole lot about. Sure, the morphology of chromosomes have their own evolution, and there may be the perfect chromosome for the frog that lives in a certain area of the earth, which might be true of any morphological character. A chromosome is a chromosome, but it's not really a chromosome, in the sense that you can't distinguish true gene sequences. It's hard to explain because you can't identify gene-for-gene on the chromosome; you can't tell anything about homologies between different species. You can point out similarities and point out differences, which I think most people have been using for a very long time for many other characters. And I don't think it's heretical to say that similar chromosomes impart some kind of similarity, especially in related species. Earlier I showed you a slide of all those frog genera with 26 chromosomes. They all had chromosomes that in some animal groups might be considered similar—mammals, for instance, which have more karyotypic variation.

If a group of mammals all have that kind of arrangement, they might be considered similar. You can't just assume similarity if you use only the number of chromosomes, especially if you don't have the complemental or comparative material. That's one reason that I haven't published a whole lot yet; I'm trying to get the fill-in group of some of the large species-groups or at least members of each species-group before I can say anything more definite. I think I'm getting pretty close to it; I don't know.

Etheridge

If I could make one observation. I think that if we are to have any hope of an integrated approach to the problems of anuran phylogeny and classification, it's incumbent upon those of us who are using a variety of characters and information from a variety of sources to show that there are correlations. No one so far has suggested an evolution of frogs on the basis of color pattern, and I don't think that Ian [Straughan] would propose revising the classification of frogs or constructing a phylogeny of frogs on the basis of vocal behavior—at least not now. So I think that I'd like to reask the question: Do you believe that modifications in chromosome number and visible structure are correlated with evolutionary change in the anatomy of the animal, and if you believe that, why do you believe it?

Bogart

I hate to hedge on this, but I do believe that chromosome structure is related to the general evolutionary phylogenetic sequence in many cases. For instance, the 30-

chromosome hylids all probably have a common ancestor that had 30 chromosomes, an ancestor that was different from the common ancestor that had 24 chromosomes. Many 30-chromosome species could have independently arrived at that number from 24-chromosome species, but I don't think that's right, and it's my judgment in this case, which you can take or leave. Without further characters to work with, I wouldn't propose a new genus for these species; I won't propose a new family or split it that way—I don't think it's time. We have to use many characters, but now we can start looking at this group with an eye that is a little bit more or a little bit less jaundiced.

Duellman

I would like to comment that first I agree with Jay [Savage] on two points: one that Holly Starrett's presentation on the evolution of anuran larvae is a very big step forward, and second, that his splitting of the old family Hylidae is correct. But I would prefer that he used the older name, *Litoriidae*, instead of *Nyctimystidae*.

I do have one question for Savage. I'd like to know if he has any evidence for the divergence of independent evolutionary lines of bufonids, ranids, and leptodactylids from one basic stock.

Savage

I'd like to take exception to the correct name for what used to be the Australian tree frogs. After I made the slides and after Ian [Straughan] had been calling them *Litoriidae* just to get my goat for a couple of weeks, I went and looked it up. The name for these frogs that has priority is *Pelodryadidae*.

Your question is a very hard one to answer, Bill [Duellman]. I have some bits and pieces of evidence and impressions, of course. You see we've been hung up so long on the idea that leptodactylids, as they exist now, are some kind of primitive group of higher frogs, primarily because that's the way Cope and Boulenger and Noble saw them from the kind of characters that they utilized. As we start to move away from that dogma and look at different kinds of characteristics, it seems to me that a more probable explanation is that these various stocks came from some base ancestral stock that we don't have much left of currently. It is my impression that if we follow back the evolutionary lines of bufonids, leptodactylids, the base ranoid stocks—something like *Sooglossus* or *Hemisus* or *Arthroleptis*—they tend to indicate discrete lines that you can't pull out of one of the other three groups. That's my impression at this particular point, using all the characters. While it seems very easy to say bufonids are leptodactylids with Bidder's organs, it doesn't work out that way when you start looking at the characters.

Rabb

One thing I want to do is defend the poor old pipids. I can tell you what you can do with them. You can use them as a parallel experiment because their sending apparatus is radically different from the vocal-cord mechanism in the other frogs, and therefore the kinds of information they communicate provide very nice parallel situations for us to explore and compare with what you turn up elsewhere.

Another point I have is more directed toward Jim Bogart because, out of my ignorance of the field, I was wondering if the term *nombre fondamental* is obsolete. I thought Morescalchi still used it, and I'm certain Goin and Goin still base their large-scale taxa comparisons of chromosomes on the number of arms.

Bogart

Well, it has been used for a long time; I think too long. In some groups it works satisfactorily—for instance, those *Eleutherodactylus* to which you can give a *nombre fondamental* of 36—but in many cases, it doesn't work at all because you must work with a submetacentric chromosome. I mean, are you going to break it and say that it is really two chromosomes that came together? In many cases in which we know that chromosomes have been gained or been lost, it isn't due to a simple fusion or fission. I think this term is rather archaic and must be used with discretion.

Craig Nelson, Indiana University

I felt that perhaps Jay Savage was a little too conservative, particularly with regard to the Microhylidae. For example, I've been considering the hypothesis that maybe the New World ones were not all related, that they might in fact represent three or four lines, each of which could be more closely related to some of the Old World ones than they are to each other. I was wondering if he had really considered whether there were six different radiations of Microhylidae or if that was just a convenient number drawn out of Parker's classification.

Savage

In my paper I pointed out a number of areas in which I believe that Parker's classification at the subfamily level, based primarily on the characters of the palate and of the pectoral girdle, probably are misleading. And so when I spoke of six major groups, that was based on a reanalysis of the available data that I have on total morphology, including jaw muscles and some other characters. It's an attempt to re-evaluate his classification, and my conclusion is that on each of the major southern masses—Africa, India, and Madagascar—in the Australian region, and in the New World Tropics, there is a distinct group of related genera that are more closely related to one another than they are to the forms in other areas. There may be an additional lineage as well, but that's very tentative. I think that it's a beautiful problem for someone to really work on the major classification of this particular stock. The group is so diverse. We're conditioned by looking at *Gastrophryne* in this country or even looking at the several genera in Latin America that have a similar morph. The Madagascar things are tree frogs and ranids. When you look at them, you'd never guess they are microhylids, and when you begin to work on their morphology, it's just as diverse. They go back a long, long time. In the jaw muscles, Holly [Starrett] has discovered that the range of variation is greater within the so-called family Microhylidae than within the vertebrates. The kind of jaw-muscle characters that are used to distinguish reptiles from mammals from birds from bony fish—these are all encompassed in microhylids. So I would say that there are about six lineages that I have a feel for; certainly five of them are very distinctive and geographically consistent, contrary to Parker, who put things that were from different geographical areas together because they lacked the clavicle or because they lacked bones in the palate. If you go back and look at microhylids, you can see the trend going from a full girdle and a basic palate through intermediate genera down to the end point. So it looks like the lineage is geographically consistent.

Inger

I'm a little concerned that in this symposium we're likely to not see the forest for the trees. We are focusing so closely on what has happened to one group of animals that we've done very little generalizing, and I think that in some cases this has led us into some trouble.

The first point I want to make is that phylogenetics comprises a variety of measures and relationships. For the most part, speakers at the symposium have been focusing on one of those, namely, cladistics—the grouping of species into major phyletic lines. Now equally part of phylogenetics are such things as the study of the branching sequence of characters in the phylogenetic tree, the estimation of phylogenetic divergence or phylogenetic distance, and the estimation of the time of appearance of various stocks.

Now, the last two things are important if we're interested in rates of evolution. I think from each of these different kinds of measures we get not just different kinds of information, but also different kinds of insights. One of the things that's happened here, which bothered me a bit, was a certain looseness of language, which has gotten phylogenetic systematics into trouble for years.

Well, we've been talking about phyletic lines that go back into the Jurassic . . . they're all ancient. We heard people describing some character states as primitive and others as derived. We heard some people use "derived" and "specialized" as though they were almost synonymous. Now, I would like, if I may, to ask two people who spoke yesterday afternoon—Linda Trueb and John Lynch—to tell us how they determined which character states were primitive and which ones were derived.

Lynch

In determining primitive and derived states, I feel there's a certain hazard in using as a means of determining what is a primitive state a fossil that is reasonably recognizable as a member of an existing family. I find it also unwise to survey anurans in general and assert that the most common character state is the primitive character state. That sort of assertion might lead us to argue that the presence of wings is primitive in insects. I will attempt to summarize my methods and reasoning.

The best method of deducing primitive and derived character states can be employed with only a part of the characteristics used in the classification of frogs. That method requires that the character states be present in more orders of Amphibia than just the frogs. The primitive character state is that state present in several amphibian orders (usually, Anura, Urodela, the fossil orders, and in some cases, the Gymnophiona), whereas the derived state is one that is restricted to the Anura. These character states I term first-order characteristics, and they are primitive or advanced for the Anura. Obviously, not all of the characteristics important in anuran classification are of this kind or can be treated in this manner, for we have many characteristics in which all character states are restricted to the Anura (second-order characteristics). If one uses only first-order characteristics, the classification must be based on relatively few characters, as was Dr. Inger's study of the phylogeny of frogs. I agree with Inger that first-order characteristics are the best sort to use; however, a number of other workers have argued that second-order characteristics can be employed. To determine primitive and derived states of those characteristics, one observes the degree of correlation of character states with the primitive and derived character states of the first order and assigns primitive and derived states for second-order characteristics on the grounds of high correlation of those character states with the character-state distributions for the first-order characteristics. It is realistic to utilize second-order characteristics, as well as first-order characteristics, inasmuch as the character states of second-order characteristics are likely (and usually are observed) to be distributed in essentially the same pattern of primitive and derived states as are the primitive and derived states of first-order characteristics in the organisms under study. I realize that there is a certain hazard to this method because it allows a certain degree of reinforcement.

When fossils have been available I have used them, although the fossil material, as has been pointed out by some of the speakers here, is in general not terribly useful because of its scarcity and the condition in which frog remains are generally preserved.

Trueb

I'm not going to have very much more to add to what John Lynch has said. He didn't discuss any specific characters, but one that I can offer as an example is the number of cranial nerve foramina in anurans. On the basis of available paleontological and comparative morphological evidence, the presence of eleven cranial foramina in an anuran is generally regarded primitive. It's also generally agreed that there is a tendency toward reduction and simplification of parts. Consequently, we can infer that the presence of eight, seven, or six foramina is a derived situation; it is less primitive, more advanced.

I also think there's something to be said for the association of some characters. The reduction in the number of the cranial nerve foramina tends to occur in frogs specialized in other ways. The monotypic *Argenteohyla* is a phragmotic, fossorial, neotropical hylid that is obviously very specialized. The number of cranial foramina is reduced to six. There is a certain correlation, and I can see that there are inherent problems here, but used with caution, it can give us some valuable insights.

Inger

I was just interested in having both of you say what you mean, just in the interest of unambiguity.

Trueb

I think I might as well point out something else, too: There's always a certain amount of "scientific" intuition involved.

Lynch

Assuming for a moment that the Orton-Starrett scheme of an anuran phylogeny is the correct one, as Jay [Savage] rather emphatically said in his opening remarks, we are left with an astounding array of—if you'll allow me to use a couple of other terms loosely—parallelisms and convergences in the so-called classic characteristics. I think you'll have to admit that that is the situation. Thus, I'm curious to know how Dr. Savage subdivides frogs beyond four groups, the same groups that Orton proposed in 1953. What sort of characteristics does he use inasmuch as he has already dismissed them, in a sense, by insisting that the larval characteristics were the only ones that were of any consequence?

A second question I want to ask relates to Dr. Duellman's comment about what sort of evidence Dr. Savage had for a simultaneous divergence of bufonids, leptodactylids, and ranids, an idea which is not so terribly foreign to me. The question I want to raise is, Why were you so emphatic?

Savage

Well, I don't want to renege on anything I said at the beginning of the talk about the larval characteristics, but I don't believe I ever indicated that there weren't other characteristics that could be used to subdivide the four groups. And I believe I mentioned something like forty-three families or subfamilies of frogs, which obviously have to be based on combinations of characteristics. The point I'm making, and the one that I think is of so much importance, is that the basic correlated characteristics in the larvae give us a real clue, a real handle to the major splits within the frogs and that these features are so essentially consistent as to indicate major phyletic groups.

This morning I was bothered—and I wanted to say something, but I didn't—about how we sort of had two conversations going on in this panel simultaneously. It seems to me that we keep getting confused about evolution. We have one course in the evolution of anurans and other groups in which the appearance of major organizational patterns provides a basic adaptive life style that's fundamental to that group; then superimposed on that through evolutionary divergence, we get other characters that modify the appearance, the behavior, and so on, of the descendant stocks within that line. So it's the familiar story: Most vertebrates have a vertebral column, and it's a major evolutionary shift, but that doesn't preclude the fact that there are different kinds of vertebrae that can also be utilized, if we wish, in classification and study, or that there aren't other characteristics of significance in terms of further divergent adaptation. I think Holly [Starrett] was talking about the origin of really basic key adaptations that opened up new worlds for frogs, and then other people were talking about evolutionary trends, characters that show trends within those base stocks, and still other people were talking about the origin of species involving isolating mechanisms. So, I would not suggest for a minute that the kinds of characters, for example, that you use with leptodactylids cannot be used to discern evolutionary trends. What I'm saying is that in the past we haven't had much to go on. If we go on vertebrae, we know that doesn't work.

Blair

Bob Inger made most of my concluding speech for me, so I will make these remarks very brief.

I think we have seen here the exposure of many, many approaches—many, many different ways of looking at phylogenetic questions. Whether we use the word evolution or not, I think phylogeny has dominated the talks of this meeting. It is very possible that we are going to see some major changes in classification and major refinements, changes for the better, as we get rid of an archaic classification based on methods that have now been supplemented by so many other recent approaches.

On the pessimistic side, I have been impressed by, to me, the apparent inability of many of the people who have participated in the discussions to heed Bob Inger's pointing out the differences between the many lines of approach to the problem and what is actually a multidisciplinary approach, an integrated approach. If we're ever going to understand the diversity of the anurans or of any other group, we're going to have to bring all kinds of evidence to bear, to look at all these lines of evidence objectively, and not to take the attitude that larval characters are all sacred, or that morphological characters are all sacred, or that biochemical studies or studies of chromosomes or hybridization must necessarily coincide.

INDEX TO ANURAN TAXA

Numbers in roman type indicate the pages upon which each item appears in the text; those in italics, the pages upon which the item appears in figures; those in parentheses, the pages upon which the item appears in tables.

S

T

SUBJECT INDEX

Numbers in roman type indicate the pages upon which each item appears in the text; those in italics, the pages upon which the item appears in figures; those in parentheses, the pages upon which the item appears in tables.

F

QL
668
E2
E93

Evolutionary biology of the anurans; contemporary research on major problems. Edited by James L. Vial. Columbia, University of Missouri Press, 1973.

xii, 470 p. illus. 27 cm. $20.00

Based on papers presented at the Symposium on Evolutionary Biology of the Anurans held during the 13th annual meetings of the Society for the Study of Amphibians and Reptiles, held at the University of Missouri, Kansas City campus, Aug. 27-30, 1970.
Includes bibliographies.

1. Anura—Congresses. 2. Amphibians — Evolution — Congresses. I. Vial, James L., ed. II. Symposium on Evolutionary Biology of the Anurans, University of Missouri, Kansas City, 1970.

270781 QL668.E2E93 597'.8'0438 72-87836
ISBN 0-8262-0134-2 MARC

Library of Congress 73 [4]